CLEAR SIGNS
OF TROUBLE
AND GREAT JOY

**What Everyone Should Know
About The Bible, History, and
the Future,
From the Sky to the Earth**

Steven R. Corey, C.P.A.

Copyright 2020 and 2021 by Steven R. Corey, CPA

All rights reserved. No part of this book may be reproduced or utilized in any form or by any means, electronic or mechanical, without written permission of the author, except by excerpts limited to teaching and evangelism of Jesus Christ, with credit given this book's author and your Author.

Printed in the United States of America

Cover production by Montrose Printing Center, LLC.

All scripture quotations are from the Revised Standard Version of the AD1611 translation from the original languages as revised AD1881-1885 and 1901, then compared with the most ancient authorities and revised AD1946-1952; second edition of the New Testament AD1971.

ISBN: 978-1-7365833-0-2

Independently Published.

To my father, Robert D. Corey, who taught me love for the truth, and to my beautiful, beautiful, beautiful wife, Charlene, whose truthfulness compels me to love her dearly. And to the greatest of truth seekers standing upon solid evidences to proclaim truth to a public awash in shifting deceit, Rush Limbaugh and Donald J. Trump. But, above all, to truth itself, light made available to any who honestly search, Jesus Christ.

I thank my wife, Charlene, for her encouragement and patience through my difficult, eight year task of writing this book, and for her critiques of my work, without which it could not have been better. I thank many others, also, for their critiques, without which this book would have been insufficient. And I thank the great men and women who dare explore, with integrity, outside the box of normal attention, sharing their discoveries with the world in spite of its consensual disregard for their work. These are not only the Rohls, Wilsons, and Pattons, but they are also the Limbaughs, Hannitys, Levins, Becks, Savages and the millions of other curious, careful, honest observers of life. Above all, I thank our loving Creator for *information*. By it, every honest, diligently searching person can clearly perceive the nature of reality's core truth, Jesus Christ, King of kings, Lord of lords, the quickly coming and definite end of all deceit; "…every eye will see…" (Revelation 1:7)

Foreword

Reality is consistent. A cup of coffee left undisturbed on your desk will sit there until removed. A monument buried in the ground will stay there until dug up. Its symbols, as discovered by archaeologists, will be the same ones its buriers knew. Today, Earth orbits the sun, and the Milky Way spins around Earth's axis, just like they have from their beginning, so consistent is reality.

Human nature is a reality. Individuals desire peace and goodness in their hearts. They design peace and goodness into their ways, so far as they are able to achieve their objectives. But when those objectives hit life's speed-bumps, as they always do, people cut corners by enlisting aid from others. The aid of others aggregates into governments; governments aim to achieve, promising no failure. So people collect into governmental powers by which they collectively cut corners across other individuals, leveling countless souls in their wakes of governmental malfeasance and murder. Nimrod led a governmental collective to The Tower of Babel. The Babylonian collective defeated an Assyrian collective at the Battle of Nipur. Roman legions collectively marched through North Africa, Europe, and the Middle East bearing the fasces symbol of collective threat before their legions, axe included. Napoleon painted Europe with its own blood for peace in accord with his collectively backed opinion about that Roman fasces. So did Hitler. America thinks it is different. Yet, the American collective has extinguished fifty-million helpless innocents, one by one, just because each was a speed-bump to another individual's comfort. Reality is consistent.

We think somehow human nature can escape old realities just because it is now us whom the world's been waiting for. But it isn't us the world's been waiting for. The entire universe has been groaning out for the return of Jesus Christ. Scoffing at that reality is another corner cut by today's human collective. The Holy Bible was once culturally regarded as the real words of God. Things weren't so peaceful and good then, either. Therefore, social reformers replaced its truth with a bunch of theories about some god-forsaken ape-man and buried Christ like an old monument. They cloaked their new "reality" in scientific, academic, governmental, and artistic deceit for collecting a consensus of unwary, common folk.

But consensus doesn't alter reality any more than rewriting history changes the past. All things real play the same tune of existence regardless of consensus' towering opinion. Reality is known by its own tune, not by man's consensual ruse. Objective science calls reality's tune "empirical observation". And now, for the first time since the consensus of normal science buried Jesus and shelved the Bible, nature is being empirically observed to have been playing science's beloved tune to the lyrics of God's Holy Bible ever since the beginning of time. Imagine that! And clearly set in each refrain of this amazing harmony between the physical universe and The Holy Word of God are praises for Jesus Christ's coming to bury human nature's age old rebellion against God's Holy Words of reality.

This book is my report of that discovery. You might decide to chalk its observations up to unimaginably immense coincidences. Yet being empirically observable by whomever invests sufficient effort to come and also see these wonders, their reality is irrefutable.

Steven R, Corey, CPA

TABLE OF CONTENTS

Foreword   vii

Introduction   13

Chapter 1: *Information*   29
     The nature of *information* and its necessity for discerning truths from falsehoods.

Chapter 2: The Rest of the *Information*   75
     Reconsidering science's disqualification of very much real *information*.

Chapter 3: Love and Message   115
     A proposition: *information* carries God's communication to all mankind.

Chapter 4: Mazzaroth   149
     The mysterious origin of the Zodiac might become clear upon considering the rest of the *information* in light of God's nature of communicating with mankind.

Chapter 5: Apophis and the Blood Moons   199
     Patterns in the occurrence of total lunar eclipses corresponding with epic events and an asteroid calls one to wonder if God directed it to get our attention.

Chapter 6: The Rest of the Blood Moons   239
     Discovering a unique set of four total lunar eclipses and their surprising implication.

Chapter 7: Measurements   273
     Observations of patterns in the chronologies of events and various astronomical phenomena corresponding to Biblically significant narratives and prophecies.

Chapter 8: The Late Great 1948   353
     An amazing testimony to the year of Israel's national independence is made by a particular pattern amongst twelve total lunar eclipses repeating throughout history.

Chapter 9: Holy, Holy, Holy   385
     The most mysterious pillar of mankind's most ancient temple and four epic con-

stellations verify that God has communicated by means of The Bible, astronomical phenomena, and Biblically significant history measured into patterns.

Chapter 10: What Then Shall We Do? 453
A rational consideration of what such extraordinarily patterned history, Biblical prophecy, and corresponding astronomical phenomena might tell us about God's communication of His salvation plans to our current generation.

Appendix 1: Analyzing Tetrads 467
Basic instructions for how you can discover the patterns total lunar eclipses have been making throughout history.

Appendix 2: Having Fun with Astronomy Software 473
Basic instructions for how you can discover the "Christmas star" and the amazing testimony mankind's first temple has made to the Bible and its clear signs in the sun, moon, and stars.

Appendix 3: Information You Should Peruse 483
Some books and videos that will aid your consideration of the implications made by Clear Signs of Trouble and Great Joy: What Everyone Should Know about The Bible, History, and the Future, from the Sky to the Earth.

Bibliography 491

Table of Illustrations:

Figure 1: The Hurrian Sala, a woman bearing the seed promised to mankind	126
Figure 2: Think Bridge!	146
Figure 3: Virgo	152
Figure 4: Sumerian seven-headed dragon	153
Figure 5: Libra	154
Figure 6: The Desire of Nations	154
Figure 7: Centaurus, Victima, and the Southern Cross	156
Figure 8: Sagittarius	158
Figure 9: Scorpio	158
Figure 10: Aqila and Sagitta	159
Figure 11: Ophiuchus, Serpens. and Corona Borealis	159
Figure 12: The Kneeler (Hercules) and Draco	161

Figure 13: Lyra	161
Figure 14: Delphinus	162
Figure 15: Capricorn	163
Figure 16: Aquarius and Pisces Australis	163
Figure 17: Pegasus	165
Figure 18: Cygnus	166
Figure 19: Andromeda	167
Figure 20: Pisces and the Band	168
Figure 21: Cetus	169
Figure 22: Aries	169
Figure 23: Perseus	169
Figure 24: Cepheus	170
Figure 25: Cassiopeia	170
Figure 26: Taurus	171
Figure 27: Auriga	171
Figure 28: Orion	172
Figure 29: Eridanus	172
Figure 30: Lepus	174
Figure 31: Gemini	174
Figure 32: Canis Major and Canis Minor	175
Figure 33: Cancer	176
Figure 34: Ursa Major and Ursa Minor	177
Figure 35: Bootes	179
Figure 36: Ara	182
Figure 37: Leo, Hydra, Crator, and Corvus	183
Figure 38: Argo	184
Figure 39: Venus and the Sorcerer	190
Figure 40: Coma Berenices	195
Figure 41: Hollywood Churches	210
Figure 42: Why Are We Here?	233
Figure 43: In the Beginning	293
Figure 44: The Jubilee and Shemitah Paths viewed together	317
Figure 45: See BS News	323
Figure 46: Overcrowding	327
Figure 47: Pigs in a Blanket	332
Figure 48: Liberal Goals	333
Figure 49: An Interesting Five Year Pattern	340
Figure 50: America's Great "X"	349
Figure 51: Signs of God Witnessing Biblical History	380

Figure 52: Aratta petroglyph — 402
Figure 53: Gobekli Tepe's Pillar 43. — 403
Figure 54: A Monument to Noah's Ark — 406
Figure 55: The constellations on Pillar 43 — 410
Figure 56: Meaning of Pillar 43 symbols compared to Mazzaroth meanings — 411
Figure 57: The sighting stone — 414
Figure 58: Some examples of Gobekli Tepe symbols found around the world — 415
Figure 59: Victory Story Told in the Milky Way — 416
Figure 60: Comparing Venus and the Sorcer to Gobekli Tepe's sighting stone — 418
Figure 61: Kulkulkan, Quetzalcoatl, and Gobekli Tepe — 419
Figure 62: The Sagittarius/Scorpio/Ara overlay — 427
Figure 63: The Sagittarius Timeline — 430
Figure 64: Israel Raised Above All Nations — 432
Figure 65: The Scorpio Timeline — 434
Figure 66: Polaris 9600BC — 440

Introduction

^{14}And God said, "Let there be lights in the firmament of the heavens...and let them be for signs..."
^{25}"And there will be signs in sun and moon and stars, and upon the earth distress of nations in perplexity...^{28}Now when these things begin to take place, look up and raise your heads, because your redemption is drawing near."
Genesis 1:14
Luke 21:25a,28

What would the world look like if science were found to be subjective and The Holy Bible was found to be objective? Would the observations evidencing reality be any different? Or would the evidences we now know only be perceived differently? Would there actually be empirically observable signs in the sun, moon, and stars today, like the Bible said those celestial lights were made for being signs? Or is any sight of such signs really only pareidolia? If extensive patterns intricately intertwining each other, history, and Biblical prophecy, intertwining far beyond any concept of mere coincidence, have actually been observed in the celestial bodies, then might they truly be the signs Jesus said would indicate the approach of our redemption at His return? Should we not look to see? Wouldn't looking to see actually be the scientific approach to these questions? Or is the scientific approach merely a salve to dull our eyes against seeing what we really should fear? If science refuses to honestly look, is it being subjective?

On the other hand, how many predictions of Jesus' return have already been made? And have been wrong! Wikipedia lists fifty-one. The website, www.bible.ca/pre-date-setters lists two-hundred-forty-two. Haven't their failures sufficiently disproved the Biblical proposition of signs in the sun, moon, and stars? If those predictions were actually based on extensive signs found by thorough searching, then the failures of their predictions would have indeed disproved that old, Biblical proposition. But none were based upon a search for signs approaching any true definition of "thorough". So, if there is an extensive system of correlating patterns amongst the celestial luminaries, Biblical prophecies and fulfillments, and other Biblically relevant history, wouldn't we face the toughest question of our modern existence? Should we honestly search for signs? Or should we dishonestly guess they don't exist? If we search, we might find trouble. But if we guess, the trouble might be double.

Even before the New Testament was completely written, predictions of Jesus' return were being made. The Apostle Paul wrote to reassure the Thessalonians that reports they heard of Jesus' return were false (II Thessalonians 2:1). Nobody had mentioned signs in sun, moon, and stars. As early as AD400, a six-day theory about mankind's feckless

Clear Signs of Trouble and Great Joy

endeavors lasting for only six one-thousand-year "days" was made the reason for expecting Christ's return then. They believed fifty-five-hundred years had passed between the creation of man and the death and resurrection of Jesus. This supposedly made their times "the end of evil in the world". But nobody noted any signs in sun, moon, and stars like what Jesus said would be there.

Another flurry of predictions set the date of His return at AD1000, or shortly thereafter. Their reasoning was mostly based on an unusually high number of crises having occurred during the tenth century, as we often hear in our times about plagues, firestorms, volcanoes, earthquakes, 9-11, and nineteen years later, Covid-19. A millennium had passed since Christ's first advent, so, of course, they thought their days were the completion of six thousand years of human history, too. They even pointed out a few signs in the heavens to hang their predictions on: a total solar eclipse, a comet, and the appearance of a new star (a supernova). But their examination of signs in sun, moon, and stars extended no further than those three.

More than half of the two-hundred-forty-two predictions set dates for His return in the twentieth and twenty-first centuries. In spite of their search being a bit more extensive, none pointed to any systematic set of signs. And we're still here. Therefore, should we be surprised everyone is tired of hearing predictions about Christ's second coming? And just how objectively is the public now willing to consider the Bible? Maybe it's time for a thorough search.

However tired we are of these predictions, the problem remains that, if the Bible is indeed God's presciently informing, objective Word to mankind, then not only will Christ certainly return, as the Bible says He will return, but His return will be preceded by clearly discernable signs and events, as the Bible says His return will be preceded. Jesus told His disciples unambiguously there would be discernable signs in the sun, moon, and stars before His return. We can know He meant those signs would be unmistakably apparent because He said we should look up *when* we see them, not *if we might be able to recognize* them. We can know He meant they would communicate clearly because He said, by them we would know our redemption draws near, not that we might be able to guess they are maybe about His possibly coming, someday.

But knowing when Jesus' coming draws near would be tantamount to predicting at least His approximate time of arrival. Therefore the Bible presents to us the sticky problem of continuing to do what we are tired of hearing.

> [35]Watch therefore—for you do not know when the master of the house will come, in the evening, or at midnight, or at cockcrow, or in the morning—
> [36]lest he come suddenly and find you asleep. [37]And what I say to you I say to all: Watch.
>
> Mark 13:35-36

The attitude of today's church is that we should not watch, since nobody but Jesus' Father knows the day or the hour of His return. As I wrote this book, many people, especially preachers, objected to my task. They said Jesus will come when He comes, implying, yet

Introduction

denying -que sera. Knowing that is good enough, they insisted. We won't know, they demanded, so don't look. Que sera.

But Jesus didn't say we won't know, que sera. He said watch because we don't know, so when we see we will know. He said watch to be forewarned. He understood that if we knew which day and hour it would be we wouldn't need to watch. Therefore, our not knowing the day or the hour is why Jesus told us to watch.

The closer the possibility of Christ's return draws near, the worse are the errors caused by never-a-date-setting. Watching is about preparing. Christ did not want His brothers and sisters to be overtaken by the same surprise that will overtake unwary guessers.

Watching envisions a mental preparedness far beyond just anticipating when and looking for signs. We watch by drawing our behavior into better accord with His. He didn't teach us to just anticipate when He might come, but to affect the world around us all the more as we see that day approaching. Watching is the same reason Noah preached before the Flood. It is the same reason prophets were sent to Israel and Judah before the destruction of those nations. It is why John the Baptist was sent before Jesus' ministry. It is why Jesus said signs would precede His second coming. When returning will be what He is doing while we have not been watching, then, by refusing to acknowledge the real signs of His coming, we will be less prepared for His return than the Pharisees were prepared for His first coming.

They ignored the signs of their times, too, as Jesus often pointed out before they crucified Him. Their failure to watch left them unable to recognize His coming. At His second coming, many nations will fail to fully understand the nature of their situation because they refused to watch. (Will America, too?) They will train their weapons on Him and suffer double for it. What makes us think our guessing will be more effective than the Pharisees' denying?

Think rationally about the reason for signs. They are for elevating effectiveness by the consideration of objective evidence for truly knowing real situations. Not just evidence of what has been, such as the truth of the Bible. But signs also evidence what will be, such as what the Bible prophesies. If the Bible is true, then, so also are its prophesies. If the Bible's prophesies have been observed to have come true, then the Bible is speaking truly. And if the Bible is true, then its Author is the Master. So if thorough searching truly discovers signs, as the Bible prophesied there will be signs, then we know the Master who inspired the Bible is coming immediately enough for us to look up and know objectively, by evidence, that our redemption draws near.

One of the common metaphors Jesus used for teaching the relationship between Himself and His followers was that of a master and his servants. If the time is coming for the Master's affairs to happen in the field, should His servants remain busy in the barns? No. They will need to busy themselves where their Master will be coming. If the Master is now ready to harvest the wheat, should the servants continue irrigating the fields? No. The servants will need to be preparing for the harvest. The closer Christ's return draws near, the more sober about the harvest His servants must become. The error caused by never watching, and therefore not knowing, begins with His followers not engaging the nature of the times, and therefore, leads to not engaging the essence of what needs their service. It isn't about knowing a date. It is about observing what our Master said would be observable for focusing

Clear Signs of Trouble and Great Joy

our behavior accordingly, which entails both aspects of watching. It's about knowing where you're at within a given process.

Had the first century Jews been more aware, they would have more focused their social communications and interactions around the probability of their times being those of the Messiah's coming. Had their communications been more focused, an elevated persuasiveness would have drawn people to more soberly consider John the Baptist's preaching. Their prayers would have more engaged the spiritual conflicts escalating at the intersection of God's Christ and the world of man's evil.

For a servant, knowledge means responsibility. Have the Master's servants no responsibilities while His second coming draws near? Should they all be asleep on the floor dreaming of how brightly their little lamps do shine? Did Christ really mean us to be the Laodicean Church at His second coming, partying in thoughtless celebration, patting ourselves on the back in recognition of what good girls and boys we have become? Or did He mean us to start buying the refined gold of truth and the salve of repentant humility for our eyes to see what we must do next? The servant takes part in the Master's plan, not visa versa.

When the signs of the Master's return are seen, the servant needs to change his current activities to "reception", whether or not such a reaction is tantamount to the setting of a date. The purpose for which a servant is commanded to watch is both to see the right thing and then to do it. If he refuses to acknowledge the sight of his Master's lamplight in the driveway because he does not believe in approximating the date of His return, should his error be any less egregious than if he refused to even watch at all?

The Bible tells us Christ's second coming will be preceded by signs, which are evidences, which are seen by watching, which purpose is to know by having seen. Should we take watching for the coming of our redemption any less seriously than we took accepting His torture for the forgiveness of our sins? Should we deny the clear sighting of a lamplight in the driveway just because two-hundred-forty-two times other servants announced His return without having actually seen any light in the driveway?

History is littered with people who honestly thought they saw some kind of lamplight. They sent warnings. People came to Jesus. Prayers were increased. Alas! They only saw swamp gas. Others fraudulently reported seeing a lamplight in order to grab the pride and power of being "spiritual" leaders, or to wring some wealth from unwary followers, or to attract a lusty harem of bedazzled ladies. No matter what those before did or did not do or see, or why they said what they said, the Master's directive to watch makes His servants watchmen. They must be watching. People need warned. So the sight of a real lamplight clearly appearing in the driveway becomes a watchman's commitment to speak up, regardless of the whining of never-a-date-setters.

> ...if the watchman sees the sword coming and does not blow the trumpet, so that the people are not warned, and the sword comes, and takes any one of them; that man is taken away in his iniquity, but his blood I will require at the watchman's hand.
>
> Ezekiel 33:6

Introduction

Watchmen must either speak of what they saw or bear the blood of those they didn't warn. Consequently, we can not heap judgmental denunciations upon everyone who has set a date. Regardless of maybe being a little short of discernment about what they thought they saw, their willingness to speak up while yet vulnerable to being impugned and castigated is laudable. They were sure they were watchmen. And they were. But only for a misguided world unappreciative of their efforts in spite of their errors.

Therefore, what might set the proposition of this book apart from being the two-hundred-forty-third error? Its author's ambition? Hope? Change? No! Only signs will distinguish it from the others. Signs like those Jesus said to watch for "in sun moon and stars". Observable, undeniably real, clearly communicative signs. Signs like none of the other two-hundred-forty-two watchmen saw. Perspicuously coherent signs like have never before been observed. Signs now being clearly visible and empirically observable. An enormous system of signs in sun, moon, and stars so purposefully arranged around history as to astonishingly communicate a particular point without words in a symbolic language developed by its complete coherence with Biblical prophesy.

> [3] There is no speech, nor are there words;
> their voice is not heard;
> [4] yet their voice goes out through all the earth,
> and their words to the end of the world.
>
> Psalm 19:3-4

Some of the previously failed predictions of Jesus' return were based on a sign or two in the sky. But thinking the time of Jesus' return can be determined by a total solar eclipse here, or a comet there, is like trying to identify the make and model of a car only by its red color and four tires. Total solar eclipses happen every few months. Comets are somewhat rarer, but not much. Even supernovae occur occasionally. They all share the same nature of being empirically observable. But there is something more an astronomical event or two needs in order to be a sign of Christ's return as much as there is something more the letter *"r"* needs in order to be meaningfully communicative. That "something more" sets this book's proposition as far apart from the previously failed predictions as it sets *"already"*, *"arrived"*, and *"are"* apart from the single letter *"r"*. This book will expose not a few signs, but an enormous system of very real, astronomical alignments intertwined with historical events of Biblical prophecy you can empirically observe for yourself. The entire system forms, in and of itself, a coherent expression of God's plan to end evil and restore righteousness for a remnant of mankind.

From the time I was an adolescent I have watched, because Jesus said, "Watch!" I noted every sign, now here, and again there, proclaimed by various other watchmen to mean one aspect or another of His soon return. Most of their proclamations didn't make any more sense to me than *"rr"*. But I remembered the signs they presented, because at least they spoke *"rr"*, which, after all, was a bit more than nothing. But their conclusions were rationally avoidable.

Clear Signs of Trouble and Great Joy

It seemed to me signs should clearly speak their own piece without human assistance. To do so, they would all have to participate in the communication like very many intricately ordered letters of the alphabet r participating together to make this a sufficiently coherent sentence. The state highway department doesn't put back-seat riders in every car on the road to read highway signs to the driver. Instead, they make the road signs coherent enough to speak for themselves. In the same way, I thought "signs in sun, moon, and stars" must adequately communicate in and of themselves to truly be signs.

I had been watching through many years for self-communicative signs when another watchman discovered an unusual phenomena and spoke up. He found four total lunar eclipses which had occurred on four important Jewish holydays in the two years after Israel became a nation in 1948. Israel is Jewish. The holydays were Jewish. One holyday commemorates God's destruction of Egypt while He passed over His Israelites to free them from Egypt. The other commemorates His dwelling with them after He freed them. Eclipses on those holydays at the reestablishment of that nation were interesting. Then four more eclipses occurred on those same Jewish holydays in the two years after the reestablished Israel won possession of Old Jerusalem through the 1967 Six-Day War. Biblical prophesy states that Christ will reign from Jerusalem after His return. That's doubly interesting! These eight eclipses stood on their own clarity far above all of the curious, little comets and supernovae before them.

> [1]For behold, in those days and at that time, when I restore the fortunes of Judah and Jerusalem, [2]I will gather all the nations and bring them down to the valley of Jehoshaphat, and I will enter into judgment with them there, on account of my people and my heritage Israel, because they have scattered them among the nations, and have divided up my land,
>
> Joel 3:1-2

Those eight eclipses on those four pairs of Passovers and Tabernacles occurring at the anciently prophesied restorations of Judah and Jerusalem's fortunes hoisted two, very clear, self exclamatory ensigns high into the skies. Should they be read in the context of Joel 3:1-2, since the Bible calls us to watch? Or should they be read in the context of scientific denial because no one but the Father knows the day or the hour? But even the fundamental nature of science is to seek knowledge by discovering order amongst nature's chaos. These eclipses on those holydays at Biblically prophesied events raised an incredible degree of order. The order amongst chaos called out for further examination. The Bible's directive is to watch for such order in sun, moon, and stars. Science and Christianity should both take interest in this orderly system of patterns.

So, why did science not notice them? One fundamental purpose of science is to take note of order happening amongst chaos. Why do most preachers either yawn in disinterest or snarl in denial? Jesus commanded them to watch. Maybe both are being subjective. It seems the Bible is the only one being objective. It said there would be signs. You can look up and see signs. That's objectivity.

The rebirth of Israel and its capture of Jerusalem form an unmistakable pattern with

Introduction

the Bible's prediction of their restored fortunes. It's a simple pattern called fulfilled prophecy. The Bible says God instituted Passover to commemorate His formation of Israel as a national people while He dismantled Egypt with plagues. It says He instituted Tabernacles to commemorate His dwelling with the Israelites through the desert and foreshadow His dwelling in Jerusalem during His Millennial reign of righteousness and peace. Think about this sensibly. How could the purely happenstantial motions of the sun, moon, and earth occur on the Jewish holydays commemorating those two concepts at the very events fulfilling the Bible's significant prophecies of Israel becoming a nation and capturing God's Holy City for its capitol? These are existing realities you can observe for yourself in the annals of history, the Bible, and NASA's lunar eclipse catalogue.

If coincidence is a sensible explanation for this pattern of order, will coincidence suffice to explain the enormity of the rest of the order between related history, Biblical concepts, and astronomical patterns extending throughout millennia? Even this small pattern in our days is too enormously meaningful to be rationally considered as just coincidentally formed amongst general chaos. Still, science will not study the possibility that the Bible accurately defines The Almighty God. Nor will cold-nor-hot preachers consider the patterns which science's mechanistic material universe itself forms in correlation with the Bible and The Lord God's affects amongst man.

Regardless of such blindness and myopia, could the Bible's proclamation of being The Holy God's communication to man actually be true? Two centuries ago science undertook the overturning of that Biblical proclamation. By now, science has become certain it has dispelled the old "God's Word" nonsense. But has the scientists' beloved mechanistic material universe now been found sassing their own conclusions by aligning the motions of its celestial bodies with historic events predicted by God's Holy Bible? Is the mechanistic material universe beginning to raise a crescendo of order in beat with the major historical developments of God's plan for your salvation? We can beg only so much of coincidence before we must rationally acknowledge the purposeful design of highly ordered and intricate patterns.

Since those extraordinary eclipses aligned with the anciently prophesied restoration of Israel and Jerusalem's fortunes, does the prescient correspondence between them further verify a nearing occurrence of the rest of Joel's prophecy about the Valley of Jehoshaphat and the horrors awaiting there for every nation? If so, is it possible that a similar ensign might be waving in the skies over those horrors in that valley? Would not such an ensign also be observable today in NASA's eclipse list as some peculiar pattern of future eclipses found to correlate with Biblical prophecies describing the essential elements of the end of this evil age? We will explore these questions.

We must understand that the public's anti-God bias instigated by science's incredulous denial of everything Biblical is not necessarily the correct perception of reality just because our culture enshrines science at the core of its own collective knowledge. What is knowledge but objectively meaningful order? Does objectivity not discover meaning in orderly patterns? If not, then why did the objective study of chemical reactions persist until the observations of their patterns produced the periodic table? If so, then why have the orderly interactions of history and the heavenly bodies begging for objective study been

Clear Signs of Trouble and Great Joy

entirely ignored by science? Is it only because the latter involves Biblical narratives and prophecies, but the former doesn't? Is not science's essential goal to discover any order amongst chaos anywhere? Does science explore with a bias? Is it subjective?

Christians are also guilty of minimizing God's patterns. They read the Bible while ignoring the significance of a seemingly inconsequential adjective here, or an obscure adverb there. And so, they read into the Bible constraints popularized by old theologies. They misread out of it their personal proclivities, while they often miss its own, subtler focus. And they treat the import of mundane events and circumstances around them the same way. Yet the Bible tells them everything serves God's purposes (Proverbs 16:4), therefore, in everything there can be found something informative about God. So I've always been as wary of Christian myopia as I have been wary of science's subjectivity.

Then it occurred to me. This entire system of empirically observable signs speaks more pointedly, and considering our late hour in God's transpiring plans, more importantly about the completely reliable, objective truth of God's publicly ignored, "scientifically" impugned Holy Bible. These signs clearly speak against the institutionalized deceit and foolishness eclipsing The Holy Bible's message from today's public square. Even science's dearly beloved, mechanistic universe seems to be warning against modern accusations about the Bible being mythological babble, as well as it seems to speak against Christianity's myopic, neither cold, nor hot watching.

"Sure," scientific minds happily admit, "the Bible is full of niceties. But it's still babble."

"Yah," myopic Christianity agrees, "it's utter nonsense to think anything might show when Christ will return."

> ...when these things begin to take place...look up...your redemption is drawing near.
> Luke 21:28

How exactly did babble predict Israel's return to nationhood with Jerusalem as its capitol fully two-and-a-half millennia before that history happened in front of our own scientifically glorified eyes? How did nonsense hang confirming, total lunar eclipses upon Jewish holydays at the very time of Israel's reestablishment? How possibly did babble and nonsense do it again when Israel won control of God's Holy City, Jerusalem, nineteen years later? Why does empirical observation, science's own beloved principle, now flirt with showing Biblical "babble" to have been correct about the sun, moon, and stars being purposefully designed to make signs (Genesis 1:14; Luke 21:25-27; Joel 3:1-2)?

We can now empirically observe science to have been wrong in its biased assessment of the Bible as being babble, and cold-nor-hot preachers to have been blind to Jesus' prediction of signs showing our redemption draws near. Empirically observed to have been wrong, not assumed to have been wrong, like science has only assumed snail-paced, epochs-long geological processes of gradualism. Nor theorized to have been wrong, like science has only theorized evolution. But observed to have been wrong, unlike science has ever empirically observed the Bible to have been wrong. (It's only theorized, assumed,

Introduction

guessed, and accused of being wrong.) These signs clearly show science to have been babbling about origins, guessing about evolution, and pontificating about the nonexistence of the Almighty and Holy God throughout the last two-and-a-half centuries. These signs clearly call us to relive the Biblical culture which an idolatrous worship of science destroyed.

The close of the eighteenth century opened a new "understanding" of life, history, and the physical universe: No design. No purpose. No God. No inspired Bible. No evidence necessary; theory works fine, thank you very much. Consequently, the public theatre in which the truth of God's Word had been for centuries displayed (although not so well obeyed) became hidden underneath many thick coats of nineteenth century, theoretical whitewash. Yet, no empirically observed evidence had ever falsified God's existence, or contradicted the Biblically implied purpose God made for the physical universe, or denied the universe's created design. Not a single shred of evidence falsified even one Biblical proposition. Not a sliver. Not then. Not now. Yet cold-nor-hot preachers still laid their twentieth century churches at the feet of science in order to mellow out the Holy Spirit.

So it was that I woke up in the middle of one night realizing why these clear signs correspond to Biblical themes, prophecies, and histories. They are empirically observable evidences scientifically answering the question of whether or not this universe was designed to specifically serve God's purpose. The Bible claims the sun, moon, and stars were made for signs. Signs are meaningfully ordered patterns. Meaningful patterns formed by the stars, sun, moon, Earth, history, and Biblical prophecy are now being empirically observed to be far more orderly than happenstance alone can explain. Were the Bible's authors merely better guessers than was Darwin? Or was their knowledge inspired by The Holy Spirit of God, as the Bible states it was inspired?

> …let them be for signs…
>
> Genesis 1:14

> [1] The heavens are telling the glory of God;
> and the firmament proclaims his handiwork.
> [2] Day to day pours forth speech,
> and night to night declares knowledge.
> [3] There is no speech, nor are there words;
> their voice is not heard;
> [4] yet their voice goes out through all the earth,
> and their words to the end of the world.
>
> Psalms 19:1-4

> …there will be signs in sun and moon and stars…
>
> Luke 21:25

The entire system of these signs is too enormous and too consistently coherent with Biblical narratives to be the purposeless, coincidental result of science's assumed, happenstancial, mechanistic processes. Therefore, science's mechanistic only, no-God theories can

Clear Signs of Trouble and Great Joy

not honestly form the core understanding of reality. The heavens themselves *are* displaying patterns of intricate and precise correlations, not with Darwin's chin-drooling ape-man, but with the Lord God's Biblical prophecies. If such correlations were discovered in any field of research other than The Holy Bible they would immediately be received as definite communication from some outside entity and pronounced to be undeniably real. SETI would be jubilant about a discovery of patterns in the universe's background radio noise having even one-percent as much of the coherent correlations we will discuss in this book. The utter silence SETI has found amongst the spewing radio noise has said nothing to anyone. Yet Psalms says the heavens tell, proclaim, speak, and declare without words, although their voice is heard around the world. And behold! We see precisely that in these very real patterns! This is the exact condition science relies upon for gathering knowledge: evidence "speaking" to observers by simply being what it is. This vast, orderly, intricate network of celestial alignments with human events strung like beads along the lines of God's Biblically revealed plan needs no words to express their two cents about reality's core concept, The Holy Bible. They brightly shine through thick layers of eighteenth century, "scientific" whitewash reflected by cold-nor-hot preacher puff. God's inspiration of The Bible is now being illuminated by science's own beloved process of empirical observation.

...and let them be for signs...
Genesis 1:14

Will these signs affect any academic theories of origin? Will the universities, committed as they are to empirical observation, now blow the dust off their old Bibles? Will the world rise to its feet at the discovery of the Biblical message written by alignments between the sun, moon, and Earth in correlation with historical events fulfilling Biblical prophecies? Would the scientific guessers repent even if a prehistoric monument buried long ago to warn our generation about Christ's reality was discovered during our generation? Surely science, being "the epitome of honest, independent, non-subjective research, thought, and consideration", would abide by the rules for discovering knowledge! Surely it will at least acknowledge the reality of these patterns. Won't it? Certainly objective science guides and constrains its knowledge by the sight of real evidence. Doesn't it?

So, might we expect the incredible analytical abilities of objective science, those superior talents which assembled the periodic table by merely noting interactions between materials, that great prowess which mapped the human genome without directly seeing it, the very ingenuity which proved gravity's affect on light by observing just one star peaking out from behind a total solar eclipse, might we expect these ingenious skills of objective discovery to revise scientific theories about the Bible being irrelevant mythology now that the signs the Bible said existed have been empirically observed to indeed be there?

No. Science is subjective.

Nor will cold-nor-hot preachers end their pandering to evolution and other materialistic assumptions.

The same science which hides evidence in order to propagate guessing will never acknowledge the Bible to be informative truth. Cold-nor-hot preachers will never revere the

Introduction

authority of God's Word over their own pride. This kind of science and preaching is neither objective nor observational. It is built upon subjective and pragmatic assumptions instead of clear evidences. The real processes for knowing and believing things and ideas are all objective processes. For knowledge comes to man, not from man.

But a remnant of scientists do investigate in accord with objective observations rather than guessing in accord with the consensual constraints of theories and imagination. A remnant of preachers preach God's Holy Word without embellishing or denuding it. A remnant of thinking people respects, honors, and uses Biblical information to live their daily lives. They're all laughed at and impugned by scientific guessers, scorned by cold-nor-hot preachers, and accused by collectivist politicians. But their conclusions and beliefs correspond far more with what is observed to be real than with the guesses and theoretical fancies of scholastic education and Laodicean churches.

Reality itself sets the rules man must use for discovering it. It would be nice if reality had written its rules for us like God etched the Ten Commandments onto stone for Moses. But just as we must grow, harvest, and process our own food to set on tables we had to also make, God left us to discover for ourselves reality's rules for how to know what is really true. And naturally, philosophers have argued and fought about even the simplest rules for how to know what's real. Consequently, their debates have produced three general theories about what to think from having seen, when to believe what is thought, and, in general, how to perceive that our beliefs have reached at least the same level of actual awareness as that of a common mouse.

Not so surprisingly, the mouse uses the same rules for knowing reality man should use just as naturally. Mice don't argue about those rules. Mice abide by them because they've worked before, which is nature's own attestation to traditionalism. For the stakes of a mouse misperceiving reality have long ago been shown to be the same steaks of a cat's perceiving it correctly. So, when the wise mouse sees a cat, he simply knows its a cat. He doesn't ponder whether the colors and shapes he perceived of what he saw were the same as colors and shapes other mice perceived when they also saw. Especially if those other mice are now cat poop. When he sneaks out his mouse hole, he even anticipates the presence of a cat from having seen cats hang around mouse holes. Wise mice don't argue about perceptions. Everything is coherently sensible to them by the observation of history's affairs. Therefore, humble mice desire precaution and patience more than they desire cheese. But self-discovered, stiff-necked, brass-foreheaded mice, craving cheese regardless of risk, have no need to ponder any correlation beyond the delight of cheese. Their desire to believe in their own perceptions above what can be observed casts them out, at the end, as cat poop.

Reality sets its own rules for discovery. Are those rules, by observation of this massive system of Biblically prophesied signs, now showing the Bible to actually be another one of reality's rules for discovering truth? Is simple observation showing the universe to be patterned around definitions the Bible provides for us to discover the knowledge of God?

Correspondence theories of knowledge, belief, and truth say you can know and believe something by seeing it. You know the chair by sitting on it. You know the table by putting your food on it. You know the food by eating it. You know the signs by observing them. Correspondence theories say truth is the squaring of your perceptions to accord with

Clear Signs of Trouble and Great Joy

what you see, hear, smell, touch, or taste.

But science tries to limit the testimony of observations. Some observations are too revealing of Biblical truth to be acknowledged. So it narrows the rules for how to see, hear, smell, touch, and taste, as far too many observable things, like the coherence of these pesky signs in the sun, moon, and stars with The Holy Bible, contradict science's own, lofty assumptions, guesses, and imaginings. Science must filter from consideration all observations falsifying its lofty assumptions by using "special rules" to "qualify" what it will consider an observation to be. Therefore, science calls upon its own, pruned version of reality's rules of observation to avoid acknowledging enormous masses of evidence illuminating science's anti-Biblical biases and theoretical falsehoods. Science's sneaky alteration of the rules for discovery will eventually drag it to a stare down with the same Cat it pragmatically ignores. At the end of that matter, reality cares not a whit about either man's abridgement of its rules nor the imaginative whitewash science has brushed over God's existence. Reality will have its way with mankind, The Holy Bible assures us.

The Holy Bible fills its pages with calls to simply come and see. Its entire narrative is constructed from testimonies of men and women who observed real events, circumstances, and a variety of other phenomena. It does not pretend there is a dire necessity to scientifically measure, quantify, or qualify anything of its reports before they can be regarded as true. The Holy Bible is the ancient book of empirical observations guided by the Holy Spirit towards the development of a knowledge of God, not by measure, but by experience. The word "testify" and its variant forms are used one-hundred-eighty times in the Revised Standard Version of the Bible as the proper response to the simple observation of a thing being what it is. We will discuss how science breaks this first rule of knowledge, denying real observations found the Biblical narrative, even though the Bible's observations are soundly verified by its undeniably fulfilled prophecies.

[3] The former things I declared of old,
they went forth from my mouth and I made them known;
then suddenly I did them and they came to pass.
[4] Because I know that you are obstinate,
and your neck is an iron sinew
and your forehead brass,
[5] I declared them to you from of old,
before they came to pass I announced them to you,
lest you should say, "My idol [Nature] did them,
my graven image [science] and my molten image [secularism] commanded them."
 Isaiah 48:3-5

[1]In many and various ways God spoke of old to our fathers by the prophets;
[2]but in these last days he has spoken to us by a Son,
 Hebrews 1:1-2a

And now we can see in our skies and our histories what the Son told us to watch for:

Introduction

signs of His coming! Beginning with Enoch and continuing through Moses, Elijah, Isaiah, Ezekiel, and Malachi to Jesus, every one of the prophets were verifiably linked to the times and places in which they spoke prophecies regarding events and circumstances that later occurred just as the events and circumstances were prophesied to occur, some in the days of the prophecies, some after, some in our days, and others yet to come. We now hold in our museums many archaeological artifacts bearing the names, inscriptions, and identifying symbols of those prophets. The Bible embeds itself in history as much as its prophecies clarify history. Fulfilled Biblical prophecies point to a reality science is not able to measure by yardstick, beaker, or scale, but could measure by spirit, if it weren't too stiff necked and brass foreheaded to acknowledge coherent patterns in the heavenly bodies correlating with The Holy Word of God.

The simplest, most basic nature of reality is that every real thing belongs to it. Everything fits coherently with everything else because reality itself is coherent. Therefore, anything we think we know must coherently interrelate with everything that has been verified to be true, else what we think we know is actually a misperception rather than true knowledge. This is the coherence theory of knowledge. In order to accurately reflect reality, a theory or perception must accommodate all of reality's evidences; no evidence can be discarded. If a perception or theory can not accommodate a piece of real *information*, then the perception or theory is in error, and it will need adjustment before it can truly reflect reality. Or it will need to be totally rejected if it is bad to the core.

The Bible is an undeniable, physical reality regardless of scientific theories and presumptions to the contrary. Its very existence demands an explanation of its origin, since it displays the unique ability to have clearly predicted the globally relevant events we know, today, e.g., the nation of Israel now existing in Palestine with Jerusalem as its capitol and the lunar eclipses on Jewish holydays at the formation of the nation and repossession of Old Jerusalem. But regardless of the Bible's uncanny prescience, regardless of its volumes of time tested wisdom, and regardless of its offering what everyone desperately wants (eternally blissful life), science has trained society to incoherently impugn, castigate, and dissemble (pie) the Bible's message as being merely human literature in order to free a consortium of scientists and politicians to believe they are the originators of truth and the purveyors of your right to exist, speak, or even move about.

For most people, personal desires and emotions arrange, groom, guide, and define the coherence of their beliefs. They might recognize empirically observed realities. They will talk about science and its "objectivity". But they do not necessarily adjust their thinking for what they see. They do not allow independent observations to affect their "life picture" by influencing the coherence of their forming thoughts. Basically, the shaping of perceptions in accord with emotions and the fear of or the desire for immediate circumstances is the pragmatic theory of belief. It is belief in what works the best for "my" situation, or in what I desire from it the most. It is the fundamental concept behind the modern belief that everyone has his own reality.

But reality continually shows itself to be mostly unaffected by our desires. Its most basic process of cause and effect is an ubiquitous, microscopic latticework of integrity truly constraining us by its own demands. In an abstract sense, everyone does have his own

Clear Signs of Trouble and Great Joy

reality. Personal choices do effect the next moment of each individual's emerging life. The real pathways stepped onto through those choices are often widely divergent from the hypothetical pathways that could have been traveled by alternative choices. For example, I chose to study accounting, therefore I am a CPA, else I could have chosen to study metallurgy and have become a welder. But every personal alteration of reality must be channeled through an empirical cause in order for reality to be affected as "yours", e.g., I had to study, I couldn't just lay on the couch and wait for talent and knowledge to fill me up. I had to work hard. We must direct our desire for bread through the channel of labor, whether we labor as thieves or farmers. We can sit on the dining room chair and desire bread all day, but it will not just appear on the table. Reality does not belong to man for each to shape by just making things up. Reality must be engaged, else it exacts a price.

    We concretely understand and obey that fundamental principle of reality every morning by crawling out of bed. We don't just lay there daily and dream our way through life. We get up and do things; we cause effects. Although too many of us have been absorbed in a broader whitewashing of "individual reality", we all are clothed (effect) by each putting on his own britches (cause). We may individually effect reality by the color of britches we put on. But we must, all the same, pull them on. And once they're on, their color can change only by the effect of another cause. Although we each have an aspect of reality that is our own, we each have had to obey reality's cause principle to effect it. In other words, reality is more a correspondence/coherence thing than it is a pragmatic thing.

    Yet for some reason, we think the correspondence to reality of what we see is not meant to affect the coherence of what we desire to think when it comes to contemplating the reality of God. Even though we must labor hard to affect physical reality, we think we can just lay on the couch and dream up whatever spiritual "reality" we like. We even think we can make God to not be at all by just thinking He isn't. For some reason, while our physical bodies are laced into a system of constraint, we think we can lace God into a system so unconstrained that He will be whatever we want to believe. Could spiritual reality possibly be constrained, like physical reality, to being as existentially defined as is physical reality? Could spiritual reality be as concretely what it is as is the mechanistic material universe, both being what they are regardless of what we make up about them? Is there some coherence of correspondent observations to suggest the existence of a spiritual realm as truly constrained by every aspect of its reality as the physical realm is by its aspects?

    In a sense, God does give everyone the personal right to think about Him however they desire. It is why the world is full of Buddha's, saviors, and culture heroes, none of whom rises any more coherently above the rest. But God does not give anyone the right to make Him up. What people think about Him is judged by what He actually is.

    The Bible claims God is I AM WHO I AM while people have made every imaginable judgment about who He is. How can the Bible not be correct in its assessment of God being who He is and man not be just arrogantly imagining? It may not be a very detailed assessment, but it is the most informative assessment. Even if God is not, then that is who He is. Does the Bible's claim about God not sound coherent with the first principle of physics, the identity principle, everything being what it is, i.e., reality is what it is regardless of anyone's personal perceptions? Science's identity principle is not a stop sign

Introduction

for thinking about your surroundings. It is the green light freeing you to study, calling you to study, and assuring that the only way you will know physics is by study. So why do we expect the spirit realm's identity principle, I AM WHO I AM, is not also a call to study, the assurance that you approach God by humble observation as certainly as you approach the edge of a cliff by very careful observation rather than wildly running imagination?

So, who is right, you or the Bible, and how can we know? Can we really just choose God to be whatever by just wanting Him to be that? We can't even choose the color of our britches by just desiring it to be different than it already is. Irregardless, everyone makes their own "God choice". Many even live the "God-is-not" choice. Occasionally some change their choice. We judge God to be only a myth by subjective, pragmatically chosen beliefs about Him. Then we have the audacity to expect God to obey our myth.

If God is the objective reality that His name implies, then the Bible's claim is true; the rest of the claims are false; and we have a lot of honest studying to do.

> Let God be true though every man be false, as it is written, "That thou mayest be justified in thy words, and prevail when thou art judged."
> Romans 3:4

If the Bible is right, does Romans 3:4 not tell us to be careful in judging who and what God is? Does history show mankind being careful? It shows him being badly messed up by pragmatically crafting his own gods. Has science judged any more carefully? Has it thoroughly studied? Or is it just making up no-God.

Science claims to have objectively told us that we can throw our Bibles away since God is only a very bad, cultural dream. It claims to have objectively considered all of the evidence, the geologic layers, the fossils, and all comparative anatomies, and to have found man raised out of the many purposeless, non-designed, happenstances of epochal time, having traveled through even the being of a worm on his way to having been an ape now become inhumanely and fool-heartedly human.

But if objectivity were allowed to form and groom all of the perceptions behind academic and scientific belief by employing everything that has truly been observed, then evolution would lose its precious poster child, the dinosaur. The "fact" that dinosaurs died out millions of years ago is meant to induce a perception of old life forms dying out so their updated versions can affectively change the gene pool until eventually T-Rex evolves into a hummingbird. But many peoples of the prescientific past depicted their observations of and experiences with a variety of dinosaurs in artwork that we can all see with our own eyes and touch with our fingers today. Nor would rock layers mean millions and millions of years since fossils, tools, and artwork of homo-sapiens-sapiens actually have been found in layers supposedly laid down earlier than when science says man evolved. Such evidences must not be seen! They do not correspond with what science tells you to coherently believe. Therefore academia throws such contradictory evidence into dumpsters full of forbidden observations. They are then able to pragmatically rearrange the coherence of "scientific" perception, making it only from the evidences they're willing to acknowledge. That is subjectivity, not objectivity. Hiding *information* is the subjective fly spoiling the scientific ointment.

Clear Signs of Trouble and Great Joy

By both his science and his religion, mankind is carelessly judging God. Objectivity allows evidence to guide knowledge to ever more accurately reflect reality. Therefore, wherever evidence goes, understanding must follow. But neither man's science nor his religion has wanted to go where Biblical evidence leads.

The Bible is actually one of reality's many evidences. Something must have endowed it with the ability to discuss events long before they happen. Something must have given it the audacity to claim there is a pattern in the positions and motions of the heavenly bodies sufficient to be communicative signs of current events the Bible discussed millennia ago. Undeniably clear, actually observed patterns amongst the heavenly bodies at events fulfilling major Biblical prophecies coherently endorse the Bible's claim about I AM being Who He Is. That should cause at least a little circumspection on the part of all God's judges.

Still, normal science and academia will try to deny these observable signs just because they correspond with Biblical themes and history to communicate the reality of Jesus Christ's soon return. They will say the signs don't exist without looking for them just because they don't want them to exist, therefore they will refuse to see them. They will throw them into the same "scientific" dumpster hiding the ancient depictions of dinosaurs.

The conclusion is logical. Scientific endeavor is biased. Human contemplation is unconstrained. They search for desired truth, not for real truth. They sort between acceptable and unacceptable evidences rather than following all evidence to wherever it all leads. Science desires to know The Holy Bible is false. Therefore it discards every evidence of the Bible's truth, thinking that will make God's Word false. All the same, Truth will return to deal with deceit. Signs now show it empirically. Observable signs show that objectivity's deniers will be cast out reality's end, quite empirically.

Hence, is it not rational to conclude that the same Book which correctly described the purposefully designed nature of empirically observable signs in the heavens also correctly proclaims the deeply spiritual nature of God and what He is doing? If the Bible is right about this vast and intricate system of celestial signs corresponding with historically fulfilled Biblical prophecies, then isn't it reasonable that the Bible's description of the God who made those signs is also accurate?

Don't hold your breath waiting for academia, science, and other subjectivity fanciers to catch up with us. Our search for the answers to these questions will take us on a journey beyond the veil of scientific slight-of-mind. Only know it is perfectly ok to draw conclusions from your personal observations instead of bowing down to the unsubstantiated theories of academic arrogance. This you should do, since the Bible warns of a final exam and emphasizes the importance of not getting your answers from cheaters. These signs in the sky confirm the approach of examination day. They graciously give unambiguous clues as to what is the correct answer to that final exam's one question: Who is Jesus Christ to you?

Chapter 1
Information

*⁶ A voice says, "Cry!"
And I said, "What shall I cry?"
All flesh is grass,
and all its beauty is like the flower of the field.
⁷ The grass withers, the flower fades,
when the breath of the LORD blows upon it...
...but the word of our God will stand for ever.
¹⁶ The LORD has made himself known, he has executed judgment;
the wicked are snared in the work of their own hands.*
Isaiah 40:6-7a, 8b
Psalm 9:16

"God in Heaven, thank you for information. Amen."
You wouldn't think a five year old could grasp a concept like "information". But one February evening of 2016 my granddaughter, Sadie, offered that blessing at our dinner table. And I doubt she did completely understand the word, although she seemed to sense its importance.

A few weeks earlier, I had been babysitting while her mommy worked an evening shift. The Denver Broncos had just beaten the Pittsburgh Steelers on their way to winning Super Bowl 50. My grandkids were in bed, and I was snoozing on the couch, basking in the radio program's post-game glee when Sadie toddled out of her room and asked to lay down with me. After she made herself comfortable, she heard the faint chirping of my radio's ear buds.

"What are you listening to," she asked with childlike curiosity.

"Information," I replied.

"What's information?"

Now, exactly how does one explain *information* to a five-year-old? I thought for a moment, then answered, "*Information* is the most important of all the things God made for us, because it shows what's true. And by seeing what's true we know Jesus is real and saves us to be happy forever."

She remembered!

Children arrange ideas into simplicities missed by adults' complex ways of thinking. Adulthood's mundane tasks of survival and constant struggle for accomplishment weave disasters and challenges into complicated veils of chaos hiding *information's* simpler interconnections. Children's minds are not so burdened by such that they miss simple con-

Clear Signs of Trouble and Great Joy

nections. They do not overrate their agendas so much that they choose not to get those connections. Their hearts are not so hardened by daily struggles that they refuse to recognize them. Sometimes a child will even call out an adult for hiding simple connections, usually the ones exposing an adult's scheming plots. And maybe that's why adults used to say a child should be seen and not heard.

Since we don't often hear the adage anymore, are we adults now seeing things more simply, as children see them? No. We're not more carefully observing *information*; we've just found more effective ways to hide it.

The 1986 Challenger disaster is an excellent case in point. The Challenger launch had been postponed five times over the previous six days. It was scheduled again for launch on January 28. But that morning arrived with its own problem: cold. The engineers at Thiokol, the contractor supplying the solid fuel boosters, knew the rubber O-rings sealing the joints between the shuttle's booster sections were barely retaining the hot, high pressure gases of the burning fuel. O-rings had been found leaking in seven of the nine previous launches. Then, adding to the menace, rubber loses elasticity when cold. Elasticity is necessary for an efficient seal. Consequently, Thiokol engineers would not certify the O-rings for launches in temperatures below 54°F.

The engineers and Thiokol management teleconferenced with NASA management the day before the launch. Thiokol management supported the engineers' recommendation not to launch. But another delay was unacceptable to NASA. So Thiokol and NASA managers teleconferenced again, this time without the engineers. Of course, the engineers' information was silenced in the latter meeting. Thiokol management empathized with NASA's delay problem. Eventually, they subordinated their information guided judgment to NASA's agenda driven judgment. Launch was recommended.

But the cold that morning was as real as the non-pliable properties of frozen rubber. The Challenger launched into the face of reality for seventy-three seconds. It fell back to reality for one-hundred-sixty-five seconds. Had they thanked God for *information*, maybe they would not have launched the disaster.

Much *information* is deliberately not seen nor heard, even though it surely exists and is definitely observable. But when the *information* unavoidably contradicts cherished concepts, people usually deny, irrationally repudiate, or sneakily cover it up instead of simply thanking God for it. The Challenger disaster will always attest to the existential nature of *information*: *information* is always what it is. However much it is ignored or hidden, *information* remains the true expression of definite reality. And just as Abel's blood cried out from the ground, disregarded *information* cries out for reasonable consideration, sometimes harshly.

What do I mean by *"information"*. Everything expresses *information*. It is every aspect of anything's state of being, every trait, characteristic, and property. Dictionaries tie the word "information" to communication in a manner that mixes its sense with data, facts, and knowledge. The technical meaning of the word involves measurement, symbols, and communication systems. But all of those are aspects of conveying concepts between concept users. My proposition needs a word expressing something more fundamental than measurements, data records, and conveyance of ideas between people. The thesis of this book regards

Information

the conveyance of concepts from reality to people, whether the reality is a pattern, a thing, another person, or God. The *information* addressed in this book is more useful than data, and it is most natural.

"Noumena" might come closer to denoting my targeted concept:

> A posited object or event as it appears in itself independent of perception by the senses.[1]

But I need to convey more than just an object or event and more than just its appearance, since *information* is more than just appearance. Maybe "noumena" is a better term, but I want to write "noumena" less than you want to read "noumena". It has no rhetorical value. It pays no metaphorical dividends. Moreover, I mean to express an aspect of reality "noumena" fails to connote. By *"information"* I mean all of the intrinsic characteristics of something expressed by its simply being what it is. Although dictionaries are not yet articulating this particular connotation of *"information"*, it is gaining ground in scientific usage:

> Many scientists…justly regard information as the third fundamental entity alongside matter and energy.[2]

This definition is too involved with measurement and communication. I need to connote something more fundamental than measurement and communication and less involved with them. Physicists deal with a concept they call "physical information".

> In a somewhat general sense, the information of a given entity can be interpreted as its identity. As such, its information can be perceived to be the representation of the specification of its existence and thus, to be serving as the full description of each of the properties…that are responsible for the entity's existence.[3]

That is closer to what this book will be referring to by the term *"information"*. But "physical information" is a technical term. The purpose of this book is not to turn its readers into techno-geeks. Even though the given definition of "physical information" is precisely my target concept, to avoid the dullness of repetitive techno-jargon, throughout this book, I will refer to the intrinsic characteristics, properties, and shape of a thing by the word *"information"* italicized. Wherever the word is not italicized, I will be referring to perceptions of things reduced to expressions of measurement and descriptions called "data information", "facts", and "knowledge", the stuff of books and reports, a very different thing than *information*. Information belongs to man, however accurately or not it reflects *information*, which belongs to its thing.

Here's some examples of what I mean by *information*. I have a rectangular, steel paperweight on my desk. Its being rectangular is *information*. So is its being steel. Its having sides is *information*, and the flatness of each side's plane is more *information*. The length of each side is *information*, even though it is unmeasured. If I were to measure a

Clear Signs of Trouble and Great Joy

side, the measurement would be data information, and it would only approximate the *information* of its real length. Its real length cannot be measured because of the microscopic imprecision of measuring devises and the inability of the human eye to see at the molecular level. If we could see to that level, we would find many significant chips and dings in the molecular structure of its edges, each chip and ding expressing its own *information* about the shape and condition of the edge. If we had a ruler accurate to the molecular level, we would need to make a set of measurements all along a side to know the actual *information* about the side's true length, for it will have many microscopic variances. My paperweight contains an exact number of molecules composed of specific amounts of iron, carbon, and maybe some other elements, such as nickel, chromium, cobalt, etc., arranged into a lattice having a precise number of unique flaws and impurities at particular locations. These are more *information* of my paperweight.

From the smallest whit of existence to the vastness of the entire universe, *information* is present wherever anything exists, whether or not it has been observed, measured, and communicated (or maybe we should say estimated and communicated). *Information* is every trait of a thing; it is intrinsically expressed merely by anything's being what it is. Therefore, *information* is enormously important. Since *information* is the product of a thing's existence, whenever any bit of *information* is observed, its thing can be certainly known to exist. As such, the first significance of *information* is not so much its precise expression of something's identity. Its bigger significance is its verification of that thing's definite existence. Where there's *information,* there is always an observable, estimable reality. *Information* is reality's smoking gun.

Information is not only present wherever anything exists, it is also present wherever there are interrelationships between existing things. Interrelationships of things form sort of constellations of *information*. For example, the total lunar eclipse of February 25, 1362BC is part of a constellation of *information,* although some of its *information* was a perception recorded by human hands. It's data in the NASA lunar eclipse catalogue is calculated from a related constellation of *information* intrinsic in the motions of the moon and earth around the sun. As the moon emerged from the earth's shadow that morning of 1362BC, the *information* of the event ended, except some Babylonian astronomer-priests saw and recorded it. Their record became data information to them, man's description of observed *information*. But to us, thirty-four-hundred years later, their data has become *information* of an eclipse observation. The clay tablet on which they recorded their observations testifies to us about the past relationship between a couple astronomer-priests in the right place at the right time to see the earth's shadow being cast across the setting moon as the morning sun rose on the opposite horizon.

But wait! Isn't there a possibility the astronomer-priests were fibbing? Maybe they didn't see an eclipse that morning as the sun rose on the opposite horizon. Maybe they only made it up to serve some political agenda. NASA's calculated lunar eclipse in Babylon's morning sky of February 25, 1362BC combined with the astronomer-priest's record forms a constellation of verifying *information*. Although its realities did not all occur at the same time or in the same place, the interrelationship between the eclipse, the astronomer-priests' record, and the NASA lunar eclipse calculation is a constellation of events expressing *information* by its pattern, truing any misperception about whether or not the astronomer-priests were just joshing. We will meet another such constellation in Chapter 9 expressing one of the world's most widespread and ancient narratives.

Observed *information* is the most basic element of analysis. It is where the awareness of any aspect begins. Although this point seems irrelevant and trivial, it is greatly important. If even one bit of *information* is observed, then the existence of something (or of an aspect of something) becomes knowable by it. Maybe the *information* is insufficient for that thing's characteristics and properties to be known, but the slightest amount of *information* attests to the sure existence of its thing, e.g., one photon reveals the existence a light source, and one Timex watch reveals that there was watch maker.

Observed *information* is the essential concept of empirical observation, the primary element of scientific investigation. It is claimed to be the starting point of all knowledge. At one time, empirical observation tightly intertwined with logic. Together they ruled the scientific roost. By the 1960's, the late, great, science historian and philosopher, Thomas Kuhn, revealed that even empirical observation had become secondarily important. He discovered that paradigms were actually being treated with more importance than observations.

A paradigm is whatever picture scientists think reality's jigsaw puzzle will look like once it was been mostly assembled. It is what they would expect to know if all of reality's *information* were to be seen at one time by a single observer. But science's problem is that reality's puzzle does not come in a box with a picture on its top (or so they claim). What science thinks an assembled puzzle might look like may not at all be what the full sum of reality's properly fitted puzzle pieces would actually show. Thomas Kuhn discovered that scientists project imaginary pictures onto their research, using imagination tainted perspectives as guides for directing further research. That injects bias into their studies. Bias shapes the paradigm, and the paradigms, in turn, further defines the bias.

But what if the puzzle picture as expressed by their paradigms doesn't look like the puzzle's picture as expressed by reality? Might they miss finding real puzzle pieces because their biased picture hasn't implied either the shape or color of those pieces? This was Kuhn's point. Each missed piece of *information* leaves an unfilled hole in the puzzle. Yet, instead of suspecting their paradigmatic picture might need correction, they enhance the problem by using their own paradigm as a template to fashion data into imaginary puzzle pieces called "theories" and "models" for plugging the holes in their puzzles left by the *information* they ignored. They could have found reality's puzzle pieces meant for those holes by simply researching without paradigmatic bias.

Thomas Kuhn discovered that, throughout the epochs of science, many times new empirical observations contradicted long held theories and paradigms. Eventually the observations' need for proper inclusion fundamentally changed (or sometimes completely eliminated) a theory or paradigm. The researchers would struggle against such mounting clues until the weight of the evidence was too much for their imaginary pictures to bear. So finally their accepted paradigm would shift, sometimes drastically, to account for the new *information* conflicting with their misassembled puzzles.

Perception is merely a reflection. It belongs to humans. It is man's reflection of reality's *information*. Reflections express more than reality. They also express characteristics

Clear Signs of Trouble and Great Joy

of the reflecting surface. A lady wondering how a dress looks on her does not stand in front of a circus mirror. Well polished brass reflects too, but it interjects its own. brassy hue. Your reflection in a pool of water is broken apart and twisted up by the ripples on the pool's surface. *Information* is similarly distorted and broken up by the agenda of its observer, his mind's pool of accumulated perceptions, desires, and fantasies, the paradigm of his soul.

The more a heart is inclined towards an agenda, the more its mirror is bent. The more the mirror is bent, the more its paradigms become unrealistic. Consequently, the less a person is able to recognize reality's actual *information* reflecting in him, the more he employs theories and models to form his perspectives. This is when perception ends up being a dangerous mess. We wonder why an ocean of deceit now drowns our news and politics, why people spit gunfire into crowds, fly jetliners into buildings, and snuff out children before they've even been born, yet we publicly encourage our surviving children to desire whatever they want, dream whatever stirs them, and worst of all, create their own realities. Perception truly belongs to humans. Mankind's mess reflects their ownership.

When we fail to realize the difference between reality and perception, we fail to realize the importance of accurately reflecting *information* in order to approximate reality's emerging picture. A more faithful reflection of *information* straightens and polishes the mirrors of our souls.

The simple point is that reality is one thing, but the human mind is something else. There is no automatic and direct link between the two. People think and live as if their minds automatically know what's real. We do know the general realities of our immediate situations: the poisons we avoid, the food we eat, the clothes we wear, the tools we use, the traffic we encounter on the road. Certainly our perceptions of those seem very automatic, rather direct, and pretty much accurate. But not all realities are so immediately engaged with our present situations and accurately discernable at a glance. Not all realities are even empirically observable. Most *information* is more scattered and seemingly disconnected. Many important matters require deliberate and directed investigation rather than the automatic sight of whatever is in front of your eyes at the moment.

Misperception, on the other hand, is very automatic. Accurate perception must be guided by the constraints of carefully observed *information*. But misperception bubbles up without constraint from the many erroneous paradigms we maintain. Automatic perception is always distorted by the self. But the perception constrained by deliberate investigation of relevant *information* eventually arrives at truth.

> Generally speaking, what you and I believe reality to be like has no effect on what reality really is like. Likewise, what our best scientists believe reality to be, or what the majority of the population believes reality to be, or what a yoga master in an enlightened state of mind believes reality to be, has little effect on what reality really is. As used in the correspondence theory of truth, "reality" is not "your reality," "my reality," "Timothy Leary's reality," the reality of an acquaintance under the influence of strong hallucinogens, or any such thing. Instead, "reality" refers to "real" reality: a reality that is completely objective, generally independent of us, and generally speaking in no way depends on what people believe that reality to

Information

be like.[4]

Noam Chomsky argued that intellectuals should be responsible for searching out truth, exposing lies, and revealing reality. He wasn't far from the truth, although he was a world away from it. Intellectuals should carefully engage *information*, not rejecting any of it, in order to form and communicate perceptions accurately. Chomsky was right about the responsibility of searching for and communicating truth. However, he was wrong about this being the duty of only intellectuals. Indeed, it is everyone's duty. Searching for and communicating truth free of distortion is a responsibility everyone owes to one another. If everyone was intellectually responsible, entire societies would not be led into tragic social disorders by the mental errors of an overly influential few.

But unfortunately, Chomsky took this useful term captive to describe a general mass of people turning their intellectual duties over to a few elitists eager to be considered superior. So, with all due regard to Mr. Chomsky, we will recapture the term to convey every individual's responsibility to accurately investigate, recognize, understand, and reflect reality through his own, unbiased quest for *information*, as well as his responsibility to accurately communicate what he's learned to others.

Intellectual responsibility is neither automatic nor passive. It requires persistent, deliberate, and disciplined investigation and reasoning. These are processes within the abilities of even less than moderately intelligent men and women. They are natural processes even animals employ. They are practiced more than learned. But they no more happen automatically than do washing your hands, sweeping the floor, or mowing the lawn. Intellectual responsibility demands personal involvement for discovering truth rather than resignation to preachers, teachers, mentors, politicians, pop-icons, Hollywood stars, or even close loved ones. Intellectual responsibility dissolves cultural bias, frays the harnesses of political correctness, and corrects the errors of group think. It isn't because other people's perceptions are necessarily invalid that we individually need to employ intellectual responsibility. It is because critical thought medicates your own perceptual errors at least, even if everyone else yields their beliefs to the shifting, whispering, imaginative drifting of the great unwashed masses. Life is full of dangerous ideas that come riding out of the desert on unnamed horses.

Let God be true though every man be false.

Romans 3:4b

So why pass around distorted beliefs like the common cold? We were perceptive enough to panic over Covid-19. Where's the panic over misperception, itself, gone viral? Only you can verify the accuracy or discover and correct the distortions of popular misconceptions. It is your responsibility. Being intellectually responsible is first to do the critical analysis, the verifying, and the corrective reasoning to your own knowledge base, thoughts, beliefs, and attitudes, and then to the cloudy perceptions blowing over you those great dunes of deceit, ever shifting and whispering erroneous ideas for burying good lives. Intellectual responsibility requires analyzing *information* honestly, measuring it carefully, categorizing

Clear Signs of Trouble and Great Joy

it accurately and consistently, then reasoning with it rationally.

But most importantly, it never casts controverting *information* aside. For intellectual responsibility always knows there is a definite reality wherever there is *information*. Wherever information is seen, a reality must be considered. Having a mind willing to consider what the eyes do see is to have a mind which impartially engages all of the *information* it meets.

Nor is intellectual responsibility complete until you begin behaving in accord with the ideas it has verified and corrected. Every implication made by *information* must be accepted in order to reflect reality faithfully. By faithful reflection, your search for further *information* becomes better directed. It is like truth has awareness. The more you responsibly correct your knowledge base, your thinking, beliefs, and behaviors, the more ability develops in you to distinguish reality from myth, truth from deceit, and *information* from perception. As intellectual responsibility corrects your paradigm by humility, respect, and honor paid to *information,* your paradigms will point you in more realistic directions for the discovery of more *information*. That is truth's awareness.

Although some people are more intellectually responsible than others, most are pitifully careless, allowing their consciousness to drift with the undercurrents of their past feelings and impressions, never measuring thoughts and feelings by reasonable comparisons with *information*, never evaluating their own sensibilities, never aware of the risks raised by unrealistic mental conditions, always impugning others for their own misfortunes, and blaming bad fortune for the constant march of disappointment their lives have become. I know this by personal experience. I once was it. My life became a collection of chaotic and destructive mental patterns causing self defeating habits and dismissive attitudes. I was the moth drawn to the flames of carelessness. I flew so close to insanity that deliberate, intellectually responsible action was my only remaining option.

We know conditions, whether tangible or intangible, by their patterns. I was able to untangle my bad habits, attitudes, ideas, and emotions by noting patterns they were making in response to each other. Patterns are *information*. They are important to note because they indicate the occurrence of processes. It is usually favorable to know what process might be developing effects in your life. So we study the patterns around us to know what processes are developing which circumstances. For example, if the news reports no longer present to us today the same perspectives they presented only months ago, either some reality has changed, or the news has become lies. Either is important to know. If it was cloudier yesterday than the day before, and it is cloudier today than it was yesterday, then, when the breeze grows strong and churlish, get your umbrella ready for rain. Patterns are informative about how you should think and what you should do.

The Bible proposes that certain conditions will arise when the process of life is prepared to end the chaos of death. Some of those conditions will arise by the deceitful, destructive processes of evil people engaging a chaotic spirituality of self service. But other processes will arise by the righteous Lord of Spirits (as the ancient prophet, Enoch, referred to God)[5], stirring within those who always work order in consideration of every reality, including the Bible. Jesus said there would be patterns at the end of the age. Technically, He called them signs in the sun, moon, and stars (Luke 21:25). But to be signs they must

Information

make patterns, such as the alignments of three heavenly bodies producing an eclipse at a certain time when something particular is occurring on the earth. Jesus proposed that God's process of eliminating evil would entail signs, i.e., patterns.

If you've been paying attention to these kinds of ideas over the past couple decades, you then know of several, rather interesting, truly real patterns amongst various heavenly bodies. Greeting any news of such signs with some amount of reservation is understandable. In fact, it is proper to investigate before believing. Yet, if reservation derails investigation instead of guiding it, then reservation is not being the operative process, but rather, prejudice is masquerading as rational thought. Intellectual responsibility will engage reservation for steering investigation towards sensible coherence. Reservation does not close one's eyes to research; bias does that. Investigation is also necessary before disbelieving something.

A great number of patterns in the recent past and near future motions of the sun, moon, earth, and stars have been empirically observed and reported by many people. The reality of their occurrences are not debatable; you can find them yourself through patient investigation. They are real. Their patterns are discoverable within the NASA logs of lunar and solar eclipses. Of course, it wasn't NASA which found the patterns. But, thank God, scientists do not have to find things for things to exist. Other people are equally capable of finding what scientists could find if scientists were unbiased enough to look. Nor do the objections of closed minded preachers make these patterns disappear. Many preachers rail extensively at them. But the last time I checked NASA's website, the patterns were still there. So I've included a couple appendices in this book to show you how to find these patterns yourself, regardless of scoffing scientists and preachers.

While you are researching, note how these patterns consistently correlate with Biblical history over thousands of years. The correlations are very different from astrology. Astrology interjects its own meaning into the entire mass of common, chaotic planetary motions to suggest what you should be and what might happen to you. But the patterns discussed in this book are defined by themselves. They consistently occur around times in which history fulfilled Biblical prophecies. Astrology theorizes possibilities about anything. But these patterns stand like testimonies only at events prophesied in the Bible and at historical processes developing into the future fulfillment of Biblical prophecies. They clearly testify to a reality producing those processes, that history fulfilling prophecy, and these patterns. They testify to the work of The Holy Bible's I AM WHO I AM, and therefore, to His reality.

By noting past correlations of such patterns with historical events, it is sensible to understand that similar patterns coming in the near future might be testimonial to whatever will be occurring then. Isn't it David Hume's principle that you can expect things of the future to occur the way we've observed things of the past to occur? Or do we now throw Hume under the bus because we have observed the Bible engaging his proposition? When considered as a whole, these patterns articulate a message which is unmistakably coherent with Biblical themes.

Most people will tend to deny this without further investigation. After all, hasn't science informed us that the Bible is hooey? Yet these patterns in the sky, history, Biblical themes, and both fulfilled and yet to be fulfilled prophecies are *information*. Where there is *information*, there is a reality generating it. These patterns are far too extensive and interre-

Clear Signs of Trouble and Great Joy

lated to be rationally cast aside as mere coincidences. Since anyone who looks for them can find them, they are not pareidolia (the mind's natural and strong tendency to see familiar patterns amongst the chaotic mix of everything). Many people will accuse them of being pareidolia simply because they are consistently cohesive with Biblical themes, narratives, and prophecies. But this array of patterns remains empirically real. It will exist in spite of accusations to the contrary. It forms a systematic network of imagery rising of its own accord to the level of discernable communication, rather like a sign language. Pareidolia does not do that.

The Bible says there will be signs. There are signs.

As such, the existence of these consistent, extensive patterns demands an intellectually responsible explanation. Science will think of them as irrelevant anomalies to be utterly ignored and deliberately discounted because its scientific paradigm fails to explain them. In fact, these patterns destroy their scientific paradigms. "Anomaly" is no explanation; it is the whitewash on science's dumpsters filled with controverting *information*. On the other hand, the Bible will thoroughly explain these signs. And recent history justifies the Bible's explanation.

Atheists insist they have never discovered any evidence (*information*) of God. But that statement and science's accusation of the Bible being only human literature can not be made scientifically. They are religious statements. They are statements of belief rather than observation. The Bible stands in our physical realm as a very real object demanding an intellectually responsible treatment of its propositions in accord with the *information* of its existence, at the very least.

If indeed God exists as the Bible describes Him, we would have for His evidence precisely the public circumstances we see today: some people denying His existence by their failure to individually observe evidences and other people professing His existence by their wealth of individually observed evidences. For the Bible proposes that God shows Himself to whomever He wills and hides Himself from whomever He wills, until He wills to act. Having the eyes to see and ears to hear the evidences of God, as Jesus often stated, comes from the willingness to pay attention to the sense *information* makes of itself more than to the sense people desire to make of *information*. A person's willingness to not pay attention to something is the most fundamental form of bias: out of sight, out of mind.

The Bible is replete with concepts such as some having eyes to see while many have eyes but see not. If the reality of "a God" accords with Biblical propositions, then His existence will not be constrained by the necessity of empirical observation. It is actually arrogant to think everything of reality must first check in with one of mankind's five senses before it can even exist. Even common sense expects an enormous number of actual, physical entities truly existing far beyond mankind's ability to sense them. It is entirely possible that one or more completely nonphysical realms do exist and are defined by whatever laws of processes their realms might entail, not requiring so much as a tweaking of mankind's nose in order to be what they are. Therefore, the Biblical God is indeed an objective possibility, and the Bible's proposition about His "undetectable", spiritual realm of entities is a completely logical construct. Moreover, the proposition is frighteningly well evidenced in the history of our physical reality.

Information

Science has fetched *information* from many unobservable places of our physical reality by observing actual affects theorized about unobservable entities -photons, electrons, and neutrons for example, and black holes, too. Why should the Bible's unobservable spiritual entities be treated any differently just because they do not share in our material substance? Could it be because belief in the black hole does not entail accepting a bona fide moral code running counter to most of mankind's basic intentions? The Bible proposes, and millions of mundane observers testify, that spiritual entities interacting with man and beast have indeed left empirically observable marks on our physical realm. Their affects usually gravitate around points of interest involving a bona fide moral code.

For example, it is undisputed history that Sennacherib besieged Jerusalem one afternoon in 701BC. Sennacherib was not the sweet little boy from next door. His ways of dealing with rebellious vassal kings left many hides and heads adorning neighborhood walls. The smarter of his vassal kings never dared to rebel. But history's question ever since Sennacherib's siege of Jerusalem has been why he went back to Nineveh the next morning in quiet resignation, leaving Hezekiah, his rebellious vassal king, happily unharmed. The Bible says God sent His death angel into the Assyrian camp, wiping out nearly the entire army. Some historians say plague infested mice sent Sennacherib packing. Plagues don't happen overnight, neither The Black Death nor even Covid-19 acted so fast. Maybe the camp chef served bad burgers for dinner that fateful evening. Whatever happened, historians admit that it occurred far outside of Sennacherib's bloody normal.

> At the end of the day, all accounts -the Assyrians, the Bible, and Herodotus, interpreted events. They didn't invent them.
> Something unexpected happened to the Assyrian army, which the people of the ancient Near East attributed to divine meddling.[6]

Maybe the people of the ancient Near East knew it was divine meddling, while today's scientists and historians whitewash their ancient knowledge with assumptions about what they would rather believe. However, that isn't the point. The point is that whatever was the cause of Sennacherib's flight left a devastated army and Sennacherib's fleeing response as marks in history. It was ascribed to an unobservable entity by the historians of its day. Today, it is assumed to have been caused by an observable entity. The ancients were closer to the event, and therefore, would know better; the moderns are deeply invested in theories, and can, therefore, distort and cover up. Today's paradigm keepers do not wish to apply the same investigative attention to those marks in history as they've willingly applied to other trace evidences for discovering such entities as the equally unobservable black hole. Holding out for accurate *information* pays dividends in truth's currency.

Thomas Kuhn demonstrated how researchers construct data and facts by manipulating *information* into the essence of their own paradigms. "Data", "facts", and other terms regarding "scientific" knowledge diminish life's billions of mundane *information* observers and users into irrelevance, however little or much *information* they might actually be observing. Those ancient historians were mundane *information* observers. But data and facts are used to obscure vast horizons of *information* which can not be quantified by the

Clear Signs of Trouble and Great Joy

paradigm-conformed methods of "qualified researchers". Data and facts are not *information*. They are interpretations of *information*. Every living individual observes a lifetime of *information*, most of which is never interpreted into either data or facts.

Therefore, it is only natural that today's science sees no data or facts supporting God's existence, although plenty of His *information* undeniably exists. God is not part of generally accepted science's paradigm. Thus, in order to be considered normal, members of our current general public, in spite of their individually observed *information*, must resign their better judgment about God to science's there-is-no-God paradigm without investigating the validity of either that paradigm or of their own better judgment. After all, hasn't science produced airplanes that fly, bridges that stand strong under tremendous loads, and the precise procedures for the gentle handling of nitroglycerine? Why would science be wrong about excluding from reality the Bible, its God, and His highly effective death angel?

The accurate understanding of physical processes does not mean the assumptive conclusions of the soft sciences (cosmology, geology, paleontology, archeology, psychology, sociology, etc.) rise to the same level of certainty. Their conclusions are too often construed from attitudes towards *information* sprinkled with a little theory ladened data and topped with a paradigmatic fact or two. Kuhn ripped open the veil of deception hiding real biases of scientific methods, research, and conclusions. He revealed through historical analysis how research projects, theories, and the paradigms constructed by bias driven data and facts present selective criteria for qualifying what will be considered as data and facts and disqualifying what will not be considered as such, therefore prejudicing *information* to the service of theory and paradigm[7] instead of humbly following *information* to reality's own, astonishing displays.

Radiometric dating is a great example of theory qualified by only paradigm ladened data and facts. David Montgomery glorifies the radiometric process in his book, *The Rocks Don't Lie*.

> Using the right tools, the age of a rock can be read like a geologic clock because radioactive isotopes decay at a fixed rate. Radiometric dating is based on the fact that younger rocks have more of the initial parent isotopes of their radioactive elements and older rocks have proportionately more of the daughter isotopes produced by radioactive decay. If you know the half-life of an isotope -how long it takes for half the remaining amount to decay- then the ratio of the parent-to-daughter isotope now in a rock tells you how long ago the rock crystallized.[8]

Montgomery's statement that a rock can be read like a geologic clock is tantamount to the statement of a fact. Yet who states this fact, the rock, or the geologist? Even though Dr. Montgomery says rocks don't lie, rocks don't talk. Rocks merely present the *information* of their current volume, shape, texture, mass, chemical composition, inner structure, discovered location, etc. at the moment those aspects of them are being observed. If a rock could say anything, whatever it would say about its past will have been said in the past, not today, because things break down and otherwise change over time. The past is when the *information* of its past composition would speak of its past. No Babylonian astronomer-

Information

priest took diction from the rock in the past for us to read about the rock's *information* back then. But today, the geologist interweaves the rock's current *information* with his own assumptions regarding the past in order to manipulate some "facts and data" for topping off one of his favorite theories about the past.

> There are also a variety of lesser factors which hinder an objective, critical assessment of isotopic dating as a whole. One of these is the well-known tendency of scientists to see what they expect to see. Hetherington (1983) provides many examples of this, showing how previous scientific measurements and observations had been biased by theoretical expectations. ...While nearly everyone agrees that fabrication of data and/or falsification of the same amounts to dishonesty, this appears to be very rare in science. By contrast, there is a large "gray area," which includes some manner of skewing the selection of data, and this is much more common in science than deliberate fraud. This skewing of data is what is endemic to studies in isotopic dating. The uniformitarian geochronologist, in good conscience believing that certain results have no geologic meaning, chooses not to publish them. No dishonesty is involved.[9]

No consciously deliberated dishonesty may be involved. But the failure to consider the affects of one's own pressures, biases, and assumptions upon his work is a subtle, most often unconscious form of dishonesty. Promoting the integrity of one's own work without fully assessing the integrity of one's own procedures is another subtle form of dishonesty. Moreover, honesty requires data to accurately mirror the *information* from which it is developed rather than the assumptions of the data developer. Any manipulation of data beyond the aspects of its underlying *information* (or short of those aspects) is a less subtle, more conscious form of dishonesty.

If any fact is truly validated by its source *information*, the fact that the geologist wasn't present to count parent isotopes and daughter isotopes in the past when the rock crystallized would be that most valid fact. He knows that. Yet he only assumes what the initial proportions of their respective quantities *might have been* in the past, a deliberate addition to the *information* the rock only made in its past. And he further assumes none of either parent or daughter isotopes were washed into or out of the rock over its lifetime, because he wasn't present throughout the past to observe that critical *information* of the past. Nor was he there to measure what the radioactive decay rate of the parent isotope was throughout the past, more critical *information* he is unable observe. He proclaims to know its rate today, but his uniformitarian paradigm instructs him to think he knows that today's decay rate was the same as it was a thousand, five thousand, or a billion years ago. *Information* tells him nothing of the sort. *Information* only displays radioactive decay today. "Decay rate" is a concept developed far beyond the evidences our limited observation can supply.

> It is widely claimed that current lab measurements on most decay constants are accurate and unbiased. However, in view of the fact that the legacy of isotopic dating is unalterably tainted with the practice of using one

Clear Signs of Trouble and Great Joy

method to "calibrate" another prior to actual ostensibly-reliable lab determinations of decay rate, skepticism is in order. There is, in fact, definite evidence that earlier lab measurements of decay rate were, in one way or another, influenced by what then was perceived to be the correct value of the decay constants.[10]

The geochronologist assumes what he is unable to observe so he can theorize what *information* has not said. He may not think of that as being dishonest, but if a rock's *information* could talk, the dishonesty of asserting "assumption" in the place of "verified conclusion" would probably be the first point it would like to discuss.

"Assumption" and "presumption", even in science, are euphemisms for "I made it up", whether or not they are important to the processes of discovery. However carefully assumptions and presumptions are formed, their basic nature always consists of at least some imagination. Imagination is not reality. It is a personalized reflection of reality at best. Theories and paradigms are also just images in the mind. That theories and paradigms are elaborate assumptions being only probable at best, and merely possible at most, is usefully forgotten. An assumption is never a certainty. Assumption isn't *information*. It isn't even data. Theories and paradigms are not verified realities. Therefore, no theory disproves another theory any more than one guess disproves another guess, or one imagining disproves another. Every valid possibility deserves complete consideration to allow *information* to sort probabilities from possibilities, to distinguish which ones are actualities, and to discover which ones are utter BS (bad science). Such consideration is honesty.

If one desires to be honest in his scientific endeavors, the following questions are then relevant: how honestly has the processing of *information* into data and facts been done, especially when only one theory (uniformitarianism/evolution) has been allowed by orthodox science to form the paradigm used for selecting and processing *information* into facts and data? Why was certain *information* selected for processing? Why was other *information* rejected? Was there any bias employed in the sorting? Was the selected *information* processed without bias -no tweaks here, twists there, no observations stretched, shrunk, or twisted to fit a theory, no cutting, pasting, or splicing of *information* anywhere? Was the rejected *information* rejected by prejudice? Were the research processes formulated to effectively reflect no more and no less than what the *information* itself displayed, or were they designed to deform *information* into facts and data trimmed and fitted to serve currently recognized theories instead of making an accurate reflection of reality? Were honestly sufficient samples taken? For example, can we really know the rates of radioactive decay over the past "four-billion years" by observing decay processes and environments during only the last one-hundred years?

Thomas Kuhn used science's own history to demonstrate how much the very act of processing observations into data ladens the resulting data with aspects of the scientists' own theories, i.e., with the scientist's own imagination, more honestly stated. *Information* is vital to the discovery of truth. Data is vital for supporting theory. The difference is immense. Yet the immensity of the difference fits comfortably into that small, four-letter word, "bias".

Information

> ...data are invariably contaminated by theoretical assumptions. It is impossible to isolate a set of 'pure' data which all scientists would accept irrespective of their theoretical persuasion, [Kuhn] argued.[11]

The immediate testability of data and facts in the hard sciences -math, physics, chemistry, etc.- boosts their theories over the bar of confidence to raise them up to the status of theorems, principles, and natural laws. Hard sciences' immediate and undeniable test results have supplied us with explanations of phenomena useful for creating the incredible technologies that are making us very comfortable, productive, and wealthy. But this golden track record of hard science tends to inspire a similar confidence to be misplaced in the far less successful track record of the soft sciences -cosmology, geology, paleontology, archeology, psychology, sociology, etc. Consequently, the general public perceives the theories of soft science as having been successfully tested to yield solid truth, like those of the hard sciences. But really, the soft sciences are little more than enormous collections of assumptions and presumptions fashioned out of attitudes, more than conclusions, regarding only a handful of observations.

The theories of the soft sciences are not so directly testable, involving more abstract concepts of the past, or of the inner workings of the mind, or of the interactions of people, and the likes. The tests used for their propositions and hypotheses are more about how well new observations fit old, well accepted paradigms. It is a bias. Generally accepted paradigms are constructed out of the same kinds of less testable propositions and hypotheses comprising the very theories those paradigms are then used to test. So they make uncertain bench marks for supporting conclusions, to put it mildly.

This isn't to say paradigms are useless. They are necessary. They're placeholders for undiscovered *information*. But when paradigms are projected as pictures for shaping *information* into "facts", "data", and other theoretical puzzle pieces, the resulting puzzle assemblage becomes highly prone to serious error caused by the ingraining of bias, especially within that paradigms' core concept. The need for tentative attitudes towards paradigms themselves becomes critical for properly aligning both science and personal perceptions with reality, above all, in regard to whether or not God has hidden His spirit realm from our five senses for better wisdom and unbiased rationale to find.

Faith in hard science's knowledge of the proper care and handling of nitroglycerine is gsoundly verified by simply observing a dropped case of it. Faith in the soft sciences' knowledge of the proper care and handling of the human soul is thought to be as well verified. But the worsening depravity of individual, social, and governmental behaviors after these last couple centuries of practicing psychology, political science, anthropology, etc. scientifically does not suggest such a sound rationale for any similar faith in the soft sciences. Their faulty theories blow up over much longer periods of time, supplying enough time for cultural perceptions to form around their dubious principles before they are even known to be exploding. Experiments with left-wing governance have been exploding bloodily over the past two centuries, yet more than a third of today's democratic populations still support communistic and totalitarian politics, while nearly two-thirds of the people support

Clear Signs of Trouble and Great Joy

their sapling stages of socialism, which often mature into various other forms of people conquering.

The soft sciences ignore the existential *information* that God's Word is. They treat it like a library of ancient bedtime stories. But reality can only be known by the consideration of ALL its *information*. A few centuries ago, science's puzzle picture looked very different. The major part of the Middle East, being Islamic, understood the world as having been created not too many millennia ago. The Western World, being mostly Christian, had recently arrived at a date for the creation event. East of the Muslim world, the Hindus figured there was no beginning of space or time, but there were cycles of creation and destruction. Many of the world's other cosmologies involved a primordial egg, a mound in the waters, or some carving up of a defeated god's carcass. The cosmologies of those days involved what they all held in common -gods.

A god of some sort, a mix of chaos and order, and some form of existence after death has been humanity's general paradigm throughout history and across the continents. It still is the general paradigm when considering individual human sensibilities rather than publicly pressured perceptions. Individual sensibility usually becomes more relevant than public perception the closer death approaches each soul. Everyone engages a god paradigm deep inside their selves, whether they perceive their Overarching Determinant to be their own selves, the universe, mere randomness, or whatever else.

The Holy Bible not only fit's the god paradigm, but it has reshaped much of the paradigm into a far more coherent understanding of reality than has any other religious writing. The Bible's paradigm makes more sense out of all observations than do the paradigms of normal science. Coherence is about beginnings and ends tying together in accord with the meaning of everything between them. In science, coherent meaning is the cause and effect of material interactions. In history, coherent meaning is cause and effect of human interactions. In Christianity it is both of those tied together by a Spirit of Truth acting upon a deceitful race. Christianity engages all reality, even the reality of your own human spirit.

Two-and-a-half centuries ago, the Western World was deeply engaged in a "God explanation" of all things. Their idea of science was to examine things for better appreciating the God of the Bible, and for finding more uses to which we could put His creation. They would have defined their acceptance of God as scientific had they employed the word "science" and its derivatives. But in those days science was called "natural philosophy", and scientists were called "natural philosophers". They subordinated their scientific efforts to their belief in the Bible and its God. It was the generally accepted puzzle picture of their time; it was their paradigm. Human history had presented sufficient observations of God's activities within human affairs to have laid, long before the natural philosophers, a bridgework of well reasoned belief in Jesus Christ and His Holy Father. This wasn't the universal condition of science in the Western World then, although it was the most prevalent condition, especially in the United Kingdom.

> In seventeenth-century England, natural philosophers surrounded their labors with a notably devotional atmosphere. Boyle and Newton, for example, did not make the sharp modern distinction between science and

Information

> theology, and they were highly serious about both their theological and their scientific studies...Lesser scientists, men like John Ray, John Wilkins, Nehemiah Grew, and many others untouched by genius, also turned their talents to expressing the majesty and wonder of God's revelation to mankind, a message that could be read not only in the inspired writings but also in His book of nature.
>
> Writings of the scientists fell into the Puritan posture of awed confidence in the face of a universe unfolding into comprehensibility at the touch of the "new philosophy," It was a sobering prospect, certainly, because God was in it and through it, but it was a prospect to be contemplated steadfastly. Those who were investigating and admiring the Lord's work were necessarily participating in the Lord's work.[12]

The God described by the Holy Bible offers a rational explanation of everything man has observed in life. The Bible clearly proposes a rational purpose designed into this universe by God. Until two centuries ago, this was by far the predominant, Western World perception of reality. But people in positions of political, social, and religious power have not historically been so enamored with Biblical explanations. By the end of the eighteenth century, philosophers and a new strain of researchers began striving to break free from studying nature for the admiration of a Biblical God. Philosophers were blazing new trails of attitudes for science's desired escape from God's perceptual and moral constraints. In Europe, especially on the mainland, people were increasingly less willing to perceive science as being constrained by the Bible. They were not seeing the natural, physical world as another book written by God to be read in the light of the Bible, as the natural philosophers were seeing it. Their desire was to leave constraint behind and turn off the old Bible's light. So science went calling upon assumptions for a plausible revision of the natural philosophers' generally accepted puzzle picture formed by interrelating the Bible, known history, and the visible universe.

That call came knocking on James Hutton's door. The philosopher, David Hume, had recently proposed that science could most usually expect unobserved things and processes to be like all observed things and processes. In other words, past processes could be thought of as being the same as currently observed processes, because, for the most part, man can consider his common experiences to be a good standard for what he might consider "natural" to be, regardless of everything he has not experienced. Hume constructed a light switch for Hutton to flick. Soon science was basking in manmade light, calling it Natural. The people of their times had never personally experienced any massive geologic event transformational on a global scale. So Hutton's imagination of all past geological history being the same normal, slow-motion events of local transformations seen in his own, comparatively microscopic lifetime was just a short step taken beyond Hume's guess. But the ignored Bible still attested to man's past experience of unique, "supernatural" events.

In the brief conception of this new scientific embryo of thought, the theorized processes of land formation became those of the grain-of-sand-at-a-time carried by a little water trickle to then be driven on by a windy gust, maybe shook a bit by an earthquake here, and heated a little by a volcano there. So, at Hume's inspiration, Hutton subjectively

Clear Signs of Trouble and Great Joy

imagined geologic gradualism for explaining the development of the earth's surface features by slow processes crawling through geologic epochs lasting far longer than those six thousand years Judeo/Christian/Islamic worldviews had portrayed for centuries. Hume's uniformitarianism introduced to science a blacklight for exploration by imagination, a process Hutton's new geologic epochs of time needed more than they needed *information,* a process performed under a light capable of hiding controversial *information* in the dark. Uniformitarianism and gradualism were processes of crafty rhetoric. They were products of philosophically ladened data and facts fashioned from selected observations shown only under that peculiar blacklight of God-denial.

These concepts were carefully sheltered from controverting evidences while they were being constructed. They were then launched as the challenger to God's authority in order to free mankind to construct utopia on its own terms. For God, they were an idea launched without consideration of fatal flaws that will lead to mankind's demise before the God they denied, His angels, and the faithful remnant of His people.

Amidst the young-earth culture of eighteenth century England, theories about gradual changes over epochs of geologic time successfully challenged the idea of a six-thousand year old earth, not because of its correctness, but simply because that's what the intellectuals of the day desired to believe. There was no more evidence for epochs of geologic time than there was for gradual changes. There were only carefully construed "data and facts". So, it took a number of decades to accomplish the paradigm switch. Theory ladened data and facts were brewed up to support crafty speculations meant to coax the eventual acceptance of the general public. Had the general public been more thankful to God for *information*, as Sadie is, the manipulated data and facts of the new, subjective geologists may not have been so submissively received.

The Bible also supplied a very reasonable and concise implication of geologic history: The Great Flood. It was a public belief before this new flood of "great" geological "insight" washed away ancient perceptions. The Bible's implication eventually fell out of favor because it was…well…Biblical, even though it was more explanatory of observations as seen in their natural light without need of man's blacklight. The Bible's *information* was old hat, even though it was yet *information.* So geologic epochs were soon read into dirt layers, which in turn rationalized seeing gradualism in fossils, and uniformitarianism in pretty much everything else. But *information* never confirmed Hutton's guess about either gradualism or uniformitarianism in general, as even the left-leaning Wikipedia acknowledges:

> Today, Earth's history is considered to have been a slow, gradual process, punctuated by occasional natural catastrophic events…[13]

> Though an unprovable postulate that cannot be verified using the scientific method, uniformitarianism has been a key first principle of virtually all fields of science.[14]

Does this mean virtually all fields of science base their theories upon merely dogma-

Information

tic, scientifically unproven preaching, rather than *information* driven conclusions? Science of origins is constructed out of paradigm driven conclusions, even though it may not be fair to say science bases its theories only upon mere dogma. But none of normal science's data and facts seem to be coherent with Biblical cosmology like ALL *information* rationally intersects the Word of God. The main difference between normal science and The Holy Bible isn't between God and no-God, it is between uniformitarianism and supernaturalism.

Each of those concepts requires a coherence amongst puzzle pieces for its conceptual structure. The coherence of uniformitarianism revolves around the nonexistence of God at best, or His Deist, noninvolvement at least. Uniformitarianism was philosophically brewed for replacing supernaturalism in order to put a comfortable distance between mankind and any concept of a mutually necessary obedience to any Higher Authority. It wasn't a discovered "reality". It was entirely conceptualized from man's shifting, whispering philosophical standards.

The Western World was becoming ever more enamored with the idea of man governing himself, an attitude born in the struggle to escape monarchy, an attitude metastasizing towards the struggle against God's expected governance over man. Whether purposefully or accidentally, uniformitarianism supplied to the concurrently forming democratic/fascistic philosophy (also in its infancy at this time) ideological support for the soon to emerge separation between church and state, cutting the church out of public influence "for ever", constraining the church's affects upon the state's age old pinnacle of power, breeding draconian governance as the logical extension of Hume's carelessness.

Today, any construction upon the paradigm of normal science must be made of data and facts conformed to the no-god-involved, uniformitarian dogma, which even Wikipedia knows to be just as unproven as is science's proclamation that the Bible is "not provable". Such is subjectivism. Such is normal science.

The non-verifiability of uniformitarianism and gradualism slipped past science's rock solid stance upon observation before the later empiricists had set observation as the critical boundary for science. In the absence of a strict demand for observation, Charles Lyell popularized Hutton's guesses, presenting his dogma palatably enough to feed the public a well seasoned, yet unverified, non-evidenced basis for virtually all fields of the soft sciences. Thus, from the *information* of fossils embedded in dirt layers was manufactured the "fact" of billions-of-years for Charles Darwin to process into an evolutionary guess still in search of any fossil which might conclusively evidence it. Fortunately, *information's* God still lives healthily and unscathed, though no longer as well noticed by the masses.

Once Darwin's evolution became popular, he was overjoyed to announce that natural selection also eliminated all purpose and design from nature, thus freeing the universe from any concept of a Designer. But what line of reasoning could possibly conclude purposelessness and non-design simply from the similarities and dissimilarities of bird beaks and other variations amongst life forms? Even if the universe and all its organisms did gradually evolve throughout eons of environmental pressures selecting minute changes in organic forms, none of Darwin's data logically construes any demand that those pressures were not designed to purposefully steer evolutionary progress towards life as we know it. The purposelessness of Darwin's proposed mutations is neither empirically observable nor

Clear Signs of Trouble and Great Joy

logically constructible. He just imaginatively pronounced it! The purposelessness of nature is not a construct of *information*. It was simply made up to drown God in a sea of atheistic assumptions dipped from rivers of manipulated data and facts. Uniformitarianism, gradualism, purposelessness, and non-design are constellations of guesses drawn around a scant number of craftily selected observations. They are speculations based upon social ambitions rather than reason, inspired by mounting desires to abandon the historical God so scientific man could become his own, utopian authority. The entire core of evolution's paradigm is speculation. How real can speculation be? How subjective it truly is!

In recent years evolution has met rising levels of opposition from even the nonreligious ranks of science. The chilling reality is that there are alternative explanations for life and existence coming from sources of knowledge other than the "magnificent school" of empiricism (in which uniformitarianism, gradualism, purposelessness, and non-design were all given hall passes the day empiricism was taught).

> Scientific explanations are supposed to be testable, they have "empirical content," their component laws describe the way things are in the world and have implications for our experience. But almost from the outset science has explained by appeal to a realm of unobservable, undetectable, theoretical entities, processes, things, events and properties. As far back as Newton, physicists and philosophers have been uncomfortable about the fact that such things seem both explanatorily necessary and unknowable. Unknowable, because unobservable; necessary because without appeal to them, theory cannot effect the broad unification of observations that the most powerful explanations consist in.[15]

> [rationalism is] the epistemology according to which at least some knowledge we have is justified without empirical test. But if some scientific knowledge is derived not from experiment and observation but, say, rational reflection alone, then who is to say that alternative worldviews, myths, revealed religion, which claim to compete with science to explain reality, will not also claim to be justified in the same way?[16]

Explaining something is one thing. Justifying the explanation -squaring it with ALL *information*- is quite another. Empirical observation works well for justifying the concrete explanations of the hard sciences. But it doesn't work so well for justifying the more abstract explanations of the soft sciences.

Consequently, in the twentieth century, Karl Popper tried to rescue theory formation from the problem of justifying inferences drawn about things too obscure to observe in total. He used the whiteness of swans to demonstrate his point, since in his day, all swans were thought to be white. But all swans everywhere throughout time could not be observed to see if just one was ever not white. Rationalism might suppose one's existence. But empiricism was yet ruling the knowledge roost; the off-white swan had to be seen, not just supposed.

How could empirical science state that all swans were white when it could not examine all swans? Obviously, science couldn't make the statement, so it should have

Information

confessed that limitation. But science had burdened itself with the task of replacing God's authority with its own. Therefore, it had to make statements about unobservable matters. Popper's solution was to make scientific statements falsifiable. Then, until the sight of one black swan would correct the scientific statement, it could be considered scientific regardless of the missing observations. Everyone was then to go looking for a black swan falsifying the statement. But for no-God theories, practically nobody does.

So Popper's problem with Hume's guess is the same problem we should have with inferring the constant rate of radioactive decay everywhere throughout all time when we can measure its rate of decay only on this Earth only in this one, fleeting, scientific moment of man's infinitesimally short existence. Intuition says the rate is constant. But *information* has not confirmed that. Einstein perceived the possibility of some black swan in the rate of time's passage and went looking for its off-whiteness. Could the radioactive decay rate throughout the unobservable past be like swans? Some scientists are saying there is not only a black swan in this matter, but also that shades of its gray have been observed.

But Karl Popper's falsification principle became more famously expressed as, "All swans are white except black swans seen by people who don't know what they're talking about." When even one empirically observed reality, which theoretically should not exist, is discovered existing, then honesty must adjust or discard its theory in accord with the observed reality. How fitting of this metaphor it was that those philosophers chose the swan's whiteness for the of their argument! The later discovery of the Australian black swan so clearly made their point that, today, any falsifying evidences of commonly held theories and perceptions are called "black swans". Alas, Thomas Kuhn's observation was that black swans get thrown into science's dumpsters far more than science cooks them into its theories. Sorry, Karl, as Tom noted, the chefs are biased. Especially against The Holy Bible.

Darwin's paradigm about nature's non-designed purposelessness will soon break up on an extensive reef of patterns discovered in the earth, moon, sun, and stellar alignments meaningfully correlating with Biblical themes and historic fulfillments of ancient Biblical prophecies. Only an Authoritative, Higher Intelligence can organize such consistently cohesive alignments. We will explore them in upcoming chapters. The mere existence of this *information* is a definite falsifier of Darwin's non-designed purposelessness of the physical universe. Generally accepted science's only options for saving its subjectively crafted paradigm of a God-less universe from those very real, intricately designed patterns will be to either categorize them as pure coincidence, or to publicly proclaim that their observers do not know what they're talking about. Categorizing them as coincidences will be irrational, since coincidence can only rise to a certain level before it demands to be acknowledged as evidence. The shear volume of these patterns demands an explanation other than coincidence. And their correlations with Biblically prophesied history all but closes the door on any "Natural" explanation for their very real existence. To deny their reality by accusing their observers of not knowing what they're talking about will be foolishly fraudulent in the face of their actual observability to anyone who spends enough time and effort to go and see them.

Therefore, in order for science to continue proposing the purposeless, non-design of nature, it needs to further limit what *information* it will observe and to build ever bigger

Clear Signs of Trouble and Great Joy

dumpsters for trashing ever more black swans. *Information* validating Biblical narratives and themes has been surfacing through the past couple centuries. Even the rate of its discovery has been increasing more with each passing decade. Ignoring *information* will soon not be a scientific option. It will become undeniably obvious that a primary ingredient is missing from the scientific stew, for the chefs continue to breach science's own, most fundamental rule of knowledge -empirical observation- whenever anything Biblical is found to be plainly observable. True science must consider EVERYTHING observed, not just what conforms to its paradigms. Honest science modifies its theories and paradigms in accord with the implications made by ALL observed *information* considered together. Dishonest science simply ignores any *information* which falsifies its own paradigms.

Science can secure its proposition of a purposeless, design-free, evolved, mechanistic material universe only after denying the Bible is God's Holy Word to mankind. For nearly two centuries science has based that denial of Biblical inspiration upon mere speculation about what it has not, and often, refuses to observe, while it twists the few, carefully selected tidbits of what it has managed to observe into paradigmatic data. The Bible not only speaks of design and purpose, it presents the Designing Purposer in a completely rational manner, even supporting its claims with empirically observable evidence.

Consequently, falsification of the Bible has matured into generally accepted science's unstated, not very well hidden agenda. But its endeavor has failed to discover any *information* that actually does falsify the Bible's claim to be God's script for mankind. It presents a great system of theories, speculations, and some paradigmatic data and facts by which it claims to disprove Biblical truth. But, at most, its selected evidences merely allude to speculations other than the Bible's explanations. Speculations and theories don't falsify anything. Science's data and facts are only *information* colored over by perceptions drawn from the same theories they are bent to support. It is the fatal flaw noted by Thomas Kuhn.

Science's hidden agenda is now in serious threat of being washed up against the reef of Biblical evidences rising out of the murky depths of the past. The inability of science to successfully hide and deny this ocean of Biblically confirming *information* is becoming its profound embarrassment. And, with every passing year, that embarrassing mass of *information* falsifying materialistic naturalism piles up higher and higher, rising even to the astronomical level of consistent, cohesive, coherent, perspicuously pesky signs in the sun, moon, and stars. Still normal science abuses its own principles of both observation and falsification in order to continue accusing the Bible of being make believe. It elevates its own guesses to the same dogmatic height from where the Catholic Church once stuffed its arrogant foot down Copernicus' throat.

> Let God be true though every man be false, as it is written,
> "That thou mayest be justified in thy words,
> and prevail when thou art judged."
>
> Romans 3:4

> all who swear by [God] shall glory;
> for the mouths of liars will be stopped.

Information

> Psalm 63:11
>
> So have no fear of them; for nothing is covered that will not be revealed, or hidden that will not be known.
>
> Matthew 10:26

Cliffs all around the world show neatly stacked layers of successive sedimentary and volcanic activity. These are what Hutton imagined eons of time formed by slow, gradual processes. But processes other than gradualism are often needed to successfully explain some real *information* contained in those layers. And processes other than gradualism are available to explain it. Mount St. Helens demonstrated how layering occurred in a mere day or two while looking as if it happened over millions of years. Layers do not prove eons. They only indicate the past occurrence of layering processes. Nor do layers describe those processes. Paradigms do that. Scientists fashion paradigms; layers don't, since, like rocks, layers don't talk; scientists do. Therefore layers do not discredit Biblical explanations; scientists do that, too.

Polystrate fossils seem to decry the millions of years evolutionists beg geologic layers and Timex rocks to sing in harmonious measure. Polystrate fossils require processes other than what entail countless eons to explain the layers through which they protrude.

> A polystrate fossil is a fossil of a single organism (such as a tree trunk) that extends through more than one geological stratum. This term is typically applied to 'fossil forests' of upright fossil tree trunks and stumps that have been found worldwide...The word 'polystrate' is not a standard geological term. This term is typically found in creationist publications.[17]

There is little doubt as to why "polystrate" is not a standard geological term. If uniformitarianism actually met the criteria of a scientific theory, then the observed, documented, truly existing polystrate fossils would definitely falsify it. Imagine any tree standing upright for millions of years while for eons sediments slowly crept up its trunk before volcanic ash blasted in to cover the sediment before eons of vegetation mats piled up prior to eons of sedimentation pressing those vegetation mats into coal followed by more sedimentation preceding another blast of volcanic ash, then again, more vegetation mats washing in before more sediment lays down, all occurring in sequential order through millions and millions and millions of years until the layering reaches the top of that poor, old, tree trunk and beyond. Throughout those millions and millions and millions of years each different layer, by its own chemical nature, petrified one section of the tree and turned another section to coal without the volcanic layers burning up the sections of the trunk extending through them. Wouldn't we love to build our homes out of such durable trees! Where have we ever seen a dead tree stand for even a hundred years without rotting to the ground? I live around dead trees in a semi-arid desert. I watch them decay and collapse in a matter of just a few years, sometimes decades, but rarely centuries, and never millennia. No wonder the term "polystrate fossil" is used by only creationists. Generally accepted science will no more speak the word than it will openly acknowledge their existence (though secretly they must, as they can't cram them

Clear Signs of Trouble and Great Joy

into a dumpster.)

Have you ever seen a dumpster successfully hide an upright tree trunk?

Therefore, science must speak of upright fossil trees, quietly acknowledging their existence with constrained reservation. Upright fossil trees are often found embedded in several layers of geologic strata, "too". And many of them seem to be associated with coal beds and volcanism, "too". Normal science offers three explanations for the admitted phenomena of upright fossil trees. The first is the scientific explanation creationists say is not possible: these uber trees stood through many thousands of years of strata formation. But of course, dirt layers now speak of only thousands of years rather than millions and millions and millions of years! Oops, that's not very consistent! Scientific theories are supposed to be products of consistency, not inconsistency. But whether dirt layers speak millions or just thousands of years, uber trees are still necessary to form upright fossil trees even if dirt layers speak only hundreds of years.

Another incredible explanation for upright fossil trees is that these upright dead trees were buried quickly and entirely, catastrophically, after which they fossilized. Untold millennia of erosion then completely exposed the upright fossil tree. Then millions more years laid down new sedimentary layers before vegetation accumulated over more millions of years, turning those sections of the fossil into coal. Finally, volcanic flows and ash piled up around the fossil. The process repeats a couple more times over the eons. But that theory requires more imagination to believe than Earth has observable realities, because it forgets to explain the fossil/coal/fossil/coal segmentation often found composing upright tree fossils, else it must call upon some black-box process able to turn sections of previously fossilized tree trunks into coal. Oops! More inconsistency.

The third explanation calls upon catastrophe (sshhh), both slow and fast. Volcanic catastrophes occur during which mudflows very quickly lay down strata. Slow rising swamps, lakes, etc. are catastrophes creeping up the tree trunks over the years. Gentle winds during catastrophic droughts and intermittent floods lay down sedimentary layers. Mount St. Helens was a catastrophe. Forests were leveled. Anyhow, combinations of catastrophes form the layering observed in upright fossils. But we need to accept the extraordinary coincidences of numerous catastrophes happening over a few short years in order to avoid believing in either the ancient Uber-Tree or the Biblical Flood. Polystrate fossils evidence very quick layering processes. Upright fossil trees evidence paradigmatic data formation.

In the early 1970's, a small museum in the old J.C. Penny's building on the corner of Main and Fourth Streets in Grand Junction, Colorado displayed a very strange steer skull. It was discovered half buried in dirt. The unburied, sun-bleached half was carbon dated to be about one-hundred-fifty years old. But the buried half was completely fossilized. Of course, the skull is no longer in the museum, at least it wasn't the last time I visited in the 1990s, after the museum had moved to Fruita. What happened between 1970 and 1990? Generally accepted science became so jealously guarding of its evolution theory in the face of mounting Biblical evidences that it began shooting every black swan it could hunt down. I have since speculated (without evidence) that this missing skull became another black swan in the dumpster, more of science's golden silence.

Silence is that mental space cleared away for fancy to replace truth. By silencing the *information* of polystrate fossils, layers of dirt can be tagged "millions of years" even though some layers must be known as representing only centuries, or even decades for no reason other than that upright tree fossils don't fit into dumpsters. And without some pesky steer skull exhibiting quick fossilization, the rest of the fossils are used to pin "millions-and-millions-and-millions-of-years" tags onto dirt layers. Potsherds, bone fragments, tools, and other objects of human activity don't sport manufacturer tags about when, where, and who made them, either. They're all just stuff-in-the-dirt until some archaeologist, paleontologist, or geologist tags his presumptions to them by some report reviewed by his peers for making sure his conclusions properly cultivate generally accepted fancies. (Maybe I exaggerate just a little.)

> All in all, the evidence suggests that scientists, who by definition search for truth, are not immune from the use of deceit in their experimental techniques or in their reports of their results.[18]

After all, paleontologists, archaeologists, geologists, and other scientists are people, too. Life presents many conundrums to every individual. Conundrums always test moral integrity. When the boss needs an employee to fudge a report, does that employee feel no pressure to retain his job?

> ...it remains a fact that everybody lies. The process of psychological development is closely related to how we learn to communicate or miscommunicate with both ourselves and others.[19]

> Let God be true, though every man be false.
>
> Romans 3:4

Other *information* displayed by reality comes in tiny, easy to miss tidbits. These tidbits conveniently fall back into unknown obscurity when popular thinkers ignore them. It doesn't matter how much the unknown thinkers attend them, or how right their thinking might be, because unknown thinkers don't warrant the general public's time of day. What they say is rarely heard.

Therefore, "Out of sight, out of mind," is the most useful concept for supporting theories against perfectly good, truly existing *information* falsifying them. After all, belief is made from what is in mind, not from what's out of mind. Even the world's most popularly developed theories stand strong against the reality of completely falsifying *information* simply because theory guardians are able to scrub that falsifying *information* from public sight, while making certain their own, sheltered theories are very widely viewed. Why destroy your own theoretical framework by discussing obscure, forgettable tidbits of falsifying *information*? The more acclaimed are the consortium of theoreticians involved in such intellectual malfeasance, the more such intellectual malfeasance is able to demonize every controverting tidbit of reality that refuses to recede into obscurity.

Everyone knows that the unknown thinker doesn't know what he's talking about.

Michael A. Cremo, a research associate of Bhaktivedanta Institute, specializes in the

Clear Signs of Trouble and Great Joy

history and philosophy of science. He researched, documented, and revealed major scientific cover-ups in the informational structure of science's quest for early man. Richard L. Thompson, Ph.D., is a founding member of Bhaktivedanta Institute, which is dedicated to studying the universe and life's nature and origin in the light of India's ancient Vedic literature. Their book, *The Hidden History of the Human Race: Forbidden Archeology*, offers over seven-hundred pages of technical descriptions and illustrations of tools, manmade objects, incised animal bones, homo-sapiens-sapiens fossil remains, and footprints found in dirt layers evolutionists ascribe to epochs far older than they claim to be the time of intelligent, tool-making, human activity. Gasp! "Sssshhh! Cremo and Thompson don't know what they're talking about!" In fact, in most scientific circles, their work is categorized as pseudo-science, not because its documentation is impeccable, not because their research can be duplicated, but because the *information* they discuss completely destroys evolution's timelines and exposes the biases employed in its fabricated fancies.

Cremo and Thompson document the same things orthodox science busily hides.

The heart of the matter is that no framework of understanding, paradigm, or theory will guide research towards the discovery of reality unless reality's *information* is allowed to participate in every process of modifying or falsifying those frameworks. Any framework of understanding must emerge from ALL relevant *information*, no matter how intricate the *information*, no matter how obscure, no matter how disproving of favored theories, ideas, or ambitions, no matter where found, or found by whom; all *information* must participate in paradigm development. Otherwise, research will veer towards imagination more than reality until it finally crashes into the barrow pit of deceit.

Jerry Bergman, PhD is an adjunct associate professor at Medical University of Ohio. He has taught biochemistry, biology, chemistry, and physics at the college level for thirty-five years and holds nine academic degrees: one associate and one bachelor degree, five master degrees, and two PhDs. He's spoken across the USA. Many of his articles and books have been translated into several languages. After carefully weighing the mass of literature ac-cumulated on the evolution side of the creation/evolution debate, he noted

> Another factor that moved me to the creationist side was the underhanded, often totally unethical techniques that evolutionists typically used to suppress dissonant ideas, primarily creationism. Rarely did they carefully and objectively examine the facts, but usually focused on suppression of creationists, denial of their degrees, denial of their tenure, *ad hominem* attacks, and in general, irrational attacks on their person. In short, their response in general was totally unscientific and one that reeks of intolerance, even hatred.[20]

Don Patton, PhD is another whose ideas evolutionists ignore when they're not attacking him. Scoffers and skeptics love to attack his credentials and assassinate his character because he publicly presents, without orthodox review (OMG!) *information* that destroys the fundamental foundations of evolution, not contrived data and facts, but reality's own, observable, unadulterated *information*.

Dr. Patton's presentations on Ica stones are favorite targets for skeptical attacks.

Information

These stones, found amongst ancient Peruvian graves, are engraved with scenes of Peruvian life. Some were presumed to have been made by recent, Peruvian locals for cashing in on the tourist trade, according to skeptics. Those stones depict a variety of dinosaurs interacting with people. Some dinosaurs are depicted as embattled by men, and others in play with men; some are even depicted as beasts of burden. Of course, this doesn't square with evolution. So, "Dr. Patton doesn't know what he's talking about! These stones were all made by the tourist industry." After all, everyone knows dinosaurs lived millions of years before humans even began to evolve. Indeed! They must have lived while those uber-trees were becoming polystrate whatnots. Or does the consortium of scientists just teach us to think that? What does reality's *information* teach us?

A Peruvian farmer, Basillio Uschuya, is the source of the popular hoax theory. In 1973, and again in 1975, he confessed to having forged thousands of those stones. Uschuya later recanted his story to a German journalist, claiming he made it up to avoid prosecution for illegally selling ancient artifacts. But weren't they supposed to be forgeries, rather than real artifacts? In 1977 he returned to his "hoax" story after being arrested by the Peruvian government for illegally trafficking in ancient artifacts (forgeries?).[21] Although Uschuya's stories are not "reports of his finds", they demonstrate how the pressures of life reshape *information* into "data" and "facts", and then, back into truth again, leaving the rest of us to wonder which is which. Even skeptics are subject to such pressures, although their pressures never affect their valuable judgment. Or so they assure us.

The Ica stones, themselves, are definitely real. A sixteenth century Jesuit priest, Padre Simon, first recorded them for the Western World long before Uschuya's time, noting the strange beasts engraved thereon. He sent a collection of them back to Spain a couple centuries before the tourist trade even knew what a dinosaur was. Many of the depictions of Apatosaurus on Ica stones found before the 1960's display its anatomically correct head twenty years before paleontologists learned they had given this poor creature the wrong head.[22] It seems the "forgers" knew paleontology better than the paleontologists knew paleontology. Or else those images were chipped onto these stones by people who really did experience dinosaurs several centuries ago.

This couldn't be possible could it? Doesn't our knowledge of billions of years comprise reality? Or maybe it only comprises a paradigm. Regardless of *information*, skeptics are skeptical, because skepticism isn't about discovering the truth as much as it is about guarding established paradigms. Every bit of *information* must be allowed to participate in the construction of knowledge regardless of biases. Because *information* belongs to reality, *information* displays realities. *Information* always tells the truth. Ignoring even one piece of *information* assures ignorance about at least one aspect of the truth.

Skeptics would be well advised to move their efforts on to some topic other than Ica stones, since more depictions of dinosaurs are woven into clothing also found in those old, Peruvian graves. Depictions of dinosaurs are also found on a variety of their pots, bowls, vases, and gold death masks.[23] The dinosaur motif seemed quite popular in their culture. Then, if skeptics yet feel bold, let them fly to Cambodia and debunk the stegosaurus relief carved on the doorpost of the Ta Prohm temple built by Jayavarman VII in AD1186.[24] On their way, they can drop by Utah where the Anasazi Indians pecked a petroglyph of a

Clear Signs of Trouble and Great Joy

Brontosaurus type dinosaur onto the base of an arch at Natural Bridges National Monument. Or they can visit my own Western Colorado where Freemont Indians pecked a petroglyph of a triceratops onto a cliff wall. Or they could stop over at the Grand Canyon where another Indian petroglyph of a bipedal dinosaur[25] can be seen. In England's Carlisle Cathedral is the fifteenth century tomb of one Bishop Richard Bell. Two unmistakable sauropod dinosaurs are engraved on his tomb's brass border.[26] All of these depictions were made generations before mankind supposedly first discovered the terrible lizard and called it a "dinosaur".

There is a bit of musing wisdom posted on the Diogenes Club blog regarding Bishop Bell's dinosaurs. This wit perfectly illustrates the point I make about how impervious to *information* accepted scientific paradigms really are, and how those paradigms destroy the general public's ability to think rationally about any *information* which might happen to escape science's dumpsters:

> Along one of the brass strips, if you look carefully, you can see two animals. The one on the left is obscured by the one on the right that looks suspiciously prehistoric. It clearly has a long tail, an even longer neck and head. With a neck like that we might have been tempted to think giraffe if it were not for the four thick-set legs and large rounded body. It is for us, of course, a classic image that every school child would immediately recognize. However, the one animal it could not possibly be is a dinosaur. That of course is quite impossible. Dinosaurs as we all know were not discovered until 1840 and this tomb was constructed in 1496. No one even knew that there were dinosaurs 500 years ago, never mind what they looked like. So it can not be a dinosaur.
>
> ...Charles Fort had a name for such data as this. He called them "The Damned" as they were damned to obscurity by a scientific community that had no room or no explanation of such things.[27]

But science must have an explanation, especially for things as simple as an old engraving in brass, otherwise, how can we really call it "science"? These depictions are clearly of sauropod dinosaurs; nothing else looks like they look. But science can not figure out how to explain them because they crush its favorite paradigm. So let's suggest to them an approach by which they can find THE explanation.

Step one is to acknowledge that a highly skilled artist intended to depict what he actually did depict: sauropod dinosaurs, whether the artist called them sauropods, dragons, or big, ole whatnots. Somebody skilled enough to make a drawing is smart enough to make it look like what he meant it to depict. He engraved dinosaurs, and not by accident. Nor could the engraving have been the colossal coincidence skeptics and normal scientists will call upon to explain the clear signs about which you will soon read. Step two is to acknowledge that those artists became familiar with sauropods (a.k.a. dragons) the same way anyone becomes familiar with anything: by seeing it, or even by interacting with it. The artist saw dinosaurs, at least culturally he did, for the engravings indeed do depict sauropods. Step three is to reassess what we thought we knew about when sauropods were first known to humans. We thought nobody before 1840 knew of them. But these engravings display our mistake as certainly as they display the engraver's knowledge. Step four is to list all of the possible ways the artists could have become familiar with those dinosaurs (dragons). Until you have completed step four, you haven't even begun to think like scientists claim they think (but don't). Then step five is to explore each of those possibilities.

How do we know dinosaurs (dragons) were not discovered until 1840, so no one ever knew what they looked like before then? Did some scientist interview everyone who lived before 1840? That's more doubtful than a fifteenth century artist accurately engraving what he had never seen. Did scientists examine every piece of pre-1840 artwork to see if anyone ever depicted a dinosaur/dragon before 1840? Well, obviously not! We've listed a few of a large number of etchings they could have examined, but haven't. And won't. So then, does science know what it knows by some means other than observation, maybe by fiat, or by paranormal inspiration? Or, does science deceitfully refuse to know what the simple acknowledgment of real observations will show? Does science just make stuff up to plug holes in its theories? Of course it does!

Science making stuff up is called assuming, presuming, hypothesizing, theorizing, and such. Creation-science making stuff up is called lying (which it should be called). Modern cosmology is full of made up stuff, matter not what that's called. Uniformitarianism, gradualism, and evolution are made up of ideas rather than *information*. Science is willing to assume in the place of observing, refusing to observe what can be seen. It lacks consideration for visible realities as simple as those old pictures of dinosaurs. It doesn't do what Jesus meant by eyes to see, i.e., honesty to admit and humility to understand.

If many people throughout the ages around the world did experience dinosaurs before 1840, how would we know it? Is there any evidence correlating with such a proposition? Try examining their artwork. Admit the real art is real. Admit coincidence is not a considerable explanation. Why does the scientific community have no room for or explanation of all this old artwork? There is only one reason. They are interested in their paradigms more than they are interested in the truth. Theoretical science is subjective. And cosmology, a.k.a. evolution, is a subjective theory well sheltered from objective evidence.

Bearing in mind that not all claims of dinosaurs depicted in ancient art are indeed dinosaurs (some just look like them but are more identifiable as something else), the following website shows how abundant ancient depictions of dinosaurs really are: http://historysevidenceofdinosaursandmen.weebly.com/visual.html.

Before the term "dinosaur" was coined and popularized in the later nineteenth century, these animals were known as dragons. The annals of history from The Book of Job through Herodotus and on down to our more recent times are peppered with stories about dragons (dinosaurs). Of course, we think of mystical, fire breathing beasts when we hear "dragon". But most woodcuts accompanying old stories of dragons depict more familiar looking dinosaurs rather than those mystical fire breathers we've fooled our children into fancying. Ascribing fake imagery to historically useful terms is an ancient trick of liberalism.

It is important to allow a framework of belief to be realigned by *information*. *Information* is the best of all evidence. It belongs to reality. "Facts" and "data" belong to man. Once a framework is realigned by *information*, *it* redefines the relevance of previously

Clear Signs of Trouble and Great Joy

misaligned evidence. David Rohl's book, *Exodus - Myth or History?* presents a great example of how one tidbit of evidence called for the realignment of a wide network of previously misaligned *information* by popping one paradigmatic balloon. His proposition generates a completely different, abundantly evidenced understanding of Middle Eastern chronology.

The last fifty years have created and popularized an historical viewpoint of the Bible called "Biblical minimalism". It is splashed all over our TV screens by myriads of "Bible Mystery", "Secrets of the Bible", and other documentaries portraying God's intervention in human history to have been nothing more than a few extraordinarily coincidental motions of nature. Biblical minimalism wafts through our culture in a foul scent of general skepticism towards every miraculous proposition the Bible makes. Minimalist scholarship has turned the faith of many churches into little more than hopeful fantasies of cold-nor-hot preachers bending and twisting Scripture to fit the non-designed, purposeless, mechanistic materialism atheists demand we worship and adore.

The twentieth century opened with an intense scrutiny of The Creation, Flood, and Tower of Babel narratives found in the Book of Genesis. At that time these accounts were, for the most part, publicly accepted as historical narratives even while they were being bitterly embattled from the great halls of intellectual elitists. Authors, journalists, movie-stars, entertainers, professors, teachers, politicians, cold-nor-hot preachers, and anybody else who wanted to be somebody soon had to fashion their rhetoric and daily decorum out of this rising viewpoint of a Biblical "mythology". Then it began obscuring the general public's recognition of God's authority as taught in the unadulterated narratives of the Bible. The old, cultural sense of God's real being slowly withered into an atmosphere of social embarrassment. Eventually minimalists were able to detach the first eleven chapters of Genesis from publicly accepted "history", hanging them on the peg of mythology like a dusty old hat. For a few decades Abraham and the Hebrew patriarchs became the new bridge between generally accepted science's evolutionary prehistory and written history.

By the 1970s and 80s, detachment of Genesis' first eleven chapters was no longer enough to satisfy Biblical minimalists. Scholars began propagating a notion that the rest of Genesis was also not historical. William F. Albright's celebrated discoveries in the Holy Land were in a few short decades demoted to naiveté. Of course, the walls of Jericho were still known to have fallen outward, as Albright found they had, as the Bible states they did. But the more-acceptable-because-she-doesn't-believe-the-Bible Kathleen Kenyan informed the world that those walls fell four-hundred years before little Joshua was even born. It wasn't that Albright did bad archeology. It was that the new way to fame now came through discrediting the Bible rather than crediting it. The band had switched wagons, and he new one became a craze. Thereafter, any admission of agreement with Biblical narrative was the quickest way to shame.

Maybe consistency only has to do with an experiment's ability to be duplicated. Maybe it has nothing to do with the intellectual integrity of treating informative evidence the same for the left hand as it is treated for the right hand. David Rohl helps us understand how just a slight lack of intellectual integrity had misplaced all of the Middle East's early chronology by a few hundred years.

Information

Rohl tells us that centuries ago scholars knew the enslaved Israelites had built the store city of Raamses for their Egyptian masters. Since Exodus 1:11 identified this city as Raamses, then surely one of the Raamses Pharaohs was the master for whom it was built, they had reasoned. It couldn't have been that Raamses was a more recent name designating a city formerly known by an earlier name, like New York was first known as New Amsterdam. Regardless, they decided Raamses II must have been the Pharaoh of Exodus because the Bible says the Hebrew slaves built Raamses. In spite of Biblical *information* pointing to 1447BC as the approximate date of the Exodus, its generally accepted date could not be pushed back any earlier than 1250BC with the weight of this "Raamses" chained to its ankle.

Because of "Raamses", a thirteenth century BC Exodus is to this day the scholarly consensus. At least it is amongst the few scholars who still admit to any Exodus at all. For no archaeological record of any large, Semitic population (Israelite type Semitic) in Egypt around 1250BC exists. Naturally then, there is no record of a large Semitic population leaving Egypt in the thirteenth century. Nor is there any record of the Nile turned to blood at that time, nor of locusts stripping the fields, nor of thick darkness, nor of dead firstborn everywhere. Moreover, forty years later, around 1210BC, when, according to the Bible, these Israelites should have been creating mayhem in Canaan, there was instead a rather peaceful time. Jericho wasn't inhabited; its walls fell centuries before the thirteenth century, as the more-acceptable-because-she-doesn't-believe-the-Bible Kathleen Kenyan informed the world. The level of destruction caused by the Israelite conquest of Canaan does not show up in the archaeological levels assigned to thirteenth and twelfth century BC Canaan.

Therefore, orthodox historians are always sure to remind us that The Exodus is not found in archeology. It must have been a story made up by Jewish priests and scribes to assuage their shame of being held captive in sixth century BC Babylon. According to the minimalist thinking, the Jews created Genesis as a special story for resurrecting national pride. Minimalists accuse the Jewish priests of composing, for themselves, a phony, supernatural heritage. Thus, the minimalist historians judge the Bible to be a complete lie, and cold-nor-hot preachers spew that judgment all over their fancy filled congregations.

Thank God for *information*, which no consortium of scientists, historians, minimalists, or sick preachers can be. Consensus is nothing more than many minds agreeing to think alike. Ideas are not true by consensus, no matter how many think they are true. Indeed, consensus popularizes ideas, but ideas remain true only by their correlation with evidence, i.e., *information*, rather than by popularity. The first principle of logic and reasoning states that everything is what it is, not that everything is what the most scholars say it is, nor what peer reviewers most hostile to the Bible demand it to be. Something being what it is generates *information* in accord with what reality is. So, *information* is who's to say what's true, being the smoking gun of reality.

One piece of good *information* can reveal more truth than a whole library full of consensus. Consensus keepers invest careers into their theories, which they need the less learned to accept as facts, else they lose their career investments. God forbid their books, their careers, their lifetime investments should ever get ash-canned at the admission of some obscure tidbit of *information*! So they invest by the guidance of theory guardians, a.k.a. peer

Clear Signs of Trouble and Great Joy

reviewing consensus-keepers, skeptics, as they are known on the streets by us commoners. Yet, *information* speaks its piece regardless of whose consensus obscures reality, or of which guardian is famous enough to simply dictate his own fancies to the world.

But sometimes, not always, nor even often, *information* stands incontrovertibly against all the defenses of consensus keeping theory guardians.

David Rohl presents to the world *information* about a total lunar eclipse the morning of February 25, 1362BC. Against all denial, it is the informational pin popping the balloon of orthodox Egyptian chronology, and to wit, Middle Eastern chronology. Unfortunately, the complexity of this pin's thrust muffles the balloon's pop, because the general public is sure to ignore whatever is too complicated for easy comprehension. Since simplicity is not the lone criterion of truth, and since I count my readers as being more apt than most of the public in general, let's patiently consider this eclipse's rather complicated thrust into that balloon, as the evidentially sound David Rohl discovered it.

Babylonian astronomer-priests recorded a totally eclipsed moon setting while the sun was rising on the opposite horizon. Fourteen days later they recorded a partially eclipsed sun setting. This is a rare combination of astronomical events to occur at any one particular location. To the astronomer-priests, it portended something of extreme significance. So they recorded it and informed their king, Samsuditana, that the end of his reign was shown by these signs. Coincidentally (or not) Hittites raids into Mesopotamia brought an end to poor Samsuditana's reign later that year. And so ended the First Dynasty of Babylon.[28]

The significance of this great coincidence (if it was merely a coincidence) is our ability to date for certain the end of Samsuditana's reign as being 1362BC. Then by adding the length of each reign reported on the Babylonian King List, the reign of Egypt's Neferhotep I can be positively dated. His name had been found in an inscription of a Canaanite king, who was his vassal. This Canaanite king had presented a certain cup to Zimrilim, King of Mari, which was identified in a gift list discovered in the royal library at Mari. Zimrilim met his end at the hand of Hammurabi, the fifth predecessor of poor Samsuditana as listed in the Babylonian King List. Therefore, dating Hammurabi by adding the assigned years of the listed kings' reigns to the newly established 1362BC dynasty end will redate Zimrilim, which redates King Yantin Ammu of Byblos (the Canaanite vassal king), which redates Neferhotep I, since archaeological evidences show they were all, except poor Samsuditana, roughly contemporary.

So what about Neferhotep I? He was a pharaoh of Egypt's 13th Dynasty as listed in the Royal Canon of Turin, a 19th Dynasty papyrus providing the basis for most of Egypt's chronology before Raamses II. Although only about half of the canon survives, it names all but six of the 13th Dynasty's twenty-seven pharaohs. Nearly all of their reign lengths are missing. But the 13th Dynasty is known to have been a time of short reigning pharaohs -three years or so each, on the average. Therefore, observing the date of Neferhotep I as linked by evidence to the known date of an astronomical event, the dates of the other pharaohs can be estimated with sufficient accuracy to true up a popularly misaligned Egyptian/Middle-Eastern chronology.

Once the chronology is squared with reality, immediately a cascade of archaeological evidence and ancient records flood into alignment with the Biblical Exodus narrative.

Information

The long, droning denials of the Bible's historicity drown in a sea of actual evidence restored to its proper place by one point of undeniable *information* -the February 25, 1362BC total lunar eclipse.

David Rohl can now name the pharaoh who made Joseph vizier, Amenemhat III. The markings on a cliff wall, which Amenemhat III decreed for recording levels of the Nile floods, evidence seven years of very favorable agricultural conditions followed by seven years of very atrocious conditions. The origin of Bahr Yussef, the canal which legend holds to have been dug at Joseph's orders, now steps out of legend into history. The sudden abandonment of the Semitic city of Avaris in the Nile delta loses its mystery to the newly discovered reality of the Exodus, as do the helter-skelter burials of a great number of Egyptian plague victims contemporary with Avaris' abandonment. And the lamentations of Ipuwer, formerly dismissed as either a novel or the tirade of some crank, become an Egyptian record contemporary with the abandonment of Avaris and whatever great plague may have caused that mass of helter-skelter burials. Ipuwer lamented the river being blood, the food being gone, darkness blanketing the delta, everyone burying a brother, and the slaves carrying off all of the land's gold and jewelry. Documented! of all things. Each of his complaints matches a plague Moses called upon Pharaoh's land. Should we wonder why?

This constellation of archaeological evidence is precisely the kind of correlation historians use for verifying their theories other than those involving Biblical narrative. So why does the same type constellation of evidence not verify the Bible's narratives? Had the events of The Exodus narrative been written somewhere other than in the Bible about people other than the Jews meeting a god other than I AM WHO I AM, scholars would have firmly insisted Ipuwer's lamentation described a real event because of so many close correlations. Even Kathleen Kenyan would have agreed. For all scholars would then have recognized the commonality of "Raameses" to have merely been a newer name for an older city.

So also the archaeological evidences of the fallen walls of Jericho and the destruction levels at Hazor are brought into agreement with the Biblical record, now that the orthodox chronology has been adjusted by that simple, total lunar eclipse of February 25, 1362BC,[29] the pin popping a balloon full of consensually heated air. However, since the eclipse is so unknown, and since it girds up the Biblical narrative, and since theory guardians yet mischaracterize David Rohl as an unknown thinker, he is attacked, and his "theory" is quarantined from public consideration by application of that well known label, "He doesn't know what he's talking about."

But David Rohl has been researching and discovering this New Egyptian Chronology for thirty years. It seems he's as dedicated to intellectual responsibility as is Dr. Jerry Bergman, PhD. But, in spite of the fact that the very same eclipse the Babylonian astronomers recorded is calculated, dated, and logged into NASA's lunar eclipse list, the academicians stand in denial of Rohl's thesis.[30] They can not allow their books and careers to be overturned. Their names are too important. So they interlock arms to guard their theory against the correcting affect of reality's *information*. Yet they do not realize how well a century of time tills and uproots former theoreticians, save the best like Copernicus, Newton, Einstein, and Kuhn. They all hope to be one of those. Alas! David Rohl is that one, not them. For he holds to *information*, while they hold only to consensus.

Clear Signs of Trouble and Great Joy

If the experimental and practical sciences were performed the way the science of origins and ancient history are performed, scientists would consistently blow up their laboratories, bridges would collapse while under construction, airplanes would fly like crowbars, and most of our medications would be worse poisons than they already are. The very method of science involves discovery of error by testing theories against observed *information*, such as that particular total lunar eclipse. Honor for *information* must be applied in the hard sciences to avoid crowbar airplanes and crumbly bridges. But even though many historical theories fly like crowbars into the headwinds of *information*, their perpetrators insist upon riding them all the way down to their dusty end.

Let's return to poor Samsuditana's astronomer-priests. They clearly indicated the moon was eclipsed during its setting while the sun was rising. This can only happen in the morning. Regardless, the common date orthodox historians choose for the observed eclipse is February 20, 1659BC. NASA's eclipse list shows that eclipse occurred almost directly over Babylon.[31] Only an eclipse at midnight can occur directly overhead, never at sunrise, as when the astronomer-priests observed their eclipse. Otherwise the sun rises at midnight in Babylon. I doubt there's much consensus for that; there's not a big enough dumpster for it. So an alternative date is set by pointing to the solar eclipse of February 25, 1631BC followed two weeks later by a partial lunar eclipse on March 12, again occurring close to midnight.[32] Yet neither the misalignment with sunrise nor the reversed sequence of the two eclipses deter orthodox consensus-keepers from embarrassing themselves by foolishly choosing this one as the astronomer-priests' observation. Maybe the astronomer-priests didn't know how to write an accurate report. Maybe their report wasn't peer reviewed. Oh yes, we all know what was wrong with their report! It would eventually serve to support rather than impugn the Bible! And that, in any "scientific" research, is worse than leprosy.

What should embarrass our modern scholar-priests even more is that the 1631BC solar eclipse was not even visible in the Babylonian area. The guardians of theory so desperately need the earlier dates of the Old Chronology to shelter their beloved, atheistic history from The Holy Bible's invasive explanations that they bet nobody would notice their inconsistent handling of *information*. For when your own name alone is big enough to support the weight of your theories, who gives a rat's donkey about *information* or anyone else? Consensus shelters paradigm from embarrassment while *information* supports Exodus as history. Thank you, David Rohl, for honoring *information*.

By the 1980's, Biblical minimalists had completely relegated King David to mythology. Unlike ancient, non-biblical texts are sufficient *information* to support the existence of their mentioned persons and places, the detailed historical *information* of The Holy Bible is never sufficient for minimalists. They always require evidence external to the Bible, even for such a widely acknowledged figure as King David. So they demanded to see David's name in some archaeological document or inscription external to the Bible before they would admit to his historical existence. But nowhere in historical records earlier than the eighth century BC was the name "David" to be found outside the Bible. With great joy then, the minimalists concluded and taught that King David was a fairy tale, along with the Easter Bunny, the Tooth Fairy, Santa Claus, and of course, Jesus Christ.

Now, Hazael was once a Syrian court official who became king of a burgeoning

Information

ninth century BC empire to the north of Israel only a century or so after the reign Biblically attributed to David. The prophet Elisha lamented Hazael's ascent to the throne (II Kings 8:10-13), since he knew by prophecy Hazael would eventually cause much devastation to Israel and Judah. Joram and Ahaziah, kings of Israel and Judah, made war on Hazael early in his reign. Joram fared poorly; he was injured. A rude character named Jehu took advantage of the opportunity and killed both him and Ahaziah, as well as the infamous Jezebel, and all of Baal's prophets, priests, and worshippers. Maybe Jehu acted as an agent for Hazael, or maybe Hazael simply embellished upon wounding Joram, but Hazael monumentalized this minor victory on a stele boasting about his having killed Joram, king of Israel, and Ahaziah, king of The House of David. Today, we call it the Tel Dan Stele. It was found during the 1993-94 excavations in the ancient city of Dan. Wait a minute! Did that inscription say, "House of David"?

It did!

But diehard minimalists still tried to claim the stele's Aramaic word for "David" really didn't mean "David". They claimed this; they claimed that; for almost a decade they squirmed. Alas. They finally admitted the name on the stele was David's, oh my! So, indeed he existed, they begrudgingly admitted, although, to this day a few diehards offer nonsensical interpretations of that *bytdwd* (House of David) inscribed upon this stone within the unambiguous context of Hazael's minor victory.

Of course, the minimalist skeptics were sure to next advance the unfounded accusation that this now confirmed King David was just the leader of a band of merry men, somewhat like King Arthur was only a romanticized, local thug. So the notion of a thief's territory having been puffed into the legend of a Davidic empire became their new attitude towards Scripture. But, in the end, a couple very large, tenth century BC, Israelite tax collection jars were discovered in ancient Khirbet Qeiyafa,[33] a city too large and far from Jerusalem for a mere band of merry men to have controlled and taxed to the need of such large collection jars. The Biblical minimalists had to finally eat the whole crow…David indeed was king of Israel, that famous empire as described in the (yuk) Bible. They should have just believed their Bibles in the first place, like Albright did. Then they wouldn't have embarrassed themselves so badly.

And this is not the only example of I AM WHO I AM arguing His case with twentieth century geniuses who think they know more than the Bible. Forty-six years earlier, skeptics were served a piece of crow for breakfast when a young Bedouin threw a stone into a cave and broke a clay jar full of ancient Hebrew scrolls. Skeptics had long yodeled the unreliability of the Old Testament text. Until 1947, I AM's faithful people had nothing earlier than ninth century AD, Masoretic copies of the Old Testament to support their claims that the Bible's text was largely uncontaminated by errors, insertions, or deletions. Consequently, they had no direct evidence for answering the skeptics' unfounded accusations of textual corruption. But meticulous analysis of those first century BC and AD fragments from Qumran showed nearly no differences between the Old Testament text, as it was before Jesus' days, and the Masoretic text of a thousand years later. (Their text does bear differences with the Samaritan text used to translate The Septuagint.) A thousand years of nearly error free transmission is unknown anywhere else in the world of ancient documents.

Clear Signs of Trouble and Great Joy

You would think the profound and abiding flavor of crow would have put an end to skepticism against God's Holy Word by now. But *information* is not as desirable as are big names, lofty careers, and personal agendas. So crow remains on their menu.

As long as individuals can not automatically know what everyone else knows, the greatest factor of persuasion will be silenced *information* rather than paradigm infected facts and data. Skeptics search for any natural cause they can find for explaining every Biblical story which threatens their box full of scientific guesses. Science can irrefutably prove the sensitivity of nitroglycerin immediately. Science can rationally prove the dire need to keep bleach, rat poison, and pistols away from children. But the mishandling of historical *information* leads to much slower, less visible consequences. The theories and conclusions of the soft sciences do not have the good fortune of being immediately testable. Either HE IS WHO HE IS created man during the sixth day, or God evolved man by the sixth epoch, or god is just the pipedream of a has-been monkey.

But the way it was is actually what the past is. And that way produced *information*. Therefore, the intellectual responsibility of the forensic scientist is to use *information* for discovering the way it was, rather than dismissing *information* to stir up consensus for the way he wants it to have been. Regardless, the last two centuries of the soft sciences' efforts have been to prop up the way skeptics think it was in complete denial of what *information* says.

Skeptics boast of *information*. They even act as if they worship it. But anyone who honors *information* can clearly see the skeptics' abuse of it. Skepticism is not a basic reason for research. Discovering what *information* shows to be real is that. Today we have electron microscopes, Geiger counters, centrifuges, and about every piece of scientific equipment imaginable for discovering *information*. We even have museums full of more *information*. So we shouldn't take inventory of our vast means for discovering what things were in the past and then fault the fourteenth century AD Troyes' Bishop d'Arcis for launching that old Shroud-of-Turin-is-a-forgery myth.

In his day, paint, dye, and charcoal were about the only medium for making impressssions on cloth. So they were a part of his mental paradigm. As Thomas Kuhn demonstrated, paradigm always ladens data. And that means any examination of The Shroud in d'Arcis' day would have anticipated seeing paint, dye, or charcoal on the fabric. Nor did it help that his predecessor, Bishop Henri, was unable to determine from where the little neighboring church of Lirey acquired this piece of cloth claimed to be Jesus' burial shroud because it bore the image of a scourged and crucified body. But we can fault both Bishops Henri and d'Arcis for their biased, unresearched conclusion that the Shroud was an artistic production. They had no *information* for drawing such a conclusion. They only speculated. They made it up.

Neither of those two bishops could stomach the Lirey church's financial success gained from exhibiting The Shroud of Turin. It was sometime around AD1355, and war had left the area more impoverished than usual. Church coffers reflected this condition. Certainly not every mite went to Lirey for a peak at Jesus' burial cloth. But the normal flow of available contributions skewed greatly in that direction. So letters d'Arcis' wrote to the French king and to the pope in Avignon more reflected attitudes of d'Arcis' jealousy than

Information

they reflected any actual analyses of *information*...

> ...a certain cloth cunningly painted...by a clever sleight of hand...falsely declaring and pretending...so that money might cunningly be wrung from [parishioners], pretended miracles were worked...[34]

Could Henri and d'Arcis' accusations have been right? They would have known it was a forgery, wouldn't they, being at the scene where the Shroud first appeared, according to modern mythology? Not unless Henri and d'Arcis' eyes were better than twentieth century lab equipment. The 1978 STURP team came away from its first analysis of The Shroud agreeing with the sixth century AD Shroud observers (what!? sixth century?) who called the image "not made by human hands". The team went into the examination equipped to analyze paint, kind of like d'Arcis and Henri did. But it wasn't paint their lab equipment found. It found human blood. Serum. And some sort of coloration on the surface of the linen's fibers made by a process their lab equipment was unable to determine. There was no pigmentation. It was not a stain. Theories were advanced about the image being photographic. But no residues of the light sensitive chemicals necessary for making photographic images were discovered on the linen. Such chemicals would have been required if the Shroud's image were a cleverly made photograph, otherwise Kodak would have used linen for film. Eventually the image was found to have been produced by a quickening of linen's natural yellowing process entailing only the surface molecules of each linen fiber. What on earth would cause only the surface molecules of individual linen fibers in only certain areas of a cloth to yellow until precisely detailed image of the scourged, crucified Jesus was perceptible? What forger could do THAT in the C14 test's fifteenth century? The Shroud's *information* definitely precludes any possibility of a forgery. And that statement isn't strong enough to convey how impossible it would have been to forge that image onto a linen using even today's advanced technology. Nobody in the fourteenth century AD knew how to accelerate the yellowing process of linen. They knew linen yellowed; sixteenth century nuns stained the cotton patches they used to repair The Shroud in order to match the linen's aged yellowing. And even if someone could have figured out how to accelerate the yellowing process, it is simply preposterous to think anyone then (or now) would be able to control such a process well enough to render the highly detailed image of Christ as it appears on The Shroud.

Moreover, when retained samples of The Shroud's linen fibers were subjected to dating processes more accurate than the famous C-14 "test", they indicated the linen to indeed be two-thousand years old.[35] Maybe those old Romans had linen's yellowing process under control?

Maybe the greatest roadblock to understanding the genuineness of The Shroud is its "unknown" origin and history before the mid-fourteenth century, when it was "first" exhibited. Since the 1960s, Ian Wilson has been searching for history's evidences of The Shroud. His book, <u>The Shroud of Turin: The 2000 Year Old Mystery Solved</u> presents his findings in greatly documented detail. He discusses the excellent clues history gives for knowing not only its first century origin, but the history of its intervening years, as well.

Justinian II began striking an image of Christ onto Byzantine coinage in the seventh

Clear Signs of Trouble and Great Joy

century AD. How was Justinian II sure what Jesus looked like? St. Augustine wrote in the fifth century that, by his time, nobody actually knew what Jesus looked like. Until St. Augustine's time, Jesus was depicted in various ways, but mostly as a nondescript, short-haired, round faced, beardless, Roman-looking chap. In the sixth century, though, everything about Jesus' perceived features changed immediately, almost overnight. Teams of "image missionaries", as Mr. Wilson calls them, began circulating outward from the Armenian area of southeast Turkey, teaching the correct image of Christ.[36] All of the churches were eagerly accepting their acclaimed depiction. Soon, every church which was any church had to have its own painting, mosaic, or copy of this new look of Jesus. Bearing in mind how stubborn (faithful) Christians can be about their religious decor, something compelling must have happened to change Jesus' look overnight. The new image quickly became popularized worldwide, as the long faced, long nosed, long haired, bearded man with large owlish eyes and a scrunched little mouth under a wide, draping mustache. (Of course, the blue eyes weren't in those ancient depictions.)

What could account for this sudden and permanent transition in the sixth century? Ian Wilson informs us that AD525 brought a devastating flood through southeast Turkey's ancient city of Edessa. The Byzantine emperor, Justin I, regarded Edessa as important enough to rebuild. It is said that during the reconstruction, the Image of Edessa was rediscovered in a sealed niche above the old city's main gate. Legend stated that it had been hidden for safe keeping against a pagan cleansing of Christian artwork about three centuries earlier. From around the beginning of the second century until AD525 this Image of Edessa had only been a little known fable about some lost piece of cloth bearing a depiction of Christ's face, for which reason it was brought to Edessa in the early first century AD in order to cure King Abgar V of a critical illness.

In its day, the Image of Edessa was described as having been approximately twenty-one inches high by forty-five inches wide and affixed to a board of the same dimensions. It was further described as portraying a face and shoulders in landscape format. Landscape format is highly unusual for portraits. If the Image of Edessa were merely a piece of artwork, the image on it would more likely have been in portrait format. Curiously, by the early seventh century, the Image of Edessa had also become known as the tetradiplon.[37] Ian Wilson informs us that, in all ancient Greek literature, the word "tetradiplon" is used to describe only this Image of Edessa. It evidently was coined for the Image of Edessa, maybe by its caretakers. The word means "four folds", not referring to four creases, but to four, two-ply sections made by three creases, the first at the seven foot center of the cloth, the next at the new center, and then once again. The Shroud of Turin is approximately fourteen feet long and forty-five inches wide. This folding technique reduces those dimensions to a manageable twenty-one by forty-five inch stack of cloth, maybe not coincidentally, with Jesus' image, from the shoulders up, appearing on the top layer, in landscape format. This is the precise description given to the Image of Edessa, i.e., the Shroud of Turin.

It then should be no surprise that the likeness on the Byzantine coinage distinctly resembles the faint image on the Shroud of Turin. And not only the likeness on the coinage, but all of the images portrayed in the sixth century and later churches, and indeed the commonly conceived portrayal of Christ today, bear the basic attributes of the face on the

Information

Shroud of Turin. Paul Vignon, a French scholar studied the Byzantine portraits of Jesus and noted that, when considered together, they all repeated one or a few of fifteen distinct attributes of the face on the Shroud. Not every attribute is repeated in every portrait, but the main attributes, with one or more subtler details, are replicated in each.[38] This image spread throughout the sixth century world onto late seventh century Byzantine coinage. It attests to the existence of The Shroud of Turin at the rebuilding of Edessa in the early sixth century, eight-hundred years before it was forged, as the Carbon-14 "tests" were abused to imply.

Ian Wilson's hard work shows us how those common attributes between the Shroud of Turin, the coinage, the Image of Edessa, and its many copies made on church walls fill in the "mysterious" origin of the Shroud before its display at the little Lirey church. Careless correlations make fallacious history, such as equating one or the other lunar eclipses of 1659BC and 1631BC to the observations made by poor Samsuditana's astronomer-priests. The Image of Edessa, some shroud at Constantinople, and the Shroud of Turin have typically been considered as different objects, because theory guardians cannot afford them to be equated with the one and the same Shroud of Turin. But every subtle detail of their *information* indicates they are one in the same object: the cloth John and Peter found laying in the empty tomb on that first Easter morning.

How many correlating points of comparison does it take to differentiate between mere coincidence and reality? Is there some reason the number of necessary correlations should increase simply because the topic of study integrates in one way or another with The Holy Bible? Sometimes the only route bias can take around controverting *information* leads everyone into the ridiculous. The mental state of never being satisfied with enough *information* while available *information* clearly makes reality's case is truly called foolishness. The conclusions now being drawn from historical considerations and scientific studies of The Shroud of Turin are excellent examples of man's refusal to humble his mind to the import of plain, unobstructed *information*.

An Abridged Summary of the Scientific Analysis of The Shroud of Turin

Evidence for authenticity	Evidence against authenticity
practically all evaluations by medical experts[39]	
lack of 14th cent. knowledge about crucifixion	
barely visible serum stains	
high concentration of bilirubin in the serum	
presence of type AB human blood stains	
same blood type on the Oviedo Cloth	
correlation of stain patterns with Oviedo Cloth[40]	
"yellowing" nature of image production	
presence of Jerusalem specific dirt	
presence of Palestinian and Anatolian pollens	
historical evidences reaching back to 1st cent.	
1st century date by recent tests[41]	

Clear Signs of Trouble and Great Joy

None.
(C-14 tests of three samples contaminated by 15th century cotton are held up as evidence, but even normal science does not consider the results of botched tests to be evidence.)

The Oviedo Cloth is a small towel sized cloth transported from Jerusalem to Oviedo, Spain in the late seventh century AD. Until then, it had always resided in Jerusalem. Since then it has always resided in Spain. It has always been venerated as the cloth (sudarium) Peter saw when he entered Jesus' empty tomb…

> …which had been on [Jesus'] head, not lying with the linen cloths, but rolled up in a place by itself…
>
> John 20:7

This sudarium has also been subjected to extensive scientific tests of its pollen, linen, weave, and the blood stains on it, and even the wrinkles and holes in it. As usual, C-14 "tests" place it in a century too late to involve Jesus, but just in time to be carried off to Spain in the seventh century, while all other evidence places it in the first century. Therefore, "the C14 test" will be the only data skeptics acknowledge, since the results of all other analyses correlate its *information* with New Testament records and The Shroud of Turin.

For not only is the Oviedo Cloth soaked by the same type blood as what's on The Shroud, but the blood patterns on it closely match the patterns on The Shroud, as would be the case when both fabrics intimately contacted the same victim. Moreover, forensic consideration of the stain overlaps on the cloth, the sequence, and timing of the overlapping stains as determined by blood coagulation and evaporation rates, realistically correlate with the sequence and timing of Christ's removal from the cross and transport to the tomb as described by the Gospels. Even finger and thumbprints left in the blood on this sudarium reveal someone's hand attempting to restrict the discharge of fluids from the nose of a head tossed about as a body was being carried to a tomb.[42] The Oviedo Cloth and The Shroud of Turin form a remarkable set of evidences for the thirty-some hours Christ's body was lifeless, and for the burst of life at His resurrection.

Yet, in the face of such conclusive evidence, the predominant reports and comments on The Shroud insult even average intelligence with such expressions as "…could be authentic," "…doesn't appear to be a forgery," "…is difficult to refute." "Impossible to refute" is a subcategory of "difficult to refute". Why not use that term, since it more accurately describes The Shroud's *information*? Intellectual responsibility doesn't achieve its task until all relevant *information* of a matter has been sorted into its ultimately appropriate category. Therefore, until someone of the fourteenth century can demonstrate an extraordinary mastery of linen's yellowing process by thusly reproducing The Shroud's image onto another

Information

linen, making by human hands a reproduction faithful to the very last detail of The Shroud of Turin, "impossible to refute" is the only honest category of its description. Otherwise, would the "yet possible but admittedly difficult to refute" description mean The Shroud of Turin could have miraculously materialized in the little Lirey church? Or maybe E.T.'s mom left it behind when stopping by Earth to do the laundry? No? Well, then…

Science claims that it demands its theories explain existing realities. If science is honest, then *information* MUST direct theory formation, since only realities produce *information*. The Shroud of Turin is an existing reality. Science has gathered almost all of The Shroud's own *information* that presently can be gathered. All of its *information* precludes, CONCLUSIVELY, ANY possible production of The Shroud by anybody of the fourteenth century, by anybody of any preceding century, or by anybody of all centuries since. Its *information* precludes any possibility of production even by this present day's near miraculous technology. The only theoretical explanation for it doesn't just require a god, it requires I AM WHO I AM. It is the only theory which explains The Holy Bible's discussion of the historical event that produced The Shroud centuries before Jesus' resurrection happened.

> [6] All we like sheep have gone astray;
> we have turned every one to his own way;
> and the LORD has laid on him
> the iniquity of us all…
> [8] By oppression and judgment he was taken away;
> and as for his generation, who considered
> that he was cut off out of the land of the living,
> stricken for the transgression of my people?
> [9] And they made his grave with the wicked
> and with a rich man in his death,
> although he had done no violence,
> and there was no deceit in his mouth…
> [12] Therefore I will divide him a portion with the great,
> and he shall divide the spoil with the strong;
> because he poured out his soul to death,
> and was numbered with the transgressors;
> yet he bore the sin of many,
> and made intercession for the transgressors.
>
> Isaiah 53:6,8-9,12 (eighth century BC)

> [25] For I know that my Redeemer lives,
> and at last he will stand upon the earth;[£]
> [26] and after my skin has been thus destroyed,
> then from my flesh I shall see God,
> [27] whom I shall see on my side,
> and my eyes shall behold, and not another.
> My heart faints within me!
>
> Job 19:25-27 (second millennium BC)

Clear Signs of Trouble and Great Joy

Considering how meaningful Jesus' crucifixion and resurrection were, why are we so surprised these two items would have been gathered up and safeguarded by the people who witnessed the events?

Information is naturally and indelibly linked to reality. Therefore it obeys the shape and conditions of reality rather than the hearts and minds of people. We can know reality only by acknowledging its *information*. Perception is commonly mistaken for *information*. But perception is naturally and indelibly linked to the mind. The mind produces nothing more than a rather feckless reflection of reality because perception obeys the shape and condition of desires and ambitions, the seeds of bias. Around the world, 24/7, confounding and tragic events demonstrate how completely reality overarches perception. Why would life be as miserable as it so often gets if reality was whatever we wished to perceive it? We can alleviate or find shelter from the disastrous side of reality only by the processes reality itself offers. Therefore, perception must be shaped by reality's *information*, not by desires, ambitions, or other biases, else a person will never fully deal with reality until reality finally deals with him.

It is popularly claimed that truth does not exist. But that is pure foolishness. Truth is not an entity to be or not to be. Truth is the condition of a reflection. If one thing is reflecting another, such as a thought reflecting an object or a situation, then there is an accuracy (or inaccuracy) about the reflection that is measurable in terms of truth or falsehood. The better the reflection matches what it is reflecting, then the more true it is. If anyone demands that truth does not exist, then he is actually referring to his own lack of desire to know things truly.

Intellectual responsibility is among the most basic necessities of both sanity and spirituality. If truly knowing involves the mind accurately reflecting, then the building of perception must involve honesty towards *information*. Honesty extends further than just applying valid reasoning to accurate *information*. It equally concerns a completeness of the considered *information*. Honesty will not omit *information*. It especially will not sidestep *information* that tends to negate its own desires and ambitions. Intellectual responsibility understands mental frailty and the need for a continuous correction of its own perceptions in order to build its paradigms well squared with reality, i.e., honestly.

This, in essence, is the goal Thomas Kuhn described for affective science to achieve. And if you consider each individual's interactions with everything around his self, from conception to his last breath, each and every one of us individuals are scientists, for we must sometimes align with, sometimes avoid, but always discover what to do regarding every reality in our paths. The Holy Bible is a reality. It's message is a rational proposition, even though the spiritual processes it describes do not square with science's misperceptions.

We've perused a few examples of real *information* falsifying the scientific paradigms misperceiving the universe to be god-less and purely mechanistic. In other cases we've perused a few instances of *information* validating Biblical narratives. These examples are just a few of the vast swaths of *information* ignored and suppressed by orthodox science in order to safeguard its delusion of a purposeless/non-designed, god-less universe. The fact that most people do not know of or acknowledge this *information* has no bearing upon its existence, nor upon the existence of its realities, although it sufficiently evidences a

Information

"scientific" cover-up. *Information* exists no matter what anyone thinks of it or against it. It belongs to reality, not to the human mind. So, wherever there is *information*, there is a reality producing it. By simply acknowledging the reality of suppressed *information*, common folks can be brighter than even scientific experts and PhDs. Although no one can know the exact and full truth about everything, everyone can sufficiently consider the implications reality's *information* makes all around them well enough to rather accurately estimate the general truth about this life and how to respond to it.

And finally, most of the *information* briefly perused above has come into view in the last century. There's now word blowing in the scientific winds that soft tissues and complex protein molecules are being empirically observed in dinosaur fossils, and not by young-earth creationists, either. These substances are far too delicate to have survived tens of thousands of years, let alone millions. Something on a very basic level is happening to mankind's ability to estimate reality in ever higher degrees of truthfulness. With each passing year *information* ever more confounds science's most basic guess about the universe and life in it having evolved over billions of years with neither design nor purpose. And now, empirically observed patterns in the moon and earth's motions around the sun have been found correlating with Biblically relevant events and prophecies. *Information* now rationally suggests that maybe this universe really was designed for a communicative purpose.

Thank you, God, for *information*.

Footnotes

1. Merriam-Webster's 11th Collegiate Dictionary. Version 3. 2003. Merriam-Webster, Incorporated. At "noumena".
2. Gitt, Dr. Werner. In the Beginning was Information, A Scientist Explains the Incredible Design in Nature. 2005. Master Books, Inc. PO Box 726, Green Forest, AR 72638. Pg 49.
3. Wikipedia at "Physical information".
4. DeWitt, Richard. Worldviews: An Introduction to the History and Philosophy of Science. Second Edition. 2010. Wiley-Blackwell, John Wiley & Sons, Ltd, The Atrium, Southern Gate, Chichester, West Sussex, PO19 8SQ, United Kingdom. Pg. 20.
5. After I had finished writing this book, a friend warned me against the use of the title "Lord of spirits". He said it had a negative and troublesome presence on the internet. It seems that some people are developing from this ancient expression a concept about an ancient Enochian language useful for communicating with the angels, somehow in connection with the "Lord of spirits". Although that indeed is troublesome, Enoch preceded the internet by some five-thousand years, or so, and he used the title in reference to God's Elect One, the Son of Man, Jesus Christ. Paul tells us to take all thoughts captive to obey Christ. The truth that some people swerve from Biblically ordered obedience to Christ by misusing terminology doesn't mean those of us who love Christ as He revealed Himself must abandon the use of that terminology. If we did have to abandon terms because they are now used for evil, then pretty much half of our vocabulary needs to be tossed into the trash, for the world is full of word and concept twisters. I will stand with Christ and His prophets, and I will use the terminology they used to mean the concepts they meant, and I will not surrender my use of God's truths which liberals have twisted into intellectual wrecks. I will do my part in taking today's falsehoods captive to the ancient Truth from which liberals twisted them, Jesus Christ, The Lord over All Things. Jesus Christ is so much The Lord of spirits that every spirit which does not obey Him will enjoy eternity in

Clear Signs of Trouble and Great Joy

the place made for Lucifer and all of the angels who rebelled against the Lord of spirits, Hell.

6. Bostrom, Philippe. How Mice May Have Saved Jerusalem 2,700 Years Ago from the Terrifying Assyrians. April 18, 2018. https://www.haartz.com/archeology/.premium.MAGAZINE-how-mice-may-have-saved-Jerusalem-2700-years-ago-from-the-Assyrians-1.6011735

7. Kuhn, Thomas S. The Structure of Scientific Revolutions. 50th Anniversary Edition being the 4th edition. 2012. The University of Chicago Press, Chicago, IL 60637. Pgs 23-34.

8. Montgomery, David R. The Rocks Don't Lie: A Geologist Investigates Noah's Flood. 2012. W.W. Norton & Company, Inc. 500 Fifth Avenue, New York, NY 10110. Pg 19.

9. Woodmorappe, John, M.A. Geology, B.A. Biology. The Mythology of Modern Dating Methods: Why million/billion year results are not credible. 1999. Institute for Creation Research, P.O. Box 2667, El Cajon, California 92021. Pg. 15. The quote references: Hetherington, N. S. "Just How Objective is Science?" *Nature*. 1983. 306:727-730.

10. Ibid. Pg. 17.

11. Okasha, Samir. Philosophy of Science: A Very Short Introduction. Second edition, 2016. Oxford University Press, 198 Madison Ave, New York, NY 10016. Pg 81.

12. Gillispie, Charles Coulton. Genesis *and* Geology: A Study in the Relations of Scientific Thought, Natural Theology, and Social Opinion in Great Britain 1790-1850. 1969. Harvard University Press. Cambridge. Pg. 4.

13. Wikipedia at "Uniformitarianism".

14. Ibid.

15. Rosenberg, Alex. Philosophy of Science: A Contemporary Introduction. 3rd edition. 2012. Routledge, 711 Third Avenue, New York, NY 10017. Pg 142.

16. Ibid. Pg 147.

17. Wikipedia at "Polystrate fossil"

18. Ford, Charles V., M.D. Lies! Lies! Lies! The Psychology of Deceit. American Psychiatric Press, Inc., 1400 K Street, N.W., Washington, DC. 2005. 1996. Pgs 17-18.

19. Ibid. Pgs 19-20

20. Sharp, Doug & Bergman, Dr. Jerry. Persuaded by the Evidence: True Stories of Faith, Science, & the Power of a Creator. 2008. Master Books, PO Box 726, Green Forest, AR 72638. Pg. 46

21. Wikipedia at "Ica stones"

22. The Ica Stones of Peru. Tracking Ancient Man. 2016. http://www.ancient-hebrew.org/1001.html

23. Patton, Don PhD. Man and Dinosaur Co-Existed: Evidence from South America. (video) 00:25:00

24. Patton, Don PhD. The Record of the Rocks. (video) 1:00:00

25. Ibid. 00:52:00

26. Bing "Bishop Richard Bell of Carlisle" and observe for yourself who is willing to acknowledge the reality of these dinosaur images and who is not. Almost all attention paid to these unquestionable depictions of dinosaurs engraved during the fifteenth century is paid by Creation-science sites. It's like Darwin's world thinks they won't exist if they don't talk about them.

27. Dr. Phil. The Strange Beasts on Bishop Bell's Tomb. July 8, 2007. Stories from the Diogenes Club. http://storiesfromthediogenesclub.blogspot.com/2007/07/strange-beasts-on-bishop-bells-tomb.html

28. Rohl, David. Exodus: Myth or History? Thinking Man Media, 6900 West Lake Street, St. Luis Park, MN 55426. 2015. Pgs 71-74.

29. Ibid. Pgs 72-74.

30. Ibid. Pg. 13.

31. eclipse.gsfc.nasa.gov

Information

32. Ibid.
33. Dever, William G. "Hershel's No. 2 Crusade For King and Country: Chronology and Minimalism". Biblical Archeology Review. March/April/May/June 2018, Vol. 44 Nos. 2&3. Biblical Archaeological Society. 2018. Pg. 35.
34. Wilson, Ian. The Shroud: The 2000-Year-Old Mystery Solved. 2010. Transworld Publishers, 61-63 Uxbridge Road, London W5 5SA. A Random House Group Company. www.rbooks.co.uk. Pgs 220-221.
35. Bertrand, Jim. EC2016 -Jim Bertrand- The Very Latest Research of the Shroud of Turin. (video) 14:00-18:00. Linen fibers taken with the samples for the 1988 C-14 tests were subjected to infrared testing, laser spectroscopy, and tensile strength analysis resulting in a date range for the Shroud's linen between 220BC and AD280.
36. Wilson, Ian. 2010. Pgs 135-139.
37. Ibid. Pg 140.
38. Ibid. Pg 142.
39. Bulst, Werner S.J. The Shroud of Turin. 1957. The Bruce Publishing Company. Milwaukee, Wisconsin. Pg 53.
40. Bennett, Janice. Sacred Blood, Sacred Image: The Sudarium of Oviedo: New Evidence for the Authenticity of The Shroud of Turin. 2001. Libir de Hispania, Publications About Spain, PO Box 270262, Littleton, Colorado 80127. Pg 85.
41. Stanglin, Doug. New test dates Shroud of Turin to era of Christ. USA Today. Published 12:53 PM ET March 30, 2013/Updated 4:23 PM ET March 30, 2013. www.usatoday.com/story/news/world/2013/ 03/30/shroud-turin-display/2038295/
42. Bennett, Janice. 2001. Pgs. 64-89, 102-135

Dayenu.

Chapter 2
The Rest of the *Information*

¹The word of the LORD came to [Ezekiel]: ²"Son of man, you dwell in the midst of a rebellious house, who have eyes to see, but see not, who have ears to hear, but hear not; ³for they are a rebellious house.
Ezekiel 12:1
A fair result can be obtained only by fully stating and balancing the facts and arguments on both sides of each issue.
Charles Darwin

    Christmas 2016, Sadie's little sister, Rhinnie, received a simple jigsaw puzzle of maybe a dozen-and-a-half pieces. She quickly assembled it until three pieces remained. She was obviously frustrated when she asked me to help her finish the puzzle. I thought it would be easy as pie for this sixty-two year old CPA. But her partially assembled puzzle didn't agree. It had no spaces shaped like those remaining three pieces. Their color patterns coordinated well with the picture emerging in the puzzle, implying they fit there, somehow. But they weren't fitting. So I began to assume the puzzle had a factory defect.

    But, before I explored that possibility, I considered a more likely one. Maybe my discernment was defective and the puzzle was fine. So I scrutinized the minute details of her work. I noticed a very slight gap between a corner of the last piece she had placed and the piece adjacent to it. The gap was barely perceptible, less than the width of a fine hair. It required deliberately focused attention to see; it was the easily missed sort of detail. The other edges of that piece fit snuggly, deceptively suggesting this piece was properly placed. Yet close scrutiny found another, confirming, hairline gap along the side of a tab opposite that nearly well fit corner.

    Had I blamed the factory for Rhinnie's problem, then Rhinnie's grandma would have been her hero that day, instead of me. She would have simply taken the puzzle apart and reassembled it in half the time it took me to find those two very slight misfits (women are like that).

    When I removed the misplaced piece from where Rhinnie had put it, immediately its proper space became apparent, almost as if the puzzle was calling out for it. After I inserted it there, those other three pieces fell into their places with the same puzzle-guided ease.

    Rhinnie's Principle lays aside biases and assumptions to consider EVERY piece of *information* by the sense a puzzle makes of itself. Focusing on one area of a puzzle while dismissing another as irrelevant scrambles its sense. Gazing into radioactive decay for dating

Clear Signs of Trouble and Great Joy

the dinosaurs while ignoring, even lying about, ancient artwork of dinosaurs scrambles the sense of universal time. Such ensuing misperceptions invoke rejection of other *information*, e.g., those "silly" dinosaur petroglyphs. Cosmology's puzzle as scrambled by radioactive-decay has no space for human observed dinosaurs. So cosmologists fashioned make-believe puzzle pieces -assumptions- to fill the spaces reality's puzzle meant for those old dinosaur pictures. Rhinnie's Principle is to look for *information* to solve puzzles instead of blindly gazing into assumptions, theories, and human agendas. It is to allow *information* to modify, and even overturn theories, assumptions, and agendas, instead of using those to construct from *information* whatever "data" and "facts" will fit "scientific" biases. And really, Rhinnie's Principle is simply Thomas Kuhn's point, not saying that Rhinnie ever read Thomas Kuhn. It is just that, finally, one scientist was humble enough to see clearly like a child.

    No *information* can be discarded from any search for what is real (Sadie's Wisdom). Rhinnie's Principle is basic to answering liberalism's favorite charge, "Who's to say what's true?" The researcher who does not discard *information* is in the best position to say. All *information* is relevant to accurately reflect reality because all *information* is the product of reality. Each piece of *information* speaks its own part in the puzzle's proper assemblage. Empty spaces indicate either undiscovered or abused *information*. Theories and assumptions temporarily fill spaces only until the correct *information* is found to fit there. If theories or assumptions then remain in that *information's* place, a dishonest researcher has been discovered. From mankind's perspective, reality's puzzle has not totally been assembled. Therefore, as research proceeds, it is critical to maintain unbiased consideration for what newly discovered puzzle pieces might say about how correctly the puzzle has so far been assembled. If a piece of new *information* is discovered to have no fit whatsoever into a partially finished puzzle, then such lack of a fitting space indicates error in the puzzle's assemblage. The assemblage must be corrected to accommodate the *information*. That is Karl Popper's principle. Until theoretical scientists research in accord with reality's own rules for assembling puzzles, then they are definitely not the ones to say what is true.

    It is critical to understand that all of the puzzles of this life have come with an important, generalized picture for guiding the proper assembling of their pieces. Orthodox science has taught the general public to deny that any telltale picture has been imprinted upon life's box-top for directing the assemblage of culture, politics, family, and personal affairs. But how reliable could science's conclusions be, considering the heap of obvious *information* it has historically discarded in order to maintain its own, biased assemblage of the puzzle against reality's *informational* direction? Eyes to see behold reality's picture imprinted on the puzzle's very most core piece, The Bible. But, completely without confirming *information,* orthodox science accuses that piece of having been faked, rather than suspecting its own theories and assumptions of being biased. Theory and assumption only propose; they falsify nothing. All known *information* confirms the Bible to be life's guiding picture.

    Most of life's puzzles are assembled by the coherence all their pieces make together. The shapes of their empty spaces and the blanks left in their emerging pictures are necessary clues directing searches for more puzzle pieces. This is the critical point of Kuhn's dis-

The Rest of the *Information*

covery which seems to have been misunderstood from the first day the world read his work. An assembling puzzle -the forming paradigm- itself influences the search for yet undiscovered pieces. So the humility of the researcher is critical to his not forcing puzzle pieces into wrong spaces or covering up errors in his puzzle's assemblage by substituting conveniently shaped assumptions for discarded puzzle pieces. The puzzle, itself, must guide the research, not the puzzler. If the researcher does not proceed humbly, his puzzle will keep producing misshaped spaces, causing him to perpetually search for nonexistent pieces, for which he eventually must formulate more assumptions, all while his own dumpster fills up full of reality's answers to his problems in the form of all the *information* he subjectively rejected.

> [3] The integrity of the upright guides them,
> but the crookedness of the treacherous destroys them.
>
> Proverbs 11:3
>
> [6] The righteousness of the upright delivers them,
> but the treacherous are taken captive by their lust.
>
> Proverbs 11:6

The telltale blessing of reality's puzzle, though, is that it has one, core piece (the one found at the bottom of normal science's dumpster) bearing the puzzle's basic, most important picture.

A paradigm must be allowed to speak its own part through the coherence of how all *information* fits together. At the point Rhinnie brought her puzzle to me, all but three of its pieces fit well. But that wasn't good enough for her. Their fit wasn't proper. The puzzle spoke it by the three orphaned pieces. Careful, focused observation confirmed it. The researcher must not put his words into the puzzle's mouth by misplacing its *information*, however slight that *information* might be. He must not put his own paragraphs into its chapters by making up assumptions to fill in holes caused by misplacing *information*. If the warnings orphaned *information* shouts are not humbly heeded, then the errors assembled into reality's puzzle will mount up to devastating heights. Ask Challenger's engineers. Eventually, the puzzle's suggestions for what piece to look for next will reach a level of foolishness and deceit, while the misassembling puzzle won't show any coherence at all with very real, yet denied and discarded, *information*. Assembling the puzzle will change from a venture towards reality to a plummet into insanity.

Normal science's puzzle has come to reflect normal scientists more than obvious reality. The resultant world culture is displaying the affects. The picture on normal science's box-top has not been reality's picture for a couple centuries. It has been that of a cherished paradigm, a.k.a. scientific imagination.

> Instead of following where data, observation, and experiment lead, normal science dictates the direction of scientific progress by determining what counts as an experiment that provides data we should treat as relevant, and when observations need to be corrected to count as data. During normal science, research focuses on pushing back the frontiers of knowledge

Clear Signs of Trouble and Great Joy

by applying the paradigm to the explanation and prediction of data.[1]

Disbelief in the Bible became normal science's paradigm as geology was being birthed a little over two centuries ago. Such disbelief didn't arise from geological observations. It wasn't the product of *information*. It arose by early geologists attending assumptions rather than observations in order that the perspicuously pictorial centerpiece of mankind's puzzle, The Holy Bible, could be "scientifically" denied. Early geological observations could just as easily have been interpreted to mean "global flood", when in the late eighteenth century the Bible was still regarded as *information*. Geological observations made today demand a global catastrophe. But even while geology was a diapered, thumb sucking infant, it learned to cast *information* aside while projecting its own ambitions onto the space reality's puzzle reserved for *information's* centerpiece. Today, normal science's resulting foolishness only babbles in the face of real *information* mounting ever higher to its contrary, such as evolution's dinosaurs stammer in the face of brontos engraved on Bishop Bell's grave, and like "only one life to live" stutters at the not-made-by-human-hands image upon The Shroud of Turin. The fulfilled and currently fulfilling prophecies of the Bible demand an explanation of the Bible's ability to prophesy successfully. But normal science must consider this kind of *information* irrelevant, otherwise, its entire scientific assumption of a mechanistically evolved, non-designed, purposeless, god-less universe would be waved off like a foul flatulent. Yet, just because David Hume assumed the naturalism which led to Hutton's presumed gradualism, Darwin's evolution dogma painted over all the spaces in science's puzzle meant to receive evolution-controverting *information*. All of the world's incurious, non-thinkers will never know the difference.

Surely, while deliberately humiliating *information* that much normal science will never discover reality. To normal science, Ica stones mean forgeries while Bishop Bell's grave means nothing; Raamses II was still the pharaoh of the Exodus (just hush up about the February 25, 1362BC total lunar eclipse); and E.T.'s mom must have left that Shroud thingy behind. The Bible must be quietly branded as mythology so normal science can avoid humbling itself to any actual *information* about the God whom little Sadie thanks. No way the scientists messed up their puzzle, Rhinnie! Never mind the gaping holes scattered around its cosmological puzzle. Those are only factory defects, Granddad. So everyone just stop thinking about all of that *information* crammed into those scientific dumpsters, and pay attention to the wonderful theories science-priests are offering the general public for thoughtless consumption.

Adults should be honest enough to admit everything that exists. We shouldn't be leaving that up to our children, or eventually, they will grow up to be the same kind of deceivers adults became. The search for truth requires an explanation of all *information*, no "anomaly" tags allowed, no trips to the dumpster granted. For, wherever there's *information*, there is reality. Reality's priests don't deceive; they don't hide; they don't make things up, because if even one piece of *information* undoubtedly falsifies a theory or paradigm, then that theory or paradigm is undeniably false, and so are any priests who maintain it anyway. The truth is not reached by ignoring *information*. Ignoring *information* reaches only ignorance. Nothing assuages ignorance, although made up assumptions can hide it.

The Rest of the *Information*

> [6] My people are destroyed for lack of knowledge;
> because you have rejected knowledge,
> I reject you from being a priest to me.
> And since you have forgotten the law of your God,
> I also will forget your children.
>
> Hosea 4:6

Still, science's cosmology fills a large dumpster with "non-fitting puzzle pieces" in order to fit Darwin's purposeless/no-design guess into the core space meant for *information's* centerpiece. That is intellectually irresponsible. Still, science presents its transgressions against *information* as if its cosmetologists are priests to be honored and followed without question. Their discarded puzzle pieces not only exist, they even indicate the corrections necessary to true these priests' puzzle with reality. Even little Rhinnie knew her puzzle was not finished until the last of its pieces were in their proper places. She would have thrown a fit had her old Granddad tossed any of its pieces into some dumpster out back.

> Truly, I say to you, unless you turn and become like children, you will never enter the kingdom of heaven. [4]Whoever humbles himself like this child, he is the greatest in the kingdom of heaven.
>
> Matthew 18:3-4

The humility of children acknowledges what they see. The proposed objectivity of science is to discover reality by building knowledge out of observations, that is, to assemble the puzzle out of reality's pieces. As such, we would think science is as deeply invested in what it has seen as was little Rhinnie. But if science were to be so invested in actual observations, then its dumpsters wouldn't contain any *information* at all. Instead, they would be full of theories and assumptions. Reality doesn't supply dumpsters for *information;* it supplies a puzzle for that. Reality's dumpster is for mankind's guesses. But science doesn't abide by reality's rules, therefore, it brings its own dumpster to the cosmological puzzle.

> Mopping up operations are what engage most scientists throughout their careers. They constitute what I am here calling normal science. Closely examined, whether historically or in the contemporary laboratory, that enterprise seems an attempt to force nature into the preformed and relatively inflexible box that the paradigm supplies. No part of the aim of normal science is to call forth new sorts of phenomena; indeed those that will not fit the box are often not seen at all. Nor do scientists normally aim to invent new theories. Instead, normal-scientific research is directed to the articulation of those phenomena and theories that the paradigm already supplies.[2]

Nobody wants to call normal science subjective. But does Kuhn's research imply anything about normal science other than subjectivity? If normal science were objectively

Clear Signs of Trouble and Great Joy

constructing its theories and paradigms, then it would use contradictory *information* to modify them until all *information* would fit into its theories and paradigms coherently. Its paradigm would then more closely reflect reality by containing more of reality's own *information* . *Information* is reality's smoke. Normal science can not honestly disregard any of it as long as reality's fire yet burns. But subjectivity admits only to *information* that builds paradigms towards its own desired conclusions and disregards the rest. Objectivity acknowledges and uses all evidence for how it coherently affects a paradigm, whether or not objectivity likes what it has discovered or how it must adjust its thinking. Objectivity is interested in what is real more than in what some normal scientist desires to theorize.

 Granddad became Rhinnie's hero because mopping up a mess wasn't good enough for a little girl's humility. To her, success was adjusting her puzzle until all of its pieces properly fit as they were created to fit. Failure would have been to use only the pieces she liked and to have placed her make believe into the spaces meant for the other pieces. But no piece of reality's puzzle is less relevant than any other of its pieces. The tiniest piece can show that something big might be found in a direction different than what the researcher had imagined, if he will only follow. No filament of smoke is any less from the fire than is the rest of its smoke. Not a single one of reality's pieces is without its precise fit into reality's puzzle, nor is it without a specific purpose for fitting there. Since science discards real puzzle pieces, doesn't science's boast seem to be more about authority to proclaim what is real instead of humility for observing what reality proclaims about itself? Science boasts of its authority to say what is or isn't knowledge while throwing out mopped up, left over *information* that won't fit in with its presumptions. Smoke can't be mopped away from its own fire. No *information* can be mopped away from reality's puzzle. True knowledge reflects reality. If science desires authority, let it be had through humble resignation to reality's *information*.

 The Bible is a reality. You can hold it. You can read it. It has a history. In it is significant observations about today's troubles written thousands of years ago. The Bible is puzzle piece. It's mere existence is *information*. But normal science has a mess to mop. It subjectively refuses to treat the Bible as *information*. Doing so would require accounting for the existence of the Bible's message. Accounting for that existence requires acknowledgement of God, the very thing atheists and skeptics twisted the meaning of scientific discovery to deny. So they accuse the Bible of being merely religious literature, nothing more.

 But that explanation (actually, an accusation) fails to account for the Bible's unique characteristics. Millions of people over the millennia have used the Bible for what it states itself to be. They've been satisfied with the affects of following its prescriptions. This can be said in spite of the fact that centuries of atrocious mayhem has been caused by masses of people abusing the Bible by doing what it never itself said to do. The testimony of humble, childlike Bible users classifies it as useful literature, at the very least.

 But normal science refuses to give the Bible even that respect. It wants no explanation of the Bible at all. So normal science accuses the Bible of being the falsehood it shows its own self to be. And that's somewhat humorous. For by accusing instead of observing, normal science refuses to be what it claims to be: objective.

 Philosophers of science haven't even developed a unified and complete explanation

The Rest of the *Information*

of science. They've just proclaimed it to be the epitome of knowledge by the same fiat they accuse the Bible of being ordinary literature. So they've left the general public holding a bag of misconceptions about how well defined and authoritative science is and the Bible isn't. Of course, science is as authoritative in describing the interactions of matter as the Bible is in describing the interactions of spirits. By knowledge of science bridges rarely crumble, most rockets launch without exploding, and everyone who knows anything about nitroglycerine knows to handle it gently. But in exploring such matters as from where our material existence came so we can know to where it goes, about what most basically constitutes matter so we can know how to keep healthy, about the reason for matter's existence so we can know why we are, and even about what forms our consciousness so we can know who we are, modern science has thrown away the Bible without thoroughly analyzing its messages or honestly contemplating how those messages could have been delivered considering what they accurately said about today thousands of years ago. The faint consideration science has given the Bible has only been to twist its messages into conformity with the purposeless/non-designed, uniformly gradual, God-less assumptions fancied during modern science's thumb sucking infancy, but never corrected to correspond with all of the controverting *information* discovered in science's maturity.

Both objective science and the Word of God speak reality in unison, one expounding upon matter, the other expounding upon spirit. Neither casts away the other's *information*. This unity is the simple reconciliation of religion and science, which the philosophers of science have avoided like leprosy.

How contorted subjective science's puzzles have become regarding such issues is exemplified by the great variety of theories its puzzles offer. Life was sparked in lightning struck ooze, one says. No! Molecules meeting on clay began life, another objects. Oh, phoo! It started in deep sea vents, proclaims some other. Nay, chopper! Ice encrusted oceans harbored life beginning reactions. Nuts to you! RNA just spontaneously happened. By golly, that doesn't add up. Esoteric nucleic acids just spontaneously happened. Poppy cock! Encapsulated, smaller molecules cyclically interacted. By Jove! You're all wrong! Life came from outer space.[3] It probably rode in on that Shroud-thingy E.T.'s mom left behind when doing the laundry.

At least somebody's getting closer. Eternal life rode in on the resurrection of whom The Shroud wrapped. Maybe science is horseshoes and close counts. But reality isn't horseshoes, so only dead-on ringers count. Just think of how confusing culture would be if history was built upon a core of assumptions instead of being built around evidences of what happened. Think about what a mess politics would be if three-quarters of the news and entertainment media did nothing but accuse everyone and everything they didn't like while successfully duping most of the population into believing nonsense. If scientists hadn't kicked the Bible out of their nineteenth century clubhouse, they could have just read from the Primordial Historian how life began and to where it flows, what we have become, what we were meant to be, and how to get back to that. But that wouldn't have made Hume, Hutton, or Darwin famous. And isn't fame the gravy of being a great scientist?

Science seems to have lost (or thrown away) the key to understanding existence. Evolution was rational before the enormous complexity of the simplest cells was discovered.

Clear Signs of Trouble and Great Joy

Today, the complexity of DNA and the mechanisms for processing its code is well known, leaving evolution to beg incomprehensible coincidences for explaining how such unimaginable complexity evolved by mere chance. Yet science still holds itself up high enough to strike down not only much of reality's *information,* but also any sense of a deeper being which most of us mundane individual's notice quite naturally (subjectively, according to scientific accusations.)

Consequently, many people keep their deeper perceptions hidden from public view in order to shelter their core beliefs from scientifically inspired ridicule. Normal science teaches our culture how to feign objectivity while maintaining subjectivity. Nobody likes it, but how do we complain? And to whom can we complain? That, my friendly reader, is normal science's greatest need for keeping the Word of God veiled from the public mind. Sufficient complaints to the Bible's God might attract so much of His involvement in man's affairs that His existence would become evidenced far beyond what subjectivity could hide. Normal science is not inclined to replay that old tune.

How has science been able to enforce such a strong bias towards the purposeless/non-designed, uniformly gradual, God-less guess in the face of so many, obviously controverting, well noticed realities? In other words, how does subjective science get away with trashing objective *information* in front of God and most everyone else?

The theory guardians of normal science -the researchers, the professors, the K12 teachers, and almost everyone whose careers interweave and depend upon scientific explanations- scrupulously distinguish between scientific and nonscientific thought and expression in order to safeguard the subjectivity of their precious imaginings (theories, assumptions, and presumptions).

The rules for distinguishing between scientific and nonscientific expressions have been popularly taught at the K12 levels as CONPTT, an acronym for Consistency, Observability, Natural, Predictability, Testability, and Tentativeness. If a statement survives the narrow gauntlet of these criteria, then the paradigm guardians salute it as being scientific. It isn't the only set of rules for guiding scientific discussion, but it is the one most commonly taught. Although these six concepts are given to assure objectivity in scientific discourse, while implying the shameful subjectivity of any other discourse, their actual and real affect is to guarantee the subjectivity of scientific paradigms and assure that reality's explanation of itself shall never be discovered by scientific processes.

Reality pays no regard to man's criteria for discourse. Reality obeys its own rules. Reality is what it is. And really now, doesn't that idea rather reflect the Bible's claim about God being I AM WHO I AM? He obeys only His rules. He pays no regards to man's criteria for discourse. Weren't we supposed to continue in His image, knowing Him as we experienced Him rather than as we theorized Him?

CONPTT is a blueprint for a box in which to shelter scientific discourse from paradigm-popping *information*. Science considers its box to be a fortification against subjectivity. But that's merely a perspective from a different angle. Working within a box is reasonable to science because science proclaims the physical universe to be a closed system, that is, a box. But observation of any closed nature of our material system can not be made from the confines of this bewildering, tiny speck of dust adrift around a little spark of sunlight within any fleeting moment of an unimaginably immense, comparatively endless universe. The closed nature of the universe is just a presumption made by science for plugging up the hole left by what it refuses to observe.

The Bible portrays the universe as being quite open to the affects of a different realm, a spiritual reality which our physical universe objectively evidences. Even empirical observation is able to discover many affects of spirits upon our physical world. Rational thought is able to discern the spirit realm through acknowledging those affects. But CONPTT is science's excuse for refusing to think about them, or even to acknowledge them. It is the blindfold science drives behind. And we wonder why the world is a wreck.

Our brief analysis of CONPTT will consider the logical extensions of its six criteria as they are described at https://www. Indiana.edu/~ensiweb/lesson/conptt.pdf. For clarity we will not discuss these concepts in the order of their abbreviation in the acronym, although we will begin with the first.

> Consistency: the results of repeated observations and/or experiments concerning a naturally occurring event (phenomenon) are reasonably the same when performed and repeated by competent investigators. The event is also free from self-contradiction: it is consistent in its applications. The weight of the evidence is also compatible with well established observations and limits.

A statement about an observation must be about an event, or object, observable by someone else in the same way the first, competent investigator observed it. For example, even though fire caused by rubbing a match quickly across a rock can be observed by anyone who rubs a match across a rock quickly, only the results of a competent investigator rubbing a match quickly against a rock qualifies for scientific discussion. So, does that mean we mundane observers should have no objective discussion with the fire department after our toddler has tried it? Is there some special difference between us common rabble and scientists? Seeing is done more by opening the eyes than it is done by higher education, and reality is more everywhere than it is in the laboratory. Rubbing matches against rocks produces easily observable results any one can do and see, and sometimes, had better see when he's done it. Of course, toying with a cyclotron requires a bit more competence, but the world is enormously more full of observable occurrences which don't need one of those to be seen.

Maybe science's prescription for competence adjusts downward with a phenomena's simplicity? Could Moses have been a competent enough investigator to observe a bush on fire but not burning up? He proved himself to be intellectually responsible by turning aside to further examine the phenomenon. In fact, his competency rose to the level of writing a rather thorough lab report about its most relevant Aspect, the same Aspect our "competent" scientists refuse to look for with their university trained, closed eyes.

But still, scientists say he wasn't competent because none of them have ever seen a bush burn without burning up, which is what they mean by demanding the weight of the evidence to be consistent with established observations and limits. Observations established

Clear Signs of Trouble and Great Joy

by who? Limited by what? And just how universally representative can be the limits found here on this speck of dust? In all due regard, Consistency is a euphemistic way of saying that if you have seen something no scientist has ever seen (or forever refuses to see,) then you don't know what you're talking about. It might be a scientific attitude, but it is not a very wise one.

Reality remains far bigger than what all mankind has ever seen, let alone than what a bunch of scientists have selectively observed in their boxed-up laboratories. Reality has a penchant for interjecting itself into human affairs wherever, whenever, as rarely, or as commonly as it so chooses, and with no regard for any scientific snobbery. Shouldn't we rationally expect such behavior from someone called I AM WHO I AM? And if Reality once brought an angel along with it, who on earth is the scientist to say it did not? Far more phenomena has occurred than what all scientists throughout history have even begun to empirically observe. And we mundane ones, who've individually observed far, far more than all those scientists have seen boxed-up in their tiny laboratories, don't know what we're talking about? So we who have seen more of what scientists have never seen often refuse to believe our own eyes so as to not be considered chumps by subjective science? The Consistency principle makes a fine floor for supporting this box of scientists and their limited considerations.

The fabric of the Consistency floor is woven from the inconsistent application of Tentativeness.

> <u>Tentativeness</u>: Scientific theories are subject to revision and correction, even to the point of the theory being proven wrong. Scientific theories have been modified and will continue to be modified to consistently explain observations of naturally occurring events.

Since all reality is far more vast than what mankind has time or the ability to empirically observe, it logically follows that there is always more *information* which needs to be considered than science has yet considered (or has yet been willing to consider). Somewhere in all of this still-to-be-observed *information* is the possibility of an observation capable of definitively refuting even science's favorite, core guess, the purposeless/non-designed nature of the physical universe. Tentativeness is important because all of reality's *information* can never be observed in total. Therefore, assumptions must sometimes fill *informational* gaps, but only after all available observations have been duly considered, and then only until correct *information* has been found to fill up those gaps tentatively painted over with assumptions. When *information* is observed to fit the puzzle where an assumption has been currently holding its place, then that assumption must be replaced by the appropriate *information*. The puzzle will need to be reshaped in accord with new *information's* implications. But normal science refuses to humble its theories even to *information* we have possessed since long before science was any practice at all. Normal science's CONPTT box is actually for limiting undesirable observations from affecting those sweetheart paradigms by which man propagates his favorite evils.

All of man's statements and theories, being incomplete in *information*, and not just

The Rest of the *Information*

his scientific ones, must be held subject to revision, correction, or even falsification by newly discovered *information,* or even by previously misunderstood but now better understood *information*. ALL assumptions and theories must be held tentatively. The discovery of unconsidered *information* obligates objective science to analyze new *information's* impact upon existing statements, theories, or paradigms, else science becomes intellectually irresponsible, and even quite deceitful. Only when *information* demonstrates the immutability of statements, theories, or paradigms do they become enduring principles.

The biggest example of the inconsistent treatment of Tentativeness to date has been Darwinism. We all know Darwin proposed that the fossil record would eventually support his theory. He predicted innumerable discoveries of transitional life forms. But by 1859 he admitted that their continuing absence was the most serious objection to his theory.[4] (At least he paid lip service to Tentativeness.) Since then, Darwinism has only stepped further away from the scientific principal of Tentativeness.

> Darwin concluded that the explanation lay in the extreme imperfection of the fossil record (1859, p. 280). As we will document, though, the millions of marine fossils discovered since 1859 have not documented these "innumerable" transitional fossils that evolution requires. The absence of fossil evidence actually is overwhelming.[5]

Darwinism holds fast to its assumptions in the face of this overwhelming absence of evidence, desperately clinging to its yarns spun in the face of many controverting evidences. Yet when it comes to The Lord God and Judeo/Christian beliefs, a presumed lack of evidence is always put forth as the greatest reason for disregarding the Bible and refuting God's existence. But actually, there is no lack of evidence for supporting The Holy Bible's explanations of the nature of all things; there is only a lack of willingness to observe the wealth of such existing evidence. This book presents the most astonishing evidence of the Bible's explanations made since the light-burst of Christ's resurrecting body imprinted The Shroud of Turin. Normal science's failure to apply Tentativeness to EVERY theory is the best evidence of science's normal subjectivity.

ONPT forms the ends, sides, and lid for this tiny box of science. Normal science is sure ONPT puts religion in a box for throwing into a dank, basement closet beside the kids' old toys. So it uses ONPT to deny what the common mind can actually discern about a spiritual existence beyond detection by the five, physical senses. ONPT implies that only what can be seen in the sterile confines of controlled laboratories is relevant, as if there is nothing else to consider.

> Observability: The event under study, or evidence of the occurrence of the event, can be observed and explained. The observations are limited to the human senses or to extensions of the senses by such things as electron microscopes, Geiger counters, etc. If the phenomenon cannot be reproduced through controlled conditions, natural <u>evidence</u> of the event's occurrence must be available for investigation.

Clear Signs of Trouble and Great Joy

A sort of sixth sense extends empirical observation beyond what only the five senses can see. Without it, everyone, including all of the animals, would be dumb as fence posts. We're not talking about Extra Sensory Perception. We're talking about simple pattern recognition. The sixth sense is the mental process of sensing patterns and perceiving their meanings. Pattern recognition is not to be mistaken for pareidolia. Reality is so full of patterns and the mind is so made to recognize patterns that sometimes the mind sees patterns where there really are none. That is pareidolia. It's all in the mind. Assumptions are rather akin to pareidolia in that they are current ideas where no actual *information* has been found to express those ideas. Pattern recognition is a very important sense. The dandelion hungry bunny had better perceive a meaning in the hawk's shadow crossing the meadow.

The best example of actually existing patterns and pattern recognition is what you are right now looking at: writing. It is the patterning of twenty six unique symbols into words, each assigned a meaning. Those are arranged into yet higher patterns -sentences- expressing ideas. Then patterns of ideas form paragraphs constructing chapters expressing the complete message of a book, like maybe the Bible, indeed, a giant pattern of ideas.

Defining and fully supporting pattern recognition as being the sixth sense is beyond the scope of this book. However, it is an important concept to both intelligence and this book's theme, for pattern recognition is the starting point of observation. Therefore, let it suffice for us to notice that all *information* enters the brain as stimulated patterns of electro-chemical signals carried along neural pathways. The brain processes the stimuli for recognizable patterns which have been assigned meanings from past experiences of similar patterns, maybe important meaning for the bunny, or for meaning to be learned if the bunny survives. Patterns of neural networks wiring and firing together in the brain, forming consciousness and subconsciousness alike, are unimaginably immense. But they are still nothing more than patterns. Physically speaking, our entire personality and character are just patterns of ideas, emotions, and stimuli occurring throughout the body and imprinting recognizable patterns onto the brain. Maybe our spirits are complete and exact records of all the patterns comprising us and their interrelationships with other patterns. Maybe our spirits cause intricate and basic stimulations of our neural networks. The Bible is filled with stories of the Holy Spirit and demons directly causing people's actions.

Let your neural networks consider the periodic table and note how the shells of atoms are formed by patterns of electron states, and how by one pattern of energy or another (covalence or electrovalence) atoms cling together to form molecules, and how molecules clump together into objects, and how objects comprise the common tools and utensils we use as well as the earth we stand upon, the solar system we live in, the spiral and cluster galaxies of innumerable solar systems, and everything else beyond the observable edges of the universe. Nature is a symphony of patterns perpetuated by the arrangements of a seemingly infinite number of fundamental particles. Then what we express by either word or deed makes patterns amongst the stuff around us which we more commonly call evidences of action. So why should it be so unbelievable that a spirit realm might be discernable by patterns it may have left in the arrangements of historical events? Actions affected by spiritual influences upon people's intentions leave evidences within the stuff of our physical realm. Does this not better explain The Shroud of Turin than does E.T.'s forgetful mommy? It clearly ex-

The Rest of the *Information*

plains the Bible. Only it can explain what you will be reading about in Chapter 9.

But wait! Now that we're talking about the spirit realm, the Bible, pattern recognition, and such, Observability no longer applies, according to normal science. There's just something not very Consistent about that. If pattern recognition is not part of Observation, then you have no idea what this sentence just said; you only see what its patterns look like without recognizing a thing about them. Thank God for pattern recognition.

Noting that the Observability criteria is for limiting consideration to the five basic human senses isn't to say science pays no attention to patterns. Ancient cultures are distinguished in part by the patterns they inscribed or painted on their pottery. Movements of peoples and ideas are discerned by interrelated patterns found throughout regions and around the world (e.g. the swastika, the spiral, the "blessing bucket", etc.) Psychology and sociology both employ recognition of behavioral patterns and patterns of cultural interaction. But normal science demands there be no pattern recognition whenever the Bible is involved. The Search For Extraterrestrial Intelligence is tolerated by normal science although it also focuses intently upon a search for patterns within the universe's chaotic background noise. Yet normal science spits in Consistency's face by ignoring any pattern related to the Bible.

Since patterns are abstract, with a little influence from a paradigm they can be waved off as meaningless or misapplied to ideas they actually have nothing to do with. But reality differs from paradigms. Because actual patterns are reality's *information*, they bear specific relationships with and hold precise places within reality's puzzle. But when a pattern illuminates what normal science does not wish to consider, normal science doesn't have to cover it up, whitewash it, or brand it as meaningless, as it must do with more concrete *information*. Silence hides patterns quite well, since most people don't look for or notice them beyond the more obvious ones of their daily experiences.

That we can not see through the box's lid with five of our senses does not mean nothing is outside the box. That's the major fallacy of subjective science's logic. It only means we have denied ourselves any way of seeing what might actually be there, such as by observing patterns made on the box's lid by what truly is outside the box.

By rejecting *information* noticeable by this "sixth sense" merely because it has to do with a very historically patterned spirituality, is normal science soon to be disgraced by the pattern of Jesus' feet descending from outside the box onto the Mount of Olives? Such an occurrence would form a very strong pattern of Consistency with The Shroud of Turin, which already has a very strong pattern of Consistency developed with the Bible and two-thousand years of Church history. Mankind's other five senses will note the answer when His feet get there. But His feet descending there is observable today by noticing conspicuous patterns of correlations made amongst positions of the heavenly bodies (eclipses, conjunctions, etc.) timed with historic events and prophecies of the Bible. Those patterns evidence the existence of a reality far more meaningful than what science's minuscule box full of subjectivity can contain. It suggests what subjective science meant its box to hide.

Normal science demands that an observation be explainable, "...the event under study...can be observed *and explained,*" as it was expressed above. But many observations won't be explained by science. Late one morning and into the early afternoon of a winter day, about 1997 or so, I listened to ongoing news casts about a light phenomenon being

Clear Signs of Trouble and Great Joy

reported by the control tower at Denver International Airport, then by the Stapleton Airport tower, and eventually by the Colorado Springs tower, as well as by various observers (including the police) as it moved southwest. It is undeniable that something was seen, though it was not explainable by our current knowledge. But what it was as much a part of reality as are the pens with which scientists refuse to note such phenomena. The notation of something is the first step towards a knowledge of it, whether or not the knowledge entails any immediate explanation. The first step towards an explanation is treating the phenomenon like the reality it actually is. But normal science's refusal to note anything it has not yet acknowledged is its next step towards the abject ignorance reality's nature.

By this same myopia, science shows no respect for billions of mundane observers. Reality happens everywhere at all times in a wide variety of ways which leave clues as to what happened, whether or not the event was explainable. All of the world's scientists are so few that they are hardly anywhere whenever reality makes inexplicable phenomena. But is that any reason to invalidate what one of the billions of mundane observers actually and truly did see, accusing them of not knowing what they're talking about? Observability is not only CONPTT's principle of subjectivity, it is arrogance. All that is required for knowing something exists is the sight of either it or an affect of it. Millions of people throughout history, and even to this very moment, have seen and reported affects of I AM WHO I AM. Of course, far more misreporting and fake reporting than true reporting occurs. (We Americans are very familiar with fake reporting.) But amongst the chaos of error and deceit can be found gems of truth being consistent amongst themselves and coherent with similar kinds of reports made in the Bible. Just because subjectivity refuses to explain such consistently observed phenomena neither means the phenomena is ignorable nor that the Bible does not explain it. It just means science has closed its only non-myopic eye, while many millions of observers are willing to see with eyes wide open.

Finally, normal science has turned its blind eye against the most important form of Observability. It insists there is no such thing as revelation from I AM WHO I AM. It has never seen it with its own five senses, so it must not exist. Moreover, there have been so many diverse, contradicting ideas throughout time claiming to be revelations that normal science refuses to believe any of them.

The reason normal science has never seen revelation from God is quite obvious. It refuses to be what I AM WHO I AM is interested in associating with: humble. God reveals Himself to the humble. The humble will follow evidence anywhere it leads. Because normal science has convinced itself there is no God; it won't even consider moving in the directions God's evidences indicate even if it undeniably sees such evidence. Consequently, not only does God not lead it to any of His evidence, normal science recognizes no need to look for any evidence of God, because it is sure He doesn't exist. Its own paradigms lead its research away from His evidence. Normal science won't even accept easily available natural evidence of God. And it certainly won't consider or honestly investigate the evidences of God "abnormal" science shows to it. Nor will it investigate the evidence you will read in this book. Normal science defines observation in a sense that renders revelation to be nonsense and narrows its own scope of knowledge. In short, normal science is closed minded science; its attitudes towards God are the epitome of bigotry.

The Rest of the *Information*

Reality reveals genuineness through the verifications its *information* makes. The Bible is a fundamental part of a pattern of *information* verifying the reality of revelation by revelation's actual correspondence with known history. Correspondence with something as natural as known history defines both revelation and God as Natural. He is more a part of everything around us than we are. His ideas, personality, and ambitions are a part of the way everything is and how it functions, because He not only designed everything, He created everything for the purpose it is presently serving. That statement is solidly evidenced by empirical observations which can be duplicated by anyone with enough patience to do the research. The only reason science has not observed God to be Natural is that it defines Natural in a manner which excludes God, acknowledging only those aspects of nature which do not involve God on the surface. All other aspects, such as Bishop Bell's sauropods, simply get thrown into a dumpster. However genuine that is not, it is truly sly.

> Natural: A natural cause (mechanism) must be used to explain why or how the naturally occurring event happens. Scientists may not use supernatural explanations as to why or how naturally occurring events happen because reference to the supernatural is outside the realm of science. Scientists cannot conduct controlled experiments in which they have designed the intervention of a supreme being into the test.

Therefore,

> Scientific explanation is no longer permitted to appeal to the deity...[6]

The criteria, Natural, forms strong, opaque sides for science's box, no windows allowed. This criteria basically requires all observations to be explained by what science is willing to acknowledge as Natural. That sounds arrogant, but let's give science a chance to explain itself.

Hume said not all nature can be observed, therefore no one can fully define what Natural is. On the other hand, he said what we are able to see can be relied upon to represent everything that is natural, whether we can see it or not. Hume's problem of how to define stuff which can not be entirely observed became science's excuse to impute whatever it imagines to anything which has not been observed in total, or to anything it refuses to admit has been observed. Normal science decided it did not want to permit God or the Bible to explain whatever they could not or would not observe. So it didn't. Uniformitarianism is a product of that gem of manipulation made into scientific criteria.

But without observing everything, no one can define Natural. Honoring the Natural criteria is then like kissing David Hume's left foot while nailing his right foot to the floor. Hume's right foot states that all of nature can not be observed, therefore, all of what is Natural can not be known, since knowing comes from observing. But science would not allow that truth to interfere with its ambition to limit the definition of nature short of including God. So, clinging to Hume's assumption that the scanty observations science is willing to permit in its deliberations can fully define Natural, science simply denies the

Clear Signs of Trouble and Great Joy

admission of God into their concept of Natural even though His effects have been observed by millions of mundane observers throughout history.

Every one of CONPTT's criteria references the unobservable imputation normal science makes about what Natural entails in order to constrain any talk or even consideration of the rest of the *information* packed tightly into normal science's dumpsters. There's actually more smoke of reality in normal science's dumpsters than there is in normal science's paradigms and theories. That observation is made in good accord with the word of Kuhn; we didn't even need to go to the Word of God.

Simple logic says normal science knows nothing about the entire nature of Natural. It is also rational that any entity which does know the whole of all reality can more clearly define what Natural is than a whole world full of people who have comparably not seen diddly-squat. For such an entity to communicate its definition to man by some means of revelation would enter revelation into man's empirical experiences, which by such experiences would make revelation to indeed be quite Natural. As we will later see, history and current events verify the Bible's claim to be that sort of Natural. And all of normal science's dumpsters are filled with evidence of that very thing.

Indeed, the Bible is a library filled with many reports of numerous consistently coherent, historically verified revelations. It even reports about revelations yet to come, such as the seven thunders of Revelation 10:3-4. What are you scientists to do? Lie? Even though Moses is the only human who was allowed to see I AM WHO I AM in person, everywhere throughout time people have been seeing I AM WHO I AM through His affects upon their personal lives simply because they are too humble to trash evidence they don't like.

God is Natural; He is much more Natural than are we. He is even normal. We humans are the abnormal ones. We are the unnatural ones. We are the ones who deny what our eyes can see until we even refuse to look where they could see. It isn't that there is a supernatural which science refuses to consider. It is that our normally experienced Natural is truly subnatural. And the most subnatural thing of our experiences is science's boast of knowing what is Natural.

But still, extraordinary claims require extraordinary evidence. Normal science denies Biblical claims because normal science claims to have found nothing about them that it can touch, lick, smell, see, or hear. But neither can normal science touch black holes, lick earth's iron core, smell Dark Energy, see Dark Matter, or hear the Big Bang. Yet normal science considers all of those to be quite Natural. Indeed, their evidences came by means of noting patterns, testing predictions, making assumptions, and constructing models, all of which can be done for the Bible more successfully than against it.

> Predictability: The natural cause (mechanism) of the naturally occurring event can be used to make specific predictions. Each prediction can be tested to determine if the prediction is true or false.
>
> Testability: The natural cause (mechanism) of the naturally occurring event must be testable through the processes of science, controlled experimentation being only one of these. Reference to supernatural events or causes are

The Rest of the *Information*

not relevant tests.

Science has no idea about what happens inside black holes. And although it knows the numerous mechanisms of the inner cell and its amazing processes, it has no idea about what makes the cell live. But since black holes and cells both exist and do what they do, there is obviously some mechanism at work in each. It is just unobservable by the five senses. Unknown mechanisms are called black boxes so that the reality of their hidden processes, observable only by their affects, can be acknowledged as being Natural. So why has I AM WHO I AM not been accorded the same respect of the black-box consideration regarding His quite apparent processes? Subjectivity? Bias?

Simply because the spirit realm is also unobservable by the five senses does not mean there is no spiritual mechanism at work there. Its processes are just as considerable by the same black-box acknowledgement. I AM WHO I AM is as much a spiritual mechanism as the black hole is a physical one. Our tests may never reveal how His mechanisms work, like our tests of black holes do not reveal how their inner mechanisms work. Regardless, the Bible's prescribed tests just as easily reveal His reality as science's predictions and tests have revealed the reality of black holes and Earth's iron core. I AM WHO I AM obliged Israel to test Him and bear witness to the rest of the nations regarding His reality.

> [58]If you are not careful to do all the words of this law which are written in this book, that you may fear this glorious and awful name, the LORD your God...will scatter you among all peoples, from one end of the earth to the other.
>
> Deuteronomy 28:58,64

Even though this test prescribed for Israel's performance was as simple as "listen and obey", it was very greatly consequential. Their history would have been far different had they performed the test by the means of obedience rather than by that of disobedience. Instead of being devastated by their enemies, cursed by their God, and sent wandering without a homeland for centuries, I AM WHO I AM would have raised them to the head of the nations in great wealth, prosperity, and honor centuries ago, blessed in all their cities and fields. Their children, crops, livestock, industry, and all of their ambitions would have been blessed, and He would have subdued all of their enemies (Deuteronomy 28:1-9).

> [10]And all the peoples of the earth shall see that you are called by the name of the LORD; and they shall be afraid of you.
>
> Deuteronomy 28:10

> [1] May God be gracious to us and bless us
> and make his face to shine upon us,
> *Selah*
> [2] that thy way may be known upon earth,
> thy saving power among all nations.
>
> Psalm 67:1-2

Clear Signs of Trouble and Great Joy

But those honors await God's merciful actions, since Israel tested I AM WHO I AM by disobedience to Him before the eyes of all mankind, and consequently, all mankind's eyes have seen them scattered from one end of the earth to the other. For centuries mankind has observed Israel pitifully wandering without being destroyed. Regardless of the means Israel called upon for doing their experiment, the results of their experiment showed I AM WHO I AM to be quite Natural. Go find another people thusly treated and thusly surviving, even having recently come back into their homeland to speak the same language they spoke twenty centuries ago, as I AM WHO I AM also promised He would do for them.

> [20] He has not dealt thus with any other nation;
>
> Psalm 147:20a

Even through the disaster of their disobedience Israel bears witness and testimony to I AM WHO I AM's reality. For these things were predicted about Israel's history before Israel first became a nation. And Israel exists today because the experiment is yet ongoing. Watch what will eventually happen to the nations rising against Israel! Christ's feet upon Israel's Mount of Olives is part of God's experiment for differentiating between the ever observant and the subjectively blind.

Although the simple test of disobey/get-scattered did not explain the spiritual mechanics of what scattered them, it indeed verified that they were scattered by a very Biblically coherent, black-box mechanism. So it is not surprising to find science accusing Biblical authors of writing "predictions" after having seen the "predicted" events occur. But if the Bible's authors had done this, then cheating would have been the black-box mechanism by which the Bible's prophecies were fulfilled, rather than by God just naturally being what He is (I AM WHO I AM). If cheating were that mechanism, then there would have been nothing of God to see, and Israel existing today would be another phenomenal testament to the absurd boundaries of extreme coincidence for no purpose at all. The assumptions of normal science were deliberately concocted to replace the revelations of the Bible, making it appear as though all of the Bible's authors cheated. But Israel's existence today says we should think a minute before agreeing with that. The core of reality is described by either nonobjective, scientific assumptions, or by Biblical revelations, but not by both. Israel being one nation in their own land, just as twenty-five-hundred years ago the Bible said it would be, is a successful test result science can not accuse of having been achieved by cheating. In fact, this test result finds scientific assumptions to be the cheats. It finds their dumpsters packed full of discarded evidences of God's reality. Israel today bears witness to the reality of God's revelation by being the successful Test of the Bible's Predictability.

> Thus says the Lord GOD: Behold, I will take the people of Israel from the nations among which they have gone, and will gather them from all sides, and bring them to their own land; [22]and I will make them one nation in the land, upon the mountains of Israel; and one king shall be king over them all; and they shall be no longer two nations, and no longer divided into two kingdoms.

The Rest of the *Information*

Ezekiel 37:21-22

Almost twenty-five-hundred years after Ezekiel wrote that prediction, the test results are proclaiming it to be completely accurate. Our generation has empirically observed, with its own eyes, Israel come together again in their land as one nation, exactly as Ezekiel said, twenty-five centuries ago, it would come together. And this same Bible predicted that, as their King prepared for His second coming, there would be signs in the sun, moon, and stars. The full moons of Israel's two most meaningful holydays about this Lord of Lord's wrath passing over them and His dwelling with them forever were totally eclipsed two years in a row beginning the year after Israel again became a nation. And eclipses on those holydays happened again when this new nation of Israel won control of what the Bible calls God's Holy City over which He is hotly jealous.

> [14]Thus says the LORD of hosts: I am exceedingly jealous for Jerusalem and for Zion. [15]And I am very angry with the nations that are at ease; for while I was angry but a little they furthered the disaster. [16]Therefore, thus says the LORD, I have returned to Jerusalem with compassion; my house shall be built in it, says the LORD of hosts, and the measuring line shall be stretched out over Jerusalem.
>
> Zechariah 1:14b-16

These eclipses happening on the full moons of those two holydays of Israel at those two epic events is not natural in the way normal scientists would define Natural. But would you expect a sign to be as natural as the butterflies and flowers on a pleasant summer afternoon? Or would you expect a sign to be something a little more beyond nature's ability, like a pillar of neatly stacked rocks by an intersection of two paths? Because those eclipses on those holydays at those events are just not Natural, does it mean we should not look at them even though they were very orderly in nature? that we should not think about them? that we should continue at ease like they just never existed? Or should we wonder who stacked them beside those events to mean what? Should those two events, fulfilling core prophecies of an ancient Book highly honored and revered for millennia but now castigated and impugned, lead rational minds to reconsider what they have been impugning? to reconsider what is Natural? and to consider just what in reality might have the ability to stack such holyday ladened eclipses above man's spiritually relevant activities? It makes one wonder if maybe we are what's being tested. How will your testing work out? Will you stop and think? Or will you proceed as if these wonders were merely butterflies flitting around a sunny Summer meadow; que sera?

Israel is not God's only laboratory. The Bible is full of Predictability/Testability for personal testing, too. Every person born is a laboratory for testing God's truth. Will denial of I AM WHO I AM for the acceptance of whatever else you want to believe delude your mind into serving the holy cathedrals of normal science, or whatever other paradigm that excludes God? Or will you test God's Word and examine His Book of Propositions in your laboratory according to the lab instructions He prescribes, humbling yourself to find reality by its *information* rather than lifting yourself up to proclaim reality to be whatever you desire

Clear Signs of Trouble and Great Joy

it to be? You are a laboratory for either testing truth or presuming deceit.

My favorite test, which you can perform in the laboratory of your life, is…

> [1] My son, if you receive my words
> and treasure up my commandments with you,
> [2] making your ear attentive to wisdom
> and inclining your heart to understanding;
> [3] yes, if you cry out for insight
> and raise your voice for understanding,
> [4] if you seek it like silver
> and search for it as for hidden treasures;
> [5] then you will understand the fear of the LORD
> and find the knowledge of God.
>
> Proverbs 2:1-5

This test gives a giant clue as to why normal science doesn't know God. If we were going to test for chemical reactions between two substances, we would need to closely follow the precise procedures dictated by the physical nature of the chemicals comprising those two substances. CONPTT is the appropriate criteria for discovering and discussing the aspects of those physical chemicals and the procedures required to successfully understand and work with them. Some test procedures are deceptively simple, for example, nitro-glycerine's unstable composition of oxygen, nitrogen, and carbon needs only a jolt for successfully testing its explosiveness. Fortunately, most chemical reactions require a more complex set of conditions and procedures to test their reactivity. Requisite conditions and procedures must be set up and maintained for any test to provide real *information*.

The Proverbs 2 test also requires a very particular set of conditions in which precise procedures must be followed in order to successfully test for realities beyond the physical. For example, the lab instructions for its test require the Bible be accepted as the revelation it claims to be, which can be done by using Tentativeness; it requires Biblical guidelines be treasured by the tester as the most valuable of all possessions, for which value they are followed carefully; it requires wisdom to be closely attended; etc. No one would accuse a science experiment of being subjective just because its procedures require precise performance of exacting lab instructions, especially when one of the instructions might be the very point of the test. So why are the precise conditions and procedures of all the Bible's tests accused of being subjective? Where's Thomas Kuhn when you need him? The Bible's tests work the same way science's tests work: by prescriptions from instructions.

The simplest, most successful testing procedure for the truth of the Bible, its nitro-glycerin test, is denial. See all of the Bible denial in today's world? See all the hell blowing up in today's world? Take the true, prophecy-speaking, prophecy-fulfilling, eclipse-stacking God out of your children's training and watch them bring guns to school and shoot each other. See the results of denying simple correlations? And normal science is objective? Since the Proverbs 2 test requires precisely followed, Biblically prescribed procedures, what possibility does normal science have of ever knowing the truth about reality? Should you really follow normal science into its exploding world? Some people know very well what's going

The Rest of the *Information*

on in this world because they strive hard to follow the Proverbs 2 lab instructions. But most don't. Most won't even read the Bible, though they are the first to accuse it. Nor will they look at signs. Has their precious normal science ever taught them subjectivity!?

Another of the Bible's condition dependent tests is:

> [6]And the Lord said, "If you had faith as a grain of mustard seed, you could say to this sycamine tree, 'Be rooted up, and be planted in the sea,' and it would obey you."
>
> Luke 17:6

The mustard seed test is probably the most scientifically impugned test of the Bible. Who have you seen lately command trees to uproot into the sea, and they uproot into the sea? Of course, we see everywhere Christians, conservatives and liberals alike, demanding all sorts of absurdities which do not happen. Maybe the reason is because most everyone we know today has supplanted mustard-seed-like faith with a more scientifically described, CONPTT-like faith.

The mustard seed is defined by its DNA. DNA is a very precise prescription. Faith is defined by the Word of God. Faith is also a very detailed prescription. Only Biblically defined faith uproots trees into the ocean and moves mountains, not just tiny faith. You don't get a mustard tree from planting an apple seed. This is why even Christian dirt contractors still use bulldozers; we're not knowing spiritual DNA so well anymore because normal science has taught us to throw the Bible's prescriptions into its dumpster where normal science has trashed the rest of reality's *information* evidencing truth about God's existence.

But the most important test I AM WHO I AM encourages for everyone to try in their laboratories does result in immediate, real, earth-moving test results, although those results might not be immediately detectable by any of the five senses.

> ...if you confess with your lips that Jesus is Lord and believe in your heart that God raised him from the dead, you will be saved.
>
> Romans 10:9

The non-Christian, scientific mind considers the immediate results of this Romans 10:9 test to be subjective because its results seem mostly, and to some people, entirely perceptual. Modern science was able to inculcate disbelief in the reality of spiritual regeneration just because its aspects can not be seen, heard, smelled, tasted, or touched immediately. K12 teaches us to interact according to the belief that everything is physical and mechanical, and that whatever might seem spiritual about our mechanistic physical environment is only a perceptual effect of deep psychology. Most of our universities demand students to study life with malevolent attitudes of opposition to every Biblical proposition. Therefore, today's scientifically trained culture pressures people to socially kick buckets of spiritual nitroglycerine rather than to socially acknowledge even the simplest of Biblical concepts.

To help the normal scientist understand the reality of the *information* he is missing,

Clear Signs of Trouble and Great Joy

let's use N^2 as a reference for that part of Nature normal science has boxed out of its consideration. N^2 is every entity, event, circumstance, and cause science refuses to explain by the honest and humble acknowledgment of "the supernatural", and "appeal to the deity". When the Biblical propositions about N^2 are honestly considered in relation to all of reality's *information,* it becomes quite apparent that the supernatural is more natural than is our subnatural, and that we should be appealing to its Deity in regard to everything.

Understandably, science can not design the intervention of a supreme being into its materialistic tests any more than it can receive signals from a probe sent into a black hole. Regardless, science refers to the black hole as part of Natural because we can observe the black hole's affects on its surroundings. But science precludes resurrection from the dead from being a Natural phenomena even though The Shroud of Turin is scientifically explainable only by the effects made upon linen from a resurrection event. This inconsistency exposes science's "inability to explain by the supernatural" as being a biased attitude rather than a scientific proposition. Science does not get to choose what it will and won't consider to be real and be truly objective at the same time in the same sense. If *information* leads to "the supernatural" for an explanation, then science must objectively go to the supernatural for that explanation, otherwise it must continue to subjectively cower in its own little CONPTT box of subnatural. Dogmatically denying *information* science has for two centuries left off its test and research schedules is to approach reality with eyes wide shut. But to have eyes wide open, CO^3N^2PTT is the criteria for determining honest scientific statements and objective considerations regarding cosmology, ancient history, and the Bible. CONPTT is little more than a lid for science's dumpster.

Let's designate Observability, the criteria the experimental sciences use for studying the properties of purely physical realities, as O^1. We'll let O^2 represent the observability of "the supernatural" availed by millions of mundane observers. Books and legends and stories throughout history abound in what these people have witnessed from the results of their simple call-on-Jesus tests, from their further faith-like-a-mustard-seed tests, and from their many other Biblically prescribed tests. They report multitudes of N^2 caused events which perchance, or by prayer, have happened for their observation.

It is the normal scientist who has no evidence for claiming none of these witnesses know what they're talking about. Science is right about many O^2 claims being at least half baloney. And some are completely cheesy. But the same mix of fraud and integrity can be observed amongst all the claims of "objective" scientists, especially in the theoretical sciences. Observations of the experimental sciences have the repetition of the Consistency criteria for combing out deceit and error. The mundane observer with integrity has experience and the Word of God by which to test claims of O^2. The Consistency criteria tests O^2 claims of mundane observers much the same as it tests claims made in the theoretical sciences, although, not by repetition of experiment, but by a coherence of observations, including Biblical observations. For you must gain the experiences necessary for understanding a paradigm in order to test claims made about it, or against it. You must accept the reality of Jesus Christ and follow His prescriptions, desiring righteousness, and endeavoring to live by God's Words in order to adequately test the O^2 claims, else you will not have the background necessary to know what you're talking about. Just like science,

The Rest of the *Information*

you must use the laboratory that your life is, living it by The Instructions God wrote for it.

Many of these witnesses (the genuine ones) have sufficient evidence showing they know what they're talking about. It's observed in the affects of their lives and the beneficial affects they cause in other lives even though everyone's been false before (Romans 3:4). But the nature of normal science's observational limitations fully evidences that science refuses to observe what its talking against, because it refuses to Tentatively consider the Bible's paradigm about the King of kings and The Lord of lords.

Many O^2 claims do float around in the ozone. Everyone's a bit ozoney, including scientists. So, wherever an incredible, but true, story is told, soon many spurious little tales sprout up around it. Man is a copycat because he is a twisted reflector. The rebellion against reflecting God has left people reflecting culture. As such, many people fake reflections because Biblical prescriptions are too spiritually invasive for them to engage publicly for making real reflections. But all of the world's spurious O^2 tales no more disqualify genuine O^2 observations than a world full of children beating on tin cans with sticks wipes out even one stroke of a Ringo Starr drum roll. No field of observation and knowledge is negated by its own brand of charlatans, quacks, and fallacious assumptions. Not of science. Nor of religion.

Finally, let's designate revelation received from the spirit realm by the term "O^3". Not misperceptions or fraudulent claims of revelation, but actual, real revelation, *information* poked into a mind by the inspiration of a spirit.

Science will not bear any thought of revelation. Revelation is amongst the rest of the *information* which science subjectively drums out of its laboratories simply for being what it does not wish to acknowledge, especially, if it is Biblical. So we can ignore science when it drums on its favorite tin can with an authoritarian stick. Whatever science refuses to acknowledge is not necessarily dismissed from reality just because they are *the* scientists.

Only observations and successful experiments falsify erroneous and false claims. Below, we will consider some observations which will validate the claimed realities of revelation and N^2, the nature and realm of the spirits. Those observations falsify Hume's claims of Natural being limited to only material things and their interactions. They invalidate science's refusal to consider God as an explanation of anything, let alone, of everything.

It is important to not ignore the reality that spirits other than I AM WHO I AM have often given revelation when considering this third kind of Observability. The witch of Endor called the spirit of Samuel up from the dead, which then imparted to Saul a tidbit of truly unnerving *information* (I Samuel 28:7-20). That was Saul's O^3 received through an entity other than I AM WHO I AM. The angel, Gabriel, revealed to Mary God's plan for bringing God's Son into the world (Luke 1:26-38). Gabriel was I AM WHO I AM's messenger, but he wasn't I AM. The snake revealed both *information* and misinformation to Eve by O^3 (Genesis 3:1-5). St. Paul once heard O^3 from a spirit possessed, slave girl; it was not only true *information*, but it was also *information* heard by everyone in the vicinity over the many days she revealed it (Acts 16:16-18). The Bible treats O^3 as a very natural part of its narratives. Moreover, the Bible claims to be the entirely inspired revelation of the Holy Spirit of God who exists in His N^2 and interacts with our N^1, making all of the Bible to be

Clear Signs of Trouble and Great Joy

O^3, observable *information* having come by means of revelation from God.

Let's pause a moment now for the scientific laughter to die down.

The mind works similar to the way the eye works. Of course, it doesn't see by the stimulus of photons striking a retina. It sees by the stimulus of ideas striking a knowledge base. We are personally responsible for what we've opened our mind's knowledge base to receive and for what we've closed it against. But just because the eye needs to be open to see doesn't mean it's a good idea to gaze into the sun. We must be responsible with our eyes, or we burn up the retina. We must also be responsible with our minds, or we pollute our knowledge base, distorting its ability to accurately reflect reality.

Everyone knows the sun is there; we've all caught a glimpse of it. We've even glanced straight at it to verify the glimpse. Even though we need scientific equipment and processes to gaze into the sun's properties, it takes only a glance to discover its existence. CO^3N^2PTT opens the mind to glimpses of the spirit realm, glimpses that God has been allowing mankind to have from the day He found them hiding in the Garden. Scientific lab tests will never be able to explore the spirit realm; as scientists have correctly pointed out; you can't design physical tests to engage spiritual processes. Spirits don't comply with lab equipment.

I'm not suggesting CO^3N^2PTT replace CONPTT in the laboratory for discovering the nature and properties of physical substances and how they interact. CO^3N^2PTT discovers the spiritual realities CONPTT covers up. It opens the knowledge base to evidences of existence which CONPTT blocks out. CONPTT is for discovering the properties of physical realities while blocking the evidences of spiritual realities. CO^3N^2PTT is for acknowledging the rest of reality's *information*. It isn't research criteria. It is acknowledgement criteria. It is criteria for opening the mind to what open eyes are able to see. The Bible is for studying what CONPTT can't handle and for prescribing a relationship to know the God and His spirit realm CO^3N^2PTT acknowledges.

The plausibility of these ideas has been scoffed at and ridiculed for nearly two-hundred years now. During that time, science has had no reasonable evidence for refuting the observations made by myriads of mundane witnesses throughout the centuries. Nor has it had any evidence for contradicting the processes of revelation, or disqualifying the spirit realm of God. So while some people around the world were observing revelation in their lives (even though many more were faking it,) science chose to close its mind to all reports of those observations. It pontificated dogmatic accusations against the Bible exactly like the Catholic Church dogmatically pontificated against the discoveries of Copernicus, Galileo, and early science. Without evidence, science made up the nonexistence of God and proclaimed the physical universe to be only a purposeless, non-designed happenstance. Therefore, if we find inexplicable, patterned order in the material universe correlating with revelatory order in the Bible's narrative and fulfilled prophecies, then N^2 will be the only available explanation for that discovery. It will become a verified reality. If N^2 exists, then O^2 and O^3 also exist, and scoffing becomes the delusion.

Clarity of thought depends upon what the mind opens to. CONPTT is a rule of thumb for exploring physical processes with a mind closed to spiritual realities. CO^3N^2PTT opens the mind to spiritual effects within the physical realm. CONPTT has successfully

The Rest of the *Information*

directed many discoveries of physical realities by directing attention to physical principles within scientific laboratories. CO^3N^2PTT successfully discovers spiritual realities by admitting Biblical *information* into exploratory thought. Spiritual principles can not be left out of scientific acknowledgement merely because they do not respond to scientific lab equipment. That is not a matter of subjectivism. It is a matter of allowing ALL observable evidences to lead you ALL the way to their conclusions. It is a matter of allowing observation to see evidence. CO^3N^2PTT offers Predictability and Testability for discovering realities CONPTT refuses to even look at.

The Biblical prediction we will shortly test is the most difficult one for scientific subjectivity to dismiss. Its proposed revelation involves the core purpose of God's activity amongst human affairs: ending evil and leading a remnant of people into eternal perfection. If this core prediction is really revelation from a spirit realm, then very significant events will be discernable in actual history essentially correlating with what the prediction states they will be. Since we will need to observe essential correlations between the prophecies and their fulfilling events (not just tangential curiosities), then we must consider some historical background in order to understand the nature of the Biblically presented activity as being core to what God is working.

The Bible proposes human history began with a very intelligent man and his sweetheart living in obedient fellowship with God, who was, at the time, empirically observable to them. They were free to interact with Him because they were yet faultless. His N^2 was a natural part of their N^1. But they colluded with a snake and broke fellowship with God by eating what He told them to not eat. The snake assured eating it would make them wise as God. They probably ate the forbidden fruit thinking they would receive an enormous library of essential knowledge, such as science thought it would discover in the dirt. Alas, their denial of God's warning only added the new experience of doing wrong to that of their having done right. Doing wrong closed the door of their minds against what God had formerly revealed to them, sort of a "scientific" result. So God withdrew their ability to interact empirically with Him, and all humanity thereafter went empirically blind to God, the scientific result of denial, kind of a spiritual lab explosion. How boring their new found science must have seemed. But eventually they acclimated to its dullness.

God didn't change; people did. And their blindness was increased by their ever growing faultiness. It would soon lead to their complete inability to see God of their own volition. Yet God retained His ability to reveal Himself to whoever of them at His volition. And by that ability He has been communicating to mankind His plan to restore a remnant of people to faultlessness. That He only communicates to those who open their eyes and minds to His reality is His choice, not their subjectivity. The rest of the people hardly have a clue, or give a rat's donkey. So, they scoff.

"And who is this remnant?" the scoffer jealously demands.

The remnant are those who desire to appropriately interact with God by His prescriptions because they desire to know Him instead of "knowing" who they think He is. They are the people who desire nothing more or less than being compliant with God's principles. They endeavor to please Him for no reason other than that He is the point of pleasure, the epitome of effectiveness, the essence of good, and the orderliness making life

Clear Signs of Trouble and Great Joy

live and things exist. They walk by His word. The multitudes who are not the remnant are simply those who could care less about God and His desires. Since God is a giver, to those who do not desire Him, He gives not Himself. Missing this point is eternally tragic. Getting it makes you one of the remnant.

God's revelation to fallen mankind began soon after Adam and Eve ate the forbidden fruit. Because of their shame, the man and his sweetheart hid themselves from God. Today, we hide from God in our science. God still sees us hiding inside our laboratories as ably as He found them hiding in the bushes. He wrote to us the same prophecy He spoke to them.

In cursing the snake for deceiving Eve, God assured the human race that the woman's offspring would crush the head of the snake (Genesis 3:15). It was a very concise communication, the accurate revelation of a complete plan expressed in a nutshell. Promises to repair what has been destroyed assuage the guilt of those who worked destruction. Reconciliation with the damaged party mends the guilt of the errant. The overcoming of error fixes broken souls. The first generations of mankind were given those hopes all packed within that one prediction of the woman's seed promised to crush the destroyer of their goodness. God left it to their understanding that this meant the blissful conditions of goodness would then return. People clung to those hopes throughout the following ages.

But as mankind's generations regressed ever deeper into falsehoods, the communication became blurred and distorted through their careless ambitions multiplied by centuries of wandering imaginations. The earliest of ancient artwork hints at ideas of an ever present snake, a lost tree, a forbidden tree, the son of a woman, a "bruised heel", and a "crushed head". Although the full demonstration of that proposition is beyond the scope of this book, we will demonstrate some of it -the most astonishing of it- in a later chapter. Yet, this proposition is in full agreement with the Biblical narrative: God's communication to mankind became ever more intermittent, and it came to only the most truth-desiring of people, while everyone else distorted God's traditional communications, sometimes only somewhat, but more often beyond recognition. Through God's prophets and witnesses thirsting for truth, I AM WHO I AM reintroduced Himself to a mankind liberally shrouded in shadows. To a small nation over the next few centuries He fleshed out that first, nutshell message about the woman's offspring, a bruised heel, and a crushed head,

> [19] He declares his word to Jacob,
> his statutes and ordinances to Israel.
> [20] He has not dealt thus with any other nation;
> they do not know his ordinances.
>
> Psalm 147:19-20

Is this proposition real? Did a God, invisible to the human senses because of human faultiness, pass notes of His plans to certain people who were deeply concerned with overcoming their own faultiness in order to discover His truth and live? The Bible emphatically says it happened that way. But nineteenth century scientists emphatically demanded it could not have happened that way, regardless of the truth that they could not see the far past to know what did or did not happen. Were these nineteenth century, guessing scientists more

The Rest of the *Information*

correct than the Bible and millennia of mundane observers just because they never saw evidence of God? Or, did none of them ever see God's evidences because of they were the ones who institutionalized blindness against spiritual realities? Already the faultiness of science has been attested by the gradual failing of many of its theories and assumptions. But time has never discarded those ancient observers' reports gathered into The Bible. In fact, with every passing year, time is attesting more and more to the truth of their reports.

The Bible is still here. And so is the ancient nation the Bible said I AM WHO I AM would eventually reign from: Israel. It's either evidential or coincidental that both the ancient reports and the nation are yet with us today, each one attesting in its own way to the other. Scientists and religionists alike agree that no man has ever had the ability to travel into the future and return to show anyone what things will be like then. But the Bible is filled with reports of I AM WHO I AM informing faithful men and women of the way things will happen far into the future. Many things having come to pass as they were prophesied have become credentials of the Bible's claim to be The Words of God.

Whittling down those Biblical credentials has become a primary goal of normal science and Biblical minimalists.

But one particular credential has been dulling every whittling knife scoffers have wielded against it. It is a verifiable revelation (O^3) confirmed by empirically observed evidence (O^1), therefore, scientifically verifying the reality of N^2. It is a prediction observably made at least two-and-a-half centuries before the historical events it predicted actually occurred precisely when and as the prediction said they would. This is how science knows stuff: fulfillments of predictions, a.k.a. successful tests. And it was not some tangential curiosity, some parlor game equivalent to Nostradamus' quatrains. It was a prediction regarding the core of God's efforts to establish a remnant of mankind in a righteous place forever. It hangs the plan God summarized for Adam and Eve upon real events having occurred in real places at actual times.

But maybe that core prophecy and its fulfillment were just happenstance, like some of Nostradamus' quatrains rather came true coincidentally. His quatrains lacked any coherent theme as a whole. The Bible's predictions, on the other hand, are greatly interrelated with the revealed theme of an Offspring who would crush the snake's head, and with the nation of people from whom that Offspring's mother was born. The Bible's predictions coherently interweave around a core theme.

> [6] Thus says the LORD, the King of Israel
> and his Redeemer, the LORD of hosts:
> "I am the first and I am the last;
> besides me there is no god.
> [7] Who is like me? Let him proclaim it,
> let him declare and set it forth before me.
> Who has announced from of old the things to come?
> Let them tell us what is yet to be.
> [8] Fear not, nor be afraid;
> have I not told you from of old and declared it?
> And you are my witnesses!

Clear Signs of Trouble and Great Joy

> Is there a God besides me?
> There is no Rock; I know not any."
>
> Isaiah 44:6-8

Many of the Bible's predictions were fulfilled by events recorded in its narratives of Israel's history. But according to Biblical minimalists, scoffers, and skeptics, the same hands recording those events wrote the prophecies. The prophecies then could not be said to be prophecies at all, but rather, fibs. Out of proper respect for the minimalists, scoffers, and skeptics, let's hold any conclusion about those fulfilled predictions Tentatively, while we consider a prediction/fulfillment which can only be explained by either extraordinary coincidence, or by I AM WHO I AM being precisely what The Holy Bible claims Him to be. If this test strips the edge off the scoffers' last whittling knife, then we will know all of the Bible's revelations belong at the core of mankind's knowledge as O^3.

The Bible says an angel named Gabriel gave Daniel an important, specific prophecy in the late sixth century BC. The events which fulfilled the prophecy are well known today. Obfuscators tried to lie them out of history. But so much validating evidence has been exposed by now that there is no longer any sane doubt they indeed occurred. Coincidence is the only possible explanation remaining for use by the scoffers and skeptics. But to call this example a coincidence will require a nearly infinite stretch of the imagination because of the seismic scale of the events predicted. Their coherence and cohesion with the rest of the Biblical narratives are too extensive to be mere coincidence. It would be somewhat like accusing the greatly coherent periodic table of being nothing more than a pareidolic fable in spite of all the chemicals reacting in accord with its illustrated orderliness. Some of the events the angel predicted to Daniel are even in the process of occurring today just as coherently with the overall Biblical narrative as the threads of your shirt are coherent with its cloth. And the events are intimately interrelated with all of the Bible's earlier prophecy/fulfillments which science accuses (without evidence) of being made up, circular fictions. If the nature and fabric of this prediction and its fulfillment tests positive, then the test will verify the reality of the rest of the Bible's predictions being accused of having been made up after the fact. If there is a thorough coherence between the entirety of the Bible's prophecies/fulfillments, Biblical narrative, and history, then the bar for explanation-by-coincidence will rise too high for coincidence to explain this stuff rationally.

What Gabriel told Daniel is the skeletal prophecy upon which hangs the details of I AM WHO I AM's plan for crushing the head of the snake and restoring blissful fortune to mankind's remnant. We will closely examine it as a practical test of CO^3N^2PTT -the proposition of a discernable, spirit realm capable of delivering revelations to humans blinded by the subjectively false science of an obviously sub-natural realm. If we find this to be real, then we will find the entire Bible hung upon the actual existence of N^2.

The Israelites were faithless towards the God they were meant to proclaim. As I AM WHO I AM led them out of Egypt into being a nation in possession of the Promised Land, He emphasized their need to remain faithful to Him. Idolatry would cost them their land. He warned that, as surely as He led them there, He would drag them away if they strayed into idolatry. They strayed. He drug the ten northern tribes away first in 722BC. Then Judah and

The Rest of the *Information*

Benjamin were taken captive to Babylon by 586BC. God didn't promise that the northern tribes would return soon. But He promised they would return before the end of the age (Zechariah 10:7-9). And He promised Judah would return to rebuild Jerusalem seventy years after it fell. To King David He had promised a lineage of descendants reigning from Israel's throne, forever, in Jerusalem.

Biblical minimalists and other scientific guessers accuse the prophecies of Judah's return to have been faked. But assumptions can not prove or disprove anything except that somebody's been guessing. The Bible asserts that there is a God. Verifying the revelation the Bible alleges He sent to Daniel would validate the allegation and establish a foundation of spiritual reality supporting the authenticity of all the other accused prophecies.

Daniel's prophecies have been obfuscated by the minimalists, too. But that obfuscation, at the end of the analysis, collapses under the weight of only a little, undeniable, historical evidence. Most of Daniel's prophesies came to pass after they had been included in the Septuagint (the third century BC translation of the Old Testament into Greek). Its translation is known to have preceded the prophesied events of Daniel 9:24-27. That should verify the unique, supernatural nature of Daniel's message like gold glittering through the veils of deceit-ful accusations and scoffing.

Daniel was led away to Babylon with the rest of Judah when he was a young man. He was faithful to I AM WHO I AM. So he was greatly concerned about God's people and when they would return to Jerusalem, God's Holy City. Many of the Psalms written by the Jews while in captivity indicate the Jews had been talking amongst themselves and praying about their eventual return home for some time. The prophets, with whom their generation had grown up and knew personally, had foretold their release, particularly Ezekiel and Jeremiah. Daniel also was praying and fasting in Babylon to learn when their release from captivity would begin. The angel Gabriel came to him with God's response:

> [24]Seventy weeks of years are decreed concerning your people and your holy city, to finish the transgression, to put an end to sin, and to atone for iniquity, to bring in everlasting righteousness, to seal both vision and prophet, and to anoint a most holy place. [25]Know therefore and understand that from the going forth of the word to restore and build Jerusalem to the coming of an anointed one, a prince, there shall be seven weeks. Then for sixty-two weeks it shall be built again with squares and moat, but in a troubled time. [26]And after the sixty-two weeks, an anointed one shall be cut off, and shall have nothing; and the people of the prince who is to come shall destroy the city and the sanctuary. Its end shall come with a flood, and to the end there shall be war; desolations are decreed. [27]And he shall make a strong covenant with many for one week; and for half of the week he shall cause sacrifice and offering to cease; and upon the wing of abominations shall come one who makes desolate, until the decreed end is poured out on the desolator.
>
> Daniel 9:24-27

The Hebrew word, *shabua*, used for "weeks" in this prophecy, is also used at

Clear Signs of Trouble and Great Joy

Genesis 29:27 and 28 to denote the seven years Isaac's son, Jacob, served his Uncle Laban for Leah, and then again to marry Rachel. It is not a shadowy, vague term. The usage of *shabua* in Genesis and here in Daniel is a Hebrew idiom meaning a concretely real measure of seven years. Although the word is used mostly in the Old Testament to mean a week of days, the context of both the Jacob/Laban narrative and Daniel's prophecy demands the idiomatic meaning of seven year periods.

Gabriel said the seventy weeks of years were for the processing of six objectives concerning "your holy city" and the Israelites being held in Babylon. How do we know "your holy city" refers to Jerusalem instead of some place of Nostradamus-like ambiguity? Eleven times the Bible refers to a "holy city" in contexts clearly meaning only Jerusalem. The expression is used in the Bible for no other city. Moreover, Jerusalem is the only city the Bible presents as having a most particular relationship with God. Well, let's just be straight forward; it is the only city God refers to as specifically being His city. It is His possession in a more particular sense than are all other things. It will be the City of His Throne. That's special. "Your people" refers to the Jewish people because the angel is speaking this prophecy to Daniel, who was a Jew. Also, the Jewish people were those who had worshipped at Jerusalem, and they were the Israelites, the remnant of which other prophecies said would be brought back to Jerusalem.

The six objectives of the prophecy are also clear rather than ambiguous. They were specifically the central effects of Jesus Christ's restoration work:

1) To finish the transgression. A transgression is an infringement upon a command, or a violation of law. Israel's major transgression was idolatry. God commanded them not to do it. But their idolatry eventually became more than just worshipping Baal, Ashtaroth, Molech, and other gods of the Canaanite peoples. What it became is why they continued to commit idolatry even though they had truly learned to never worship any entity other than God. Even worshipping I AM WHO I AM by false concepts or self contrived principles is a transgression, for it ascribes vanities to God's name. God demands service in submission to His words, nothing more or less. When the Messiah/King came to the Jews in the person of Jesus, their understanding of God and His Word was dull because they had assumed far more than God had commanded. Consequently, they rejected Jesus because they found Him to not be what they had imagined. Guessing is disastrous. Theorizing is dangerous. Yet, to this day Judaism imagines more about the oracles I AM WHO I AM gave to them than they actually obey. It isn't very funny that Christians do the same. It is why there are so many denominations of both. Obviously, finishing that transgression is yet to be done.

2) To put an end to sin. Sin is any and every misconception or error in thought, emotion, or deed, whether deliberate or not. The New Testament uses a Greek word for "sin" that means to miss the mark of your aim. Intent is not the point; bull's eye is. This concept shows how hopelessly imperfect man is, and will be, to the very day imperfection ends. Revelation 20:1-3 prophecies that this age will close with the binding and quarantining of the Accuser/Deceiver/Destroyer for a thousand years. It will end the reign of sin over mankind. Sin will still exist, but its reign will be over. Sin will end when Satan is freed from his quarantine to be captured and hurled into the lake of fire.

3) Atone for iniquity. Transgression is an infringement. Sin is as simple as a mistake.

The Rest of the *Information*

But iniquity isn't a categorical concept as much as it is an attitudinal one. Iniquity is the deliberate employment of either transgression or sin. It is sin with a mission; it is transgression for a purpose. If righteousness is the intentional aim at the bull's eye, iniquity is the intentional aim at anywhere but the bull's eye. Both arrow's may strike the target in precisely the same place, but one is there by aim, while the other merely missed its mark. Missing the mark conveys the extensiveness of imperfection; iniquity conveys the nature of lawlessness. Atonement for iniquity is the preparation for our separation from evil. Atonement came through the cutting off of the anointed one of verse 6. The Bible proposes Jesus never so much as sinned; He was a perfect marksman hitting the bull's eye with every shot of His life. It is an enormously important proposition. Although death is the wages of sin, Jesus died without earning those wages. He didn't do any work payable in death. We did. Yet He was paid the wages we earned. So the natural order of doing the spiritual accounting was broken. Jesus had to punch out on our timecards, since He worked no sin on His own. He took the pay we earned. Then He gave each of us the choice of leaving our paycheck on His cross or receiving our own paycheck in disbelief of His cross. His receipt of our wages was His atonement for our iniquity. And this is why He was perfect to the very last shot.

4) To bring in everlasting righteousness. The end of transgression, sin, and iniquity will not leave a vacuum. Everlasting righteousness will fill the void. Righteousness has become somewhat of a bad word anymore. Our concept of it is overly related to religious activities, rituals, and other forms of religiosity, and to attitudes of political superiority. These always seem to settle quite naturally into bigotry. They are not producers of actual righteousness. The Bible uses the word to mean thinking, feeling, and doing what's right. The right thing to do always benefits everything it affects. It produces no detriment to anything. It doesn't try to control what it affects; it only benefits it. It is that simple. Therefore, the Hebrew concept extends further than just doing right things. It entails right reasoning. Right reasoning is a forensic concept. It is what the scientific process was meant to be before it devoted itself to assassinating God. Right reasoning explores reality to shape the mind in accord with what it finds. Righteousness involves intellectual responsibility for discovering and implementing deliberate actions of mutual benefit to everything they affect. Intellectual responsibility does not cut off and throw away any *information*, not a scrap of it, especially not the fundamentally basic and coherent *information* which the Bible is.

5) To seal both vision and prophet. Sealing a document affixes to it the authority of the one sealing it. The event fulfilling a prophecy affixes God's authority to the prophet and his vision. Fulfillment is one of two Biblically prescribed criteria for discerning between true and false prophets and for knowing God did something instead of its happening by pure chance.

> [21] And if you say in your heart, 'How may we know the word which the LORD has not spoken?'—[22] when a prophet speaks in the name of the LORD, if the word does not come to pass or come true, that is a word which the LORD has not spoken; the prophet has spoken it presumptuously, you need not be afraid of him.
>
> Deuteronomy 18:21-22

Clear Signs of Trouble and Great Joy

When a book filled with specific predictions is handed down from generation to generation while its predictions are coming to pass in the same manner as they were predicted, then we know for certain the book's authors were getting their *information* from something transcending the normal flow of time. The Bible reveals this transcendence to be I AM WHO I AM. So it is very natural to propose that the fulfillment of the Bible's prophecies affixes the seal of God's authority to the Bible. By that authority comes our ability to treat the Bible's concepts as the empirical observations of ancient witnesses, even for scientific purposes. Natural philosophers knew this in the eighteenth century; scientists packed dumpsters full of evidence to deny it. And Nostradamus was not as well sealed, yet he gets more public curiosity than the perfectly well sealed Bible. Many available books about its fulfilled prophecies demonstrate that truth. Scoffers object. But scoffers offer no evidence. They only accuse and scoff about who's to say what's true. But vision and prophet offer tested results, seals of authority, to say what's true and Who's right.

6) To anoint a most holy place. By the end of these seventy weeks of years, when the world has become the kingdom of the Lord and has been horrifically cleansed of self-centeredness, Jerusalem will be anointed as Christ's throne on earth -a most holy place. This is a prophecy in the process of happening before our own, modern eyes, as Chapter 7 will demonstrate. The sealing of this prophecy's truth is presently occurring around us as current events are drawing Jerusalem ever closer to surviving a clash with a world full of anti-Semitic hatred before it is anointed as the most holy place of The King of Kings and Lord of Lords, Enoch's Lord of the Spirits, The Holy Bible's I AM WHO I AM.

The Book of Daniel had been included in the Septuagint close to three-hundred years before Jesus became the "cut off" anointed one. And not only did this passage of the Septuagint rightly predict His being "cut off" before it occurred, but it accurately indicated when the cutting-off would happen. Gabriel told Daniel this time period of seventy weeks of years would begin running with the issuance of the decree to rebuild Jerusalem. Artaxerxes I issued that decree to Nehemiah in 445BC,[7] ordering the rebuilding of Jerusalem. It started the seventy weeks of years ticking like a Timex watch.

The prophecy said an anointed one would be cut off after sixty-nine weeks of years (7 weeks plus 62 weeks). If we search for the cutting off of an anointed one in AD40, four-hundred-eighty-three years after 445BC (7X69), we will not find Christ's crucifixion happening then. The accepted range of historical dates for His crucifixion is between AD27 and AD33. Our first test did not work because we think in terms of solar years (365 days). The Hebrews, to whom this prophecy was written, thought in terms of lunar years (360 days). Converting lunar years to solar years is easy: 360X483/365=476.4 solar years. Just over four-hundred-seventy-six solar years after 445BC is AD33. That test is successful.

The most significant event to any and every human being, Christ's crucifixion to pay the debt of our sin, occurred precisely when Biblical prophecy indicated it would occur. It hangs the Bible's authenticity upon a nail of evidence driven into the wall of reality by the hammer of God's authority. In the eyes of humble and honest people it does, anyway. But skeptics, scoffers, and normal science need neither evidence nor good sense to accuse the Bible of being only man's words instead of God's revelation. Accusations need only

The Rest of the *Information*

mouths to vent air well heated in the flames of rebellious hearts. "Coincidence" is just another put on air, sometimes, thoroughly heated. Regardless of deceitful accusations, honest people note the historical coherence of this prophecy written at least two-hundred-seventy-five years before the event it predicted (giving skeptics, scoffers, and accusers the benefit of what little doubt might be remaining), but more likely written four centuries before then.

Empirical evidence shows Christ lived. It shows He died. The scientifically tested and found to be otherwise inexplicable Shroud of Turin displays His resurrection from the dead like a photograph without the possibility of having been photo-shopped. Those who believe all of the *information* about Him know beyond a doubt He lives today because they experience the affects of the resurrection in their lives (O^2). That is testimony. We predicted that if the Bible is The Holy Bible, The Holy Word of God, the revelation of basic truth, we could find an empirically observable fulfillment of a Biblical prophecy, and we found it. And not just any one, but THE one, THE most core prophecy of God's work. THAT is evidence.

So when we step a few decades further into the first century AD, we find the city and the sanctuary being destroyed by "the people of the prince who is to come", just as it, too, was prophesied. That stacks more weight against explaining Daniel 9:24-27 by reference to some, extraordinarily grandiose coincidence.

Therefore, some people try to explain away this prophecy/fulfillment by denying its reality. The oldest copies of the Septuagint containing Daniel date only to the fourth century AD. It is natural for skeptics to hope the prophecy was inserted into the text of Daniel by Christians trying to manufacture evidence after these predicted events occurred. And The Dead Sea Scrolls seem to support that theory. Those tatters of a Qumran library contain no less than eight copies of Daniel. Two very small fragments date from the second century BC. They would seem to date the writing of these prophecies to nearly two centuries before their fulfillment. But they don't do that, since the only fragment of Daniel 9 amongst The Dead Sea Scrolls contains a few portions of Daniel's prayer, but not one letter of the prophecy we're discussing. This prophecy is amongst the missing fragments. Or, as skeptics would rather state the situation, this prophecy never was for there to be any fragments of it. So maybe miscreant Christians really did pen the prophecy after Jerusalem fell.

Judaism failed to acquire the same fondness towards Jesus as the Western World eventually did. Jewish scribes and scholars in Medieval Tiberius, Jerusalem, and Babylonia were the keepers of the Old Testament's Hebrew text. These were the Masoretes. The Dead Sea Scrolls evidence how accurately the Masoretes passed down the Hebrew text over the thousand years spanning from the first or second century BC Qumran library to the date of our earliest Masoretic copy. The Masoretes all but worshipped the Biblical text. Every page they copied was scrutinized, letter for letter, against the original. The copy had to be a perfect match. The letter at the exact center of the copied page had to be the same letter of the same word at the exact center of the original page. The text was worshipfully maintained by these careful, extreme deniers of Jesus. And their centuries long, meticulously worshipful work contains Daniel 9:24-27, the supposed Christian insertion diabolically meant to validate Jesus as the Jewish Messiah, an extreme blasphemy, according to these Masoretes.

Clear Signs of Trouble and Great Joy

Copying an inserted, blasphemous corruption of their holy literature would not be the nature of highly religious men paying great homage to anciently venerated texts. That this passage was supportive of their ultimate nemesis would make copying it even a worse anathema if it was a Christian insertion. That these extremely conservative Masoretes scrupulously retained Daniel 9:24-27 in their tedious work, even though they knew it correlated well with Jesus' crucifixion (obviously a "cutting off" of an anointed one), pretty much ends any possibility of this prophecy being a later, Christian insertion. Yet no amount of valid reasoning upon sound evidence ends cheap, subjective accusations.

The rationally verified authenticity of Daniel 9:24-27 and the later, well known events fulfilling its prophecy point singularly towards the existence of a spirit realm from where the Almighty I AM WHO I AM had been poking ideas into Hebrew minds by revelation for quite a few centuries. But the destruction of the city and sanctuary of verse 26 is not the end of this prophecy. It discussed seventy weeks of years. We've only analyzed through the sixty-ninth week. There is one week of years left. And it is foretold at verse 27.

> [27] And he shall make a strong covenant with many for one week; and for half of the week he shall cause sacrifice and offering to cease; and upon the wing of abominations shall come one who makes desolate, until the decreed end is poured out on the desolator.

One school of eschatological thought (the study of end-time prophecy and themes) assigns these last seven years to the destruction of Jerusalem in AD70. But additional details prophesied of these seven years in the Book of Revelation correlate with prophecies throughout the Old Testament describing a time far different than AD70. In fact, much of Revelation's terminology and imagery flows directly from the scrolls of the Old Testament prophets. Those correlations place the week much later in world history than AD70. Most errors of interpretation occur by unconsidered *information* more than by invalid reasoning. When all Biblical prophecies are considered together, verse 27's descriptions anticipate a time when Israel has become a nation unto which its Messiah/King comes and reigns. Especially Zechariah's prophecies indicate Israel will be lifted to the head of all the nations at the end of the seventieth week of years. Israel was destroyed in AD70. It remained destroyed for eighteen-hundred-seventy-eight years. So, verse 27 can not be about AD70 and remain coherent with either the rest of Biblical prophecy or actual history.

Israel's nonexistence for the eighteen centuries after AD70 compelled many theologians to think Christianity had now replaced Israel in all of the Biblical prophecies regarding the close of this evil age. They claimed God forever discarded Israel and chose Christianity to enjoy its prophesied events instead, while ignoring the distinctly different roles scripturally assigned to each. The Church supposedly became the Israel of all end-time prophecies after the faithful I AM WHO I AM went back on His word to Abraham, Isaac, and Jacob. Of course, these theologians do not actually consider God to be unfaithful, although their theology implies it. They say He told Abraham, Isaac, and Jacob that their descendents would possess the land forever while meaning every believer in Jesus would be a member of Abraham's family instead. This bait-and-switch-prophecy is called replacement

The Rest of the *Information*

theology.

But the establishment of Israel as a nation in 1948 raised an ensign over the death blow to the replacement misconception. The continuing prosperity of Israel still throws dirt on its grave today.

> [11] In that day the Lord will extend his hand yet a second time to recover the remnant which is left of his people, from Assyria, from Egypt, from Pathros, from Ethiopia, from Elam, from Shinar, from Hamath, and from the coastlands of the sea.
> [12] He will raise an ensign for the nations,
> and will assemble the outcasts of Israel,
> and gather the dispersed of Judah
> from the four corners of the earth.
>
> Isaiah 11:11-12

The Church isn't a remnant of Israelites. It is a remnant of mankind. And although it was initially driven from Jerusalem, it spread throughout the world because it was Christ's witnesses sent to the ends of the earth, to Assyria, Egypt, Pathros, Ethiopia, Elam, Shinar, Hamath, and to the coastlands of the sea. The entire world was the land of the people who became the Church. The Church wasn't cast out. It was greatly persecuted in its first two-and-a-half centuries, but it never was "the dispersed of Judah". It was "the sent of Christ". So it did not have to be gathered home as if it had been ran out of its place.

Not only was the current nation of Israel prophesied in those ancient scriptures, but later, the New Testament even prophesied God's correction of replacement theology. It is found in The Book of Revelation's letters to the seven churches. Among other purposes, those letters play a distinctly prophetic role. They metaphorically portray seven epochs of church history quite coherently. The major characteristics of each epoch are symbolized in a few traits presented about each church to which a letter was addressed. The letter to the church at Philadelphia characterizes the recent three centuries of a world evangelizing church by the symbolism of an open door "which no one can shut" (Revelation 3:8). Late in this evangelistic epoch, generally considered to be the eighteenth through twentieth centuries, Israel became a nation again. And this wasn't some quiet event in an obscure corner of the world. God's chosen people were led out of Hitler's infamous death camps while the entire world gazed in horror at what it had allowed to happen. Three years later, those tortured souls became the nation God promised to Abraham, Isaac, and Jacob. A towering, neon sign could not have made it more obvious to the world. The nation of Abraham's descendents, named through his grandson Jacob (a.k.a. Israel), promised by God to be a nation for ever, twice driven off their land, and now having become, for the third time, a nation on that very land in our own scientific times to receive its Messiah as King in Jerusalem, Israel's capitol today, is within the very core of all the Bible's prophecies. God's faithfulness to His promise was raised up like an ensign as He guided the return of nationhood to His people in 1948, then hung in that new nation's skies four total-lunar eclipses on its most holy Passover and Tabernacle ceremonies.

Consequently, the deep love God has for everyone who abides in His Word shined

Clear Signs of Trouble and Great Joy

forth brightly from this promise well kept. It cast revealing light upon the deceit of those who said they had permanently replaced Israel in prophecy, claiming to be Jews while they truly were not.

> [9]Behold, I will make those of the synagogue of Satan who say that they are Jews and are not, but lie—behold, I will make them come and bow down before your feet, and learn that I have loved you.
> Revelation 3:9

Jesus addressed Peter as Satan when Peter objected to hearing about Jesus' impending death (Mark 8:33). And Jesus again used hyperbole in this letter to the church at Philadelphia, calling the people of replacement theology "the synagogue of Satan". Replacement theology has no scriptural basis. The Bible neither teaches nor implies it. So advocating that the Bible teaches it is to lie. Lying is Satan's thing, not God's. Would we expect a synagogue of lies to be God's synagogue? Especially a synagogue of such an egregious lie as advocating that I AM WHO I AM would abandon the people He promised to revive, bless, and comfort after His harsh punishment of their unfaithfulness? Replacement theology is a shameful indictment of God, worse than even the indictments raised by atheists. Israel having become a nation again during the Philadelphian church epoch greatly illuminated that synagogue of Satan, teaching them how special to God are those who hold fast to the Bible, having their faith sealed by God's evidence.

We predicted that, if communication of the way things are and what will soon take place has been made to us from a spiritual nature overarching our physical one, then we should find at least one significant historical event occurring just as it was foretold beforehand. That is Predictability. We then turned to the world's most unique, literary resource, said to have been given from an overarching spiritual nature, while we also searched known history for an event fulfilling just one of its prophecies. That is Testability. We found not one, but four undeniable instances of significant, well known historical events happening just as The Holy Bible prophesied they would happen. We even verified the certainty that those prophecies were written before the occurrences of their prophesied events. The Bible's successful prophecy of Christ's death, of Jerusalem's destruction, of Israel's existence at the close of this age, and even of a false theology's humiliation validate the spirit realm's reality as N^2 and its revelations to us as O^3. Subjectively accusing by neither evidence nor rationale is not Tentativeness. It isn't even scientific. Consistency is understanding that these four instances of clearly fulfilled prophecy evidence the very real probability that the rest of the Bible's prophecies are also realities. Our test confirms the objective reality of Biblical prophecy and revelation. But should we expect this confirmation to correct the myopia of CONPTT? Should we expect science to climb out of its blind box and see more clearly by use of CO^3N^2PTT as the more complete scientific criteria?

Of course not! It wouldn't be normal science if it went where evidence led. But reforming science, humble science, Tentative science will go wherever evidence leads. Even though science could test the Bible and find it true by every real scientific analysis honestly performed upon it, such is not the treatment God purposed for His Holy Word. It was the

The Rest of the *Information*

treatment He purposed for The Shroud of Turin. But The Bible's purpose is to test the truthfulness of every individual heart, and especially to test the truthfulness of the scientists, since they claim to be the most objective of us all. Science is neither necessary nor useful for testing the individual. Only contact with the Truth is necessary for that. A heart's reaction to Truth will reveal the nature of a heart's substance every time that test is run.

Truth has awareness as surely as magnets have field. Truth and sincere hearts attract each other. That isn't to say any one man's heart is all sincere and everyone else's are completely insincere. St. Paul, filled with the Holy Spirit, told us all men are false, not that they are all totally false, but everyone is erroneous about something or more. Any heart set upon finding its own "truth", or upon finding the "truth" of the best scientist, or of the majority of the population, or of the yoga master, or of yours, or of mine, or of Timothy Leary, or of hallucinogenic distortion will no more respond to reality's truth than a bottle of beer will respond to a magnet. But a heart set upon knowing reality's truth is made of reality's nature, humility, the acceptance of what is, the acceptance of I AM for who I AM is. The sincere heart and reality's truth attract each other like nails and magnets. This attraction is the essence of what Alex Rosenburg implied about honest science -to follow data, observation, and experiment wherever they go. Normal science does the same thing religious bigots do; they follow their hearts desire to wherever they can find some evidence for making up a theory or a theology. Normal science and normal religion are each a self directed search producing equally deadly train wrecks. The conflict over truth is not between science and religion. It is between sincerity and bias; sincere science and biased science; sincere religion and biased religion.

Considering this, let's ask again, did Predictability and Testability correct CONPTT by demonstrating the reality of O^2, O^3, and N^2? Is the testimony of millions of witnesses of God's revelations over the millennia Consistent with the findings of our test? Or should the objective mind hold the demonstration Tentatively while searching for more conclusive evidence? How does your heart test out?

My heart takes up faith only at the end of evidence. Before I entered High School, my witness-mom, my witness-dad, my witness-step-mother, and many other witnesses of God's reality presented to me better evidences in their lives than did the skeptics and deniers of Jesus Christ. In my fifteenth year I set my faith upon Christ. But only tentatively. For the "great" manuals of science government schools were cramming into my little "skull full of mush"[8] implied the Bible was a lie. They taught that man was progeny of apes rather than Adam's offspring. They taught The Great Flood as being just poppycock of fools, and that Jesus was an emotional crutch for weaklings, just another member of the Santa-Claus/Easter-Bunny/Tooth-Fairy crowd. I'm sure this humiliation rings familiar to many.

I became a certified public accountant to the chagrin of my High School art teachers. Art is for imagination. But mine is a career for the independent, diligently seeking mind, humbled by what it finds. Through a picture, any idea can be correlated with any other, matter not their real differences. But in reality, the Bible and humanist science can not both be right at the same time in the same sense, just like a financial statement can not show both a profit and a loss in the same sense for the same year. Reality is one thing, or it is the other. It can not be two different things at once. Therefore, my faith demanded a search for which

Clear Signs of Trouble and Great Joy

idea most fundamentally defined reality, generally accepted science or The Bible. My faith desired to leap only from the end of as much evidence as I could gather for consideration. So, I became engaged in a life long search for which idea provides the best jumping point for faith, publicly perceived science, or The Bible.

Reality's puzzle pieces are shaped and colored for a proper fit into its puzzle by its own nature. David Hume suggested we can not know for certain anything we have not observed in total, but only what we have observed can be relied upon to indicate what ought to be considered natural about everything we haven't seen. It is a rather fair proposition, as long as ALL, real observations are indeed used to assess what Natural is or is not. However, Copernicus was every bit as dead and gone as was Isaiah at the time David Hume was willing to consider Copernicus' observations for defining Natural, although he was not willing to consider any of Isaiah's observations. The Shroud of Turin is another observation he never considered, although it was available for his consideration. So was the mass of anatomically accurate, ancient dinosaur art found around the world. The total lunar eclipse of February 25, 1362BC, and the chain of evidences it links to the long hidden archeology of the Exodus are considerable, today. The Bible itself is an observation about reality. And so are the consistent testimonies of millions of eternal life's mundane observers. But none of those have been allowed to participate in defining what Natural is because normal science doesn't want to think about them.

Attention allocation is the primary cause of bias. Natural will never be known without proper attention paid to ALL observations. Hume proposed Natural to be known by what has been observed, therefore, disregarding any observation skews its definition.

We can forever call upon coincidence to explain away observations we don't like. But eventually one should remove his ear from the rails after sufficient rumbling has sensibly evidenced the approach of a train. Somehow, sufficient evidence must be acknowledged in a world where perfect evidence just does not exist. We demonstrated by the prophecy of Daniel 9:24-27 that God does raise rumbles of warning before train wrecks. But will He raise enough rumble that every person will get their heads off the generally accepted track before the Tribulation train comes wrecking through? Or will He raise just enough evidence to convince honest thinkers, leaving the rest to the fate of insincerity?

The Bible expressly proposes that everything is made for a purpose (Proverbs 16:4). This extends to the material universe and life in it also being purposeful and designed. The Bible proposes that in the beginning the sun, moon, and stars were made for the purpose of being signs (Genesis 1:14). This Hebrew word for sign at Genesis 1:14 means

> a *signal* (literally or figuratively), as a *flag, beacon, monument, omen, prodigy, evidence*, etc.[9]

Every sign establishes the existence of a purpose designed by a communicator. An orderly level of purpose, design, and process, if discovered within the inanimate realm of the sun, moon, and stars, far beyond the reach of mankind's biased, tampering hands, would solidly evidence the reality of God's Word. Therefore, if we find orderly relationships amongst the motions and positions of the sun, moon, and stars being consistently cohesive with historical events meaningfully relevant to the core prophecies and themes of the Biblically proposed plans of God, such that those relationships themselves rise to a coherent level of rational expression, then science's core assumptions will be invalidated by means of Popper's principle of falsification as Darwin's non-designed purposelessness of the universe will have been empirically observed to be false. The guesses and theories of normal science regarding a cold, dead materialism will have to be replaced with the Biblical explanation of the eternal realm of spirits from which The Lord of Spirits designed a spiritually malleable universe for a particular purpose.

Even if we do find communicative signs in the heavens, not all science will be impugned. Math, chemistry, physics, engineering, etc. are not effected by these issues. They humble themselves to reality, else bridges wouldn't stand and laboratories everywhere would consistently explode. Moreover, many hundreds, if not thousands of Biblically minded scientists offer cosmological propositions consistent with the Bible to explain phenomena which mechanistic materialists can only categorize as anomalies to be carried off to dumpsters. Their propositions explain physical realities far more coherently than do the uniformitarian dogmas. Their propositions about the cause of dirt layers answer questions normal science can only ask. Their propositions guided by Biblical *information* find a coherent fit for every observation made without a need for appealing to anomaly or sneaking stuff into normal science's dumpsters. Rhinnie would be proud of them.

Normal science's puzzle can be completely rearranged around the core understanding of a creation intricately designed for a very specific purpose. Normal science's dumpster full of puzzle pieces can empty into every space of reality's puzzle wherein normal science substituted an assumption. Every piece can take its proper place and participate with all of the other pieces in a unified presentation of a completely sensible understanding of the universe if the Bible is simply accepted as the *information* empirical observation finds it to be.

Oddly enough, normal science's puzzle is itself loosing coherence amidst all of the new discoveries being made. It is what Thomas Kuhn said eventually happens to myopically held paradigms. At the same time, Creation science's puzzle is beginning to show an astounding level of coherence between all empirical observations, past and present, including every observation revealed in The Holy Bible and every observation acknowledged by normal science.

Where there's smoke, there's fire. Where there's *information*, there's reality. The incredible, reestablished nation of Israel is the smoke of God's revealing Holy Bible. The Holy Bible also prophesied signs in the sun, moon, and stars. So let's do some research to see if anything of the sun, moon, and stars might possibly be considered signs. If we find obviously communicative interrelationships amongst them, then maybe we should consider the Bible's messages more soberly.

Footnotes

1. Rosenburg, Alex. Philosophy of Science: A Contemporary Introduction. 2012. Rutledge, 711 Third Avenue, New York, NY 10017. Pg. 222.

Clear Signs of Trouble and Great Joy

2. Kuhn, Thomas S. The Structure of Scientific Revolutions. Fourth Edition. 2012. The University of Chicago Press. Chicago, Illinois 60637. Pg. 24.
3. Choi, Charles Q. 7 Theories on the Origin of Life. March 24, 2016. http://www.livescience.com/13363-7-theories-origin-life.html
4. Bergman, Jerry, PhD. Fossil Forensics: Separating Fact from Fantasy in Paleontology. 2017. BP Books, Bartlett Publishing. Pg. 147.
5. Ibid. Pg 147.
6. Rosenburg. 2012. Pg. 104
7. Encyclopedia Britannica: A New Survey of Universal Knowledge. Vol. II. 1956. Encyclopedia Britannica, Inc. pg 448. Britannica states only that Nehemiah was appointed governor 445BC. But confer Nehemiah 2:4-8 and Josephus' Antiquities of the Jews, Book XI, Chapter 5, Sec. 6. Nehemiah's appointment rationally follows Jerusalem's need for rebuilt walls, the distress of Nehemiah to which Artaxerxes I responded with the appointment. The decree to rebuild the walls accompanied the appointment, it is what the appointment was about.
8. An endearing expression of the great Rush Limbaugh highlighting the intellectual vulnerability of children within our modern classrooms filled with political propaganda.
9. Strong, James. Strong's Hebrew and Greek Dictionaries. Electronic Edition STEP Files copyright 2003, Quickverse, a division of Findex.com, Inc. At H226.

Dayenu.

Chapter 3
Love and Message

> [4] *I say to the boastful, "Do not boast,"*
> *and to the wicked, "Do not lift up your horn;*
> [5] *do not lift up your horn on high,*
> *or speak with insolent neck."*
> [6] *For not from the east or from the west*
> *and not from the wilderness comes lifting up;*
> [7] *but it is God who executes judgment,*
> *putting down one and lifting up another.*
> [8] *For in the hand of the LORD there is a cup,*
> *with foaming wine, well mixed;*
> *and he will pour a draught from it,*
> *and all the wicked of the earth*
> *shall drain it down to the dregs.*
> [9] *But I will rejoice for ever,*
> *I will sing praises to the God of Jacob.*
> [10] *All the horns of the wicked he will cut off,*
> *but the horns of the righteous shall be exalted.*
> *Psalm 75:4-10*
>
> [20] *Our God is a God of salvation;*
> *and to GOD, the Lord, belongs escape from death.*
> *Psalm 68:20*

In the beginning, Eve got it wrong. And Adam was no help. They welcomed deceit into the perfect universe God created. Their perfection ended, and the universe was thrown into a state of decay, slowly stumbling towards the eventual meltdown of its elements (II Peter 3:10). The disobedience of God's two humans fired up a culture of deceit and triggering the eventual global warming.

But God did not destroy them for that. Nor did He destroy the snake for introducing His humans to deceit. Instead, He cursed the snake. Embedded in that curse, He revealed His plan to restore a remnant of mankind to righteousness. The woman would bear a Child, a seed, as God called it. The snake would replicate treachery. The snake's treachery would wound the Child. But the woman's Child would put His foot down on the snake's head. In the end, the snake's posterity will be crushed. Deliberately, within the hearing of His first two people, God pronounced this curse upon the snake and that blessing for those people.

God then clothed His first couple, maybe in fox hides, and sent them away from the tree of life so they would not eat of it again to live eternally in their newfound state of

Clear Signs of Trouble and Great Joy

falsehood. God's love is not only expressed in what He gives, it is also expressed by what He withholds.

It's commonly said that Eve's first mistake was listening to the snake and discussing the forbidden fruit with it. But she was made in God's image. God is no snob. God did not forbid her from speaking with the snake. He forbade eating of The Tree of Knowledge of Good and Evil. Nor did He forbid her touching the fruit of that tree. But when the serpent asked if God said not to eat of the tree, she replied that He commanded her not to even touch its fruit. That was Eve's embellishment of what God said.

When we either embellish or denude God's Word, or what can be observed of His Creation, we are left to reason upon our own strength, having, with our own minds, polluted the good sense of His wisdom. His Word is distorted by adding anything to it or subtracting anything from it. Since His Word is the key to understanding, since it defines the fundamental paradigm in which everything we encounter is correctly understood, twisting or distorting it twists and distorts our understanding. Trouble is the eventual product of any endeavor engaged without sound understanding. So the Bible is careful to instruct its readers to neither add to nor subtract from its words. We think this command regards the heightened duty of ancient copyists to copy the Biblical text carefully. But as enormously important as the copyist's duty was, it's the least of this command's meaning. The Bible's message is easily distorted by subtle additions and subtractions of thought, memory, and even nuance in your own mind. God's Word had been added to and subtracted from long before Christians began brewing up a plethora of theologies. In the end, numerous shoddily conceived theologies deceive vast numbers of lost souls into doubting the Bible's claims.

When Eve embellished God's instructions to not eat that fruit, her mind ventured onto a course of less sound reasoning. Sloppy thinking mishandles relevant details and misses subtler connections. It turns our challenges into Challengers and substitutes ape-man for The Loving Creator. Meanwhile, God remains precisely who He is, and His goals stand firm against our imaginations. Had Eve reasoned precisely with what God commanded, any desirability of the fruit would have been superfluous to her decision about what's for dinner. But having stepped away from the authority of God's unembellished Word, Eve was left to think for herself against a spirit entity, created only a little less superior than God Himself, leading her by the twisting words of a snake. Her embellishment shaped her thinking to attract the serpent's subtle deceit.

> [5] Every word of God proves true;
> he is a shield to those who take refuge in him.
> [6] Do not add to his words,
> lest he rebuke you, and you be found a liar.
>
> Proverbs 30:5

Why did God put such a powerful system of deceit in the perfect Garden of Eden with people of limited mentality? The entire Bible is the narrative of God's solution for imperfection. Had Eve not lost God's Word amongst the embellishments of her mind, had

Love and Message

she simply told the snake he was wrong about the fruit because she knows what God commanded, the Bible may have thereafter read far differently than it now reads. But she didn't. So the Bible is as much a warning about the mopping up God must perform as it is instructions for how to avoid being mopped up. Once the mopping is over, everyone who remains will know what all the mopping was about.

> [32] For the simple are killed by their turning away,
> and the complacence of fools destroys them;
> [33] but he who listens to me will dwell secure
> and will be at ease, without dread of evil.
>
> Proverbs 1:32-33

The most dangerous misconception about evil is its perceived, high degree of heinousness. Evil isn't evil because it is heinous. Evil is just what isn't righteous. The basic concept of righteousness is pervasive order. Pervasive order replicates itself into a perpetual system of building up. Pervasive order is every cause producing only beneficial effects to anything it affects. Eternity can work on that principle. But one imperfection spoils the system, like one fly spoils the ointment, or one pinprick pops a balloon.

A flaw in a foundation spoils the skyscraper. It's even more subtle than a fly in the ointment, since "fly" and "ointment" do not share the same nature. Flaws and the universe of manmade things share the same nature. So our flaws don't seem so evil, especially since our experience with life is steeped in endings instead of eternality. One flaw causes infinite repercussions within an eternal process. Flaws and errors seem benign to us. But the tiniest flaw destroys all perfection. The destruction of perfection is, indeed, very evil.

For eternity to ensue, God must confine evil to a place sequestered from any interaction with righteousness. He is not a mean-weenie curmudgeon for this. He is entirely the opposite because of it. He is the ultimate giver of eternity. He is the epitome of truest love. Love nourishes. It shelters. It protects. It fixes. A fix requiring destruction of imperfection doesn't make the lover mean. What parent would be a mean-weenie curmudgeon for squashing a black widow in her baby's crib? The black widow suffers its fate because mommy loves her baby. Evil will suffer its fate because God loves those who hunger and thirst for righteousness.

But everyone hungering and thirsting for righteousness are evil, too, else all their thoughts, emotions, and deeds would be perfectly and always righteous. Yet, everyone has sinned and fallen short of the glory of God.

> [2] The LORD looks down from heaven upon the children of men,
> to see if there are any that act wisely,
> that seek after God.
> [3] They have all gone astray, they are all alike corrupt;
> there is none that does good,
> no, not one.
>
> Psalm 14:2-3

Clear Signs of Trouble and Great Joy

It isn't that none want to do good. It is that the first flaw has long since worked its effect. Everyone is evil today, simply because no one is perfect. If we could peer deeply into details we could see the vastness of the chasm between our best behavior and perfection. To the perfectly righteous God, who meant us to be His baby, we are the black widow.

Maybe we should hope God's love communicates!

Eve's mistake spread to her children and her children's children. They all liberally embellished and denuded God's Word as they pleased, until there wasn't much traditional sense of His Word conserved anymore. It was a time and condition which had to end. It was a nest that had to be drowned. Yet God is no curmudgeon. He is love. The drowning of the black widow nest was for the preservation of its remnant who wanted to be His babies. So, He didn't just turn on the spigot without a word. According to The Book of Jasher (an anciently venerated Hebrew history book) Adam's great grandson, Cainan, became a wise king calling the people's attention back to their Creator. Then God rested His Spirit upon Cainan's great grandson, Enoch. Enoch also reigned, but with the Spirit of great prophecy.

> And he taught them wisdom and knowledge, and gave them instruction, and he reproved them, and he placed before them statutes and judgments to do upon earth, and he made peace amongst them, and he taught them everlasting life, and dwelt with them some time teaching them all these things...
>
> But in the latter days of Methuselah [Enoch's son, the grandfather of Noah,] the sons of men turned from the Lord, they corrupted the earth, they robbed and plundered each other, and they rebelled against God and they transgressed, and they corrupted their ways, and would not hearken to the voice of Methuselah, but rebelled against him.[1]

Finally, God rested His Spirit of action upon Noah. The Bible calls Noah a herald of divine truth -a preacher (II Peter 2:5). He preached before the Flood to little avail. His wife, sons, and daughters-in-law listened. They joined him in the ark he built to preserve a remnant of mankind through the Flood. Everyone else continued eating, drinking, marrying, and giving in marriage (Matthew 24:38), probably laughing and poking fun at the "unscientific" warnings of Cainan, Enoch, Methuselah, and now, this nut-job, Noah. But reality washed their laughter away.

Noah taught what he learned from his dad and great-granddad, Enoch, after the Flood to, maybe, a little more avail. We have solid archaeological evidence that, after the Flood, some people did closely follow the essentials of what Enoch wrote in his book. But again, almost everyone began straying from Noah's anciently taught truths, because they didn't give a rat's donkey about what their ancestors knew. Who wants to do daddy's stodgy old ways when the kids are being so delightfully liberal! Men moved to a beautiful, watered place on the banks of the Euphrates, and there they proceeded to liberate themselves from God so they could become the slaves of their own imaginations. Go figure! Their new master, Nimrod, aimed to build a shelter for their own name against what they feared God could do, or maybe against what they feared He should do.

Love and Message

It seems like nothing changes. Faithless deniers of God's providence stirred up a public panic to make a name for themselves instead of stirring up a fear of God. Their Tower of Babel epitomized the same collectivism today's demand for consensus forms. God destroyed it to divert history onto a more beneficial course of errors than what Nimrod's global fascism would have led through. Today, mankind has built its own tower of artificial security against fearing God. We've erected two-hundred years of scientific assumptions, stacking uniformitarianism upon naturalism, piling evolution onto uniformitarianism, and extracting from evolution a very pugnacious nihilism for seasoning the entire, towering heap with the politics of despair. Elitists always seek to rule others through towers of assumptions. They wall mankind's heart from God's reality to squelch any public contemplation of His ordinances, thus detaching public perceptions from His authority.

Scientific technology is not rebellion against God, as religiosity perceives. God is creative. Therefore, man is creative. To the chagrin of the Amish, God is proud of mankind's technological advancements. It is the natural progress of man made in the image of a creative, constructive God. Mankind thanking themselves, instead of God, for their technology is their rebellion. Their courageous denial of God's involvement in their technoogy is rebellion. Their scoffing at God's revelation is it. Their accusing God of their own hatefulness and destroying the foundations of truth in their own minds is rebellion. Every embellishment and denudation of God's Holy Word pours from idolatry's rebellious cup. Indeed, nothing changes. Both, the technological blessings into which God's love led mankind and the towering assumptions a world of fools praise for those blessings, will perish in an hour (Revelation 18:10,17,19) because of mankind's ungratefulness towards I AM WHO I AM and their disrespect for the empirically observable truth of all things.

Truth and love correlate in a very particular way. For example, I love my sweetheart beyond the love I have for all other people. She is the pinnacle of my earthly priorities. But if I sought to feed her the same venison and trout I enjoy, I would not be loving her, for she is repulsed by both. If she craved only woodchips and nails for breakfast, lunch, and dinner, my love would refuse to feed her those. Love engages the whole of truth in all its thoughts, emotions, and actions, or it is not love. It serves to others only what does good. My love serves to her what she might enjoy eating other than what would harm her. Love serves true benefit only. It builds up. It does not insist on its own way. The only reason love might destroy something would be to remove what actually causes damage in order to build something of actual benefit in its place. After I ran over a deer with my motorcycle, it was very loving of Char to scrub the pavement out of a hole torn into my foot no matter how much that scrubbing should have hurt. Love is not love outside the truth about what is good.

The Bible proposes that I AM WHO I AM also loves black widows. So He warns every black widow to repent of its poisonous, self indulgence and cry out to Daddy God for rebirth as one of His loving children. Even though God did not articulate a message of repentance in His curse on the snake, crushing the head of the serpent engages repentance for us. The intoxication of deceit is crushed out of every individual who searches for shelter in God's unembellished, undenuded Word. By its warning, black widow babies are given the opportunity to abandon their venom for becoming children of God.

Clear Signs of Trouble and Great Joy

> [12]"Do not call conspiracy all that this people call conspiracy, and do not fear what they fear, nor be in dread. [13]But the LORD of hosts, him you shall regard as holy; let him be your fear, and let him be your dread. [14]And he will become a sanctuary, and a stone of offense...
>
> Isaiah 8:12-13a

> "What then is this that is written:
> 'The very stone which the builders rejected
> has become the head of the corner'?
> [18]Every one who falls on that stone will be broken to pieces; but when it falls on any one it will crush him."
>
> Luke 20:17b-18

So, how does a black widow know to call on the Daddy of broken, but not crushed children? Message. Truth written in black widow language. Truth backed up by what any black widow has the eyes to see for its self. But because God's message was penned by the hand of man, it is accused of deception. The Spirit of God moved human hands to write about fearing God alone. Therefore, men's hands moved pens courageously to express thoughts attracted by God's Spirit of Truth.

> [3]Thou didst come down upon Mount Sinai, and speak with them from heaven and give them right ordinances and true laws, good statutes and commandments,
>
> Nehemiah 9:13

> [9] the fear of the LORD is clean,
> enduring for ever;
> the ordinances of the LORD are true,
> and righteous altogether.
> [10] More to be desired are they than gold,
> even much fine gold;
> sweeter also than honey
> and drippings of the honeycomb.
> [11] Moreover by them is thy servant warned;
> in keeping them there is great reward.
>
> Psalm 19:9-11

> [16]All scripture is inspired by God and profitable for teaching, for reproof, for correction, and for training in righteousness, [17]that the man of God may be complete, equipped for every good work.
>
> II Timothy 3:16

The Bible is very confident about its being the Words of I AM WHO I AM. Scoffers are very confident that it isn't. Verifying either proposition requires evidence, an extract of *information*, the product of reality. If I AM WHO I AM is the pinnacle of reality, then His Holy Word is the most fundamental of *information*, and it precisely evidences reality. What evidence do scoffers bring to show the Bible is not God's Holy Word? That the people

Love and Message

of Israel, whom the Bible predicted would be a nation forever, no longer exist? Well, no. That Christ's Church, which Jesus promised to be with forever, no longer exists? Well, no. That Jesus never rose from the dead? The otherwise inexplicable Shroud of Turin solidly evidences all four Gospel accounts of that event, and so do billions of living reflections of His resurrection. They are witnesses to the resurrection by the light of their repentant lives overcoming every shade of their rebellious inclinations; they are O^2. So, is the Bible not God's Holy Word because scoffers have no eyes to see the evidences of God? No. Abundant evidence for the truth of God's Word indicates the scoffer's lack of honest research.

God's evidence is everywhere buried under mankind's layers of deception. Scoffers are delighted to think their philosophies impugn the Bible's truth. But their philosophies impugn only their own lack of honest investigation, their settling for the paper currency of popular assumptions rather than mining golden evidence, their theorizing rather than observing. Truth is discovered, not made up; it is observed rather than theorized. So scoffers are only able to accuse, but never able to prove. The Bible stands as evidence. Its prophetic nature is not refuted by scoffing assumptions.

In the last chapter we considered an example of simple evidence -the irrefutable case of the fulfilled and yet to be fulfilled prophecies of Daniel 9:24-27. But, for the sake of discussion, we will pay some respectful heed to the possibility of it being only an enormous coincidence, even though its coherence with every other Biblical prophecy/fulfillment is yet more evidence. The consistency, cohesiveness, and coherence of the Bible's message, from its beginning to its end, is evidence. It is not in man's liberal nature to retain such detailed, consistent focus upon such a complex theme for the fifteen-hundred years through which the Bible was written by over three dozen different authors, or even over the five-hundred years in which scoffers accuse it of having been forged by priests and scribes. The coherence of its message with known history from the time of the ancients through today's fulfilled prophecies is evidence. The correlation of Biblical narratives with archeology is evidence. The Bible is God's warning for each of us to take refuge in the Messiah, who was tested and groomed for mopping up the nursery to receive His Daddy's cherished children.

The Bible often uses understatement as a literary device. For example, only once in the Bible is Jerusalem referred to as the mountain in the far north (Psalm 48:2). Only once in the Bible is the Accuser/Deceiver/Destroyer quoted as proclaiming that he will sit on the mount of assembly in the far north (Isaiah 14:13). This is a tidbit of quite extraordinary Biblical *information*, considering that Mt. Zion, Jerusalem, is not in the far north. What significance does "the far north" have for describing such an enormously important place as Jerusalem, which is actually south of most Israel? Another example of understatement is the one time only that the Bible directs the worshipful and celebratory singing in the Christian gatherings to one another as well as to God (Ephesians 5:19). Are we supposed to count how many times the Bible mentions something in order to weigh the importance of its meaning? No. God's Word is all His words. Any one of His ideas needs to be spoken only once to have been spoken sufficiently. The duty to pay enough attention to grasp it is our responsibility. Ask Eve. Understanding a simply spoken word is the test of our desire to hear. One word of a communication is as important as are the rest.

Our next test of the Bible examines a concept it presents only once. Yet this concept

Clear Signs of Trouble and Great Joy

is not at all insignificant, even though it is only once mentioned. Still, it is a most important characteristic of the universe as described by the Bible in Genesis 1:14, which we have touched upon before...

> ...and let [the lights in the firmament of the heavens] be for signs...

Without addition or subtraction, these words simply state that the stars serve a purpose of signifying. Signifying what? Genesis 1:14 doesn't say. But Luke 21:25 proposes that they will signify the time of Christ's return. We will discuss that later. But for now, there is only one other direct prophecy in the Bible about a star signifying a particular event, although the Bible claims that an historic star led some wise men to a particular child that had been born in Judea. Interestingly, the one prophecy and the historic star both concern the same birth.

> [17] I see him, but not now;
> I behold him, but not nigh:
> a star shall come forth out of Jacob,
> and a scepter shall rise out of Israel;
>
> Numbers 24:17

Astronomy has had a long history in the Middle East. It was an ancient practice even by the time poor Samsuditana's astronomer-priests observed the evidential lunar eclipse of 1362BC. And as their case demonstrated for us, the early peoples of the Middle East ascribed very anciently perceived meanings to the constellations, planets, sun, moon, and their alignments. Speculation in those times, and even by some people today, was that their meanings were delivered to mankind by the gods. Many Mesopotamian cultures, as well as the ancient Egyptians and Greeks, thought the gods delivered their ideas through an "Enoch" type person. The essential consistency of many (but not all) meanings ascribed to the constellations and planets across various cultures and over the centuries reflects the possibility that some sort of "Enoch" was the original source of knowledge about signs amongst the heavenly bodies.

Then in 607BC, the young prophet, Daniel, was captured into the court of the Babylonian king, Nebuchadnezzar. The Babylonians were certainly not unfamiliar with the Hebrew people and their religion. But until Daniel arrived, their knowledge of Hebrew traditions was sketchy. Daniel was one of those men who neither embellished nor denuded God's words, but drew close to God by carefully obeying each one of them. So God opened up to him like He opened up to Enoch. At the occasion of a deeply perplexing dream, Nebuchadnezzar discovered how real and closely Daniel was in communication with God. Daniel was able to reveal a perplexing dream the king had and its meaning. Consequently, he developed a good standing with Nebuchadnezzar and his succeeding kings; he was commissioned to serve high in the Babylonian, and then in the Medo-Persian governments. Daniel became highly regarded in Mesopotamia for his prescience, wisdom, and knowledge of God throughout his life there.

Love and Message

The speculation that Daniel taught his Hebrew knowledge to the Babylonian and Persian astronomers is most likely true. By the time the three wise men of Matthew's gospel saw the famous Christmas star, its eventual shining would have been anticipated for over four centuries because of Daniel's influence. The wise men probably knew what they were watching for, being of the Persian astronomers who may have been watching for The King's star throughout the four centuries passing from Daniel's time to the little scene at Bethlehem.

Additionally, the idea of a child being the desire of the nations born to a special woman was another widely, but rather vaguely perceived, religious element passed down from the deep shadows of the Middle Eastern past. Of course, as it goes with most concepts passed through multitudes of generations, including any possible meanings passed along from some "Enoch", their eventual imagery becomes only misty expressions of denuded truths repeatedly embellished until their primitive meaning becomes only a vague hint obscured by layers of imagination.

These wise men coming to Jerusalem seeking the whereabouts of the baby king born to the Jews, reckoned that this baby king was highly significant to their people, as well. Otherwise, would they have journeyed hundreds of miles over bandit-filled highways while bearing gifts fit for royalty to pay Him kingly homage (Matthew 2:1-11)? Surely Daniel's expectation of the coming Messiah, the desire of the nations, spread through the ranks of Mesopotamian wise men (astronomers) and most likely effected what they watched for in the sky.

Frederick A. Larson is another of the many mundane, O^2 type observers. He became greatly curious about the star St. Matthew claims led the wise men to Bethlehem. If it were other than merely an apparition, Larson reasoned, then it surely would have been some noteworthy, highly unusual motion of a planet, or something of the nature. He describes his search for what the special star might have been and the discovery he made in his video, <u>The Star of Bethlehem</u>.[2]

I watched his video in amazement. I am a Bible believer who does his best to neither denude nor embellish what the Bible says. I had little trouble believing that Frederick Larson discovered what could have been the star which led the wise men to the baby Jesus. But my "show me" attitude demanded I download astronomy software for checking Mr. Larson's work before I could accept his proposition. And although what I am about to describe to you is from notes I took journeying through my astronomy software, I would know nothing about this without Mr. Larson's work.

Royalty has been associated with the lion from times long before written history. Some of the early figures of The Mother Goddess are seated between lions. Early Mesopotamian depictions of Inanna, a later somewhat denuded and greatly embellished conceptualization of the ancient Mother Goddess, show her standing on the back of a lion. The Assyrian cherubim were winged lions. Whatever else the constellation Leo may have represented to the ancients would have been in addition to its essence of royalty. And particularly pertinent to our discussion, the little district of Judah had long been represented by the lion and Leo, the constellation. None of this would have been news to St. Matthew's wise men.

Jupiter, the planet understood from Mesopotamia through Greece to Rome as the king planet, passing through Leo would stir familiar, traditional interests, since, over the

Clear Signs of Trouble and Great Joy

centuries, it has passed through Leo once every twelve years. And maybe those ancient perceptions regarding royalty and the lion would have been more heightened for the wise men because they were the products of a culture influenced by Daniel's knowledge and the prophecies of his scriptures about Judah and its coming Messiah. "The star that would rise from Jacob" may have meant Jupiter to them. But Regulus was anciently regarded as the king star. It is the brightest star in the constellation Leo and the twenty-first brightest star in the night sky. It is actually not one star. It is a system of four stars. These four stars appearing as one have long been thought to rule the affairs of the heavens,

> ...a belief current, till three centuries ago, from at least 3000 years before our era.[3]

The wise men most likely were intently watching Jupiter approach Leo in 3BC. They were as able to calculate planetary motions as we are today, so I doubt they were surprised by what they soon saw. Most likely they anticipated the conjunctions (a planet passing within one-and-a-half degrees of a background star). Probably they were watching specifically to see them.

Millennia earlier, the ancients carefully aligned Enclosure D of the astonishing Gobekli Tepe monument with the motions of Cygnus in the north and the sun's passage through the Southern Gate. Secular history concludes that mankind had been tracking the heavenly bodies with at least some level of mathematical precision for nine-and-a-half millennia by the time the wise men were intently watching for this spectacle in Leo. I don't think it a bit absurd to suggest that they knew Jupiter was nearing three conjunctions with Regulus in 3BC-2BC.[4]

Wait now. If a conjunction is when a planet passes within a degree of a star or another planet, then how do you get three of those out of the one time Jupiter slips through Leo? Surely the planet doesn't back up, or something strange like that? Well, yes, it does something like that, but not really so strange, since it doesn't actually back up. Because of a too-complicated-to-describe-here correlation between Earth and Jupiter's orbits around the sun, Jupiter twice annually appears to stop in the sky, backup, stop again, and then proceed on its normal course. Earth and Jupiter both maintain their actual motions without variation. But in appearance only, Jupiter seems to back up against the astronomically distant backdrop of stars. It is called retrograde motion.

And any first century BC wise man would have been very familiar with not only Jupiter's retrograde motion, but also with when it was about to occur, and therefore, where it would occur amongst the background stars. They whiled away their evenings watching stars and spinning yarns about their motions. A few centuries of that made them wise enough to know how often and when Jupiter's retrograde motion would occur while passing through Leo.

3BC to 2BC was to become extraordinarily meaningful. I am confident the wise men did not just happen to see the star of the newborn King. I propose they were experienced enough in astronomy to know Jupiter cyclically conjuncts with Regulus three times by retrograde motion. Therefore, they expectantly watched the king planet, Jupiter, go into

Love and Message

near conjunction with the ruling star of the heavens, Regulus, on September 13, 3BC, in Judah's constellation, Leo. Jupiter then stood still to begin its retrograde motion approximately December 3, 3BC. On February 17, 2BC it was again in near conjunction with Regulus, and on March 31, 2BC, it again stood still before returning to its normal motion. Then for the third time, on May 5, 2BC, this king planet was in near conjunction with the ruling star of the heavens in Judah's constellation for the third time. I respect the mental skills of the ancient astronomers enough to be certain the wise men were awaiting this moment of astronomical speech, for the following words of David were most likely known to them through Daniel.

> [1] The heavens are telling the glory of God;
> and the firmament proclaims his handiwork.
> [2] Day to day pours forth speech,
> and night to night declares knowledge.
> [3] There is no speech, nor are there words;
> their voice is not heard;
> [4] yet their voice goes out through all the earth,
> and their words to the end of the world.
>
> Psalm 19:1-4

Speech delivered from the Divine number (three) of these near conjunctions in Judah's Leo must have spoken enormous, long contemplated meaning to the wise men. So it was that even during this ancient, astronomical show in the sky, most likely their camels were being fed and their gold, frankincense, and myrrh (3) were being purchased and packed for the journey to Jerusalem.

Nor do I think they would have been surprised to observe Jupiter enter conjunction with Venus thirty-three days later. Although this conjunction was not as spectacular as Jupiter's triple conjunction with Regulus during the preceding year, it most likely bathed in the essence of being special.

From the times of ancient Sumer, three-thousand years earlier, Venus was regarded as Inanna's star. Inanna was Ishtar to the later Akkadians and Babylonians and Isis to the Egyptians. She was the queen of the heavens. But her queenship came not by marriage to some king, although many cultural representations of Inanna/Ishtar/Isis indeed portrayed her as married. The essence of her queenship actually rose from the prehistoric semblance of the mother of all the living. The ancient Australians portrayed this great mother in a very nearly identical depiction that she was also found to have been portrayed at Gobekli Tepe: giving birth like an overflowing cornucopia. Therefore, we should not be surprised at where the concept of such a Mother-Goddess/Inanna/Ishtar/Isis originated before it was denuded and embellished almost beyond recognition...

> [20] The man called his wife's name Eve, because she was the mother of all living.
>
> Genesis 3:20

Clear Signs of Trouble and Great Joy

Considering today's disregard for the Bible, our surprise at (or rejection of) that most probable origin is to be expected. Most people today think of Eve as being the first ape-lady of Darwin's animal book, not the original beauty of God's Holy Book. The Great Mother Goddess began her hallowed popularity as the much more svelte Eve.

The planet Venus shared in a bursting, cornucopia concept that had been carried down and applied to the well dieted Inanna/Ishtar/Astarte/Isis from the more ancient, Great Mamma Eve. But I AM WHO I AM doesn't participate in star yarns spun upon ancient rooftops. His words are neither denuded nor embellished in His mind, no matter how much men once crafted figurines of a porky lady seated between lions. If Genesis 1:14 indeed meant that signs would ring like bells in the heavens around events I AM WHO I AM directed upon the ground, then ring out loud without a voice to be heard they did!

> [3] Praise him, sun and moon,
> praise him, all you shining stars!
>
> Psalm 148:3

"Queen" was not God's point in either Venus or the birthing concept carried away from Gobekli Tepe to the Australian aborigines as sort of a cornucopia imagery. The restoration of life filled with blessing and goodness, righteousness instead of evil, and prosperity instead of poverty were. God prophesied the woman bearing the seed of salvation as He cursed the snake in the Garden of Eden. To a people freshly expelled from their blissful existence into the chaotic troubles of wilderness survival, the restoration of what they lost would be an enormous concept. The Mesopotamians twisted God's concept of the seed bearing woman into "the queen of heaven" and applied that to the planet, Venus.

Before leaving the constellation of Judah, the planet of the King (the seed promised to crush the head of the serpent), came into conjunction with the star of the woman to whom God promised the conquering seed. In Leo, Jupiter and Venus joined in melodious harmony with the voice of God's Word to speak of a humble birth in Bethlehem.

At this sight, Frederick Larson presumes, the wise men mounted their camels and set off for Judah. Four months later, Jupiter and Venus again approach near conjunction on October 13, 2BC, again appearing as the brightest star in the sky. But now it shined in the head of Virgo. Virgo is even more conspicuously the woman of the promise than is Venus. She's been given the name of a virgin from Roman times. From far more ancient times this constellation was associated with a goddess bearing a head of wheat. The name of the goddess was neither Akkadian nor Semitic. Its normalized form is identical to the Hurrian, *sala*, meaning "daughter". Currently regarded as the goddess of wheat, she is depicted as holding a head of wheat in her hands. A head of wheat is represented by the star, Spica, in Virgo's left hand. New Testament imagery presents The Bread of Life ground from the head of wheat, the seed of the woman. Virgo was known as Bethulah to the Hebrews,[5] the same Hebrew term for "young woman" found at Isaiah 7:4,

Figure 1: Sala

Love and Message

[4]Therefore the Lord himself will give you a sign. Behold, a young woman shall conceive and bear a son, and shall call his name Immanu-el.

Immanu-el is Hebrew for God-with-us. The connection is obvious. Skeptics like to accuse the Bible of borrowing these memes from ancient Mesopotamia. But if they are wrong and the Bible is the true words of God, then sound evidence will show Biblically correlating, prophetic imagery being more ancient than the Mesopotamian concepts which minimalists claim to have been most of the Bible's source. Seventy-three days after its conjunction with Venus in Virgo, the king planet began retrograde motion. It had traveled from near Virgo's head to near her heart. And there, by the heart of the virgin, Bethula, in the morning of December 25, 2BC the king planet would appear to stand still for several days. About 6:00AM it would seem to be standing over Bethlehem as observed from Jerusalem, although it was rather high in the sky.

We can't be certain these motions and conjunctions of those planets, stars, and constellations are what the wise men followed, although they most likely were aware of them. Regardless, Frederick Larson revealed to our 21st century, TV-blurred eyes this wonder in the sky at the time of Christ's birth. They were obviously signs of His birth, the heavens telling the glory of God. They monumentalized mankind's greatest delivery without speaking a word.

Familiar with Jupiter and Venus' motions through the constellations, the wise men would have known that every twelve years Jupiter passes through Leo close by Regulus. It orbit's the sun along a path through the stars and constellations called the solar ecliptic. So does earth. This planetary pathway rests on a plain extending through all twelve main constellations, including Leo. This pathway tracks very close by Regulus.

The ancient wise men also would have known that every year Jupiter goes through retrograde motion twice. I am sure they were, and many thousands of astronomers today are fully aware that every 11.86 years Jupiter travels through Leo, passing close to Regulus at least once. And I am sure most of today's astronomers are aware that every sixty years a twenty-four year period begins in which Jupiter triple-conjuncts Regulus each time it passes through Leo. So these triple conjunctions occur twice in a row every eighty-four years. And maybe most of today's astronomers understand, as those ancient wise men probably understood, that at least one of the two years in which Jupiter "crowns" Regulus three times, it will enter retrograde motion within a few days of December 25. Moreover, somewhere amongst the constellations, Venus will conjunct with Jupiter at least once a year.

This is just the geometry of the solar system. It is how it was made, according to those who neither embellish nor denude God's Word.

But to skeptics, it is how it coincidentally evolved.

Although these interrelationships between Jupiter, Regulus, and Venus may not alone have been the Star of Bethlehem signaling St. Mathew's wise men, I am sure the astronomer-priests farther back than even those of poor Samsuditana were fully cognizant of these motions, and therefore, were also expectant of their predictable occurrences. The motions were simple to observe. Hundreds of ancient, astronomical observatories around the

Clear Signs of Trouble and Great Joy

globe were not constructed just to make men sweat under the burden of massive stones. They were built to observe the motions of the lights in the heavens. Whatever the wise men saw made sense to them, even if it entailed some amount of apparition.

And so, the wise men wound up in Jerusalem giving gifts to the newborn King.

None of our current historians, scientists, astronomers, or astrologers have shed clear and revealing light upon the origin of the Zodiacal meanings ascribed to the constellations. They all assume Neanderthallic grunters breeding their way through Paleolithic times ascribed great imagination to star patterns while picking their nuts and berries. That those meanings were held for tens of thousands of years through hundreds of generations of critters able to communicate only by grunts and hand gestures is not rational. There must have been more reason for them to stick than just "we hunt bulls", "lions are mean-weenies", and "women have babies". Mankind's most ancient artwork, e.g. the Hall of Bulls in the Lascaux caves, demonstrates not only sober attention paid to these lights in the sky, but also a perception of the constellation figures, somewhat similar to how we perceive them today.

By the time the wise men rode for Jerusalem, the motions of Jupiter and Venus through Virgo, Leo, and Regulus were not only well known, but they were perceived as the king planet, the king star, the star of the queen of heaven, the constellations of the seed bearing woman, and the constellation of the king, possibly because of these interrelated, cyclically repeating conjunctions and retrograde motions.

Often truth and falsehood lay only a few degrees apart. Secular history sees Neanderthallic grunters ascribing cultural aspects onto the lights in the sky, maybe for transmitting their perceived cosmology to their descendants. Yet it is rather amazing to think such a complex concept as cosmology could have engaged the thoughts of grunters so bereft of mental prowess as to not improve the way they chipped stones into tools for millennia. On the other hand, most young-earth creationists are certain the ancient Sethites passed down oral histories and cosmologies through only a couple dozen generations before the Hebrews eventually inked them onto parchments and papyri. Their literature describes how the prophet Enoch was shown all the lights of the heavens by a messenger sent from The Lord of Spirits (Jesus). Both stories involve the darkness of the millennially deep past. Both stories involve the concept of a cosmology written in the lights of the sky. But one story evolves from deliberately blindfolded guessing, while the other story involves the passing down of a very widely and long acknowledged history, as well as a great variety of ancient images bespeaking some shattered and scattered puzzle of old. Did people ascribe meanings to those planets, stars and constellations because of their motions? Or did God create them and their motions to deliver us the meaning His Spirit ascribed to them?

Leaving this question about origins to the unfolding of more *information*, let's for now just understand that the ancients of the Middle East perceived these planets, stars, and constellations to be representing kingliness, queenship, and some sort of propagation, and that every eighty-four years, from time immemorial through today and forward to the bitter end, their motions have been blinking, and will continue to twice blink, the Christmas story.

Maybe these conjunctions were "the star" that alerted the wise men, while the star they followed, which eventually settled over the place where they found the baby Jesus, might have been an apparition. We must remember that the inexorable knot, cinching our

Love and Message

every thought to the necessity of "mechanistic material" for the explanation of all sights and sounds, was tied for us by a bunch of nineteenth century, Bible-denying, not-all-that-well-educated guessers.

Would the birth of I AM WHO I AM into human form not be the most awesome act of The Most Almighty Being? Would God becoming man to bear every person's sins for the eternal salvation of anyone who chooses to believe not be the greatest act of bedrock humility throughout all of the ages of eternity? Would it actually, then, be surprising that the God responsible for such a thing would adorn it with this continuously blinking praise throughout earth's time? And is it not noteworthy that the general cycle of this blinking praise for Christ occurs twice within every seven sets of twelve years, seven being the number of spiritual perfection and twelve being the number of complete order, a.k.a. righteousness, and two being the number of emphasis?

The secular explanation of this phenomena must call upon extraordinary coincidence, not so much extraordinary because of the celestial motions and cycles, but more extraordinary because of the coincidence of an early, gruntingly idiotic ape-man presciently attributing characteristics to the very stars and planets that would eventually symbolize the birth of mankind's Savior, blinking a discernable message in the sky when He was born. And adding immensely to that coincidence is the further coincidence of a "Holy Bible", concocted entirely from human imagination, that actually prophesies His birth centuries beforehand, as well as prophesying the exact year in which the One who was that-year-born would be cut off, and even more, prophesying the following destruction of His beloved city. But good sense sees the infinitely talented I AM WHO I AM working the order of message into all these motions. Coincidence, or good sense? That is the question.

The coherence of such *information* as that February 25, 1362BC eclipse, Ipuwer's lamentations, the quickly abandoned Avaris, ancient men drawing clear pictures of actual dinosaurs around the world, new geological observations implying tsunami sequences of The Great Flood, and the Daniel 9:24-27 prophecy nailing to the year its prediction of the crucifixion of this King born under one, particular occurrence of these cyclical, prescient, astronomical patterns pushes coincidence beyond absurdity to parking it at abject foolishness. But for the worship of atheism, over-stretched coincidence becomes ordained explanation rather than good sense. Otherwise, we're left to suspect Enoch prophetically revealed God-assigned meanings of the constellations and stars, such that their patterns and interrelationships correlate with historical events to tell the most important story of all times to whomever is not too arrogant to listen. If mechanistic materialism is reality, then acknowledged history presents a network of absurd coincidences for no purpose at all by no design of anyone. If I AM WHO I AM is reality, the explanation of all things is trying to tell us something pretty darned important.

It was important for the Jews of the first century to recognize their Messiah when He came to them. Not recognizing Him because He was not what they were expecting is to live the idolatry of worshipping a form of God contrived by imagination rather than observed in reality. His name, I AM WHO I AM, demands observation of His reality instead of making it up. Accordingly, the Daniel 9:24-27 prophecy and the Jupiter/Regulus cycle would both be observations of Him erecting signs along history's highway to shout, "Israel! Your

Clear Signs of Trouble and Great Joy

King is Coming! Get it right!" Malachi, the last of the Old Testament prophets, prophesied in the closing words of his book:

> [5]Behold, I will send you Elijah the prophet before the great and terrible day of the LORD comes. [6]And he will turn the hearts of fathers to their children and the hearts of children to their fathers, lest I come and smite the land with a curse.
>
> Malachi 4:5-6

Amidst this cycle of Jupiter/Regulus motions was also born John the Baptist, accompanied by a similar flurry of angelic announcements as what ushered Jesus into the world. Of course, these flurries of angelic announcements were no more directly evidenced than are uniformitarianism, gradualism, or even Darwin's evolutionary ape-man. But they have vastly more indirect evidence. The same good sense that assembles a picture of God from all of these strange, astronomical events understands these meaningful patterns to also evidence those angelic flurries. It is simply a matter of consistency, cohesiveness, and coherence, things any mundane mind can recognize. Assumption alone imagines Timex rocks for perceiving the geologic epochs ape-man needed to climb up the tree of evolution and ascribe lasting meaning to the stars for no ultimately important reason even while not yet being smart enough to wet on a seed buried in the ground. Else, in the beginning, The Almighty Loving God wrote with the stars an astronomically important message for all existence, and then revealed it to Enoch.

> ...let them be for signs...
>
> Genesis 1:14

> [3] There is no speech, nor are there words;
> their voice is not heard;
> [4] yet their voice goes out through all the earth,
> and their words to the end of the world.
>
> Psalm 19:3-4

> [6] The heavens declare his righteousness,
> for God himself is judge!
>
> Psalm 50:6

> [4] He determines the number of the stars,
> he gives to all of them their names.
>
> Psalm 147:4

> [1] The Mighty One, God the LORD,
> speaks and summons the earth
> from the rising of the sun to its setting.
>
> Psalm 50:1

> [3] Praise him, sun and moon,

Love and Message

> praise him, all you shining stars!
>
> Psalm 148:3

And every eighty-four years Jupiter, Regulus, Venus, and Virgo repeat the story of Immanu-el, God With Us, even to this day!

It was imperative for the Jews of the first century to get the message instead of discounting its coincidences. The King of Kings and Lord of Lords did not force Himself upon them. He presented Himself for their acceptance or rejection. It wasn't His test. It was theirs.

Forty is the Biblical number of testing. Jesus was tested in the wilderness forty days. The Hebrews wandered in the wilderness forty years. Moses was forty years old when he fled Egypt after striking dead an Egyptian taskmaster, and forty years later God called him to lead His people out of Egypt. Forty years transpired from the beginning of Jesus' ministry to the destruction of Jerusalem and The Temple.

Would the Jewish leaders direct this nation to accept its crucified King, or would they lead it into continuous rebellion against Him during its forty years of testing? It would have helped their case immensely had they not hung Him on the cross. But, above all mankind's atrocities, God is forgiving. So, would they repent and call on their resurrected Messiah? Daniel 9:26 answered that question before it was even asked. A comparative few repented. But the majority didn't. Forty years after they crucified their King and failed to repent, their city and prized Temple were destroyed. Jesus even foretold its destruction. And while Jerusalem was burning in AD70, Jupiter was again crowning Regulus three times. Moreover, eighteen-hundred-and-ninety-seven years later, when Israel won control of Old Jerusalem in 1967, the city from which that crucified and resurrected King will reign, Jupiter again began crowning Regulus three times. Coincidence? Or message for good sense?

We don't understand such things today, because our minds have been formed into towers of assumptions about unbreakable laws of physics and the nonexistence of anything nonmaterial. We've been "educated" to accept without question unfathomably colossal coincidences necessary to believe in the evolution of an almost infinitely complex ecosystem by nothing more than crapshoot chances of genetic goof-ups somehow assembling near perfectly patterned, seemingly unending DNA instructions for each one of billions of different life forms. No wonder coincidence will be invoked to explain these obvious messages written in the stars by an infinite Creator. Skeptics stretch "coincidence" over endless enormities. The magic of coincidence aside, surely it is sensible to believe that the Almighty Creator of physics could and would cause a few scientifically abnormal experiences for the sake of attracting attention to a most critical message meant for the dwellers of first century Jerusalem.

> Let God be true though every man be false, as it is written, "That thou mayest be justified in thy words, and prevail when Thou art judged."
>
> Romans 3:4

This is the biggest reason behind God's communication to all mankind: He is going

Clear Signs of Trouble and Great Joy

to prevail over every judgment against Him. Since God loves everything, even those judging Him must be warned to accept Him for what He is rather than to judge Him for not being what they'd rather Him to be (or rather Him to not be at all). Mercy and message for anyone who will incline their hearts to His communication is a part of His basic nature of working only good effects. He is a giver. To those who seek Him He gives Himself. To those who seek not Him, He gives not Himself. Of this He austerely warns.

But responsibility for communication has a flip side. Successful communication is the accurate formation in one mind of what another mind expresses. The same meaning must be received as what was delivered. To know what was expressed requires a full and honest regard for the communicator. The receptor mind will receive the communication only if it is faithful to follow all indications given by the expressing mind. If the expressing mind indicates the use of signs in the sun, moon, and stars for conveying some of its messages, then attention must be paid to the sun, moon, and stars to receive those messages.

Genesis 1:14 indicates this very purpose for the heavenly bodies. Luke 21:25 calls attention to signs in the sun, moon, and stars for knowing that Christ's return draws near. If God uses prophecy for communication, then attention must be paid to prophecy. If He uses inexplicable events and circumstances to communicate warnings, then we must not snub reports of seemingly miraculous abnormalities. We can not discount evidence merely because it topples our twenty-first century tower of assumptions. We can not disregard *information* merely because it contradicts our most desired conclusions. We can not disrespect coherence simply because it presents a picture we choose not to see. If God has revealed through signs, prophecy, and miracles, then we must carefully consider the reported observations of signs, prophecies, and miracles.

God sent Jerusalem its King, and forty years later that city paid a high price for having rejected Him. Josephus was one of the Jews inside the city when the Romans besieged it. He was wise enough to dismiss the hope influential Jews were touting about God changing their plight in this treacherous situation. He soberly considered the example of Jeremiah's call for an orderly surrender to the Babylonians when Jerusalem had been under siege six-hundred-fifty years earlier. To Josephus, that was *information* warning everyone to get out of Jerusalem. So he went missing from the disaster. Also conspicuously missing from the disaster of AD70 were the Christians. They had some *information* of their own. Jesus warned:

> ^{43}For the days shall come upon you, when your enemies will cast up a bank about you and surround you, and hem you in on every side, ^{44}and dash you to the ground, you and your children within you, and they will not leave one stone upon another in you; because you did not know the time of your visitation.
>
> Luke   19:43-44

Nor was it only Jesus' prophecy spreading this warning like ominous writing on the wall. Many strange things were reported as AD70 approached. Regarding some audacious

Love and Message

hope and change being propagandized about Jerusalem's bleak besiegement in spite of the more prescient clues of impending disaster, Josephus wrote:

> 2...Now a man that is in adversity does easily comply with such promises [of delivery]; for when such a seducer makes him believe that he shall be delivered from those miseries which oppress him, then it is that the patient is full of hopes of such deliverance.
>
> 3. Thus were the miserable people persuaded by those deceivers, and such as belied God himself; while they did not attend, nor give credit, to the signs that were so evident, and did so plainly foretell their future desolation; but, like men infatuated, without either eyes to see or minds to consider, did not regard the denunciations that God made to them. Thus there was a star resembling a sword, which stood over the city, and a comet, that continued a whole year. Thus also, before the Jew's rebellion, and before those commotions which preceded the war, when the people were come in great crowds to the feast of unleavened bread, on the eighth day of the month Xanthicus, [Nisan,] and at the ninth hour of the night, so great a light shone around the altar and the holy house, that it appeared to be bright daytime; which light lasted for half an hour. This light seemed to be a good sign to the unskillful, but was so interpreted by the sacred scribes as to portend those events that followed immediately upon it. At the same festival also, a heifer, as she was led by the high priest to be sacrificed brought forth a lamb in the midst of the temple. Moreover, the eastern gate of the inner, [court of the temple,] which was of brass, and vastly heavy, and had been with difficulty shut by twenty men, and rested upon a basis armed with iron, and had bolts fastened very deep into the firm floor, which was there made of one entire stone, was seen to be opened of its own accord about the sixth hour of the night. Now, those that kept watch in the temple came thereupon running to the captain of the temple, and told him of it; who then came up thither, and not without great difficulty was able to shut the gate again. This also appeared to the vulgar to be a very happy prodigy, as if God did thereby open them the gate of happiness. But the men of learning understood it, that the security of their holy house was dissolved of its own accord, and that the gate was opened for the advantage of their enemies. So these publicly declared, that this signal foreshewed the desolation that was coming upon them. Besides these, a few days after that feast, on the one and twentieth day of the month Artemisius, [Jyar,] a certain prodigious and incredible phenomenon appeared; I suppose the account of it would seem to be a fable, were it not related by those that saw it, and were not the events that followed it of so considerable a nature as to deserve such signals; for, before sun-setting, chariots and troops of soldiers in their armour were seen running about among the clouds, and surrounding of cities. Moreover, at that feast which we call Pentecost, as the priests were going by night into the inner [court of the] temple, as their custom was, to perform their sacred ministrations, they said that, in the first place, they felt a quaking, and heard a great noise, and after that they heard a sound as of a great multitude, saying, "Let us remove hence." But,

Clear Signs of Trouble and Great Joy

what is still more terrible, there was one Jesus, the son of Ananus, a plebeian, and an husbandman, who, four years before the war began, and at a time when the city was in very great peace and prosperity, came to that feast wherein it is our custom for everyone to make tabernacles to God in the temple, began on a sudden to cry aloud, "A voice from the east, a voice from the west, a voice from the four winds, a voice against Jerusalem and the holy house, a voice against the bridegrooms and the brides, and a voice against this whole people!" This was his cry, as he went about by day and by night, in all the lanes of the city. However, certain of the most eminent among the populace had great indignation at this dire cry of his, and took up the man, and gave him a great number of severe stripes; yet did not he either say anything for himself, or anything peculiar to those that chastised him, but still he went on with the same words which he cried before. Hereupon our rulers supposing, as the case proved to be, that this was a sort of divine fury in the man, brought him to the Roman procurator - where he was whipped till his bones were laid bare; yet did he not make any supplication for himself, nor shed any tears, but turning his voice to the most lamentable tone possible, at every stroke of the whip his answer was, "Woe, woe to Jerusalem!" And when Albinus (for he was then our procurator) asked him, Who he was? And whence he came? And why he uttered such words? He made no manner of reply to what he said, but still did not leave off his melancholy ditty, till Albinus took him to be a madman, and dismissed him. Now, during all the time that passed before the war began, this man did not go near any of the citizens, nor was seen by them while he said so; but he every day uttered these lamentable words, as if it were his premeditated vow, "Woe, woe to Jerusalem!" Nor did he give ill words to any of those who beat him every day, nor good words to those who gave him food; but this was his reply to all men, and indeed no other than a melancholy presage of what was to come. This cry of his was loudest at the festivals; and he continued this ditty for seven years and five months, without growing hoarse, or being tired therewith, until the very time that he saw his presage in earnest fulfilled in our siege, when it ceased; for, as he was going round upon the wall, he cried out with his utmost force, "Woe, woe to the city again, and to the people, and to the holy house!" And just as he added at the last, "Woe, woe to myself also!" there came a stone out of one of the engines, and smote him, and killed him immediately: and as he was uttering the very same presages, he gave up the ghost.

Now, if anyone consider these things, he will find that God takes care of mankind, and by all ways possible foreshews to our race what is for their preservation; but that men perish by those miseries which they madly and voluntarily bring upon themselves.[6]

Josephus informs us that Titus began the siege of Jerusalem at the time of Passover.[7] Jerusalem was full of worshipers who became imprisoned by the siege. Starvation swept so freely through the city that the healthier were too repulsed to bury the dead. The weaker of them collapsed on the piles of bodies they were attempting to bury and died, too, Josephus

Love and Message

reports. Houses became graveyards for their occupants. Dead bodies so overwhelmed the living that they were thrown over the city wall in astonishing masses onto the valley floors below.

> ...Titus, in going his rounds along those valleys, saw them full of dead bodies, and the thick putrefaction running about them, he gave a groan; and spreading out his hands to heaven, called God to witness that this was not his doing.[8]

Did God indiscriminately pour this horror onto the city because He wanted the good and bad alike to suffer terribly? Over the centuries, Christian reflection on the destruction of Jerusalem presents it as punishment for the Jews rejecting and crucifying their very own King. But there was a certain synchronicity plainly visible for any one who looked at things through eyes to see, as Christ preached and Josephus noted. The Jewish leaders collectively rejected Jesus' message from the beginning, which rejection the majority of the Jews eventually followed, even though they had some four decades of signs and prophecies begging for their repentance. But those who came to believe Jesus during those forty years recognized the mounting dangers. They were willing to view portentous events in the light of what Jesus preached. They were willing to see an overwhelming number of "coincidences" speak a message. They fled.

> [1]There were some present at that very time who told [Jesus] of the Galileans whose blood Pilate had mingled with their sacrifices. [2]And he answered them, "Do you think that these Galileans were worse sinners than all the other Galileans, because they suffered thus? [3]I tell you, No; but unless you repent you will all likewise perish. [4]Or those eighteen upon whom the tower in Siloam fell and killed them, do you think that they were worse offenders than all the others who dwelt in Jerusalem? [5]I tell you, No; but unless you repent you will all likewise perish."
>
> Luke 13:1-5

At the time Jesus dropped this clue, He had not intimated anything of Jerusalem's destruction. But He took occasion in this conversation to warn about the importance of repenting, turning around, fundamentally transforming your own behavior instead of trying to force fundamental transformation upon others. The "likewise" of the perishing at Jerusalem, AD70, bore an ominous parallel to the miseries of those Galileans and the eighteen who died under the tower in Siloam. Titus would soon bring all of Jerusalem's occupants parallel to these disasters, but on a far greater scale. The people caught in the mess were those who failed to give Jesus His due attention. They didn't repent. They scoffed. So they didn't flee when they saw the Roman legions drawing near.

> [6]A man had a fig tree planted in his vineyard; and he came seeking fruit on it and found none. [7]And he said to the vinedresser, "Lo, these three years I have come seeking fruit on this fig tree, and I find none. Cut it down; why

Clear Signs of Trouble and Great Joy

> should it use up the ground?" [8]And he answered him, "Let it alone, sir, this year also, till I dig about it and put on manure. [9]And if it bears fruit next year, well and good; but if not, you can cut it down."
>
> Luke 13:6-9

For three years Jesus went through Judea, seeking the spiritual fruit of His people. From one end of the country to the other He called people to leave their sin and mend their ways because the kingdom of heaven was at hand. Although His hearers could not have extracted the sense of an unrepentant Jerusalem's destruction from this parable alone, it's dots should have connected near the end of His ministry when He added:

> [20]But when you see Jerusalem surrounded by armies, then know that its desolation has come near. [21]Then let those who are in Judea flee to the mountains, and let those who are inside the city depart, and let not those who are out in the country enter it; [22]for these are days of vengeance, to fulfil all that is written. [23]Alas for those who are with child and for those who give suck in those days! For great distress shall be upon the earth and wrath upon this people; [24]they will fall by the edge of the sword, and be led captive among all nations; and Jerusalem will be trodden down by the Gentiles, until the times of the Gentiles are fulfilled.
>
> Luke 21:20-24

By the end of His ministry, Jesus had supplied two pieces of crucial information to His people: 1) God destroys the fruitless, and 2) Jerusalem would be destroyed for their fruitlessness. His crucifixion was the key *information* about their fruitlessness. In the forty years from the beginning of His ministry to Jerusalem's destruction, the Jews rejected Him and the warnings He gave them; they failed to repent; they failed to call on I AM WHO I AM as He showed them to do it; they failed the test; their fruitlessness was revealed.

But Jesus' followers listened. They knew something was afoot because they believed what Jesus said, and they saw happening what Jesus prophesied. They didn't need hypotheses, scientific measurements, or paradigm guardians to simply hear and think for themselves. For them, the decision to believe was the best scientific response to their having seen everything Jesus said would happen actually happen the way He said it would happen. Having observed Jesus' life, truthfulness, and prescience, they watched with minds open to new and relevant *information*. Having accepted Jesus' warnings and noting the nonresponse of most Jews thereafter, their eyes were able to see where the course of history was about to flow.

It was not likely that all of the Christians individually perceived what was about to befall Jerusalem. Their lives were lived even more by social interaction than we live ours. Some knew much; some knew less; others knew little. But those who knew shared what they knew, and those who didn't know accepted what was shared. We should try that; we might like it.

In the last several years, a few godly folks with eyes trained on *information* are perceiving correlations between current events, signs they have seen in the heavens, and warn-

Love and Message

ings they have read in the Bible. Jonathon Cahn presented to America a clear warning about America's current denial of its historic establishment by the providential hand of God. Mark Biltz presented the signs of the blood moons occurring at the defining events of the modern nation of Israel (and at the discovery of America.) Joel Richardson has compared for us the uncanny similarities between Islam and the Antichrist. Even Edgar C. Whisenant, Hal Lindsay, Tim LaHaye, and many others have stirred up an ominous sense of our current history's trundle towards a time more frightening than even Jerusalem's destruction. These men, and more, have for several decades been raising the alarm about God's determined victory over evil and Jesus Christ's return to reign in Jerusalem. God's voices of warning resound throughout all avenues of life. Every aspect of culture and politics is affected adversely by deceit and beneficially by the truth. So even Rush Limbaugh, Shawn Hannity, Mark Levin, Michael Savage and many other intellectually responsible voices have for thirty years been shining bright, well documented and evidenced educational lights on the festering growth of deceit within our current politics, fake news, biased education, and subjective culture. Many who are concerned about the observable, objective truth today are seeing, listening, considering, and talking.

This same condition of sobriety and awareness existed in the Christian communities leading into AD70. As people of fellowship, they deliberated and discussed the import of what they saw. According to Eusebius, the fourth century AD church historian, Christians did not perish at Jerusalem. They fled before the siege occurred.

> The whole body, however, of the church at Jerusalem, having been commanded by a divine revelation, given to men of approved piety there before the war, removed from the city, and dwelt at a certain town beyond the Jordan, called Pella.[9]

"Dear God in Heaven, thank you for *information*. Amen." If little Sadie had been at Pella, they would all have known exactly what she meant by that prayer. The fruitful were not God's target. God warned everyone to leave Jerusalem by appealing to eyes trained to see and ears practiced to hear. He gave signs portending the approach of this impending disaster foretold by Jesus four decades earlier and prophesied by Daniel five centuries before that. The Christians of Jerusalem perceived those signs as *information*. They left. They lived. Amen, Sadie. Amen!

Yet scoffers say, without evidence, that Eusebius doesn't know what he's talking about; Josephus was just bearing religious tales; and what does little Sadie know? But no scoffer of today was there to compare what actually happened to what Josephus reported. Even if they were there to see what Eusebius and Josephus said hold true to what happened, scoffers would still accuse and impugn their reports, because scoffers are not about knowing the truth; they are about guarding theory at all costs, even at the cost of their own eternal lives. Their minds are locked up in towers of assumptions; their eyes are yet blinded by drawn curtains. Their mouths are masked against speaking the truth. They are pinnacles of subjectivity.

God calls people to become His and know reality by its evidences, the chief

Clear Signs of Trouble and Great Joy

evidence being The Holy Word of God. In it He repeatedly calls people to come and see what He has done and will do: empirical observation and Testability. He is traveling a path to where only righteousness will exist. And He is leading His people there. This is the answer to the often asked question, "Where is God's love in all this world's suffering?" His love is in the demonstration; it is in the *information*; it is in the message, "Hear and follow;" it is in the nursery to be cleansed of the black widow. God's love is in defeating evil for the joy of those who love right. Therefore, His love is in the warning to depart from the evil which He will horrifically defeat. He doesn't scoff at us; He warns us. His love is in the warning.

Today's public mindset scoffs at the idea of God's message and refuses to move in its direction regardless of evidence. His warning is hard to receive. It threatens the way people want their things to be. So today's public obfuscates God's Word and ascribes all of the fingerprints and tracks He has upon history to poppy-cock stories. "Sure Jerusalem was destroyed," they bleat, "but come on, now! A heifer giving birth to a lamb?!" They scoff at such reports even though the possibilities of those stories are no more impossible than is the image not-made-by-human-hands appearing as it does on the Shroud of Turin. Scoffing does not arise from *information*. Scoffing does not rise from observation. It is the product of a mentality emotionally stirred up to reject anything that might controvert what it lusts to believe.

If God truly is love, and if man has truly been allotted a time to effect how his own spirit will be considered at Judgment Day by The Ultimate Judge, then God would certainly communicate man's need to appear acceptable, as well as, how to acquire that appearance. Everyone who knows God by His evidences knows He has communicated exactly that in His Holy Bible. He sprinkled The Bible with prophecies regarding all phases of His work towards crushing evil and raising up the repentant ones of mankind. One of those prophecies regards signs in the sun, moon, and stars before the scoffer-silencing, paradigm-altering return of Jesus Christ. Whether or not we are willing to think deeply enough to accept the prophecies and signs as truths by their evidences, we would be wise to believe them at least for the desirability of the more glorious propositions they make.

Nobody scoffs at the reality of death. No mortal man is alive today who lived through Moses' time. This we know for certain. But science wishes us all to believe that it also means the mental lights simply go out at death; game over; experience done. Science sells that idea, without evidence, to an enormous majority of intellectual consumers. Yet, it is a complete and utter guess. No evidence whatsoever logically bears such a conclusion. Science can only observe the decaying body. The scientific thesis of mechanistic materialism denies all spiritual realities simply because the normal scientist can not see them with his own eyes at the moment of his own choosing. Their underdetermined guessing is bought up like gold, today, for secularism comforts souls who fret about meeting God while apparently being black widows. The lights-out idea is believed, not because it is evidenced, nor because it is desired, but because it supplies emotional shelter from a God determined to eliminate evil.

Our postmodernist community casts off the empirical constraints of science and is discarding any impression that science is the anointed story-teller of all narratives, the bard

Love and Message

of all beliefs, the only process for knowing reality.

> The traditional view of science of course favors a "totalizing" narrative, one in which either the whole truth about reality is ultimately to be given, or in which the complete tool kit for predicting our future experiences can be constructed. Both of these versions of the totalizing narrative seek to subsume all stories (the "total" narrative) by employing words like "universality," "objectivity," "essence," "unity," along with "truth" and "reality." Of course these expressions are merely sticks with which to beat into submission those who dissent from scientists' (and their philosophical fellow-travels') orthodoxy. Once we recognize that these inscriptions and noises ("the truth, the whole truth and nothing but the truth") have no fixed meanings, the claims science employs them to make are open to contestability. It is only by wresting the power to influence the audience from the totalizing narrative of science that it can be replaced by other narratives, ones that will emancipate those social groups whose interests are not served by science or at least science as it has hitherto been practiced.[10]

Science's egg is beginning to crack, near to being poured onto the ultimate testing griddle -the arrival of the future. David Hume founded science's most basic claim to all knowledge upon manipulating the honest admission that most everything can not be known entirely. Folks of Hume's day thought all swans were white even though they hadn't seen them all. But, eventually the day came when that "white-swan" egg was fried by the discovery of the Australian black swan. If there were any "whole truth and nothing but the truth", it would be that we have not even begun to observe every detail of all there is to see. We can see one star rather closely, but the overwhelming rest of the stars we can not even distinguish individually. Yet we pompously assume to know everything important to know about stars. We can see the fine details of no more than what is within a few feet of us; we can hear no more than what is within a couple hundred yards; we can not touch, taste, or smell what does not contact our bodies directly (odor is caused by chemical contact with the nose). Moreover, none of those sensations can be received outside the short moment in which the sights, sounds, odors, etc. briefly existed. Empirical observation is immensely constrained to time and place. Even though we can build equipment to see far beyond the body's physical limitations, that equipment sees far short of reality's immense set of cohesive, coherent details because of those limitations. So what we've observed to be natural around us is indeed a very, microscopically small sampling of all there is remaining to be experienced -the rest of the griddle.

Everything we know about what we have not ourselves observed we know only by the testimony of others who saw, or by the principles of sound reason. But not only has nobody seen everything, everybody hasn't really seen much of anything, yet. Hume sidestepped this problem by proposing that we can rely upon our microscopically tiny experience of reality to represent the immense realms we have not yet seen. Truthfully, folks, who was Hume to say such a thing about infinitely more than he could never even have imagined experiencing? If philosophy of science were honest with its terminology, his idea would be

Clear Signs of Trouble and Great Joy

known as Hume's Subjectivity instead of Hume's Principle.

Just because we of today do not normally see what folks of the Bible times said they normally saw does not mean they didn't know what they were talking about. It only means we do not desire to know what they were talking about because they were talking about what we don't want to acknowledge as being true. Man's refusal to acknowledge what he doesn't like is the fire burning under reality's griddle coming to test the whites of science's eggs.

Messages are made of contrasts, like the printing on this page. The only reason you are able to view it is because there is enough contrast between the black of the ink and the white of the paper for the words to be seen. Then also note that, not only in this message I am writing to you, but in almost every message presented to anyone, there are contrasts between ideas which set their messages against a background of opposites. What message would have stood out by the normality of heifers bearing calves if on that ominous day the heifer led to sacrifice bore a calf? But the heifer birthing a lamb would provide enough contrast to draw the attention of anyone who saw it. Indeed, at that moment a few observant humans were close enough to see the lamb born of the heifer.

They testified.

We denied. Since no scientist was there to see it, those witnesses must not have known what they were talking about.

Right?

Regardless of our scientific standpoint, they bore witness to what they saw, and Josephus recorded it. Others bore witness to the noonday light shone upon the midnight altar, and others to the temple gates opened of their own accord, and to Jesus' seven-year rant about Jerusalem's end, and to the chariots and soldiers dashing around in the clouds. We smug mechanistic-materialists, who have witnessed none of their times or perils, we whose senses are confined to no more than the circumstances of our own few days of these present times rant against the reports of those events, since we never saw anything of that sort ourselves. How humble is that?! We have no empirical basis for denying what they reported. So we apply Hume's Subjectivity to history, too.

That you can know the nature of all human history by having directly experienced only one-third of a percent of it is not a sensible proposition. The sample size is much too small. Moreover, the sample is drawn only from the last fleeting moments of current history rather than having been randomly selected from the entire range of human experience. Only randomly selected samples are valid for honest analysis. Josephus' report could just as well have been included in the sample of real history drawn from his times. The Shroud of Turin evidences Jesus' resurrection to be another sampling of history reported in The Gospels. Chapter 9 will discuss an amazing sample of very ancient history. Honesty requires the same adjustment to the definition of Natural to be made for Josephus' report which it demands to be made for the Shroud of Turin. Science's sampling of real history is as invalid as was its C14 test of Jesus' burial shroud. Hume's Principle remains unverified in the face of controverting evidence, making its demands about Natural to be entirely subjective.

But those witnesses of Jerusalem's destruction saw what they saw. They were there. And Josephus reported what they said they saw.

So, if there is any truth that stands completely upon its own accord, it is that today's

Love and Message

observers have no say at all about what yesterday's observers saw in yesterday's moments. Today's observers can accuse yesterday's observers of not knowing what they were talking about. But they can't prove that accusation any more than they can empirically observe their "having not seen" what they reported. Mechanistic materialism only has Hume's Subjectivity for its support, nothing more. Even Hume confessed that not all of what is real can be observed. Yet, we are all K12 trained to cower before science's great "objectivity" through which the universality of Nature is "known" to be nothing more than happenstancial processes of mechanistic material's purposeless interactions. Worse yet, our postmodern community is casting off science's beloved constraint of knowledge by use of observation. Science is now little more than a bad-boy whip for herding mindless masses into political beliefs.

Maybe there is something more fun to believe than the old "eat, drink, and be merry, for tomorrow we die" narrative. We can choose to believe the fifteen centuries of testimonies written into The Bible narrative just as validly as we have chosen to believe science's hollow boasting. Or we could believe almost anything else, for that matter. Pick this. Pick that. Pick the other. We're all postmodernists now! If science has failed to prove what's true, then maybe we really can select our own reality. Maybe we really do live lives scattered through many parallel universes, or we all make this one ourselves by just believing whatever we want.

But, if we could simply change reality by what we believe, don't you think most every one would? If anyone could have his own reality by simply believing whatever he wants to believe, wouldn't the world quickly fill up with multi-billionaire kings of the universe? Or at least with independently wealthy ones? Wouldn't traffic lights always be green? But nobody can change their immediate circumstances by just changing their beliefs. Nobody can step into a different universe before disaster strikes. Nobody can still the winds by his own command. Nobody can tell death to take a holiday, although anyone through Jesus Christ can tell eternal death to get lost eternally. Something else is bigger than all of us.

The Bible speaks rationally about mankind being subject to the decisions of an Almighty God, whether it likes it or not. Signs in the sun, moon, and stars agree, whether anyone chooses to look or not. So why not believe the Bible is The Holy Word of that God since it says eternal, blissful life is ours for the believing? That would enormously change your future reality merely by how you believe today. Science has found nothing to prove the Bible's claim isn't real. Billions of mundane O^2 observers attest that it is real. Normal scientists only guess. Even though critics and skeptics accuse The Bible of deceit, their accusations are made entirely without evidence. So why not observe the infinite love and splendid plan of a God who did not count Himself above His imperiled, forlorn creation, but took part in it, lowering Himself into mankind's circumstances, taking upon His perfect self the creation's faultiness to break imperfection's power and to lead into eternally blissful life everyone who will simply believe His report on reality? It is no more preposterous a belief than is the belief that a lightning-struck mud-hole brought forth a fully functioning cell filled with unimaginably complex DNA, fully equipped with the intricate bio-machinery needed to process DNA's precision codes, a cell even given some sort of prophetic knowledge that it must reproduce because it soon will break down. Without such prescient knowledge about

Clear Signs of Trouble and Great Joy

what "that first cell" never had experienced, the one chance in a zillion lightning-struck mud-holes which did produce a cell will have been totally wasted by the time of the cell's inevitable death. The beauty of a world full of life forms deliberately created with enough intelligence to process the infinitely intricate messages of God's Word is a much more rational belief, if rationality still means anything. So, why not choose a rational belief that gives more hope than a logically impossible belief which delivers only despair?

God's Word has a great cloud of witnesses. Normal science is a cloudy bandwagon of Bible accusers. Which would you like to join? The scientific way should decide by empirical observation. But it doesn't. The forensic way looks for evidence. Science throws evidence into dumpsters. When observations and evidence are in short supply, the philosophical way is to either honestly consider what a situation needs, or to dishonestly consider what we want it to need. Normal science withholds many observations and much evidence from public view in order to create a perceived need for its pragmatic philosophies. The general public doesn't clamor against such practice because they don't know the difference between sufficient and insufficient evidence. Far too few of them think carefully enough to appreciate the concept of sufficient evidence. So, how does the general public choose between joining the cloud of witnesses or climbing onto a cloudy bandwagon?

Influence. Fad. Peer pressure. They rely on believing others either know truth or spread falsehoods. We imagine one guy speaks truthfully while we assume the other guy doesn't know what he's talking about. So how do we discern which guy knows what he's talking about and which guy doesn't? By our empathy for the guy who's been talking good about what we like and bad about what we don't? The fundamental nature of pragmatic epistemology is that simple.

Pragmatic theories wear all shades of rouge. Some ascribe the nature of a situation's circumstances to be truth's criteria: whatever is needed to make a situation work will be considered its truth. One situation might call for a "truth" completely opposite of another situation's required "truth". Who cares?! Each situation determines what's true for itself; nothing has to actually correlate with anything else; so don't ask questions! Just believe! Other theories ascribe the nature of the people involved within a situation to be truth's criteria. Whatever works for them is their "truth". But for an alarming number of people, whatever they want to believe becomes "the truth for everyone else", too, if they can fool enough voters into giving them power (or just cheat the vote if they can't). Pragmatism forms a central core for draconian governance. And pragmatists are ambitious to control all governments using "science" as the handy-dandiest corralling tool.

> Among the most influential students of science committed to the improvement of science as a social institution have been feminist philosophers of science. Some of these philosophers begin their examination of science from an epistemological insight, sometimes called "standpoint theory." This theory begins with the uncontroversial thesis that there are certain facts relevant to the assessment of scientific theories that are only detectable from certain points of view -standpoints. Sometimes the point of view or standpoint in question involves using a certain apparatus; sometimes, these philosophers argue, it requires being a woman, or the member

of a social class, or racial minority, or having a certain sexual orientation. To be interesting, the thesis needs to be given strong and potentially controversial content. It needs to be understood as claiming not merely that if a male, or a Caucasian, or a corporate executive, or a heterosexual, were in the same epistemic position as the woman or the minority or the relevant social class, the male would detect the same fact; rather, it must claim that they cannot detect such a fact for the same reason they cannot be female. The fact must evidently be a relatively complex, perhaps historical, certainly a theoretical fact not open merely to one equipped with the five senses. And feminist standpoint theorists have not been reluctant to identify such facts.[11]

Watch out men! The Mother Goddess is back in town! And she ain't smilin' no more! But wait, ladies! Rosenberg published his wit on standpoint theory in 2012, years before draconian governance legislated a man's right to be a woman if he damned well chooses to be one!

"I can be anything I want to be! And I want to be a girl!" the little boy is trained to say at school.

And you had better call him a her, too, for he…oops…she can turn you into a jailbird if you don't. Therefore, ladies, can a man-women, or can a man-women not, detect the same facts a woman-woman can? If they can't, then they're not really women. But if they can, then so can the rest of us men.

Coherence is important to the true reflection of reality.

Yet authoritarian political systems, a.k.a. draconian governments, thrive by pragmatism instead of coherence. If they need something to be true, they merely represent it as true, and their subjects had better agree if they know what's good for them. The scarier the authoritarian, the less its subjects will resist his/her demands. The Hitlers, the Stalins, the Gueveras, the Mao Tse Dongs, and the Maduros of the world made people disappear (killed them) to scare the masses into subservience. The Obamas, the Omars, Schiffs, Pelosis, Muellers, and Comeys are presently too hindered by America's Second Amendment to simply make people disappear (except for "disappearing" fifty million of your prenatal neighbors). These Progressive, authoritarian, dracos find character assassination to be almost as effective a manipulation device as Guevera and Tse Dong's pistol shots to the back of many millions of heads. By threat of character assassination, American dracos stir up bandwagons of thieves to steal not only everyone else's sovereign right to practice their own thinking, but also to censor and conceal the *information*, observations, and evidences a population needs for discerning truth themselves. They are the precise image of Draco, the snake in Eve's Garden.

In the end, the pragmatic method of "knowing" things stands intimately near everyone's need to make a decision, enticing them to choose what they desire even though good reason operating upon verified evidence might show that what they desire to be right might actually be wrong. Pragmatism is at the heart of human nature; it is the "all men are false" thing. It makes draconian governance possible. Humanity's pragmatic following after desires is how Progressivism can be so obviously incorrect even while it is being wildly popular.

Clear Signs of Trouble and Great Joy

If everyone had Mr. Spock's ability to mind-meld, if everyone knew what everyone else knew, deception would lose all of its utility. Deception thrives by hiding *information*. The utility of pragmatic "truth" selection extends from the belief that somewhere there is someone who knows all of the evidences which verify a chosen idea to be true, even though evidence of its truth may not exist at all. The belief in a woman's right to kill our prenatal neighbor rests upon the "knowledge" that everyone understands a child to not become human until it is born, regardless of the reality that everyone knows "fetus" is only a cover word for a living, prenatal human being. Another good example is belief in evolution. All of its believers are sure that someone else knows the evidence proving it. The intellectually responsible Dr. Jerry Bergman testifies from his own research that this principle operates among evolution's academicians and researchers.

> Many evolutionists acknowledge that the fossil record in their specialty lacks evidence for evolution, but maintain their faith in the theory because they believe that other specialists have shown evolution to be true.[12]

That is similar to blowing a whistle from only second hand knowledge. Dumber yet, the cultic nature of pragmatic belief systems actually blocks the individual's response to empirical observation and destroys his objectivity. Standpoint theory illustrates the cultic nature of every human social institution. It denies the possibility that anyone else can make objective observations on the one pragmatic proclamation that only the members of its exclusive club can know what affects that club's members, even though everything indirectly affects everyone, and even though every club is always "the exclusive club" of its own members. Standpoint theory is the epitome of group-think led by the pinnacle of arrogance. Dr. Bergman offers an experienced perspective on the same cultic nature of science.

> At the time I was going through this transition from skeptic to believer, I was a professor at Bowling Green State University in Ohio. My colleagues were very unhappy about the direction my research was taking me, and they made it very clear that they disagreed with my conclusions. I had many long discussions with them and again experienced the same response that I experienced from the [Jehovah's] Witnesses -many were "true believers" in the full sense of the word and were simply unwilling to look at the evidence. In truth, most knew *little* about either creationism or evolutionism except that evolutionism was "true" and creationism "only religion."
>
> Furthermore, they were not interested in learning much about the evidence. Why should they? They already knew evolutionism was true, so why read about it? My critics invariably questioned my intelligence. The fact that I am a member of MENSA, as are a number of creationists, had a 4.0 GPA for both my PhDs, have close to a 4.0 GPA for all five of my masters degrees, and scored in the 98 percentile for the GRE in my area did not impress them much. They stated that I was like Isaac Newton, a genius, but the last of the magicians.[13]

Love and Message

It is amazing that even the scientific community casts reliance on empirical observation out of the debate in order to dodge the obvious truth about the loving Creator. Then they have the temerity to claim empirical observation casts God out of reality. The world's mess of cultures are quickly becoming Alice in Wonderland's magic rabbit hole simply because it is human nature to deny any observation which might counter one of its uninhibited desires.

Don't go down that hole, especially when there's a bad moon on the rise. Correspondence between observations and coherent understanding drives better stakes for discovering what is real.

But observing and thinking are hard to do. Searching and learning are time consuming, and often boring. Nobody has the time or mental prowess to search and learn everything. So we must tentatively trust others to know what we have not yet discovered for ourselves. Still we must vet what they tell us by noticing how reality deals with it. Reality is like Rhinnie's jigsaw puzzle. Its pieces fit together only in one way. Getting any one of them out of place destroys the puzzle's coherence. Coherence overarches pragmatism, serving a corrective function in concert with *information* gained by observation. When pragmatism is allowed to shape coherence, the perception of reality becomes distorted and ultimately declines into the depths of insanity; eventually no pieces of evidence will find their fit within a totally pragmatic puzzle assemblage. In accord with everyone's limited ability to observe, coherence becomes somewhat of a joint endeavor. But that does not dismiss the individual's responsibility to observe, reason, and correct cultural misperceptions, at least for himself. The trust we place in group and cultural perception must be tentative, otherwise the individual risks a cultural insanity eventually overcoming the carelessly associating community of individuals.

Sadie and Rhinnie's little brother, Teddy, watches what they do and observes its affects on their happiness. Maybe he trusts them more than he should. Or maybe his sisters are more trustworthy because they honor *information* and like the way things fit together of their own accord. Maybe he does it just because he's a child. At any rate, whenever he sees them having fun, he is sure to come running, "Can I try? Can I try?" He thinks he will like it every time he sees them liking it. It's Teddy's Theory.

The ultimate authority of a communicating, loving God is found to be a more satisfying belief than all the nonevidenced guesses about ape-men, billions of years, and the nonexistence of any form of jurisprudence above mankind's own murderous governments. You can believe in evolution and no-God if you desire. You can believe in Krishna, Buddha, or even Kukulkan, if you so like. Or you can believe there is no authority higher than your own self. Whatever you believe, your mind will pragmatically weave all of your experiences into a seemingly coherent image for justifying that belief, as long as you keep any controverting *information* at bay.

But reality will continue all the same. Reality is what it is; your mind alone can't change it. It will run over your pragmatic thinking like a Mack truck loaded with evidence paying no regard to your own personal proclivities. Science honors reality by claiming the necessity of empirical observation, but then dishonors it by pragmatically holding to theories in the face of controverting observations. Truth goes only to where all evidence leads,

Clear Signs of Trouble and Great Joy

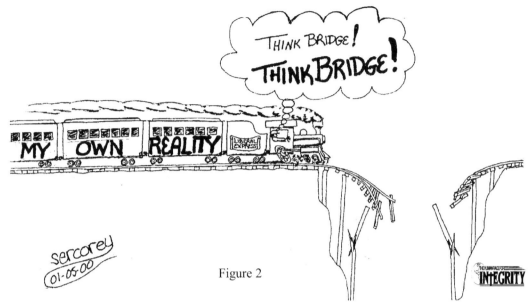

Figure 2

and therefore, its believers never plummet into the abyss of despair when theories fail to bridge reality's evidence.

Theories pragmatically denying the Bible's claims, narratives, and themes eliminate neither historical events fulfilling Biblical prophecies nor very real signs in the sun, moon, and stars. The signs remain realities of the physical universe regardless of what anyone thinks about them, correlating with fulfilled prophecy whether or not anyone cares. Both testify to the coherent warning God's Word has been delivering to a myopic mankind.

Billions of lives, having accepted the reality of His warning, evidence His love. No one knows the truth in all of its details. But anyone can acquire the ability to estimate truth well enough to live pleasantly himself and considerately towards those his life affects. Pragmatic estimation of truth isn't for making up whatever serves your self. It is for estimating what benefits everything that might be affected by a necessary guess, or by a distant approximation. Everyone can acquire the ability to discern pleasant from bitter, construction from destruction, freedom from captivity, use from abuse, etc. That old "Who's to say what's true?" is just a deceitful ruse. Ways that build up others and make life pleasant for everyone are more true than ways which destroy and hurt others to serve the self. That is Teddy's Theory at work: when you see something working well for others, try it. You might like it. His theory doesn't necessarily discover the truth. But it usually leads to a helpful clue for discovering which direction leads to the truth.

The Holy Bible has been both used and abused. Its use has worked great benefits for people. Its abuse has caused grievous destruction. Whether you believe it for the benefits its use has worked or reject it for the destruction the abuse of it has caused does not affect the integrity of God's Word. It effects your integrity. Try His Word. you will like it.

Footnotes

Love and Message

1. Johnson, Ken Th.D. Ancient Book of Jasher, Biblefacts Annotated Edition. 2008. Biblefacts Ministries, biblefacts.org. Pgs 11-12.
2. The Star of Bethlehem. Copyright by Frederick A. Larson. 2007. Presented by MPower Pictures.
3. Allen, Richard Hinkley. Star Names: Their Lore and Meaning (formerly titled: Star-Names and Their Meanings). 1963. Dover Publications, Inc., 180 Varick Street, New York 14, New York. Pg. 255.
4. Regulus lies just slightly above the solar ecliptic, the path the planets travel around the sun. Therefore, its conjunctions with Jupiter every twelve years are not precisely conjunctions, but are often just near conjunctions, i.e. they appear to the naked eye as conjunctions.
5. Allen, Richard Hinkley. 1963. Pg. 464.
6. Josephus, Flavius. The Wars of the Jews; or The History of the Destruction of Jerusalem, Book VI, Ch V, Secs 2-3. Translated by William Whiston, AM. 1960. Kregel Publications, Grand Rapids, Michigan 49501.
7. Josephus, Flavius. Wars of the Jews, Book V, Ch III, Sec 1.
8. Josephus, Flavius. Wars of the Jews, Book V, Ch XII, Sec 4.
9. Eusebius Pamphilus. The Ecclesiastical History of Eusebius Pamphilus. Translated by Christian Frederick Cruse and Isaac Boyle. Baker Book House, Grand Rapids, Michigan. Sixth printing, 1973. Book III, Chapter V.
10. Rosenberg, Alex. Philosophy of Science: A Contemporary Introduction. 2012. Routledge, 711 Third Avenue, New York, NY 10017. Pg. 262.
11. Ibid. Pg. 266
12. Bergman, Ph.D., Jerry. Fossil Forensics: Separating Fact from Fantasy in Paleontology. 2017. BP Books, Bartlett Publishing, Bartlett publishing.com. Pg. 5.
13. Sharp, Doug and Bergman, Dr. Jerry. Persuaded by the Evidence: True Stories of Faith, Science, & the Power of a Creator. 2008. Master Books, PO Box 726, Green Forest, AR 72638. Pg. 48.

Clear Signs of Trouble and Great Joy

Chapter 4
Mazzaroth

³² Can you lead forth the Mazzaroth in their season,
or can you guide the Bear with its children?
³³ Do you know the ordinances of the heavens?
Can you establish their rule on the earth?
Job 38:32-33

¹ The heavens are telling the glory of God;
and the firmament proclaims his handiwork.
² Day to day pours forth speech,
and night to night declares knowledge.
³ There is no speech, nor are there words;
their voice is not heard;
⁴ yet their voice goes out through all the earth,
and their words to the end of the world.
Psalm 19:1-4

The Mazzaroth of Job 38:32 is what we know as the Zodiac. Many Christians don't find it very comfortable to think the Bible would speak of the Zodiac in neutral terms, let alone afford it the honor of Psalm 19:1-4, since it has been the instrument of astrology for centuries. It came to us through the Greeks and to them from the Chaldeans and Babylonians. To the Greeks it was more of a mnemonic system for their mythology. The Chaldeans incorporated into it the aspects of divination. But what meaning did it have to their ancestors? Did the Chaldeans obscure a deeper, more important meaning of the Zodiac which had been passed down to them? Maybe the Greeks, enamored as they were with Mesopotamian art and culture, preserved a truer meaning of the Zodiac that was possibly better known in their times than ours. From mankind's beginning people have practiced the liberal art of twisting traditional concepts to the service of current ambitions. Is it possible that the original meaning of the Mazzaroth was twisted to serve an astrological purpose as the Zodiac? If that is its actual history, then what did the most ancient people see in the Mazzaroth?

In the early nineteenth century, Frances Rolleston endeavored to correct the record about the origin and nature of the Zodiac. She was proficient in several languages, including Hebrew and Arabic. Her admirers considered her to have been a linguistic scholar. But others think she approached her studies with subjective presumption more than objective research, and perchance, she knew several languages. She made copious notes and inferences on the constellations and stars as she searched old languages for the original meanings

Clear Signs of Trouble and Great Joy

of their names. She didn't finish her research or write her book before she died. Instead, her admirers compiled her notes into a book which is still on the market today. Soon, two renowned preachers and Biblical scholars of the later nineteenth century construed her speculations on the Mazzaroth and its meanings into a theory about the forty-eight constellations representing the gospel as God revealed it to Enoch, or Seth, or possibly even to Adam. Many other books on the gospel in the stars (as this speculation has come to be known) have since been added to those first two.

But the problem is, according to Dr. Danny R. Faulkner, hardly anyone has contrib.- uted information beyond what Ms. Rolleston first presented in her notes,

> The gospel in the stars thesis owes its origin to what we have from Rolleston, and very little scholarship has been done since, so the scholarship foundation of the gospel in the stars is Rolleston's scholarship…And Rolleston's scholarship does not favorably meet up to modern standards.[1]

Even I noticed an annoying level of similarity between the books and papers available on the topic and those of Rolleston's *Mazzaroth: Or the Constellations*, while reading E.W. Bullinger's *The Witness of the Stars*, and Joseph E. Seiss' *The Gospel in the Stars* (the aforementioned Biblical scholars). But even more annoying about these books is their common practice of presenting different meanings for the same words occasionally used as names for various stars. For example, according to Seiss and Bullinger, the Hebrew word, *Deneb,* naming a star in Capricorn means "the sacrifice comes". But of a star in Cygnus, this same *Deneb* means "the Lord and Judge comes". When I consulted a Hebrew dictionary for the meaning of *Deneb,* I found it simply means "tail" rather than any idea of "sacrifice, "Lord and judge", or "comes". Rolleston and company are at least consistent in rendering "comes" from both applications of *Deneb* in those two different constellations. But that only means they're more consistently wrong; the Hebrew word means "tail", not "comes". Rolleston and her followers also find many different star names all implying the same idea. They say the name, *Al Zimach,* of a star in Virgo means "branch". But in Bootes, *Al Katurops* is "the branch, treading underfoot". *Kornephorus* in Hercules (The Kneeler) means "the branch kneeling". *Enif* in Pegasus is "the branch" according to Seiss and Rolleston, but it is "the water" according to Bullinger. And, of course, "the branch" in Cassiopeia is *Caph*.

By the time I had read a variety of books on the Mazzaroth, I wasn't trusting anybody to tell me what those old star names meant. And I still don't. It is my nature to check the background of concepts sufficiently before considering them, let alone, before thinking with them. In this gospel-in-the-stars business, it is rare to find agreement between the meanings of star names perceived by the Rolleston school and their meanings as perceived by other sources. Faulkner also noticed this problem.

> In summary of Rolleston's methodology, she did not, in spite of what many of her supporters claim, find her meanings in ancient sources. Instead, she assumed that Hebrew was, or was closest to, the original Adamic language.

Mazzaroth

With this assumption, she proceeded to look for Hebrew homophones in star and constellation names in various languages...while claiming or at least implying that she had found these meanings in ancient texts...The earliest sources Rolleston listed are medieval; none are ancient...If she had truly found these meanings preserved in ancient texts, she utterly failed to document them. This is strange, for it was standard practice in scholarly works to document things carefully in the early nineteenth century. Thus, her scholarship in this matter is seriously lacking.[2]

But William Banks prefaces his book, *The Heavens Declare: Jesus Christ Prophesied in the Stars* with thrilling praise for Ms. Rolleston,

> Miss Rolleston was a scholar in every sense of the word. She was a classicist, a linguist, and one who worked and consulted with many of the most educated people in England during the 1800's. She also corresponded with many other experts from around the world.
>
> Frances was a gifted student who became deeply committed to her studies and began diligently studying Hebrew in 1818 and continued as a student through 1835. She read daily from the original Hebrew Scriptures even after completing her formal study of the language. Among her other exceptional accomplishments were the following: she studied Latin and Greek doing her own translations of Homer, and other Greek poetry ("just to keep my hand in"); she also mastered French, Egyptian, (Coptic and Hieroglyphics); the Semitic root languages of Arabic, Syriac, Aramaic, Sanskrit, and Chaldee; and became an authority on the origin and roots of language.[3]

Maybe she became such a linguistic authority that she didn't need to consult a Hebrew dictionary on the meaning of *deneb*. But Dr. Faulkner's analysis is more technically articulated than is Banks'. Although the possibility those old star names held meanings for the ancients similar to the ones Rolleston suggested cannot be summarily dismissed, Dr. Faulkner's proposition seems the more probable.

Regardless, the idea that the forty-eight constellation images display the gospel seems to stand on its own merit without requiring any support from those funky star names. The seamless coherence of gospel themes spanning the Zodiac's anciently developed images is far more impressive than any face of Jesus found on burnt toast, potato chips, or in spilt Cracker-Jacks. The various concepts comprising the Biblical themes of the gospel are very parallel to the essence of the forty-eight constellation images as they were perceived by the fourth century BC Greeks. Even though those images extend from Mesopotamian, Greek, and Latin mythologies, the abstracts of their illustrious characters and their portrayed actions match the abstracts of every gospel concept.

Art expresses what bewilders science, even though most cosmology is more artistic than scientific. But the gospel bewilders science far more than any art. Let's consider the gospel themes the Rolleston school sees in the artwork of the constellations as the fourth century BC poet Aratus describes them. His was the earliest definition of the complete Zodi-

Clear Signs of Trouble and Great Joy

ac as we know it today. Then we will consider whether the possible source of Rolleston's perceived themes might be The Artist Of The Constellations, or just her own imagination.

It will also be instructive to bear in mind the commonality of the rudimentary sky observatory in prehistoric cultures around the world. Stonehenge is by far not the only one. They are found sprinkled across every continent, even if their construction is nothing more than a few purposefully placed stones. Gobekli Tepe is recognized as the oldest (9600BC) and the most mysterious. And although the various zodiacs of the different civilizations that laid out those observatories are neither identical nor complete in all their symbols, they bear more similarities with one another than just their all being constructed of stone circles. We will consider the imagery of the Greek and ancient Middle Eastern zodiacs while evaluating the possibility that the Mazzaroth was developed to illustrate the story of the gospel, since those images are all elements of that area's history.

The following demonstration of the gospel in the Mazzaroth closely follows the Rolleston school of thought. I have expounded and clarified many of its attributes where *information* and sound logic have found good reason for so doing. This demonstration will show how seamlessly the essential elements of the gospel overlay every feature of the Zodiac as it has been displayed since the fourth century BC, how anciently the essence of many of those elements were perceived in the constellations long before even then, and therefore, how the Zodiac very likely originated as The Mazzaroth revealed to ancient man, much the same way Daniel's seventy weeks of years were revealed through God's inspiration. Modern science will scoff at this whole idea, but only by its own subjectivity. As you read the following, note the high degree of correspondence between the essence of the Zodiac's features and the elements of the gospel, and notice the great coherence it all makes.

Virgo proposes the idea of propagation -nature's call upon every individual for the survival of its species. The female seems always to be the metaphor of propagation. The early Sumerians and later Assyrians saw Virgo as a plowed field, "The Furrow", clearly the essence of preparation for the planted seed. But since its early days, prior to the current era, Virgo has longer been viewed as a virgin holding in her hand a head of wheat, which, of course, is seed. We met Sala with her head of wheat in association with the constellation, Virgo, in Chapter 3. The Hurrian origin of her name is quite interesting.

Figure 3

The Hurrians were an ancient culture in the northern area of Mesopotamia. To secular scholarship, their origin is vague. Their language is not of the Indo-European or Semitic groups. But it does have similarities with Sumerian, another ancient culture of vague origin. Where written history no longer shines light into the darkness of the past, historians commonly paste posters of ape-man's children, embellishing long yarns spun out of a tooth here and assuming an entire civilization out of a potsherd there. (I exaggerate somewhat, but my drift is accurate.) However, a brighter light shines into that past, a light which most

Mazzaroth

historians reject because it shines out of The Bible's pages.

The Bible ascribes mankind's salvation to the seed of woman. In cursing the serpent for inspiring chaos into His well ordered creation, God said,

> I will put enmity between you and the woman, and between your seed and her seed; he shall bruise your head, and you shall bruise his heel.
>
> (Genesis 3:15)

Virgo clearly illustrates the seed bearer of mankind's Savior enough to suggest the bruise-and-crush prophecy as being the origin of Virgo's concept. If it is, then the prophecy and the constellation were both carried across the Flood in Noah's mind, and possibly in at least one book. (Writing is anciently ascribed to Enoch, Noah's great granddad.)

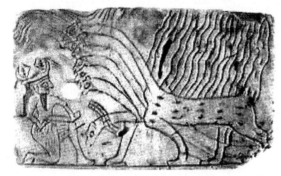

Figure 4: Sumerian seven-headed dragon with seventh head slain by a strongman.

Amongst the early Mesopotamian themes there is a strongman depicted as dispatching a seven-headed serpent by striking one of its heads. This serpent is sometimes portrayed as standing alone with a line struck through the neck of its lowest head, as if to signify the strongman's blow. We will consider the seven-headed serpent and this wound in another chapter, and we will cross its imagery again in this discussion. But, in Virgo, we can see the Seed promised to "bruise" the serpent's head. Virgo bears it in her hand, Spica, the head of wheat which provides The Bread of Life. This connection is made clear in a prophecy given to Ahaz, a king of Judah in the later part of the eighth century BC,

> Behold, a young woman shall conceive and bear a son, and shall call his name Immanu-el,
>
> Isaiah 7:14

which is a concept Revelation links to Virgo,

> And a great portent appeared in heaven, a woman clothed with the sun, with the moon under her feet, and on her head a crown of twelve stars; she was with child and she cried out in her pangs of birth, in anguish for delivery.
>
> (Revelation 12:1-2)

Virgo is the natural beginning of the Mazzaroth, since it clearly depicts Jesus' birth. The ascriptions of propagation to Virgo are also accurate, but the propagation is of eternal life rather than just fertility; it is about God's call upon one, particular individual for the survival of a human remnant.

Clear Signs of Trouble and Great Joy

The ancient concept of the weight of a soul is seen in **Libra**, the scales. In ancient Egyptian themes, scales are the prominent element of judgment. But the Egyptian and Hebraic concepts of the soul's weight are opposite. The Egyptian soul must be light as a feather to reach the blissful afterlife. The Hebrew soul must be weighty. The Egyptian scales measure for the weight of sin; the Hebrew scales measure for the weight of righteousness.

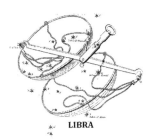

Figure 5

> [9] Men of low estate are but a breath,
> men of high estate are a delusion;
> in the balances they go up;
> they are together lighter than a breath.
>
> Psalm 62:9

According to the gospel, the weight of Jesus' perfect life sufficiently balanced the scales for all of us false humans. One of the most significant verses of the Bible compares man's soul to God's, as if measured on a scale. In accord with its significance, I cite the verse often in this book:

> Let God be true though every man be false, as it is written, "That thou mayest be justified in thy words, and prevail when thou art judged."
>
> Romans 3:4b

We all know the meaning of the Christmas' name "Immanu-el" as being "God with us". Jesus' soul weighed true to that name. The gospel identifies Him as the only man who was not Himself false, although He became everyone else's falsehood and was put to death so anyone can be trued with God through Him. He becomes weight on the scales for sufficiently paying the price necessary to eternally "propagate" our false lives thusly trued.

> Truly no man can ransom himself, or give to God the price of his life, for the ransom of his life is costly, and can never suffice, that he should continue to live on for ever, and never see the Pit.
>
> Psalm 49:7-9

THE DESIRE OF NATIONS

Figure 6

The Dendera planisphere[4] places below Virgo a depiction of a seated woman holding a child on her outstretched arm. The Rolleston school calls this constellation **The Desire of Nations**. They claim it indicated, from prehistoric times, the woman presenting her snake-head-bruising Seed to every nation of the world. But zodiacs after the fourth century BC place a shock of woman's hair, Coma Berenices, above Virgo and leave out the more ancient depiction of a woman holding forth her child. Since Coma Berenices and The Desire of Nations were not composed of the same star group, Dr. Faulkner

154

argues that Coma Berenices did not replace The Desire of Nations. But does it really matter if we call it "replace" or something else? The affect is the same: Coma Berenices is now part of our familiar Zodiac, but The Desire of Nations is not. What's up with that? We will return to that question once we've perused the entire Zodiac and gained enough perspective to note the full implication of Coma Berenices' presence and The Desire of Nation's absence. But for the topic at hand, this woman and her child on the Dendera planisphere correlates with imagery far more ancient than does the fourth century BC's Coma Berenices.

From birthing imagery found at Gobekli Tepe to the birthing leopard-lady of Catal Huyuk to the obese Mother Goddess spread across the ancient Mediterranean and Mesopotamian worlds, fertility and survival appear linked in ancient expressions when we allow the images of our minds to be painted by ape-man. But these ancient expressions do not come to us with any factory labels or user guides defining their representations to be the natural importance of fertility. Therefore, presuming they represent mere fertility is subjective. That presumption has no substantiating *information*. In fact, the later imagery developed from the ancient Mother Goddess was far more about the queen of the heavens, Inanna of the Sumerians, who becomes Ishtar of the later cultures, the Canaanite Astarte, the Egyptian Isis, and the Etruscan Uni-Astre. All of these women were about more than just fertility. Their earliest depictions were made with motifs of power, for example, the two lions seated on each side of the ancient Mother Goddess, and the leopard-lady giving birth to a horned bull. (Ouch!)

Horns became the widespread symbol of divine power throughout ancient civilization. They adorned the images of Inanna, Ishtar, Astarte, Isis, and every ancient depiction of any god. The theory about this "Queen of Heaven" idea rising from a perceived importance of make-a-baby and grow-a-crop seems a bit less intellectually responsible than perceiving its rise from the faint memory of a woman promised to bear a seed to crush (power) the serpent's evil out of mankind's affairs. We will shortly come across an even more compelling source for this systematic imagery of woman, power, and Queen of Heaven seated in the Milky Way.

Aratus wrote of Virgo in *Phaenomena*, his epic fourth century BC survey of the heavens,

> Yet in that Silver Age was she still upon the earth; but from the echoing hills at eventide she came alone, nor spake to any man in gentle words. But when she had filled the great heights with gathering crowds, then would she with threats rebuke their evil ways, and declare that never more at their prayer would she reveal her face to man...But when they, too, were dead, and when, more ruinous than they which went before, the Race of Bronze was born...then verily did Justice loathe that race of men and fly heavenward and took up that abode, where even now in the nightime the Maiden is seen of men...[5]

Aratus seems to more describe a powerful, insistent, and guiding mother than a perpetual baby machine. The woman with her son was a more ancient symbol than was Isis presenting Horus on her outstretched arm. The abstract of all these symbols is quite clear: a virgin

Clear Signs of Trouble and Great Joy

brings the King of kings and Lord of lords to mankind's remnant.

Centaurus is the well known half-man/half-horse from Greek lore. But the Egyptians were depicting the same creatures even before the Greeks existed. The Centaur came to the Egyptians from the "more ancient" Sumerians. Sagittarius, the archer, is also one of these ancient creatures come from the land of Biblical beginnings. In both can be seen Christ defeating evil, Centaurus with a lance, and Sagittarius with an arrow. But why would Christ be represented by imagery as strange as a half-man/half-horse creature? From where in Biblical concepts would such a motif arise?

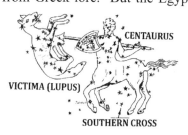

Figure 7

Ancient culture was adorned with combined creatures. The Egyptians depicted men with ibis heads, hawk heads, hippopotamus heads, etc. They borrowed the sphinx from the Sumerians, too. The Sumerians depicted lion/goats, lion/horses, lion/dragons, lion/fish, goat/fish, men-and-women/fish (mermen and mermaids), and scorpion/people, amongst many others. They put wings on lions, bulls, men, sphinxes, and centaurs. We will later comment on Pegasus, the winged horse. The Bible presents some combo creatures, too. Zechariah 5:9 describes two women with stork wings. Ezekiel saw four living creatures with bovine feet, wings, and three different animal faces in addition to the face of a man. We could generate quite a list of this stuff, but let's get back to how such a man/horse image could represent Jesus in the gospel artwork of the stars.

The Rolleston school suggests it is a depiction of Christ's dual nature. Metaphor is often ripe with multiple meanings. The ancient combo images bring the attributes of two different animals together into one entity. In the case we are addressing, the human half of the centaur sensibly depicts Christ's human nature. That leaves the horse to depict His divine nature. But from where might the connection between "horse" and "divine" come?

Consider God's ambition for Israel. He chose them to witness His eminence, providence, and protection. He brought the congregation of Israel out of Egypt "with an outstretched arm". They didn't fight their way out. God did the fighting and the lifting for them. The Hebrew word for congregation, *edah*, is the feminine form of *ed,* meaning "testimony" or "witness". God had *information* about Himself to present to the world through His relationship with Israel. It prominently shows that He defends and avenges His beloved people. He destroyed the charging Egyptian army at the Red Sea with no Israelite effort beyond Moses' raised staff. When Gideon raised an army to go against the Midianites, God trimmed its numbers back to three-hundred men who set thousands to flight. And when Sennacherib besieged Jerusalem, God disbanded his great army with a plague. God continually told Israel their hope and protection was to be in His providence. He bade them to trust His might rather than their own weapons, chariots, or horses. Yes, horses.

> [16] A king is not saved by his great army;
> a warrior is not delivered by his great strength.
> [17] The war horse is a vain hope for victory,

Mazzaroth

and by its great might it cannot save.

Psalm 33:16-17

⁷ Some boast of chariots, and some of horses;
but we boast of the name of the LORD our God.

Psalm 20:7

¹⁰ His delight is not in the strength of the horse,
nor his pleasure in the legs of a man;
¹¹ but the LORD takes pleasure in those who fear him,
in those who hope in his steadfast love.

Psalm 147:10-11

The horse was a military breakthrough for the ancients, so, it represented security in military strength. The horse served as a fitting metaphor of God's divine, protective nature of defending His people against the surges of evil, death, and despair. This concept is Christ's other nature portrayed by the military strength of the horse. I AM WHO I AM is our victory, our military, our horse. Jesus was both man and I AM. Centaurus and Sagittarius are horse/men; they both represent the I-AM-WHO-I-AM/Man.

Then, under Centaurus' flank appears the **Southern Cross**. Jesus gave His life on the cross. The strength of Jesus laying Himself down defeated death for those He lifts up. Centaurus thrusts his spear into **Victima**'s heart. Although they were Roman nails which fastened Jesus to the cross, the Jews needled the Romans into crucifying Him. At the appropriate time in His ministry, Jesus declared His divine nature, "…before Abraham was, I AM," (John 8:58). The Jewish leaders knew very well "I AM" meant the Supreme God, so they misperceived Christ's declaration, thinking it to be blasphemy, because they misperceived Him to be a mere man. Jesus understood that declaration would effect His conviction. Thus, Jesus needled the Jews into needling the Romans into crucifying Him.

Once nailed to the cross, He could have called the entire host of angels to His rescue, because, of course, He was I AM. Or He could have simply turned the cross into a Snicker Bar and smugly walked away, maybe sharing a piece of it with those taunting, Jewish rulers. But it was not His nature to shirk duty. His nature was to defeat death by the power of His perfection, which was His penultimate duty. Thus, by remaining where He could easily have left, He drove the spear through His own side; He dismissed His own spirit,

> "Father, into thy hands I commit my spirit!" And having said this he breathed his last.

Luke 23:46

In these three constellations -Centaurus, the Southern Cross, and Victima- is the portrayal of Christ sacrificing Himself on the cross. This is a vivid, compelling picture when viewed as a single scene rather than three individual, quasi-related depictions. Here, the Mazzaroth shares a characteristic with the Bible and most other literature: context indicates interpretation. A broader context of the Mazzaroth as a whole will suggest why Victima

Clear Signs of Trouble and Great Joy

might represent the slain Christ as a dog. We will get to that shortly. But here, Centaurus, Victima, and the Southern Cross depict the supreme authorship, power, and authority of all existence, I AM WHO I AM, in the form of a lowly man, subjecting Himself to the treatment of a miserable dog, hung on a Cross, the most shameful, painful way to die.

We will find this same picture carved into an archaeological record more ancient than Moses' Pentateuch, Father Abraham, and all of the Mesopotamian literature which scoffers claim the Bible plagiarized. Mesopotamian mythology indeed existed before the Bible. But it didn't inspire the Bible. The Holy Spirit existed before both, moving in the minds of humble men. Abraham humbled himself and began a great family. Then Moses humbled himself and began a Holy Library. A more ancient one was humbled to carve an archaeological record seminal to all of their concepts. The liberal Mesopotamians twisted his message almost beyond recognition. They weren't humble. But over the next several centuries, a few, humble men were conservative enough to write those more ancient concepts of verbal lore into the Bible without the subjective spin found in Mesopotamia's "liberated" literature. The Bible's consistency and coherence maintained over a fifteen-hundred year period of historical development verifies the its inspiration by God. Where have we ever found mankind being naturally humble enough to observe, maintain, and develop a seamless theme of conceptual connections over so many centuries without major cuts, pastes, whitewashes, rewrites, bickering, and infighting over every trifle? If all this indeed is the case, then the Mazzaroth also came from the Holy Spirit's purpose for the stars: signs.

Sagittarius draws His bow to dispatch the sting of death, **Scorpio**. His arrow is drawn to aim at the scorpion's red heart. Defeating evil and vanquishing the sting of death are Christ's missions. They require the supreme power of I AM's army. Thus we see in these two centaurs those two missions: 1) provide man a way to escape eternal death, and 2) eternally destroy death. As Sagittarius levels an arrow at Scorpio, **Aquila**, the eagle, is falling from the sky, having been struck by **Sagitta**, (Figure 10, next page) Sagittarius' arrow. We will see in Chapter 9 why the Bible often portrays Christ by various aspects of the eagle.

Figure 8

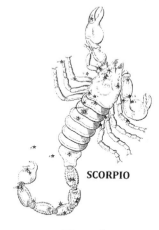

Figure 9

Here, as Aquilla, the eagle portrays Christ laying His own life down. It wasn't that Sagittarius was so bad with the bow as to miss the scorpion and hit an eagle. As Victima represents Christ's self-sacrifice at the tip of Centaurus' lance, providing man's escape from death, Aquila, struck by Sagittarius' arrow, represents Christ destroying the power of death by laying down His own life. He received the authority, right, and ability to shoot death (Scorpio) in the heart by His having accepted the most ungodly death for no sin of His own. Understanding this concept of the gospel clears up the mystery of the Mazzaroth's two centaurs. This man in all ways is

Mazzaroth

our warhorse defeating eternal death and destroying the Destroyer.

Satan, that ancient serpent, was created perfect and above all other angels. But his greatness eventually dredged up arrogance from the bottom of his heart instead of pouring his heart full of gratitude towards His Creator. Because of his beauty he lusted to elevate himself into God's place instead of praising the glories of God from his own place. He aspired to reach God's level and seize His crown of authority.

Figure 10

> You were the signet of perfection,
> full of wisdom
> and perfect in beauty.
> [13] You were in Eden, the garden of God;
> [17] Your heart was proud because of your beauty;
> you corrupted your wisdom for the sake of your splendor.
>
> Ezekiel 28:12b-13a, 17a

> [13] You said in your heart,
> "I will ascend to heaven;
> above the stars of God
> I will set my throne on high;
> I will sit on the mount of assembly
> in the far north;
> [14] I will ascend above the heights of the clouds,
> I will make myself like the Most High."
>
> Isaiah 14:13-14

Serpens, the snake amongst Scorpio's three deacon constellations, has stretched Forward[6] to seize **Corona**, the northern crown. But holding Serpens back, preventing him from reaching the crown is **Ophiuchus**, the serpent bearer. Meanwhile, Ophiuchus' foot is on Scorpio's head, hinting at who this Ophiuchus is. Aquila and Victima share a theme of death. Their sacrifice ends the sting of eternal death, the product of serving one's self, by being its opposite, the death of selfishness to live for what's right. Satan lifted himself up to put God and others under him, but his selfish function was crushed by Christ's laying Himself down to lift others up to reign with Him forever. By laying ourselves down we join Christ's nature of lifting others up. Humility defeats arrogance; selflessness dissolves selfishness. Scorpio dies by the wound of the Eagle. Satan's arrogant attempt to seize God's crown is presently restrained by Christ's

Figure 11

humble interactions with human affairs. Ophiuchus, Sagittarius, Aquila, Centaurus, The

Clear Signs of Trouble and Great Joy

Southern Cross, and Victima show Christ laying Himself down to raise us up, subjecting Himself to death on the cross for everyone who chooses to follow in His nature of a servant.

In the final analysis, God's position is that of the orderly giver. Satan's position is that of the pragmatic taker. The choice to give or take is your bet to place. Choose not to see God now and forever you will not see God, the giver of the choice. Of this He warns.

Jesus' death was that of the worst of the reviled. The religious leaders of His own people more than rejected Him. They reviled Him. They accused Him. Today the geniuses of scientific mythology, governmental tyranny, public misperception, and even many of our own friends and families often join those Jewish leaders in reviling Christ and castigating His Holy Word. It has become far more popular than worshipping Him. Christ was counted with the dogs then. He is counted with the dogs today. But the dog dying at the tip of His spear and the eagle shot down by His arrow defeated Satan's reach for the crown by doing what the revilers have no substance in themselves to do -love sacrificially.

> "Death is swallowed up in victory."
> [55] "O death, where is thy victory?"
> "O death, where is thy sting?"
> [56]The sting of death is sin, and the power of sin is the law. [57]But thanks be to God, who gives us the victory through our Lord Jesus Christ.
>
> I Corinthians 15:24b-56

Death is the wages of sin (Romans 6:23). Christ was paid wages for which He never worked. He never sinned. He was God's child born of a virgin and portrayed on the outstretched arm of the woman as The Desire of Nations. Being sinless, He could no more die than wages are paid for not working.

So the sin of all mankind was marked upon His timecard for securing a flush wage for sin paid upon a rugged old cross. Yours and my sins are tallied onto His timecard. Since wages are only paid once, should we choose to leave our sin on His card and instead work His nature of the Way, the Truth, and the Life into our being, we will have eternal life with Him rather than eternal death by ourselves. He will retain the wages of our sin if we choose to accept His receipt of our paycheck.

Yet death could not retain Him. His perfection was too powerful, righteousness being stronger than sin and order being the strength of integrity while chaos is merely a feckless scatter. By reason of truth, death only has a claim upon its earner. Christ did not earn death; we did. But we weren't paid death; He was. So death had to cough Him up. Nor is death able to keep anyone whose earnings have been paid. Therefore, from Hades He brought the keys of death for releasing everybody who leaves their sin upon His cross and their wages in His account. With the keys of death He opens Hades' gates for our resurrection unto eternal life.

> I died, and behold I am alive for evermore, and I have the keys of Death and Hades.
>
> Revelation 1:18b

Mazzaroth

This self-sacrificed and self-risen Christ is clearly seen in **The Kneeler**, known to the Greeks as Heracles and to us as Hercules. In His left hand, The Kneeler holds the prize earned by his humility and courage: Cerberus, the three headed dog, guardian of the gates of Hades. One of Heracles' twelve works took him into Hades. Christ also descended there and preached to the spirits in prison (I Peter 3:19) until He returned to life the following Sunday morning. To return from Hades, Heracles had to defeat its gate keeper, Cerberus, whom he is depicted as holding in his left hand to strike with a club raised with his right hand. Christ is described in Revelation 1:18 as having died and risen, possessing "the keys of Death and Hades". Christ is described by Revelation 1:20 as holding seven stars representing the seven angels of the seven churches in His right hand. Then maybe it is His left hand in which He holds the keys of death? Would not striking Cerberus with the messengers of the seven churches aptly illustrate the Church-spread gospel snatching souls from their fate of eternal death, The Kneeler raising the war club in his right hand to strike Cerberus in his left hand?

Remember Hercules. We will discuss this constellation more in Chapter 9.

Figure 12

The imagery developed around this constellation is strikingly parallel to the Biblical theme of Christ conquering death and then spreading salvation through the activity of His church. Both The Kneeler and Christ are not only depicted as controlling the keeper of death's gate, but The Kneeler is also depicted as crushing **Draco**'s head with His planted foot, just as God said the seed of the woman would do to the snake. With the keys of death Christ will open the gates of death for everyone who follows His Word. Then, in the final end, death and Hades will be thrown into the lake of fire with the dragon, the ultimate and final crushing of the snake's head. Truly, that is for what all the praise, celebration, and hallelujah of **Lyra** goes up!

LYRA

Figure 13

Aquila and Lyra, both being eagles, correlate with the Biblical symbolism of mankind's remnant sheltered by the wings of an eagle

> ...tell the people of Israel: [4]You have seen what I did to the Egyptians, and how I bore you on eagles' wings and brought you to myself.
>
> Exodus 19:3b-4

> [11] Like an eagle that stirs up its nest,
> that flutters over its young,

161

Clear Signs of Trouble and Great Joy

> spreading out its wings, catching them,
> bearing them on its pinions,
> [12] the LORD alone did lead [his people, Jacob,]
> and there was no foreign god with him.
>
> Deuteronomy 32:11-12

Therefore, praise for Christ the Savior mounts to heaven on the chords of Lyra, the lyre often portrayed within the body of an eagle. Although Aratus described Lyra as having been made from a turtle shell, following the myth of Orpheus, it was an eagle which carried it into the heavens. In this Greek myth it was the first lyre made. Playing it, Orpheus mesmerized everything, charming the rocks, trees, and streams.

> [11] Let the heavens be glad, and let the earth rejoice;
> let the sea roar, and all that fills it;
> [12] let the field exult, and everything in it!
> Then shall all the trees of the wood sing for joy
> [13] before the LORD, for he comes,
> for he comes to judge the earth.
> He will judge the world with righteousness,
> and the peoples with his truth.
>
> Psalm 96:11-13

The heavens are telling the glory of God without words (Psalm 19:1-4), so let's not be surprised at the earth, sea, and trees singing forth Christ's sacrificial victory, too.

The lyre was thrown into a stream when Orpheus died, goes the myth, and Zeus sent an eagle to carry it into the sky. Because of their long lives, the turtle was one of several symbols of eternal life to the ancients, therefore Lyra was strung across a turtle shell to portray the eternal nature of its melodious praise.

Because Christ's sacrifice gives us eternal life, **Delphinus**, the dolphin, is associated with Aquila and Sagitta amongst the constellations of Capricorn. Why a dolphin?

DELPHINUS
Figure 14

> In many sculptures from the East, the dolphin is associated with Atargatis, the mother goddess, goddess of vegetation, nourisher of life and receiver of the dead who would be born again. In later myths, particularly in Roman literature, and again in art and statuary, it is the dolphin that carries souls to the 'Islands of the Blest', and around the Black Sea images of dolphins have been found in the hands of the dead, presumably to ensure their safe passage to the afterlife. Taken together these references seem to point to a deeper association with the processes of life, death and rebirth, perhaps linked to the dolphin's ability to pass between the air-breathing, living world of humans and the suffocating, terrifying world beneath the waves.[7]

Mazzaroth

It has always been the dolphin's nature to accompany man's ships. So its presence was with the ancient seafarers. The dolphin became a symbol of blessing and life to the sailor, as heard in many stories and myths of them saving the lives of flailing, drowning humans.

The depiction of **Capricorn** begins as a sacrificial goat dying, and it ends as a living fish. It is one of those ancient, Mesopotamian combinations: a goat from the chest forward, and a fish from the chest back. (Nebraska Indians long ago depicted a deer being fish from the chest back.) The Rolleston school perceives Capricorn to represent the sacrifice and resurrection of Christ, His effect of eternal life for whomever will follow Him. The metaphor in the goat we can see; goats were sacrificial animals. But why now a fish to represent eternal life rather than a turtle or a dolphin?

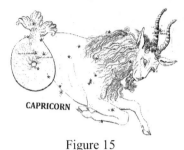

Figure 15

After Capricorn's sacrifice and resurrection are completed, **Aquarius** comes pouring a stream of water into the mouth of His fish. Aquarius is also an ancient theme about the coming Seed as are the centaur, the goat-fish, and the eagle. Aquarius' imagery reaches deep into the Sumerian past as Ea, the patron deity of Eridu, the place of the abyss. The abyss, or apsu as it was known to the ancients, was the underground, fresh water reservoir rising up to water Earth. The earliest Mesopotamians perceived this abyss to be the one welling up from the ground to water Eden, and then flowing away in four rivers to water the rest of the world. This imagery of the apsu strikingly parallels Genesis' description of both Eden and the conditions of the restored Holy City, Jerusalem, after Christ has returned to reign for a thousand years.

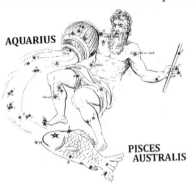

Figure 16

> [8]On that day living waters shall flow out from Jerusalem, half of them to the eastern sea and half of them to the western sea; it shall continue in summer as in winter.
>
> Zechariah 14:8

Ea is timelessly ancient. He was The Sky God mostly known as the Storm God who rode his thundering chariot through the cloudy skies. Today, he is misconceived as the progenitor of the Hebraic idea of Yahweh, I AM WHO I AM. There is no "progenitor of the idea of Yahweh". There is only a liberally twisted misconception of what was more conservatively known by the Hebrews. Sumerians, Akkadians, and Canaanites misconceived what Adam, Eve, Enoch, and Noah knew about Yahweh. The many, humble authors of the Bible were led by Yahweh's Holy Spirit to write it correctly. But our last three centuries of "astounding scholarship" guesses the Bible borrowed heavily from Middle Eastern texts.

Clear Signs of Trouble and Great Joy

If you wish to deny and scoff at the reality of I AM WHO I AM, it can certainly appear that the Bible's authors copied Mesopotamian themes. But, without a doubt, *information* evidences the Mesopotamian texts to be the embellishments, denudations, and distortions of the same ancient memories which were more accurately kept by a deeper, Sethite respect for actual history. The Mesopotamians were as consumed by their own story twisting as today's scientific revolution is consumed with making up ape-man. Both are story tellers accusing God's messages of their own falsehoods.

Ea was widely depicted with streams of water flowing from each shoulder. Fish swim up each stream. Historians typically subject Yahweh to the concept of Ea, as if Yahweh were just the next step in man's invention of some god myth. But man's falsehood was born first of simple bias, and then it grew into today's scientific stupor about the past. But God did communicate directly with Adam and Eve. He spoke revelation to Enoch. He warned Noah, commanding him to build an ark. God was not unknown to the ancients. The earliest ancestors of the ancients were more conservative than are the smarter of us today. They passed down traditions and knowledge from generation to generation to maintain a culture of truth. Their liberal descendants articulated a few aspects about God as they were passed down to them, retaining a few kernels of truth in their twisted literature, thus somewhat paralleling the truths of the Bible. But the two are not congruent. So we find the Sumerian Ea engaged in the same aspect of water production as we find Christ's Jerusalem after His return. We find Ea riding His chariot through the clouds and making rain like we find Yahweh doing the same. Although the Mesopotamian depictions are embellished and twisted from the core outwards, the myths they engage bear the same seed of truth at their core as what The Holy Bible develops into unadulterated, revealed knowledge. Such similarities are what we would expect from the testimonies of two independent witnesses about truths experienced in earlier times.

The first liberal embellishers, twisters, and denuders of Noah's known truths traveled to Eridu[8] in Sumeria, southern Mesopotamia, where they misconstrued God into this god of the abyss' streams and fishes. It really was always and still is I AM WHO I AM who owns those waters and its fishes.

> [4] There is a river whose streams make glad the city of God,
> the holy habitation of the Most High.
>
> Psalm 46:4

> [21]...there the LORD in majesty will be for us
> a place of broad rivers and streams,
>
> Isaiah 33:22

> [3] For I will pour water on the thirsty land,
> and streams on the dry ground;
> I will pour my Spirit upon your descendants,
> and my blessing on your offspring.

Mazzaroth

Isaiah 44:3

²⁴...let justice roll down like waters,
and righteousness like an everflowing stream.

Amos 5:24

As Christ raised Himself from the dead, He will raise us also, pouring forth streams of living water for those who swim in truth, as represented by **Pisces Australis**. The inextinguishable life He gives is represented by the dolphin swimming safely from the shipwreck of mutual sin upon the chaotic sea into the fish ladened streams of the Holy Zion.

The disdain in which Christians were held in the first few centuries of the current era of our Lord and the resulting persecutions heaped upon them pressed an *ichthus* out of their expressions. The *ichthus* is a fish sign early Christians used as a secret symbol of Christian identity. It is the one we see today on the trunk lids of many cars. It was a natural representation, since Christ made His followers to be fishers of men. To this day the fish symbolizes the remnant of mankind born again of the Holy Spirit into the body of Christ.

The fish was an ancient symbol of life and spirituality, as the fish swimming Ea's streams attest, as the early Sumerian goat/fish, mermaids, mermen, etc. also attest. The seven sages of Assyrian and Babylonian lore were each robed in a fish hide draped from his head. The ichthus was easy to draw in the dirt. Christ drew useful metaphors from fish. So it is easy to see why the fish became a Christian symbol. But how the fish may have risen to the representation of eternal life amongst the ancients requires a bit more thought.

Imagine for a moment you are one of Noah's family. As the Bible says, your language has few words. Your concepts have few symbols. But through your ancestors, Noah's parents, grandparents, great grand parents, etc., you know of a prospect for eternal life after death. How would you symbolically portray life which transcends death? You grew up in a very rich and beautifully forested world. It teamed with life that perished in the fleeting instant of a horrifying, global flood. All life forms had to board your ark to survive. Except the fish. They transcended the horrendous death event without any need of your ark. They offer the perfect symbol of the transcendence of death promised by the conquering Seed of the woman. And so we see Ea's fishes, and those fish skin vestments of the seven sages symbolizing concepts of eternal life.

Figure 17

The winged horse, **Pegasus**, is a deacon of Aquarius. The victorious King of kings and Lord of lords returns to earth on a white horse, descending from the heavens like Pegasus, to rule for a thousand years in Jerusalem. From the earliest depictions made by man, wings meant mobility through the heavens. Earthbound critters walk, creep, and worse yet, slither through the dust from which they came. But birds break free of the dust and soar on wings into the sky where they safely navigate the space of the heavenly host. The ancients of

Clear Signs of Trouble and Great Joy

Catal Huyuk depicted the guardians of the deceased as eagles with outstretched wings. The Bible depicts cherubim with wings. Mankind depicts angels with wings. God's victory, protection, and providence flies to the Water Bearer's fish as the likeness of a winged horse, white for the purity and righteousness it bears.

Overhead in the northern sky, the ever circling swan, **Cygnus**, soars on outstretched wings,

CYGNUS

Figure 18

> ...the lordly bird-king of the waters, in all ages and in all refined countries considered the emblem of poetic dignity, purity, and grace. By the Greeks and Romans it was held sacred to the god of beauty and the Muses, and special sweetness was connected with its death.[9]

The swan's song was a fable even before the days of Greece. It was proverbial by the fourth century BC. Ancient, urban myth believed the swan spent its life without making music until the approach of death. Then it broke into the most beautiful songs. Observation does not back up every particular of this story. But certain varieties of swans possess an additional tracheal loop in their sternum which whistles a series of notes as its lungs collapse in death. The science of it is not the metaphor. The distinction of its last song placed it in the northern sky, according to mythological imagination.

As much as metaphor is limited, it can also be vast. Jesus died when he was thirty-something. He spent ninety percent of His short life occupied with the same, ordinary, and mundane endeavors as do the rest of us: working, eating, sleeping, and laughing. But His world-changing sermons came in His final three years before death. Cryptic and enticing they were at first. Then they became gloriously revealing and pointed towards His end. His "singing" of the grand victory His Father plans for whoever follows Him did not end until He breathed its final notes on the cross, "Father, into thy hands I commit my spirit!" (Luke 23:46). Here metaphor has met its limit. Christ's swan song brought about His death; death did not bring about His song. Had He remained silent His entire life, He may have died peacefully, and alone, for the single sin of having remained silent. But He sang the glory He knew and received the company of all those who sing the same song for which the world killed Him. They are put to death with Him, but most of them not physically. Today, the characters of those who follow Christ are assassinated by a media bullyhood practiced and well prepared to slander whoever dares not join their evil dog pack. Everyone who flees the liberal character assassins into silence quietly starves on a nutrient free diet of fake belief.

But the swan ever circles in the northern sky, carrying the song of ultimate truth. Christ is eternally alive from the first to the last, and His struggle to restrain evil (Ophiuchus), conquer death (Centaurus/Victima, Sagittarius/Aquila), and vanquish the serpent (Leo pouncing upon Hydra) has continued through the centuries just as He had it planned. The more concrete meaning of Cygnus is the Seed born of woman to crush the serpent's head. But we must await Chapter 9 to be able to fully appreciate how that essence is expressed by

Mazzaroth

a waterfowl (a swan, as we know it in today's Western World).

Meanwhile, He causes His enemies to sing their swan songs. The antediluvian world did not repent at Noah's preaching; they sang their swan song. Pharaoh did not humble himself before God at Moses' command; he sang his swan song. Many prophets could not convince Israel and Judah to seek only I AM WHO I AM; those generations sang their swan song. Even America has turned to lies from the Christ of its Founding Fathers. Maybe we should be confronting those who write the fake tunes of America's swan song instead of greedily wolfing down their fake news.

To a world steeped in mystical deceit, having been drawn away from the truth by scientific stretches of the imagination and irrational passions of the heart, Christ is giving clear signs regarding the truth of His Holy Bible and the approach of His fateful and glorious return. Isn't it maybe more than coincidental that the major stars of this ever circling bird-king of the waters also form the most conspicuous cross in the heavens? Will you see with eyes to see and accept His *information* before the fullness of His reality arrives to judge fake belief? Or will His return be your swan song?

Most will sing their own swan songs while spending their own paychecks; it is profoundly sad. But the final thousand years of human history shortly to begin will be the victorious swan song of the obedient to God sung before leaving this physical universe to its fleeing end (Revelation 20:11). We should take these tunes seriously enough to sing His new song: He continuously abides with the faithful like this beautiful swan circles forever overhead.

But God's remnant of faithful people are still chained by the dust of their fleshly bodies to a physical world where natural disaster, accidents, murderers, thieves, and lying governments prowl to feed upon good hearted people and murder their children with deceit. **Andromeda** is chained to the seaside cliff where she awaits to be the great sea monster's

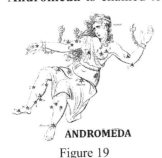

ANDROMEDA
Figure 19

afternoon snack. But God's provision still flies to her on wings grown from her belief in His reality. Our contemporaneous blessing of the Lord is knowing the future demise of this malicious, materialistic, trap of a world filled with chaotic sea monsters. While chained, Andromeda has little cause to rejoice in this material reality. Her physical circumstances are a constant state of being cornered. The church's comfort and joy is the promise and hope of an everlasting spiritual place with her Savior, not the physical conditions of a deteriorating universe. The ever returning Swan will soon release Andromeda into a time when this material Creation will reflect the orderly reign of Jesus Christ for its final thousand years.

From now to forever the Water Bearer pours His living stream into the mouth of His fish, and the Lord's provisions fly to us, Andromeda, on the spiritual wings of Pegasus while Christ ever circles over those calling upon Him.

> [7] How precious is thy steadfast love, O God!
> The children of men take refuge in the shadow of thy wings.

Clear Signs of Trouble and Great Joy

[8] They feast on the abundance of thy house,
and thou givest them drink from the river of thy delights.

Psalm 36:7-8

[2]...for you who fear my name the sun of righteousness shall rise, with healing in its wings.

Malachi 4:2

[11]Then I saw heaven opened, and behold, a white horse! He who sat upon it is called Faithful and True, and in righteousness he judges and makes war...
[14]And the armies of heaven, arrayed in fine linen, white and pure, followed him on white horses.

Revelation 19:11,14

Do you want to be a part of it? Although science failed to falsify The Word of God, science still denies the reality of I AM WHO I AM as if it had falsified them both. But that unsubstantiated denial only verifies its own subjectivity. Time is beginning to more than hint at God's truth. Irrefutable evidences of His reality and the Bible's truth are increasing with every passing year. But the ultimate way to prove the truth of God's Word is to put on its promises and try them out, to test drive belief in I AM WHO I AM and experience the resulting affects upon your life. This is not subjectivity. It is the Consistency principle applied by testing for yourself. (Funny how an objective criteria becomes subjective only because it involves the Lord God.) So what, if the PhDs and Ivy-Leaguers howl about subjectivity? They know nothing of God's things because they refuse to try the experiment according to its lab instructions. That is subjectivity. Consequently, they will forever be singing their own swan songs. Go ahead! Use Teddy's Theory. Try the experiment. You'll like it.

The Lord's remnant of Israelites who will soon turn to their King at the end of the Tribulation, and everyone who's been calling on Christ to heal the tribulation wounds of their individual lives, are the two fishes of **Pisces**. **The Band** ties Israel and the Church by their tails to the chaotic sea monster spoiling the joys and pleasures God intended for even this physical life. Who can look at the repetitive rebellions of God's Israel and the "great splendor" of The Church's Inquisition and not notice how bound to a spiritually erring sea monster they've been? When men put their heads together in order to determine which direction leads to God, most often they proceed like a pack of viscous, marauding dogs.

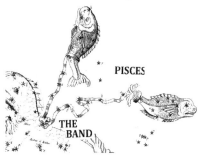

Figure 20

To the ancients the sea was a ruthless chaos. As much as Delphinus goes safely through it into eternal life, **Cetus** brings up from its churning depths a destructive force of eternal doom. The sea monster represented chaos to the ancients, the constant unraveling which only consistent, arduous effort to restore order could allay. Chaos is that stuff of physical life which refuses to work for good. It is the marauding army, the prowling beast, the sneaking thief, the lying neighbor, the contemptuous liberal, the cheating spouse, the dying child, the abortion of living children, the howling wind, the storm, the flood, the shaking ground, the city crumbling down, and the draconian government acting like it's God. It is the sprouted misperception grown through years of delusion into frightfully demented deeds. In the Bible it is called Leviathan. Although dying to sin through choosing Christ mounts our souls up to the heavens on wings of eagles, the material of the physical body yet secures our fleshy feet to the earth's shaky ground. The chaos we must presently experience tempts us to abandon future hope.

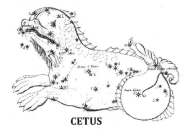

Figure 21

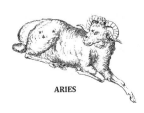

Figure 22

But **Aries**, the Lamb of God, blazed the alternate course of self-sacrifice for everyone who is no friend of Cetus. Aries dilutes Cetus' chaos with a self-controlled struggle put up by the committed pursuit of righteousness, peace, and joy in The Holy Spirit against its bands. Cetus may yet affect the lives of Aries' brethren, but he can not destroy them eternally; he has lost control over the nature and destiny of all who turn to Christ. And he's lost more than that!

Perseus makes off with Madusa's head. Christ is the promised Seed crushing the head of evil. It is the curse God pronounced upon the snake in Adam and Eve's hearing; the woman's seed would bruise the head of the snake. Christ is now the Head of all those who call on Him. That relationship frees His betrothed Andromeda, His beloved fishes, from the permanent affects of this faulty world, as Cetus dives, feet first, into the flaming Eridanus. For us who are the bound fishes and the chained Andromeda, chaos has lost its headship to Christ's perfect order, so now Christ, our Head, effects life in us through our spiritual death in Him, although our bodies yet face physical deaths, which is no problem.

Figure 23

[28]We know that in everything God works for good with those who love him, who are called according to his purpose. [29]For those whom he foreknew he also predestined to be conformed to the image of his Son, in order that he might be the first-born among many brethren. [30]And those whom he predestined he also called; and those whom he called he also justified; and those whom he justified he also glorified.

[31]What then shall we say to this? If God is for us, who is against us?

Clear Signs of Trouble and Great Joy

³²He who did not spare his own Son but gave him up for us all, will he not also give us all things with him? ³³Who shall bring any charge against God's elect? It is God who justifies; ³⁴who is to condemn? Is it Christ Jesus, who died, yes, who was raised from the dead, who is at the right hand of God, who indeed intercedes for us? ³⁵Who shall separate us from the love of Christ? Shall tribulation, or distress, or persecution, or famine, or nakedness, or peril, or sword? ³⁶As it is written,

"For thy sake we are being killed all the day long;
we are regarded as sheep to be slaughtered."

³⁷No, in all these things we are more than conquerors through him who loved us. ³⁸For I am sure that neither death, nor life, nor angels, nor principalities, nor things present, nor things to come, nor powers, ³⁹nor height, nor depth, nor anything else in all creation, will be able to separate us from the love of God in Christ Jesus our Lord.

Romans 8:28-39

When the coming trouble is finished, Jesus Christ will be the head of all the earth, restoring everything to great joy and abiding righteousness. This will be the time when Aries, the sacrificial lamb, becomes The King of Kings, **Cepheus**. He places his church, **Cassiopeia**, on His throne to reign with Him for a thousand years from His Holy City, Jerusalem. Cassiopeia sits in the sky grooming herself for her coming place of sharing the throne of her beloved Cepheus. Should we not be grooming ourselves, perpetually performing the good works God prepared for us to do (Ephesians 2:10)? Jesus said several times that those who are faithful in a little shall be given much. By dealing faithfully with what occupies us in the moment, we prepare for the throne coming soon. By being joyful in our chains, we are released into His reign. The joy of righteousness overarches the despair of evil, faith in the Truth overcomes deceit.

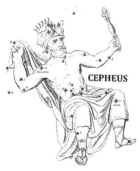

Figure 24

Figure 25

This is the portrayal Seiss and Bullinger extracted from the sense of Frances Rolleston's notes. The ancients also saw in Cassiopeia a beautiful woman enthroned.

Albumazer says this woman was anciently called "the daughter of splendor," hence "the glorified woman."[10]

But we will eventually discover the most ancient ascriptions to Cassiopeia being far less bashful than what even today's X-rated minds might comfortably perceive. Yet we will find this ascription to be the most important thing of God for man. Blessed is he who reads on through Chapter 9.

Mazzaroth

Those who lay themselves down to take up God's importance will not fear the great Day of the Lord's wrath against which nobody can stand. The righteousness of His wrath to be released upon evil will go forth like the power of a charging aurochs. **Taurus** is more than just a bull. Giant trouble will come for all Earth's little fools lifting their selves up to put I AM WHO I AM down. At the points of His horns the Aurochs will drive everything evil from the presence of God's beloved remnant who love righteousness and order instead of evil and chaos.

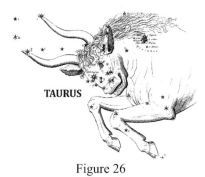

Figure 26

The aurochs stood seven feet at the shoulders. It was as fearless as it was fierce, maybe like a rhinoceros in cowhide. No evil will escape the power of The Auroch's fierce horns. Nor will any evil escape the sacrifice made of it. God specified the bull to be Israel's sin offering. So the sacrifice of a bull rolled back sin, and the horns of the bull rolled back everyone too rebellious to call upon God with a proper sacrifice. Taurus portrays the power of God over evil, forgiving the evils of the repentant and ending the unrepentant. In Chapter 9 we will see archeology itself reveal how anciently early man knew that snubbing sacrifice justified God's coming wrath.

Of Taurus, we must note before passing, from cover to cover, the Bible symbolizes power by "horns". The Hebrews were not the first to employ this symbolism. The much older Sumerian and Egyptian cultures were steeped in the idea of horns as not just being power, but being divine power. The Canaanite culture, the Hittites, and the cultures of ancient Europe also depicted power by the display of horns. We can trace this metaphor back thru Alaca Huyuk and Catal Huyuk to even the Lascaux caves of France, so ancient has this symbolism been. What happened several millennia ago to correlate such a universal theme throughout mankind which has lasted over the ages? Could Gobekli Tepe lend a clue?

God's power works for, not against, anyone who is attracted to His righteousness. Righteousness is the substance of God's relationship with His mortal followers; they desire it and He leads into it. **Auriga**, the charioteer, holds a goat and her kid in His protective relationship. Although the tip of Taurus' horn presses at Auriga's foot, he carries them safely.

Figure 27

The significance of the charioteer is surely derived in part from the importance of the horse and chariot introduced by warrior Indo-Europeans as they moved into southern Europe, the Middle East, and the Indo-Iranian land masses during the second and third millennia B.C.E. The driver of the warrior's chariot would naturally have been an extremely important figure on whom success in war depended. The chariot was the jet fighter plane of the second and

Clear Signs of Trouble and Great Joy

third millennia, and the charioteer was its pilot, a member of an aristocratic class. The relationship between the warrior and his driver was crucial and inevitably the subject of much lore.[11]

[17] With mighty chariotry, twice ten thousand,
thousands upon thousands,
the Lord came from Sinai into the holy place.

Psalm 68:17

Figure 28

The focus of God's wrath on evil can be seen in the strongman, **Orion**, raising the club in his right hand to strike the destroyer clenched by his left hand. His foot is raised to crush the head of uncleanness, Lepus the rabbit. The lion is one of a few objects from which the Bible draws metaphors representing both sides of the war between good and evil. Of course, Christ is the Lion of Judah. But Orion strikes the head of our adversary, the prowling lion.

Your adversary the devil prowls around like a roaring lion, seeking someone to devour.

I Peter 5:8b

[8] As I have seen, those who plow iniquity
and sow trouble reap the same.
[9] By the breath of God they perish,
and by the blast of his anger they are consumed.
[10] The roar of the lion, the voice of the fierce lion,
the teeth of the young lions, are broken.

Job 4:8-10

[6] O God, break the teeth in their mouths;
tear out the fangs of the young lions, O LORD!

Psalm 58:6

From under the raised foot of Orion also flows **Eridanus**, the river of God's wrathful fire, into which the chaotic Cetus is soon to be plunged. The woman's Seed did nothing to deserve a wound to His heel. By graciously accepting the unwarranted wound, He is now warranted to wound His assailant. The beast, the false prophet, Satan, and all who reject the gracious sacrifice of Christ will have been cast into the lake of fire by the end of God's Day of Judgment. By His truthfulness I AM WHO I AM prevails against all of mankind's accusations heaped upon Him (Romans 3:4).

Figure 29

Mazzaroth

[10] A stream of fire issued
and came forth from before him;
a thousand thousands served him,
and ten thousand times ten thousand stood before him;
the court sat in judgment,
and the books were opened.

Daniel 7:10

Aratus is considered the first author known to have given the name "Eridanus" to this constellation flowing from Orion's raised foot planted upon Lepus' head. The name came to him as that of the mythical river flowing into the sea north of Europe at what was then perceived to be the world's end, where there was an abundance of goods. It was the river into which the flaming Phaethon plunged after fatefully loosing control of his father's chariot, The Sun. Most of the Eridanus was dried up by the heat of Phaeton's burning wreckage, a story continuing to be told in Aratus' days of the fourth century BC.

Etymologists have not determined the origin of the name "Eridanus". The possible conjunction of Eridu, the city of the Sumerian Ea, and the Indo-European river deity, *danu* is a tempting suggestion. But the city name, Eridu, shows no trace of literary connection between Ea's time and the time of the pre-Greeks, let alone Aratus' Greeks, well over a thousand years later. By then the city of Eridu was gone, and its name was no longer written. But does this mean its name was never again spoken? Oral traditions tend to trickle through various channels over astonishing spans of time with hardly a trace but for something maybe like this river's name. Sometimes vocal conflations persist long after their original meaning is lost. But such speculation is of little help.

The Greeks construed their Zodiac from the Babylonian constellations. There was no "Eridanus" amongst those, at least none we know of today. So it is tempting to consider a possible connection with Eridu's name for another reason. The Sumerians said Eridu was the place where a band of men traveling west to Shinar (Genesis 11:2) found the apsu, the underground reservoir of water welling up to irrigate the earth. It was reminiscent of the apsu of Eden ("a mist went up from the earth and watered the whole face of the ground" Genesis 2:6), which was a Middle Eastern theme absorbed by Greek mythology as well. Under this apsu of Eridu flowed a treacherous river separating the underworld from those alive on the earth. If these concepts were mixed into the naming of Europe's mythical Eridanus, Phaethon's crash into it would seem to make some sense, disregarding its mislocation. Such an Eridanus would not be the only Mesopotamian symbol embedded in Greek mythology. The centaur, mermaids and mermen, the seven-headed serpent, and even wings on horses were Mesopotamian long before the Greeks even were.

The river of Daniel 7:10 shares a fiery essence with the lake of fire, and it flows from before the Great Judge like Eridanus flows from under Orion's foot, who is prepared to "judge" the prowling lion with His raised club. Whether or not these flaming rivers of the Bible and the Mazzaroth are imagined to flow into the lake of fire, Cetus is placed amongst the constellations as diving feet first into the river of flames with the two fishes of Pisces yet

Clear Signs of Trouble and Great Joy

attached, hoping to drag them there with him. But for every follower of The Great Judge, this band of attachment shall break at death. No fire awaits Christ's fish. The river flowing into their mouth is the refreshing, living water pouring out from the Water Bearer.

Also under Orion's planted foot is **Lepus** the hare, the uncleanness of mankind's judgments. Lepus has been represented as the Jewish sinner carrying sticks on the Sabbath.[12] We think it is wise and innocent to question the validity and straight forward sense of God's

Figure 30

Holy Word. Every question honestly seeking the truth leads to God when followed resolutely, for all things flow from God. All things are traceable to God. But the question in search of an excuse always develops another accusation of God, because all excuses flow from the self. Excuses are the product of biases multiplying like rabbits. Excuses hop down any and every bunny trail imaginable to avoid knowing God as He is. But only through pure and right purposes does careful thought find truth. Falsehood runs into the uncleanness of "any which way that works for me". Righteousness requires sharp correlation with reality. It requires a coherence of good reason. But wickedness needs only some bunny hopping down whatever trail might lead to another moment of personal gratification.

> [6]And the hare, because it chews the cud but does not part the hoof, is unclean to you.
>
> Leviticus 11:5

This constellation has also been represented as Cain sent to the moon with a bundle of sticks because he murdered Abel (Genesis 4:6-7). Lepus' association with the moon is widespread around the world. The moon is the reflective body of the two great lights; its mission is to reflect the sun onto the dark side of the earth. Mankind was created to be the reflective species; their mission is to reflect their Creator into the affairs of physical existence. But in Eden, the serpent warped their mirrors.

In the Dendera planisphere and the Persian zodiacs this constellation is a snake. We think of ourselves innocently because we reflect one another more than reflecting Christ, becoming dulled by one another's falsehood while loosing sight of His righteousness.

There at the intersection of Orion's foot and Lepus' head is the beginning of Eridanus' fiery flow.

For a thousand years Christ will reign on earth with His Church. In **Gemini** we see a twin concept. Christ and His people reigning. Or Jesus the Lamb and Christ the Lion. Adam of whom all men are and The Second Adam who is God with us, Immanuel. Peace and praise are another twin system. One twin has set down his unstrung bow. The other has set down his mace and taken up the lyre. Metaphor in these twins is multifaceted. But the striking characteristic about them is that the club we've

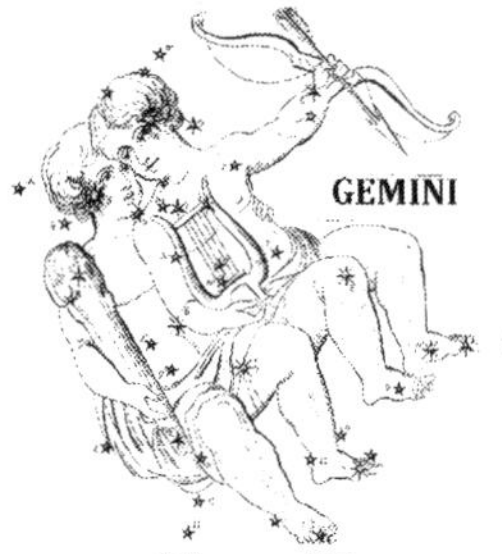

Figure 31

Mazzaroth

seen Hercules and Orion raise against death and evil is at rest, and the bow Sagittarius had loaded with arrow and cocked towards Scorpio's heart is unstrung in peace. The lyre has then been taken up. The two men are enjoying rest.

God did not create this disparaging place to continue forever. He is now working towards His predetermined goal of righteousness soon to be reached. It is the universe's purpose which Darwin denied. Christ will reign in glorious peace with the people found praising God rather than guessing against Him. St. Paul described this condition recognized in the peaceful Twins, at rest with the lyre picked up and the weapons placed down, as the kingdom of God -righteousness, peace, and joy in the Holy Spirit (Romans 14:17).

The Greeks named Sirius for its being the brightest of the stars. But this glowing star in the nose of **Canis Major** is not a single star. Consistent with the theme of the twins, it is a binary system, two stars orbiting each other. The children of Abraham shine forth the light of their Christ. They outnumber the sand which is on the beaches. We see twins in the children of his faith reconciled: Israelites and Christians. Together their testimonial lights shine for the Lord, though they are as of now treated as dogs by the majority of the world.

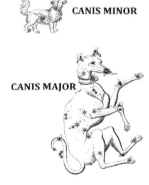

Figure 32

Procyon of **Canis Minor** is the eighth brightest star in the sky. Interestingly, it also is a binary system of two stars in mutual orbit. Again, twins. And as an aside, the Biblical meaning of the number eight is "new beginning". A new beginning is God's plan for His followers. And at the new beginning, when the kingdom of the world becomes Christ's kingdom (Revelation 11:15), Israel will have been raised to the front of the nations. The ranking of its star attests to the new beginning Israel will have when Christ takes up His reign.

The Church shines the gospel light of Christ into the world while Israel shines the testimonial light of God's faithfulness to His promises, having been brought to nationhood again in the very land God promised to their father, Abraham. Thousands of years ago God showed His prophets He would bring Israel into their land forever. Those prophecies are strung from Isaiah through Malachi like lights on a string. And they are sprinkled about the Pentateuch like stars. It is fitting, the brightest star in the night sky shines in The Big Dog, the gospel gone into the whole world, while Procyon, the eighth brightest star of the night sky, shines in the Little Dog, representing the testimony of God's faithfulness towards the new beginning of Israel's nationhood. They whisper God's purpose into our overly scientific, twenty-first century ears. Soon Christ is going to clear away the world's deceitful underbrush for Israel's new place at the forefront of all the nations in mankind's new beginning at the righteous, millennial reign of Jesus Christ.

> [4]They are Israelites, and to them belong the sonship, the glory, the covenants, the giving of the law, the worship, and the promises; [5]to them belong the patriarchs, and of their race, according to the flesh, is the Christ.

Clear Signs of Trouble and Great Joy

God who is over all be blessed for ever. Amen.

Romans 9:4-5

¹¹Then he said to me, "Son of man, these bones are the whole house of Israel. Behold, they say, 'Our bones are dried up, and our hope is lost; we are clean cut off.' ¹²Therefore prophesy, and say to them, Thus says the Lord GOD: Behold, I will open your graves, and raise you from your graves, O my people; and I will bring you home into the land of Israel. ¹³And you shall know that I am the LORD, when I open your graves, and raise you from your graves, O my people. ¹⁴And I will put my Spirit within you, and you shall live, and I will place you in your own land; then you shall know that I, the LORD, have spoken, and I have done it, says the LORD."

Ezekiel 37:11-15

Dogs are scavengers. Without regard for all they have benefited mankind, they're happy to eat garbage and roll in dead things. The Bible doesn't whitewash their nature. It usually presents dogs as licking up the blood spilt from some vile person's carcass. The dog depicts the world's attitude towards Christians and, especially, Jews. The common adjective used for a Jew is "damned". The general public doesn't think much better of Jehovah, I AM WHO I AM. They see Him as the impetuous big brat who pounded the innocent little Canaanites for burning their own children alive instead of mercifully sucking them through surgical tubes as is done in this twenty-first century full of "holy" liberals. In Lepus' way of dodging *information*, the world heaps insults on top of accusations against God and His people. Thus, the loving Christ who died so every man might live, and all those who have chosen to live with Him, as well as those to whom the Oracles of God were given are seen in these stars as the dogs, Victima (Lupus), Canis Major, and Canis Minor. And, of course, we will briefly return to the dog in Chapter 9.

God remains true regardless of every man's falsehood. At the pinnacle of truth is the faithful grasp in which He securely holds these reviled-by-the-world people who've come to Christ, courageously and humbly, amidst snorting accusations. God's clutching grasp around them is secure and unbreakable.

³⁸For I am sure that neither death, nor life, nor angels, nor principalities, nor things present, nor things to come, nor powers, ³⁹nor height, nor depth, nor anything else in all creation, will be able to separate us from the love of God in Christ Jesus our Lord.

Romans 8:38-39

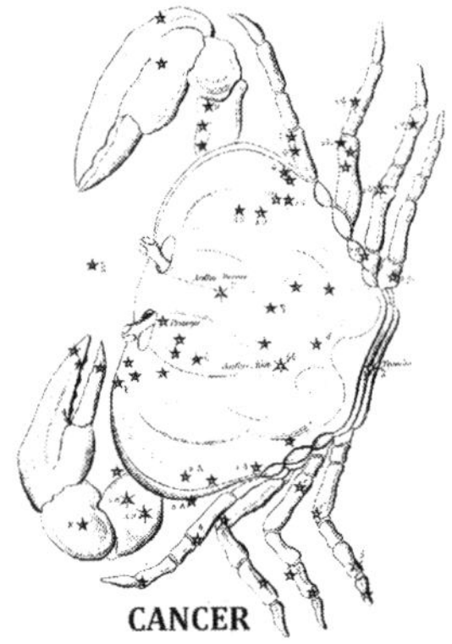

The name, **Cancer**, comes to us from the Latin for "crab". Crabs have two notable characteristics by which their name gets applied to that very bad disease: 1) they creep all over, and 2) their claws lock on very tightly, like the disease spreads throughout the body and hangs on relentlessly. By the hope of the gospel Christ creeps into

CANCER

Figure 33

Mazzaroth

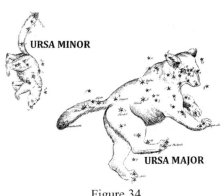

Figure 34

hearts all over the world. And He holds tightly onto those He finds.

This sense of determination is further refined in the depictions of **Ursa Major** and **Ursa Minor**, two of Cancer's deacons. Over the millennia, these two constellations have represented different ideas to various cultures. They are generally viewed as long tailed bears. But, instead of tails, the Hebrews saw cubs following mothers, or even sheep entering folds, rather than bears leading cubs. The essential idea is protection and shelter.

[32] Can you lead forth the Mazzaroth in their season,
or can you guide the Bear with its children?

Job 38:32

We all know how a bear protects her cubs. If you rob one, you had better take it far away fast. Three times the Bible uses the analogy of a mother bear robbed of her cubs to portray fierceness and resolve. One of those times regards the fierceness by which God tore Israel out of their land because they worked idolatry. The world never has understood God's wrathful resolve. It sees Him as just plain old mean and capricious. But the world looks through its own Lepus colored glasses rather than the Bible's righteous instructions. For that, its adventure of distortion will end as disastrously as did Phaethon's theft of his father's chariot.

But in the midst of God's wrath, God always reserves a safe place for those who turn to Him. It may not appear that way to us dwellers in the dust, for we consider the return to dust to be our defeat. But physical death isn't defeat; it's merely, a doorway. Death is a spirit vacating its physical body to move on to its next designated place. This place of dust is just the first of other rooms. Stepping through the door into the glory of eternal life is the safest of situations and the most blessed of bliss. Those who follow Christ through the door of death are led into eternal joy after this brief trouble of temporal life. They are glorified for willing to be seen as dogs. They win for having been willing to lose.

But those who demand their victory in the dust lose their lives for all eternity. His wrath towards accusers, liars, and destroyers is the same as our wrath towards the black widow in baby's crib. Our disposition appears mean and wicked to the black widow. She's only hungry. Baby is juicy. To the baby, mommy's wrath upon the black widow is protective, loving, wise, right, and joyful. Nor is it evil of God to give each one of us the choice of being either the black widow or the baby. It's purposeful.

God's wrath serves more than safe space to His beloved. Righteousness operates in it to procure prosperity, peace, and joy for those who cling to their Bibles. Righteousness, peace, and joy are His objective, while His wrath targets destruction, deception, and

Clear Signs of Trouble and Great Joy

unfounded accusation like a mother bear vindicating her cubs, or like a human mommy squashing a black widow (actually, like a human daddy squashing it for mommy).

God securely keeps two promises to His faithful. To those who call on Jesus Christ He promises eternal life. To the Israelites He promises a nation in this place of dust honored by all of the other dusty nations, as well as a Jerusalem with Christ ruling in it for a thousand years. Because even Christian lives, today, are also messy, and the whole world loves to hate Jews, it doesn't seem like God any longer honors these promises. But such deception is a construct of the world's paradigm, not God's. It does not extend from God's communications. Those who call on Christ (which a remnant of Jews will do by the end of the Tribulation) will outlast evil's demise and be carried across the turbulent oceans of tribulation to a thousand year haven of peace and joy, fulfilling all but the last few of God's promises. It just hasn't happened quite yet. Still, signs pointing to it are happening ever more frequently. And their clarity about kept promises increases as each passing year exposes ever more evidence of I AM WHO I AM to either convince or convict, in accord with your conclusion.

But until Christ brings His careful clingers to their desired haven, things get pretty bad. Evil is not a mist just wafting around in the air. It isn't abstract. Nor is it just a worse degree of bad. It is deceit and selfishness worked by deliberation; it is temptation deliberately not fled; it is people excusing each other of wrongdoing, excusing politicians for lying, excusing scientists for denying the incontrovertible evidences against their God-less theories while accusing truth of not being real. Evil leads away from truth with deliberation and deceives otherwise innocent people into eternal graves. It propagates exponentially within bullyhoods we foolishly call governments and throughout dog packs we ignorantly call democracies. It drives the truth about God and His pure institutions out of our schools, universities, and public spaces by hidden forces inching down slippery slopes towards Eridanus. Evil grows without suspicion until we have become the same mess we once scorned just a few decades earlier. We pride ourselves in thinking the Holocaust will never happen again. We elevate ourselves for having ended slavery. But pride comes before the fall, and elevation determines the impact. Today, public attitude embraces a sanctimonious tolerance for sin and deceit in the wake of Hitler's Holocaust, securing a parallel path for Islam to enslave the world to Allah at the tip of a sword drenched in Jewish blood. What fools are we! Islam has been proclaiming its desired extermination of the Jews for centuries. It boasts of soon scraping Israel into the sea. Yet we excuse Islam of that evil in spite of their sharing Hitler's dream of no more Jews.

> The last hour would not come unless the Muslims will fight against the Jews and the Muslims would kill them until the Jews would hide themselves behind a stone or a tree and a stone or a tree would say: Muslim, or the servant of Allah, there is a Jew behind me; come and kill him;
>
> *Sahih Muslim,*
> *Book 041, Number 6985*

All humanity will once again cave in to the serpent's deceitful hatred of God's people. The Antichrist will come with death for God's Christians and Jews, and with chains for

Mazzaroth

the world's fools.

> ⁹When he opened the fifth seal, I saw under the altar the souls of those who had been slain for the word of God and for the witness they had borne; ¹⁰they cried out with a loud voice, "O Sovereign Lord, holy and true, how long before thou wilt judge and avenge our blood on those who dwell upon the earth?" ¹¹Then they were each given a white robe and told to rest a little longer, until the number of their fellow servants and their brethren should be complete, who were to be killed as they themselves had been.
>
> Revelation 6:9-11

Flesh and blood cannot inherit the kingdom of God. Otherwise we would have no need to face the humility, pains, and often bloody mess of death. Regardless of its messiness, the deaths of those who love the truth about God and His righteousness are a reaping into glorious victory and joy. But the deaths of those comprising the bullyhood are for the worse, by far. They are reaped into the eternal flames of destruction.

The Tribulation time of trouble ends with a giant, reaping frenzy. Everywhere will be death. Only a remnant will see this trouble's end. **Bootes** has been variously seen as a hunter leading two dogs on a leash, or as a plowman leading an ox pulling the heavens around its axis, or the keeper of the bears, Ursa Major and Minor, or a farmer rushing forth to reap.

BOOTES

Figure 35

As the farmer, Bootes is depicted with a raised sickle in one hand and a spear in the other. Neither Aratus of the fourth century BC nor Ptolemy of the second century AD mentioned the sickle. It is evidently a later addition to Bootes' imagery, possibly of Christian origin. For the Tribulation will be the time of a great reaping of souls, some sent by the persecutor's sword into the joy of God's kingdom, but more gathered into the winepress of the Lamb's great wrath.

> ¹⁴Then I looked, and lo, a white cloud, and seated on the cloud one like a son of man, with a golden crown on his head, and a sharp sickle in his hand. ¹⁵And another angel came out of the temple, calling with a loud voice to him who sat upon the cloud, "Put in your sickle, and reap, for the hour to reap has come, for the harvest of the earth is fully ripe." ¹⁶So he who sat upon the cloud swung his sickle on the earth, and the earth was reaped.
>
> ¹⁷And another angel came out of the temple in heaven, and he too had a sharp sickle. ¹⁸Then another angel came out from the altar, the angel who has power over fire, and he called with a loud voice to him who had the sharp sickle, "Put in your sickle, and gather the clusters of the vine of the earth, for its grapes are ripe." ¹⁹So the angel swung his sickle on the earth and gathered the vintage of the earth, and threw it into the great wine press of the wrath of God;
>
> Revelation 14:14-19

Clear Signs of Trouble and Great Joy

The Tribulation is a time when the entire world full of anti-God rebels steps between Christ and His two, beloved cubs, Jesus' followers and the Israelites. From this precarious place they bully God's faithful to join their deceit, or die.

Much fear about the world's coming end is being spread everywhere. Every hurricane, earthquake, bloody coup, terrorist attack, and new flu virus are pumped up as signs for stocking the bomb shelter, buying, locking, and loading the firearms, and hoarding gold. It's a response coming from the fear of death, not from the fear of God. Fear of death is the trap. Fear of God is its escape. Most people will do anything to avoid the death of their living dust, except fear the Lord God. He calls us to death because it is the doorway out of this troubled world.

> [9] The coming of the lawless one by the activity of Satan will be with all power and with pretended signs and wonders, [10] and with all wicked deception for those who are to perish, because they refused to love the truth and so be saved. [11] Therefore God sends upon them a strong delusion, to make them believe what is false, [12] so that all may be condemned who did not believe the truth but had pleasure in unrighteousness.
>
> II Thessalonians 2:9-12

Placating the fear of physical pain and demise will not secure us through this quickly coming hell on earth. Security comes by braving physical threats, face to face, out of fear for an even worse and more eternal, spiritual demise.

> [4] "I tell you, my friends, do not fear those who kill the body, and after that have no more that they can do. [5] But I will warn you whom to fear: fear him who, after he has killed, has power to cast into hell; yes, I tell you, fear him!"
>
> Luke 12:4-5

Although the people of the Antichrist will be able to physically persecute God's people at will for a few years, they will personally experience a bloody place between Momma Bear and her cubs. The Tribulation is that time when the Serpent Bearer, God working the ultimate restraint of the mystery of lawlessness, will set the foolish serpent loose into the last place it should have ever desired to slither.

> [7] For the mystery of lawlessness is already at work; only he who now restrains it will do so until he is out of the way. [8] And then the lawless one will be revealed,
>
> II Thessalonians 2:7-8a

Serpens, the snake Ophiuchus restrains, will seem to have seized the royal crown which it first reached out to grasp in the Garden of Eden and has continued struggling to steal throughout history. The short spell of its seeming victory is such an important epic of

Mazzaroth

God's plan to conquer evil that a prophecy of its rise at the end was carved onto an ancient pillar. Serpen's struggle will escalate into the dark before the dawn of Christ's return. For centuries this dark spell has been perceived to be very far away. And so, today's mistake is to scoff at the clear signs of its quick coming because, for centuries, it was never more than a futuristic shadow. But Jesus told His disciples signs would indicate the nearing of this dark time before His return just as surely as He has reserved that time for Serpen's reckless abandon. The one restraining evil at the end is even prophesied on the pillar.

Separating the lovers of chaos into a place of their own consequences is this world's ultimate purpose. Dante depicted Hell to be made of levels descending into ever greater degrees of torture. Others have claimed simpler visions of Hell. Jesus likened it to the fires of Gehenna, the valley around Jerusalem where some of Judah's kings sacrificed their own children, the reason that valley was used for a perpetually burning garbage pit. Fire is a chaotic process of destruction, an apt illustration of chaos unleashed against order. But Hell's fire will not happen until the time comes for I AM WHO I AM to destroy evil and permanently imprison its source, Serpens, the accuser, deceiver, and destroyer of men.

The Tribulation is that dark before that dawn, when evil will become unrestrained before it is defeated and imprisoned.

> [3]Let no one deceive you in any way; for that day will not come, unless the rebellion comes first, and the man of lawlessness is revealed, the son of perdition, [4]who opposes and exalts himself against every so-called god or object of worship, so that he takes his seat in the temple of God, proclaiming himself to be God.
>
> II Thessalonians 2:3-4

> [27.26] Allah, there is no god but He: He is the Lord of mighty power.
>
> Koran: The Ant 27.26

> [9.29] Fight those who do not believe in Allah, nor in the latter day, nor do they prohibit what Allah and His Apostle have prohibited, nor follow the religion of truth, out of those who have been given the Book, until they pay the tax in acknowledgment of superiority and they are in a state of subjecttion.
> [9.30] And the Jews say: Uzair is the son of Allah; and the Christians say: The Messiah is the son of Allah; these are the words of their mouths; they imitate the saying of those who disbelieved before; may Allah destroy them; how they are turned away!
>
> Koran: The Immunity 9.29-30

> [8.39] And fight with them until there is no more persecution and religion should be only for Allah;
>
> Koran: The Ascension 8.39

> [27.76] Surely this Quran declares to the children of Israel most of what they differ in.

Clear Signs of Trouble and Great Joy

> [27.77] And most surely it is a guidance and a mercy for the believers.
> [27.78] Surely your Lord will judge between them by his judgment, and He is the Mighty, the knowing.
> [27.79] Therefore rely on Allah; surely you are on the clear truth.
>
> Koran: The Ant 27.76-79

The more the world population embraces the depths of deceit, the more it moves between the Bear and His cubs. Jesus did not count equality with God as a thing to be grasped even when He was in the form of God. Instead of proudly exalting Himself, instead of breathing threats of torture and beheading at everyone who refuses to accept his superiority, Jesus laid His own life down in the humiliating death of a criminal, a dog, and bade people come to Him, laying their lives down, too. By wisdom and love Christ attracts lovers of truth and humility. He leaves it to His Father to exalt Him and His followers. Every self-elevator will fall to the bottom of his own shaft and die in the deceit he spun at the rooftop of his self glory. That is trouble. But the humble will be raised to peaceful eternity for having been trampled at the bottom to where they lowered themselves in order to elevate truth. That is great joy. Because Jesus laid Himself down and dished out His perfect life as a sufficient price for everyone, He holds the right to overturn the flaming altar, **Ara**, onto a world trying to elevate itself above I AM WHO I AM.

ARA

Figure 36

The Bible is clear about Jesus' perfection. It is clear about His knowledge of the truth about God, man, and God's plans for man. The Bible is clear about Jesus working His plan for mankind's healing, and about this servant of all good and righteousness being no less than the Creator of the physical realm and the dusty life entrenched therein. Every human dies because of his own sin. But Christ had no cause of His own to die. His death was caused by the sins of others. And by His unjustified death He offers eternal freedom from sin, error, and destruction for anyone who aims his belief at what Jesus has to say about God, man, and the coming end of evil.

In a sense, then, God gives each person a chance to judge Him as being true in what He has claimed to be, or false. But the judgment every individual passes upon Christ through belief and faithfulness or disbelief and rebellion will rightfully fall upon each individual judge. If his life and thoughts judge Christ to be true, then He will receive the truth, the life, and the way of righteousness which Christ is. But if his living and thinking judges Christ to be in error, then the truth of Christ's sacrifice made on the cross for even his sins will pour the flaming wrath of that spurned sacrifice upon him for snubbing such a great and simple salvation. The fires of Ara are not poured onto Earth because God is mean, as many deceitful judges accuse. Death is a personal thing. It is earned quite personally. Ara pours out the judgment which each individual calls upon his own self by daring to deny the reality of Jesus Christ, The Savior, King, and Lord of Lords.

> [5]Then the angel took the censer and filled it with fire from the altar and threw it on the earth; and there were peals of thunder, voices, flashes of lightning, and an earthquake.

Mazzaroth

Revelation 8:5

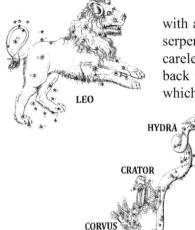

Leo, the resurrected Lion of Judah, is seen pouncing with all claws extended to slice and dice **Hydra**, that ancient serpent who brought chaos to God's Creation through Eve's carelessness. **Crater**, a bowl full of wrath, teeters on Hydra's back where it will overturn onto the arrogant nations through which the serpent slithered.

Figure 37

[8] For in the hand of the LORD there is a cup,
with foaming wine, well mixed;
and he will pour a draught from it,
and all the wicked of the earth
shall drain it down to the dregs.

Psalm 75:6

[17]Then I saw an angel standing in the sun, and with a loud voice he called to all the birds that fly in midheaven, "Come, gather for the great supper of God, [18]to eat the flesh of kings, the flesh of captains, the flesh of mighty men, the flesh of horses and their riders, and the flesh of all men, both free and slave, both small and great." [19]And I saw the beast and the kings of the earth with their armies gathered to make war against him who sits upon the horse and against his army. [20]And the beast was captured, and with it the false prophet who in its presence had worked the signs by which he deceived those who had received the mark of the beast and those who worshiped its image. These two were thrown alive into the lake of fire that burns with sulphur. [21]And the rest were slain by the sword of him who sits upon the horse, the sword that issues from his mouth; and all the birds were gorged with their flesh.

Revelation 19:17-21

Corvus, the crow, sits ready to feed on Hydra's soon to be dead carcass, craving its chicken-like flavor. Leo, Crater, Corvus, and Hydra clearly portray the theme of the Tribulation's final scene, the defeat of evil and the penultimate crushing of the serpent's head. The essence of this final scene was contained in the first prophecy given to man, the woman's Seed crushing the snake's head.

It is a theme present in most every religion of ancient origin, Zoroastrianism, Hinduism, Judeo/Christian, Islamic, and Nordic religions to name just a few, although the images expressing the theme differ. The ancient Sumerian depiction of a seven-headed dragon with one head wounded demonstrates the antiquity of this idea as eventually expressed in the Bible's Revelation to John. And here, at Corvus' feast, the concept was

Clear Signs of Trouble and Great Joy

portrayed by the Greek understanding of the ancient Zodiac four centuries before John wrote of this dragon's demise. The concept is too intertwined in world mythology to glibly claim the Bible's authors borrowed it from the Mesopotamians. Genesis' authors presented the actual history of its origin. Ancient liberals denuded, embellished, and twisted history into myths. Just because one preceded the other does not mean the other came from the one.

When all of the reaping is finished and the days of the Tribulation's storming ravages have been calmed, that good ship, **Argo**, carrying the Lord's companions safely through evil's chaotic destruction, docks in a restored world of righteousness, peace, and joy. Only a small remnant of humanity who refuse to exchange the truth about God for lies will emerge from the Tribulation's bleak trouble with life left in their dusty bodies. They will have been brought to their home port, their safe haven, as Aratus attested of this deacon constellation of Cancer,

Figure 38

> ...the ship Argo is hauled stern foremost. For not hers is the proper course of a ship in motion, but she is born backwards, reversed even as real ships, when already the sailors turn the stern to the land as they enter the haven, and everyone back-peddles the ship, but she rushing sternward lays hold of the shore. [13]

Argo is the fitting metaphor of I AM WHO I AM faithfully securing His believers through the storm into the blissful haven of righteousness. It is characteristically the ship of Jason who led the Argonauts to fetch the Golden Fleece from Colchis by which he could take his rightful place as king of Thessaly's Iolcus. The meaning of his Greek name, Jason, is "healer". Jesus is "healer" to those who go with Him aboard the Argo.

Who can deny the world is chaotic, both past and present? Even the most self controlled life is surrounded by what can not be controlled. Even the most gracious life must engage the perilous business of survival. Life carries us to joyful, spiritual heights to drop us into the depths of disaster and guilt. When bounded by chaos, our limitations are scary. Even though we who love God sail on the Argo, our houses burn down; our cars wreck. Liberals destroy our cultures and governments. We lose jobs. And we die, ingloriously sometimes.

> [23] Some went down to the sea in ships,
> doing business on the great waters;
> [24] they saw the deeds of the LORD,
> his wondrous works in the deep.
> [25] For he commanded, and raised the stormy wind,
> which lifted up the waves of the sea.
> [26] They mounted up to heaven, they went down to the depths;
> their courage melted away in their evil plight;
> [27] they reeled and staggered like drunken men,

Mazzaroth

and were at their wits' end.
^{28} Then they cried to the LORD in their trouble,
and he delivered them from their distress;
^{29} he made the storm be still,
and the waves of the sea were hushed.
^{30} Then they were glad because they had quiet,
and he brought them to their desired haven.
^{31} Let them thank the LORD for his steadfast love,
for his wonderful works to the sons of men!
^{32} Let them extol him in the congregation of the people,
and praise him in the assembly of the elders.

Psalm 107:23-32

For those who love I AM WHO I AM, He constrains chaos to the affects of only this physical place; it will not touch their eternal existence. It will not destroy their eternal government. The followers of Christ have a spiritual orderliness affecting their temporal conditions. The spiritual affects of chaos are controllable by the Lord for the good of anyone who chooses to love His ways. For those who love righteousness, He will come to calm the storms with assurances of love, relationship, and life eternal. The sea seemingly settles to a rest for everyone who is willing to believe the Lord's *Information* and put it to use, even though the storm rages. And for their belief in the victory promised, the calm of Argo's spiritual deck is carried beyond their graves into an eternal, bliss filled reality entirely without chaos. Just imagine, no more liberals.

For thousands of years wayward peoples have accused and persecuted God's faithful followers, demeaned their sentiments, and scoffed at their Father's Holy Word. You're reading this book because we who live by faith in Christ Jesus are still here, and our faithful ideas are still being expressed, and God's Bible is yet available to be read. If you wish to know by true science, then give up "your right to pronounce truth" and accept as true what is actually evidenced to be real. Give up your subjectivity and see objectively. I AM WHO I AM is the simplest statement of truth. Pick up a copy of His Bible, climb aboard the Argo, and personally observe its joyful docking in a harbor on the other side of disastrous trouble.

The best of art provides avenues in which its admirers are able to commingle their personal meanings with meanings expressed by its artists. Some Scripture is similarly artistic. The constellations are very much artistic. Their ancient admirers adorned every continent with rudimentary sky observatories. From Europe through Mesopotamia, Egypt, and India to the Mayans of Mesoamerica we find ancient representations of the sun's twelve houses. China, moreover, has anciently observed those twelve houses by peering at them through the king-planet's twelve year orbit. (We noted its affect and relevance in the previous chapter.) Among the cultures of Africa, Australia, South America, and North America are mythologies woven into commonly recognized constellations, often in subtle accord with more abstract concepts expressed throughout the ages around the world. A long lived adherence to those abstractions ascribed to the imagery of some of these constellations hints of an ancient, more traditional art, distorted a little here and somewhat washed away

Clear Signs of Trouble and Great Joy

there by generation after generation of liberals subjectively personalizing their meanings. As The Preacher said, nothing new is under the sun.

But if the Mazzaroth was designed and shown to man as being Jesus' gospel painted with stars, then it is indeed art done very, very long ago meant for expanding the message of a bruised heel and a crushed head for the generations of faithfully traditional children. Is this its history? Has the Mazzaroth's message been turned into the Zodiac by the interwoven imaginations of liberal viewers? Is today's Zodiac an eclipse of yesterday's Mazzaroth?

I have demonstrated above how the gospel's core elements naturally fit the constellations' forty-eight pictures. Many of the constellations clearly depict gospel themes as if they were designed to specifically portray them: the virgin bearing the promised Seed; the strongman holding back the snake from seizing the crown while being stung in the heal by the scorpion; the strongman returned from Hades bearing the guardian of death's gate in His hand while crushing the serpent's head underfoot; the water-bearer pouring life-giving water into the mouth of His fish; and the triumphant lion dispatching the serpent and tipping onto it the cup of wrath, then feeding it to the birds. Christ's virgin birth, His ultimate restraint of lawlessness, His victory over death, and His final dispatch of the serpent are key elements of the gospel unmistakably portrayed in the depictions of Virgo, Ophiuchus/Serpens, Hercules/Cerberus, Draco, Leo, Hydra, Crater, and Corvus. The obviously intentional correspondence of those gospel elements with these constellations illuminates subtler correlations between the rest of the constellations and other gospel imagery, such as the sufficient price measured in the scales of Libra; Christ's dual nature portrayed in the centaurs; the world's dogged treatment of Christ, His followers, and the Israelites as seen in Victima, Canis Major, and Canis Minor; and God's resolute determination to protect His people as the Great and Small Bears of the ever spreading, ever clinging crab; the victorious Perseus, who struck off the monster's head, rushes to free His fair lady from bondage; they become the King reigning with His Church as Cepheus and Cassiopeia. Even the gospel's theme of providence is present in the winged horse, as well as Christ's continuous presence portrayed by the ever circling swan, and His quick coming as the reaper, Bootes. All of the constellations together avail an unmistakably coherent representation of the gospel. All of the images described above were common depictions made by the Greeks at least three-hundred years before the gospel themes of the Old Testament were developed into the imagery expressed by the New Testament. Today, in spite of every argument raised to the contrary, the common "zodiac" imagery lends itself to the display of Jesus Christ, from A to Z, like well fit puzzle pieces, each presenting its part of a unified theme.

The details of the forty-eight constellations offer more than sufficient imagery for the world to build whatever meaning of them it desires. It's the way art works. In the Western World, we are most familiar with the Greek meanings. Muslims see different meanings. And the Hindus perceive different ones yet. But our question here is: Did God reveal this Mazzaroth to the first few generations of mankind as prophetic signs of His saving work, a message eventually denuded, embellished, and twisted by generations of liberal subjectivism? Or did the Zodiac merely emerge from man's mortal imagination with all the amazingly coincidental points necessary for it to also portray a most incredible gospel myth seemingly evidenced by an impossible to explain burial shroud and impossibly lucky guesses mas-

Mazzaroth

querading as anciently told prophecies coincidentally having come true before our own, ingeniously scientific eyes?

Without some kind of empirical evidence carved onto some relevant stone of the ancient past, or something of the sort, any belief that Enoch was given the Mazzaroth by revelation would indeed only be a pragmatic guess. All but three of the forty-eight constellations of the Mazzaroth, as we know it today, were acknowledged in Aratus' *Phaenomena* of the fourth century BC. But his *Phaenomena* is only one of many interpretations available to us; it isn't the only historical note on the constellations. Nor is it ancient, although the Zodiac's meanings came to Aratus from a deeper past.

The Biblical book of Job is far more ancient than Aratus' *Phaenomena*. It mentions the Mazzaroth, Orion, the Pleiades, Ursa Major (and/or Ursa Minor), and Bootes (the latter by the name of its brightest star, Arcturus). Many consider Job to be the Bible's earliest book written. Moses is credited with writing most of the Pentateuch, which would have been late in the fifteenth century BC. So the Book of Job is possibly older than the mid-fifteenth century BC. This would indicate not only a common knowledge of the constellations in the early second millennium BC, but also a possible connection between the Mazzaroth and God in the ideas of Job's culture. Furthermore, both Josephus and The Book of Jasher, an ancient Hebrew history, tell us Abraham's father worshipped twelve gods, one for each month of the year. Most likely, those gods were representations of the twelve main constellations. Abraham lived in the beginning of the second millennium.

Although we don't have complete representations of Akkadian, Sumerian, or earlier planispheres, many ancient symbols bear essential similarities with the figures of today's forty-eight constellations, from centaurs to goat/fish to Ea and more. Certain features on the Dendera planisphere also indicate the Mazzaroth may have had at least a twentieth century BC origin, and possibly earlier than that. The present structure of the Dendera temple is a late first century BC production with some of its additions having been completed as late as the first century AD. Although the temple was constructed upon the foundations of a much older one, many scholars believe the actual production of this Dendera planisphere embedded in its ceiling could not have been earlier than the first century. But the orientation of its four figures of Hathor, each positioned in one of the four directions of the compass, suggests a time when the constellations near those four Hathors aligned with the seasons no later than twenty-five-hundred BC. Therefore some scholars believe the Dendera Zodiac represents something very much earlier than the first century BC. It is possible that a copy of an ancient, maybe highly revered, planisphere was made for the ceiling of the later Dendera temple. There is even speculation that this planisphere might have been itself an antiquity built into the Dendera ceiling.

> The Dendera Zodiac is the only complete map that we have of an ancient sky, from Egypt in the first century BC; it shows the classical zodiac surrounded by the Egyptian constellations for the rest of the sky...But the zodiacal constellations are not shown in their Graeco-Roman forms; the shapes of the figures on the Seleucid and Dendera Zodiacs are almost identical to each other and to the boundary-stone pictographs from the second

Clear Signs of Trouble and Great Joy

> millennium BC...So the Dendera Zodiac seems to be a complete copy of the Mesopotamian zodiac.[14]

The Mesopotamian Zodiac is commonly said to be as old as the early third millennium BC. But no complete planisphere of its twelve (or eleven) constellations has been found, unless the Dendera planisphere is it. Yet there exist rudimentary pictographs of particularly interesting images from the third millennium and earlier bearing immediate similarities with the constellations.

> Prominent in these artworks are bulls and lions, and sometimes scorpions. These same animals were pictured in the sky as the earliest zodiacal constellations -Taurus, Leo, and Scorpius. We do not know when these constellations were actually defined, but it was most probably around the same time, ~3200BC. Although the main importance of bulls and lions was as real animals and as power symbols, some of these figures were decorated with stars and so may have represented the constellations. [15]

Rogers suggests the possibility of Elam being a source of these images, therefore suggesting it to be the origin of the Zodiac rather than Sumer. But this same set of images appears in sacred places which are considered to be far older than even Elam or Sumer.

Catal Huyuk was a large town in its day, some nine-thousand years ago (if you wishes to accept orthodox chronology as set by Timex rocks and Bulova bones). Scholars estimate seven-thousand people lived in this city situated in the Konya Plain of south-central Turkey. Its discovery in 1958 and excavation in the early 1960's created a stir of interest much like what Gobekli Tepe's excavation created nearly forty years later. And for the same reason. Both sites are considered Neolithic, a very important time period for evolution mythology. That mythology imagines man to have staggered through hundreds of thousands of years of blithering stupidity before tripping over the phenomenal ingenuity which landed him on the moon comparatively in the next historical moment. Evolution mythologists say the beginning of the Neolithic age transformed mankind's ignorance into brilliance when he exchanged picking his nuts and berries for sheltering his flocks and planting grain. After *information* has adjusted one's eyes sufficiently to see through this mythical fog, amongst archaeological evidence appears beads of familiar images meaningfully strung along time's thread by far more intelligence than any children of rib scratching, flea eating, nut picking ape-men could have mustered.

Catal Huyuk's buildings were replete with bulls' heads. And lions were there, too. They weren't exactly the well descript, majestic lions of later Assyria, Babylonia, Egypt, and such places. But they were large felines, leopards mostly. And The Mother Goddess was there. She is also at Gobekli Tepe, but in somewhat risqué, more abstract form. She is spread across Europe in her Catal Huyuk features, south to Malta, east to Crete,

> ...her high priestesses controlled lands, animal herds and human occupations all the way from Iraq to the Orkneys...[16]

Mazzaroth

Her symbolism of large cats shows up much later in Sumerian imagery, of which we first hear her named Inanna. Her influences even reach down to us, today, in the sense of Gaia (Mother Earth) and the woman worship of our current feminist movement. Maybe there is some sort of actual spirituality behind all of these images.

> Artists carved those statues depicting her as a naked, pregnant woman which archaeologists have found all over Europe, the Middle East, and India. The Great Mother remained imaginatively important for centuries. Like the old Sky God, she was absorbed into later pantheons and took her place alongside the older deities. She was usually one of the most powerful of the gods, certainly more powerful than the Sky God, who remained a rather shadowy figure. She was called Inana in ancient Sumeria, Ishtar in Babylon, Anat in Canaan, Isis in Egypt and Aphrodite in Greece, and remarkably similar stories were devised in all these cultures to express her role in the spiritual lives of the people.[17]

Mother Goddess figurines were found in all layers of Catal Huyuk right up to its end at about six-thousand BC. So she is not only found spread afar. She is also very ancient and enduring. Symbolism of the highly stylized statuary at Gobekli Tepe indicates a developing mythology of a great, Cosmic Mother[18] at what archaeologists consider to have been nearly twelve-thousand years ago.

Wherever this Mother Goddess went amongst the ancients went also the bull, the large cats, scorpions, and eagle/vulture/birds. It is like some intellectual net captured and drew these images together. The people imagining them exposed their dead to the vultures, (at least in their minds), to be stripped of flesh to the bone and carried to a new home in the sky. Motifs of the vulture and headless stickmen show up from ancient Gobekli Tepe in southeast Turkey to as far west as the Lascaux Cave in France. There we see a bird-headed stick in front of a stickman laying in repose, big cats, bulls, and outside the caves, a little figurine of the famous, corpulent lady. Some say Scorpio squares off with Taurus for a fight in one Lascaux cave painting.[19] And in what's called the "Venus and Sorcerer" of the Chauvet Cave is a mix of the woman, the bull (this time in the form of a bison), and big cats.

> The Venus in question is an abstract torso, hips, vulva, and legs of a full bodied woman, arguably one of the oldest known two dimensional representations of the female form in existence; it is also the only human figure to be seen anywhere in the caves. The Sorcerer is a young bison, its head overlaying the woman's belly or womb, its left front leg doubling as her left leg, indicating a special relationship between the two figures. Completing the scene is the head and upper body of a large feline, perhaps a panther or lion, which extends above and to the left of the woman's body as if the former is emerging from the latter...
>
> The strange image is unique, without anything else quite like it in the entire cave complex, and would seem to have been a central focus of the cave artists' ritual activity. Thus any suggestion that the panel might depict abstract representations of celestial objects should be taken seriously,

Clear Signs of Trouble and Great Joy

especially when we find that it could well reflect the same region of sky as Lascaux's Shaft Scene.[20]

Briefly consider that these cave paintings are said to have been made some ten to twenty thousand years before mankind developed enough mental capacity to bury a seed in the ground and wet on it. Yet these chin drooling, pre-mental mammals supposedly projected such compelling meaning onto the chaotic scatter of the stars that the images they saw in the stellar patterns are still the rustic images we see in them today?! Step aside Picasso, Van Gogh, and DaVinci! Your affects were bested millennia before your time by a bunch of flea bitten, half chimpanzees!

But the Bible says otherwise. It says mankind was brilliant from the beginning and went dumber as the years passed (to which the state of current politics bears sound witness.) Could this be why we see much of the same essence in the star patterns as the ancients perceived? Some concepts held about certain constellations must have been so important to early man that they inspired a common set of enduring images to be passed down through time.

Figure 39

Since the Lascaux Cave discovery in 1940, a possible relationship between its paintings and the zodiacal constellations has been surmised. In 1999, Chantal Jegues-Wolkieweiz, PhD teamed up with Jean-Michel Geneste to demonstrate how the last rays of the Summer solstice sunset shined through the cave entrance to light up the famous Lascaux bull painting. (We will return to this summer lighting effect in a later chapter.) Moreover, what had been noted decades before their research, what most likely had inspired their research, at least in part, were the seven dots over this bull's back where the Pleiades would be if the bull represented Taurus. Of course, those dots could be regarded as being there by mere coincidence. But there are also dots in the bull's face, too, where the Hyades adorn the face of Taurus, coincidence upon coincidence.

Assuming Carbon-14 dating is correct, which is not actually as rational an assumption as it is a pragmatic one, some researchers tell us these Lascaux Cave paintings are nearly twenty thousand years old. (Others say thirty-thousand years, which only sheds more light on the irrationality of radiometric dating.) We know practically nothing of this people's historically significant events, or so the ape-man myth instructs us to believe. We have little clue of their language, again, as we are taught to believe. And we can only imagine what they looked like, yes…taught…you get the drift…never entirely evidenced but always specifically taught. However,

> Art historians have long been delighted that the cave paintings are accurate to a minute degree in their knowledge of animal anatomy and seasonal habits of each species. But that is not what is important. What is implied is that each painting in the Hall is aligned with a corresponding zodiac constellation. "This is what we hold to be true," said Chantal Jegues-Wolkieweiz. It is the positions and relationships of the animals that indicate astronomical knowledge of the solstice positions, the constellations and the fixed stars.[21]

Mazzaroth

Rock hard evidence of that art on those cave walls points to some universal seed of meaning applied to the same constellations the ancient Anatolians, Mesopotamians, Egyptians, and Hindus all recognized. Indeed, even the North and South American Indians pointed to many of the same constellations involving the same stars, too, sometimes perceiving them in similar terms, though not identical ones. There is too much coincidental coherence between similar images spread too far across the globe amongst too many different peoples throughout too many era's to so glibly deny the Zodiac's starting point in the early history of some centralized population. The evolution myth offers no possible interconnections for explaining the spread of these symbols around the world as remnant pieces of a once, whole unit of understanding eventually broken and scattered. But the Bible offers a perfectly rational connection with a close knit culture from where such pieces of a soon broken whole might have spread across the globe: those days when the entire, human population was only Noah, Shem, Japheth, Ham, and their wives.

In the northern Aegean Sea on the island of Samothrace is a classical monument to the gods. Sailors of those times indulged an ancient Samothracian mystery religion believed to protect them from shipwreck. This monument is believed to overlay a much older, pre-Greek shrine of the Great Mother Goddess. There is a point of interest relative to her shrine.

> Also venerated were twins called Cabeiroi, believed to be protectors of sailors, but who were otherwise regarded by ancient writers with great fear and ignorance...The ancient Samothrace Sanctuary's pantheon of a Great Mother Goddess, an inferior male god and mysterious twins exhibits some striking similarities to the cult that Mellaart found in the shrines of pre-flood Catal Huyuk. Corroborating this, Cabeiroi is a non-Greek word, indicating that those who had lived on Samothrace since the time of the great sea level rise were non-Greeks and spoke an as yet unidentified pre-Greek language.[22]

The flood to which Wilson refers is that of the rising Aegean Sea bursting through the Bosporus Strait into the Black Sea. Archaeologists have found settlements established far below today's suffocating waters of that sea. Wilson proposes that the peoples displaced by the resulting flood founded Catal Huyuk, Alaca Huyuk, Gobekli Tepe, etc. In a third millennium tomb at Alaca Huyuk in northern Turkey, somewhat westerly of Catal Huyuk, twin figures were found.[23] Furthermore, Great Mother Goddess societies lived under a system of twin kings. Kings ruled through a royal lineage of priestesses, rather than through the male right of inheritance. They became king through marriage to the high priestess of the Great Mother Goddess. Both co-kings married the high priestess, and then each reigned successively over fifty lunar-month periods. (Fifty is an interestingly significant number to the later Hebrews.)

> The twinship system, obviously redolent of the mythical Cabeiroi and those twin figures found at Catal Huyuk and elsewhere, undoubtedly accounts for

Clear Signs of Trouble and Great Joy

why, in the mythology of cities and royal dynasties, it is often twins, or kings with alternating names who figure as their original founders…Such twinship always seems to be a sign of the culture having originated from a Great Mother Goddess basis, whatever the system of kingship that might later come into being.[24]

There is more confirmation of a connection, for, astoundingly, it is said of the deceased pharaoh in the Pyramid Texts: "You ascend with the head of a hawk and all your members are those of the Twins of Atum." In the later Hermetica Atum is credited with the creation of the Zodiac, in which context there would be no doubt about whom the twins of Atum are - they are Gemini! Yima in the various stories goes first into the Otherworld to become Lord of the Dead. Here we have him, in the Pyramid Texts, ascending to that place with the same bird-headed form 'the head of a hawk' that we see at Lascaux.[25]

What can be known about the antiquity of the Zodiac is underscored by not only the paintings in those caves of Lascaux, The Mother Goddess, the bull, the big cat, the scorpion, the bird, and the twins, but also by a connection the most ancient portrayals of these things have to alignments with solstices and equinoxes. At shortly before 4000BC, the Spring equinox sun rose in Taurus the Bull, the Summer solstice sun rose in Leo the Lion, the Fall equinox sun rose in Scorpio, and the Winter solstice sun rose in Aquarius the Man. In three of these constellations are the familiar faces of the bull, the lion, and the man. The fourth familiar face is that of the eagle, the nemesis of the scorpion of the Spring equinox sunrise. The interplay of the eagle and the scorpion is also most ancient. It is found chiseled onto Gobekle Tepe's Pillar 43 before it lasted through millennia of ancient Mesopotamian art. It is strikingly obvious that the cultural elements of Chauvet Cave, Lascaux Cave, Catal Huyuk, etc. were spread throughout the Mediterranean and Middle Eastern worlds to India from ancient times forward expressing the four corners of the earth and sky by these images of the bull, the lion, the man, and the victorious eagle flying the flesh of carcasses into the preserving sky after death had stung.

Should we be surprised to find these four faces on the cherubim supporting the throne of God which Ezekiel saw? They are also the faces of the cherubim John saw surrounding God's throne in heaven. What has been so compelling about those bulls, big cats, men, and eagles/vultures that they not only adorned the world's first artistic displays, but they lived on, adorning the heavens with the same imagery, and then Bible, too? If the Mazzaroth really was some form of mnemonics given by God to warn mankind of the impending defeat of death and chaos, if it did command such grand attention as to leave this set of portrayals scattered across times throughout the globe, then why has half the world, from the earliest of times, held the woman in so much religious fascination rather than the man (Aquarius, The Water Bearer) positioned in the celestial corners of the Zodiac?

Change is a generational thing, because, to children, the familiar grows old, obedience gets boring, and limits become humiliating. So tradition is for overthrowing. Conservatism is for abandoning. The inevitability of liberalism doesn't make it beneficial. Whenever Cain wants to beat Abel on the head with a rock, he'll take issue with daddy's preaching about love, calling it hatred, and accusing Abel of anything worthy of a well swung stone. As Abel lays there bleeding out, Cain will concoct his own version of love for covering the irrationality of his brother's busted head. Repenting to embrace again daddy's idea is just too passé for human nature to bear. Besides, it exposes our crimes to everyone. That's the true reason we can't turn back the clock. Liberalism is more readily apparent in history than is even idolatry. It is a major piece of the Bible's overall puzzle: God is true; men are liberals. God reveals; man twists. Since The Seed of the woman crushes the head of the serpent, surely the woman is the most significant thing to mankind! Since the woman ate from the Tree of Knowledge, she holds the wisdom of humanity. And so mankind hates the Bible and its "brat" God of The Seed.

Actually, God provided The Seed. The woman only delivered it. She was not the victor. But by the same lingering ingratitude blowing smoke at the Bible today, Adam's kids blew smoke at their daddy's experience with God then. The Genesis story reveals it. Man became so evil in just ten generations of living in accord with their own ideas that God decided a great Flood was necessary. Today, man is becoming so evil the Tribulation is necessary. Jesus said the time of evil's end would be like the times of Noah.

We can not say for certain why the ancients came away from the Garden of Eden with a mindful of Eve, but apparently, they did. Maybe she was overbearing. Maybe after the fall, Adam lost his marbles and she had to make all the decisions. Or maybe she was just able to make that seem so and usurped Adam's leadership role. Maybe being the bearer of the seed, she became regarded as the hope of mankind's return to Eden where the Tree of Life was lost to a careless bet on some snake's assumption. Adam, Eve, the Tree of Knowledge, and the snake are all represented on Sumerian cylinder seals. Where'd that come from? The ancients didn't lose their memories of God. They merely twisted them away from the truth. They stopped conserving history and began imagining liberally.

Legend has it, and ancient Hebrew histories tell of it, that Seth's descendents received the knowledge of astronomy before the great Flood. Enoch, born to the fifth generation of Seth, related how an angel showed him the constellations:

> I saw how the stars of heaven [sun and moon] come forth; and I counted the constellations out of which they proceed, and wrote down all their gates, of each individual star by itself, according to their number and their names, their courses and positions, and their times and months, as Uriel the holy angel who was with me showed me. He showed me all these things and wrote them all down for me: their names, laws, and operations.[26]

We need to be wary of the smoke blown at the Book of Enoch even by "highly knowledgeable" Christians in these modern times of endless assumptions and imagination. The Book of Enoch was well received history in the first century AD and the centuries before it. Fragments of more than one copy have been found amongst the Dead Sea Scrolls. 1 Enoch's theological reflections parallel New Testament themes. Or should we say the New Testament parallels its themes, since 1 Enoch is obviously older than the New Testa-

Clear Signs of Trouble and Great Joy

ment. Its prophecies of the Messiah closely parallel the Bible's prophecies of Him. Many of its concepts became embedded in the New Testament, and it is even quoted there. It remained popular with the Church fathers throughout the first three centuries of Christianity; they also quoted it and vouched for its authenticity. If we attribute the Bible's writing to the inspiration of the Holy Spirit, then we must honestly consider the Holy Spirit's use of 1 Enoch's many eschatological concepts as legitimizing that book's authenticity. Of course, today, we guess otherwise.

Regardless, most of 1 Enoch is likely Enoch's work. Knowledge of several constellations as early as the seventh generation from Adam probably was common. Yet the Book of Enoch speaks nowhere of any gospel spelled out by those constellations, unless that idea might be either the "laws" or the "operations" referenced in the above quote. In the forty-third chapter of Enoch (curiously), the angel referred to "pictures", but the context is not completely clear that he was referring to the constellations. The text of 1 Enoch seems to merely present their motions somewhat scientifically stated. But at least the suggestion of these constellations representing rudimentary elements of the gospel given to Enoch by angels is warranted by the imagery in which Enoch conceptualizes God. Basic gospel concepts are found in 1 Enoch: sinners weighed in the balances and found wanting by the Lord; salvation made available by a messiah born of a virgin; His blood shed for salvation; His resurrection; our resurrection; God's gift of eternal bliss.[27] Abstracts of those ideas are found sprinkled around the Middle East as if carried to us from the heavens by some Mercury/Hermes/Thoth/Enoch character. 1 Enoch acknowledges the constellations. It acknowledges the elements of the gospel. It contemplates both concepts together. But did Enoch mix those concepts into the Mazzaroth's artwork? We will eventually discover that by the time of the Flood, someone did.

Considering the sizable stakes the Bible raises for our gamble on what lies beyond death, it behooves us to play our cards only after carefully considering ALL of the available *information*. While *information* is the smoke of reality, the wager is about reality's shape in the hereafter. Allowing atheistic assumptions to wipe volumes of explanatory *information* about the past from our minds is a vastly foolish way to bet your hand in the face of stakes so immense as eternal bliss, unending agony, or even that famously assumed "lights out".

Biblical concepts, the constellations, those ancient cave paintings, the Mother Goddess, her big cats, bulls, scorpions, the eagle/vulture/birds, the twins, and of course, snakes everywhere seem to consistently step forth from a deep, prehistory. Traditional meanings surrounding these ancient symbols faintly imply a wide ranging distortion of some earlier, more coherent understanding. Every generation has had its liberal wave of rebellion against the conservative traditions of its ancestors, even if the rebellion has amounted to little more than inwardly held contempt. So we find these symbols everywhere, often having essentially similar meanings even though the narratives rather differ between cultures, just as one would expect from a once coherent theme having been twisted into an abundance of divergent perceptions.

It appears from the evidence that, as the first few generations developed after the great Flood, animosity grew against I AM WHO I AM, maybe out of the horror such a great flood would have evoked. It would have settled into the descendant leaders of a very

Mazzaroth

small world. It would have provoked in them the same self dependent arrogance our generation wallows in today, replacing truths carefully designed into ancestral imagery with its own intoxicating assumptions. They built the tower of Babel for security. We build the tower of mechanistic materialism. They ran from the God who washed their world clean. We run from the Christ who washes our souls clean. Liberalism has been the world's most enduring mental disorder, as Dr. Michael Savage aptly informs us.[28]

The Zodiac has been its most enduring anthology. Enoch has been a most ancient messenger. So it is compelling to see the gospel expressed in those constellations.

John Rogers explained the Mesopotamian symbols as progenitors of a zodiac emerging in the fourth and third millennia BC. The symbolism in the Lascaux Caves and Goebekli Tepe show elements of a zodiac existing earlier than the Mesopotamians. Whether God revealed the Mazzaroth to Seth or Enoch, complete with a gospel meaning that was later lost to overbearing mythologies, or whether the Mazzaroth was developed by moronic ape-men over a great period of time into imagery remarkably capable of portraying the gospel, the reality remains that the gospel indeed can be clearly seen in these forty eight constellations. They are Biblically descriptive right down to the consideration of that one, lately usurping constellation -**Coma Berenices**.

As we noticed earlier, the Seed of the woman, the Desire of Nations, was replaced

COMA BERENICES

Figure 40

by Bernice's Hair (Coma Berenices) in the fourth century BC Zodiac. The worship of the Great Mother Goddess replaced ancient man's worship of the Creator, our Savior. The Book of Jubilees (8:1-5) says Ham's son, Canaan, rediscovered the secrets of pre-Flood astrology and taught them to his children. Canaan's nephew, Nimrod, rebelled against God and built The Tower of Babel. Ham mocked Noah. Something went askew in Ham's family. They rejected Enoch's traditions brought through the Flood aboard Noah's ark. The ancient Australian mother goddess and rainbow serpent are revealing tracks along a clear trail of liberal flight from the earlier traditions which shaped Biblical narrative. Evidently, early in mankind's new beginning, worship of and honor paid the woman and the serpent replaced The Seed of the woman promised to crush the serpent. She became Inanna, Ishtar, Astoreth, Isis, etc.

The Bible defines worship of anything other than I AM WHO I AM as idolatry. It portrays idolatry as harlotry, wayward womanhood. The Bible makes no bones about the fact that the woman sinned first (I Timothy 2:13, if anyone still cares). The Great Mother Goddess stole God's worship. Then various other idols over the millennia obscured this reality by twisting the rest of God's anciently known attributes almost beyond recognition.

Today, the evolution myth has elevated mechanistic materialism into a twenty-first century idolatrous pantheon. The worship of science is not itself The Great Harlot riding the scarlet beast of Revelation 17:3, but it is one of her numerous expressions. The blanketing swarm of monstrous creatures released onto the earth at the blast of Revelation's fifth trumpet all sport women's hair (Revelation 9:7-11) to symbolize disdain for God's warning, like Eve's aspiration to eat the forbidden fruit, Ham's drive to mock Noah's faith, and liberalism's obsession with lying about everything.

Like Coma Berenices in today's Zodiac supplants The Desire of Nations, the Anti-

Clear Signs of Trouble and Great Joy

christ will soon set himself up in the temple and pronounce himself to be God at the midpoint of the Tribulation. The woman of Babylon (Zechariah 5:7-11), the woman riding Revelation's beast, the Biblically portrayed whore of false religion, the ancient Great Mother Goddess, and today's assumptions supplanting God's Word all proceed from the human penchant to passionately make believe rather than searching soberly for truth. Even five-year-old Sadie knew enough to thank her Creator for *information* rather than ignoring it to attack Him with her imagination.

The woman usurper is a Biblical theme. Even that lock of hair having come to the Zodiac in the fourth century BC to supplant The Desire of Nations fits meaningfully into the gospel as told by the stars. Just another coincidence…right?

The most ancient, direct, literary reference we have to the Zodiac is "The Mazzaroth", as I AM WHO I AM Himself spoke about it to Job. The order of its forty-eight constellations purpose these celestial lights to brightly shine Christ's gospel into humanity's darkness. The Mazzaroth bears witness to the heavens speaking His glory. All things will have demonstrated His Holy Word by the time He ends evil, the purpose for which God made everything.

But whether the reference to the Mazzaroth in Job is to some, incredibly ancient, astronomically coincidental concoction from an Einstein-like ape-man's mind, or whether it is more rationally a reference to a revelation given by God to Noah's great granddaddy, Enoch, must await a direct, confirmational link found within some extraordinarily ancient, archaeological reality, whatever that could possibly be.

And so the rest of what can be seen amongst all the constellations, archeology, and the Word of God is for another chapter another day. What will be presented there is even more astounding than is this amazing coherence between the gospel and mankind's ancient thoughts depicted in the constellations. About the perspicuity of the Mazzaroth's presentation of the gospel, even St. Paul attested…

> [17] So faith comes from what is heard, and what is heard comes by the preaching of Christ.
> [18] But I ask, [has mankind] not heard? Indeed they have; for
> "Their voice has gone out to all the earth,
> and their words to the ends of the world."
>
> Romans 10:17-18

Footnotes

1. Faulkner, Dr. Danny R. A Further Examination of the Gospel in the Stars.
https://answersingenesis.org/astronomy/stars/a-further/examination-of-the-gospel-in-the-stars/
February 6, 2013. Pg. 10.
2. Ibid. Pg 14.
3. Banks, William D. The Heavens Declare: Jesus Christ Prophesied in the Stars. 1985. Impact Christian Books, Inc., 332 Leffingwell Ave., Suite 101. Kirkwood, MO 63122. Pg. xi.
4. The Dendera planisphere is a bas-relief zodiac found in the ceiling of a portico to the Temple of

Hathor at the Dendera temple complex. The bas-relief was most likely carved in the first century BC, as it contains symbols representing a solar and a lunar eclipse occurring 51BC and 52BC. However, the arrangement of other elements indicate a depiction of the constellations at a much earlier time.

5. Aratus. Phaenomena. http://www.theoi.com/Text/AratusPhaenomena.html. Theoi Project. Copyright 2000-2017. Aaron J. Atsma. New Zealand. Pg 3.

6. From the beginning of the twentieth century on, "Forward" has been a code word of political movements bent upon conquering people and controlling them into socialistic and communistic subservience. The word has been used in and as titles to numerous fliers and publications deceiving the unwary, minimally thinking masses into viewing various forms of Marxism as "saving" ideologies. It is the conceptual equivalent of "Progress", a term used to deceive today's careless thinkers into believing mankind's budding knowledge will carry them forward into utopia, if only everyone will submit to draconian governance. These concepts sprout from the same ancient deception the snake employed on Eve. In a later chapter, we will see the rise of Marxism and its murderous, twentieth century governments prophesied in very ancient artwork.

7. http://www.pbs.org/wgbh/pages/frontline/shows/whales/man/myth.html. Copyright 1995-2014 by WGBH educational foundation.

8. Rohl, David. From Eden to Exile: The Five Thousand Year History of the People of the Bible. Greenleaf Press, Lebanon, Tennessee. 2009. Pgs 35-40. Rohl remolds Biblical history by modern conceptions of ancient history to fashion an account of Enoch leading the first Sumerians into Mesopotamia and building the first cities there, beginning with Eridu. Although this produces myth rather than history, it yet illustrates their passing a fading, twisted memory of God from one generation to another until we see it represented on Mesopotamian cylinder seals in cultural terms of their own. Yahweh was not copied from Ea. Yahweh is who the Sumerians poorly represented as Ea. He is who later revealed Himself accurately to Abraham, Moses, the prophets, and Jesus. Yahweh is not a copy. Yahweh is the correction.

9. Seiss, Joseph A. The Gospel in the Stars. (Originally published by Kregel Publishing in 1882.) 2005. Cosimo Classics. Cosimo, P.O. Box 416, Old Chelsea Station, New York, NY, 10113. Pg. 78

10. Ibid. Pg. 94.

11. Leeming, David. The Oxford Companion to World Mythology. 2005. Oxford University Press, Inc., 198 Madison Ave., New York, NY 10016. Pg. 71.

12. Wright, Ann. www.constellationsofwords.com/Constellations/lepus.htm 2008.

13. Aratus. Pg 6.

14. Rogers, John H. Origins of the ancient constellations: I. The Mesopotamian traditions. Journal of the British Astronomical Association. Volume 108, 1, 1998. Pg 10. This article can be accessed free at http://adsabs.harvard.edu/abs/1998JBAA..108....9R

15. Ibid.

16. Wilson, Ian. Before the Flood: The Biblical Flood as a Real Event and How it Changed the Course of Civilization. First St. Martin's Griffin Editon: March 2004. St. Martin's Press, 175 Fifth Ave, New York, N.Y. 10010. Pg 212.

17. Armstrong, Karen. A History of God: The 4,000-Year Quest of Judaism, Christianity and Islam. 1993. A Ballantine Book. The Random House Publishing Group, Random House, Inc. , New York. Pg. 5.

18. Collins, Andrew. Gobekli Tepe: Genesis of the Gods: The Temple of the Watchers and the Discovery of Eden. 2014. Bear & Company, One Park Street, Rochester, Vermont 05767. Pgs 107-111.

19. Fabricius, Karl. What the Lascaux Cave Paintings Tell Us About How Our Ancestors Understood

Clear Signs of Trouble and Great Joy

the Stars. October 10, 2009. http://scribol.com/anthropology-and-history/archeology

20. Collins, Andrew. Pg 72

21. Lima, Pedro. Lascaux Planetarium Prehistorique? The Incredible discovery of a paleo-astronomer. Translation displayed at http://cassiopaea.org/forum/index.php?topic=23833.0 Pg. .5.

22. Wilson, Ian. Before the Flood: The Biblical Flood as a Real Event and How it Changed the Course of Civilization. Pg 158.

23. Ibid. Pg 175.

24. Ibid. Pgs 210-211.

25. Glyn-Jones, William. Yima and his Bull: Gemini and Taurus in the Lascaux Caves. January 28, 2008. https://grahamhancock.com/glynjonesw1/ 2003-2017.

26. Johnson, Ken Th.D., The Ancient Book of Enoch. 2012. Biblefacts Ministries, biblefacts.org. Chapter 33:3-4. Pg 45.

27. Ibid Pg 10-12.

28. Savage, Michael. Liberalism Is a Mental Disorder: Savage Solutions. 2005. Nelson Current, a division of a wholly-owned subsidiary (Nelson Communications, Inc) of Thom-as Nelson, Inc.

Dayenu.

Chapter 5
Apophis and the Blood Moons

[10] I spoke to the prophets;
it was I who multiplied visions,
and through the prophets gave parables...
[13] By a prophet the LORD brought Israel up from Egypt,
and by a prophet he was preserved.
Hosea 12:10,13

It is interesting how closely the sense of ancient, mythological events often parallel Biblical narratives. The Babylonian creation epic, Enuma Elish, parallels the Bible's "In the beginning" with "When above the heavens did not exist, or the earth below". The Akkadian Atra-Hasis rode out a global deluge in an ark as did Noah. So did ancient China's Gun-Yu. But the Egyptian serpent, Apophis, is this chapter's particular case in point.

The Egyptians believed this serpent was present at the creation of the cosmos, like the Bible implies Satan was present when God created the universe and its physical life. To the Egyptians, Apophis represented the primary symbol of all evil and opposition to cosmic order, quite similar to the Biblical portrayal of Satan.

> Apophis was essentially only ever a bringer of entropy, disorder, mayhem, and misfortune.[1]

The Egyptian sun god, Ra, triumphantly made his orderly trek across the daytime sky. He shined daily blessing on Earth to nourish all life. The Bible says God causes it to rain and shine on the good and the bad alike.

> Of all the extremely varied symbols the ancient Egyptians used to represent divinity - water, silt, birds, snakes, crocodiles, lions, bulls, thrones and many more - the sun was the most important to them. Accordingly, the sun god - the numinous power they perceived and experienced within the sun - was their single most important deity.
> The ancient Egyptians called their sun god by many different names, such as Amun, Khepri, and Atum. But of all these, none were spoken of so highly, and so often, as Ra.[2]

Ra was the upholder of Ma'at, cosmic order. Apophis was Ra's greatest enemy, the cause of chaos. Apophis chased Ra through the night trying to destroy him and throw the

Clear Signs of Trouble and Great Joy

cosmos back into chaos.[3]   I AM WHO I AM is the upholder of righteousness, the essence of order which Satan must destroy to seize the crown of Heaven and to set chaos on the throne of God. Satan's goal and Apophis' goal are the same: breaking down the glorious complexity of righteousness' mutual benefit of everything into the decent of all things towards their lowest state of being. Chaos is everything behaving only towards the purpose of its own benefit, which in physics, eventually becomes a state totally depleted of energy, where everything is equally full, equally empty, and equally feckless, where nothing can possibly differ from anything else, rather like a Marxist dream state.

And so the main villain of Stargate SG-1's first four TV seasons was named after the Egyptian serpent. Stargate's Apophis launched heavy attacks on Earth, which for a suspenseful while, were undefeatable. Earth appeared to be doomed to chaos and destruction.

Stargate fans, Roy A. Tucker and David J. Tholen discovered an asteroid which NASA catalogued as asteroid number 99942. When an asteroid's discoverer is able to calculate its orbit, NASA allows him to name it. By 2004 Tucker and Tholen initially calculated 99942's orbit and found that it had a 2.7% probability of striking Earth on April 13, 2029, a Friday, no less. So they named their asteroid after Stargate's Apophis.

But additional observations allowed improvements to Apophis' calculated course. By 2006, Tucker and Tholen ruled out the probability of a 2029 impact. Yet there remained a possibility of a strike in 2036, when it again comes our way. Their adjustments revealed a six-hundred mile wide, very ominous keyhole that, if Apophis passed through in 2029, Earth's gravitational effect would draw it into an April 13, 2036 impact. Further observations were made, and better calculations began shrinking this keyhole until, by 2008, it was a meager two by five mile rectangle.[4]

This was Tucker and Tholen's prognostication when I first read about Apophis in 2010. I knew the heavens had recently displayed some interesting phenomena relative to the signs Jesus said would precede His return (Luke 21:25). They were even being considered by some people as signs of His return in those days. Hubble had photographed the debris field of an asteroid collision. At the head of a long stream of dust, pebbles, and boulders was the main spray of the collision. Its appearance was like a cross embedded in the Star of David (see Figure 50, page 349). A couple years earlier, in 2008, a record-sized gamma ray burst had been observed in Bootes, the constellation bearing the gospel-in-the-stars meaning "Christ quickly comes". This gamma ray burst was

> ...the most intrinsically bright object ever observed by humans in the universe.[5]

The King of Kings and Lord of Lords, the quickly coming Christ, is the most intrinsically bright truth in the universe, according to The Holy Bible. Was it merely coincidental that this outburst occurred in that constellation? Or was it purposeful?

I have always been a fan of the Bible's Revelation to the Apostle John. Any information, or even clue, regarding the Tribulation or end times catches my attention fast. The gamma ray burst in Bootes and the cross/Star-of-David debris field photographed by Hubble alerted me to the possibility of more signs coming. So I was watching. At the time

Apophis and the Blood Moons

(2010), speculation of the Lord's soon return was running rampant in the eschatological circles of Christianity. But these signs and a few others, as exciting as they were, were just not sufficiently communicative in themselves about when He returns. I was watching for something more resolutely indicating the approach of His return. It seemed rational to me that the enormous mayhem, disaster, and loss of life involved with His return would certainly compel a clear warning from a loving God known to warn before acting.

> [7] Surely the Lord GOD does nothing,
> without revealing his secret
> to his servants the prophets.
>
> Amos 3:7

Even Jesus said there would be signs in the sun, moon, and stars as the time drew near.

But Apophis' seven year cycle, and possible 2036, impact captured my attention, especially since it was said to return on a Friday 13th. To me, it smacked of Revelation's seven-year disaster known to Christians as "The Tribulation" and to the Jews as "Jacob's Trouble". By itself, though, Apophis failed to rise to the level of the objective communication required of a warning sign. The only aspects it had in common with the Tribulation were its seven-year cycle and the ominous nature of its Friday 13th appearance. But a sign of dire warning should clear a high standard. The challenge for a sign to approximate the occurrence of the Tribulation raises the bar higher yet. Jesus said He did not even know the day or the hour of His return. The Tribulation is ultimately about His return. So knowing its approach would be akin to knowing the time of Christ's return. Indeed, Apophis highlighted a very particular set of seven years. It also was named after the Egyptian version of the Tribulation's dragon. And even its NASA assigned number communicates the essential purpose of the Tribulation (see pgs. 267 and 268), according to Biblical numerology. But what I knew of it then lacked the objective confirmation needed to clear the high bar for indicating the Tribulation's occurrence beyond just "sometime soon". So, Apophis became a point of humor for me to entertain my friends.

Of course, I understand the difference between not knowing the day or the hour and not knowing the year, or even the decade. I also understand the difference between knowing the day of His return and knowing the time of the Tribulation. But even suggesting there might be such a difference raises hackles in the churches, they have so embellished "not knowing the day or hour" into meaning not knowing at all.

Christ's return will occur at a particular point in time, a particular hour of a particular day. The Tribulation is a seven year period spanning much more time than just a day or an hour. But being a set amount of time, if you can tell when it begins, then you will know when it ends. The Bible places Christ's return at the end of the Tribulation. Should the Tribulation's beginning be a bit too vague to discern, then surely the Antichrist's usurpation of God's throne at the Tribulation's midpoint will be sufficiently apparent to show the Tribulation is indeed in process and that its end is twelve-hundred-sixty days away. As the Bible describes it, the Tribulation can't be missed, I mean as far as knowing that it is occurring. This becomes quite problematic for the "won't know anything" crowd. Revela-

Clear Signs of Trouble and Great Joy

tion schedules Christ's descent to the Mount of Olives at three-and-a-half lunar years after the Antichrist's self-elevation to God's throne. And even Jesus said discernible circumstances will clearly approximate the time of His return.

> …when these things begin…look up…for your redemption is drawing near…
>
> Luke 21:28

    The Tribulation is a time of pending doom worse than any Friday 13th could bring. But it ends in the glory of Christ returning to destroy all denial of His reality and every spurn of the authority He earned by sacrificing Himself. It will break evil's hypnotic power over mankind. It will yank the ancient serpent away from God's crown and throw him bound and gagged into confinement, socially distanced far from every human heart. Then Christ will begin a righteous reign of justice in His Holy City, Jerusalem, teaching peace, prosperity, and the worship of I AM WHO I AM to a no longer brainwashed world.

    The news article which introduced me to Apophis presented both of its rendezvous with Earth as being on Fridays. This pair of Friday 13th's, book-ending a period of seven years, sent me laughing to my friends about an asteroid marking the beginning and end of the Tribulation. When I later discovered that April 13, 2036 was not a Friday, but was actually Easter Sunday, I sobered up a little. The Tribulation ends with Christ's return from Heaven. Easter Sunday was Christ's return from the grave. But what could some passing asteroid have to do with anything Biblical, especially an asteroid which seems to have been whizzing through Earth's neighborhood every seven years or so for how many millennia?

    I admit; I wanted Apophis to be a sign about the Tribulation. I'm human. Humans are subjective. Evil is painful; deceit is liberal; I want them to end. And I understand the Tribulation to be the process of Christ's quarantining evil from mankind for many centuries before evil is eternally eliminated from the place of those who cherish the truth and love doing right. Furthermore, the events and circumstances described by Revelation fit the trajectory of current history so well that it seems as if the world stage is being set for the Tribulation: Israel being a nation again with God's Holy City for a capitol; deceit rising to unprecedented heights; and even a head-hunting, "worship like I do or pay in blood" religion persistently grasping for world dominion. So these occurrences in the heavenly bodies over those previous few years seemed quite foreboding to me.

    But several good coincidences and some eerie times do not a clear message make. Maybe I am a bit more skeptical than others. Maybe I over compensate for my own biases. Whatever the case, to me, Apophis was still just entertainment for my friends. We had some good laughs for a couple years. But, little did I know, more coincidences revolved around this three-hundred meter long space rock than even my own skepticism could eventually withstand.

    When I saw an advertisement for a book about some blood moon phenomenon, I had to turn aside and read if God had another wonder to be seen, or if it was just the usual interesting coincidence being puffed up by the next preacher. The book was John Hagee's *Four Blood Moons* (Worthy Publishing, 2013). His proposition was simple. The clockwork motions of the earth and moon will occasionally produce four total lunar eclipses occurring six months apart, a sequence which NASA calls a "tetrad." Sometimes these tetrads will correlate with two of the most significant Jewish holydays: Passover and Tabernacles. But that in itself is not highly coincidental. Passover and Tabernacles both begin on the day of a full moon, and of course, the full moon gets eclipsed by Earth's shadow, sometimes, every six months over the course of two years.

Passover commemorates the night God's death angel passed over the Israelite slaves as it went about killing the Egyptian first-born, simply stated. The Christians recognize the Paschal Lamb of Passover as representing the Messiah, Jesus Christ, of whom everyone must partake to be passed over by the angel of the eternally final death, Christ's judgment of all sin. Then the Feast of Tabernacles celebrates God's dwelling with Israel through the desert after they exited Egypt. But like it is God's nature to pack His symbols full of meaning, Tabernacles also celebrates Christ's resurrected, physical presence in Jerusalem for a thousand years of righteous and just reign over all mankind: Immanu-el, God with us. Maybe God "planned for making signs" by calling for these two holydays to begin on nights of full moons six months apart. Therefore, it is somewhat circumstantial that these two phenomena, the feasts and the eclipses, occasionally coincide.

John Hagee refers to four blood moons occurring on both of these Jewish holydays over two sequential years as "tetrads". But I must adhere to NASA's use of the term for meaning any four, total lunar eclipses, each occurring six months apart whether on Jewish holydays or not. "Tetrad" was NASA's ascription before it was Hagee's term. NASA deserves its honor. We must discuss many tetrads which do not fall on those two Jewish holydays. Therefore, it will be useful to retain NASA's ascribed meaning of the term. We will refer to any tetrads occurring on all four of these two important Jewish holydays as "holyday-tetrads", or as "special tetrads", depending on the contextual mood. And we will reserve the term "unique tetrad" for a unique discussion in Chapter 6.

In his book, Hagee discusses three holyday-tetrads having occurred AD1493-1494, 1949-1950, and 1967-1968, and then another to occur 2014-2015. Two of these holyday-tetrads began the year after enormously significant events of Jewish history: 1) Ferdinand and Isabella's expulsion of the Jews from Spain by issuance of the Alhambra Decree in 1492, and 2) the Jews declaring the nationhood of Israel in 1948. The third began in the year the young state of Israel won control of God's Holy City, Old Jerusalem, as a result of the 1967 Six Day War. Of course, Hagee's book was written before the 2014-15 holyday-tetrad occurred, so it speculated an event it might mark.

These holyday-tetrads are not extraordinary because of Jewish holydays correlating with total lunar eclipses. They are enormously coincidental in that they occurred immediately after the historical events bringing to fruition God's promise made to Abraham, the Father of the Israelite people of those holydays, nearly four thousand years earlier. And so it was thought that the 2014-15 special tetrad might share in the same, prophetic mystique of the first three. But the crucially important event corresponding to it was yet a mystery at the time Mark Biltz discovered and John Hagee popularized this phenomenon.

The first three special tetrads happened either the year after or spanning a year of great concern or joy for the people to whom God's Word unilaterally promised the

Clear Signs of Trouble and Great Joy

possession of Palestine. And as Hagee points out in his book, the coincidence is even greater because these four holyday-tetrads were the only holyday-tetrads occurring from the fifteenth century to our present day within a larger group of seventeen tetrads. Most of the Christian analysis being made of the 2014-5 special tetrad attributed a message to it. Some saw it to be merely illuminating. But a few rejected the idea of a message represented by any of these four, special tetrads.

The Holy Bible promises Israel's preservation through times of trouble meant for its nemeses. Revelation and Zechariah foretell a time of tremendous doom. His Spirit will not strive against man's rebellion forever. This was true before the Great Flood. It is true today. Jesus said the day of His return would be like the days of Noah,

> [26]As it was in the days of Noah, so will it be in the days of the Son of man.
> [27]They ate, they drank, they married, they were given in marriage, until the day when Noah entered the ark, and the flood came and destroyed them all.
>
> Luke 17:26-27

The people did not eat, drink, and make merry until the Flood swept them away merely because they had never heard of its coming. The Flood narrative implies that God warned Noah of the impending Flood one-hundred-twenty years before it happened. The anciently venerated Book of Jasher said Noah preached a daily warning to the people during those years.[6] They watched him build the ark. They laughed. They jeered. They scoffed. It wasn't that they simply never knew. God is a loving God, therefore He is a warning God. They ate, drank, and were merry in spite of one-hundred-twenty years of daily warning. They didn't think old Noah knew what he was talking about.

Jesus said there would be a warning before His second coming; He said there would be signs in the sun, moon, and stars (Luke 21:25). Those signs are as empirically real today as Noah's preaching was to the people of his day. Their articulation is as perspicuous about the coming Tribulation as Noah's preaching was about the Flood. The eating, drinking, and making merry Jesus said would be happening at His second coming as in Noah's day indicates that, indeed, today's people will fail to heed these signs and know of His arrival.

Jesus Christ lived a perfect life. Man rewarded Him with a criminal's death. Two-thousand years ago, His work was calling people to grace and the truth. The end of His completely righteous life satisfied the wrathful flames of Heaven's altar awaiting to be cast upon mankind for their crimes and errors. Yet, by His unwarranted death on a torturous cross, Christ earned the right to overturn the flaming altar upon everyone who is faithless towards the truth, denying His authority over their lives, which His faithfulness sufficiently earned. But His willingness to die effected a temporary reprieve for the deniers and liars. They have been graciously granted one last stretch of time to see the light, to open their eyes. Then, at the appointed time, Christ will overturn Ara, the flaming altar, upon the rebellious people of the world who, in spite of having been warned, refused to take shelter in the King of Kings.

As in the days of Noah, a remnant of mankind will survive the Tribulation event. Their lives will be passed over. Immanu-el will tabernacle with them in God's Holy City, Jerusalem. And Israel will be raised to the forefront of the nations. For a thousand years

Apophis and the Blood Moons

Christ will lead the world into justice, righteousness, and peace while the Church, the Bride of Christ, and the little nation of Israel will be shining brightly amongst mankind. So there are these four holyday-tetrads speaking without words about the preservation of God's chosen people through a disastrous crushing before He comes to dwell with them. These holyday-tetrads monumentalize historic events setting Israel and Jerusalem onto the Tribulation's stage in the theater of reality. As in the days of Noah, today there is also a warning. As a few heeded and boarded Noah's ark, some will heed today's warning and call on the coming King.

Some argument could be made that Hitler's Holocaust was a more major event than were Ferdinand and Isabella thrusting Jews out of Spain. They didn't gas and incinerate six million Jews. So it can be asked why there was a special tetrad after 1492 instead of 1945? We will later discuss how Hitler's Holocaust is indeed part of this blood moon picture. But for now, Hagee points out how Columbus' discovery of America in 1492 also occurred at the first of these four special tetrads. America, Israel, and the Bible have a common and very special relationship which struck Hitler like a Kenworth full of reality. Those first three special tetrads marking key events of Israel and America were more than interesting enough to make the 2014-5 tetrad greatly curious. And we will find in the fourth special tetrad an essential correlation with both the Holocaust and Apophis. It will more than somewhat invoke an eerie sense of those seven Tribulation years.

So curious were these eclipses of 2014-5 that some folks proclaimed they meant Israel would gain control of the Temple Mount by the end of 2015. Others were going so far as to anticipate the beginning of the Tribulation by then. A few were even saying the Tribulation began in 2008, and Christ would return in 2015. Well. None of that happened. The laughter will eventually subside. As we discussed in Chapter 1, what reality's *information* says is far more important than what people say. Those ideas sounded good, but they lacked *information*. *Information* unlocks mysteries. People make up ideas.

I loathe preacher puff, informationless exaggerations from the pulpit. I've heard enough of it in my lifetime. I wasn't in the mood for more when I first heard of these special tetrads. Pastor Hagee's proposition about the return of the temple mount to Jewish control smacked of preacher puff, at least in my estimation it did. But I understand the hub of the Tribulation's action will revolve around the Jews, Israel, and Jerusalem. These are the obvious points of those historic events of 1492, 1948, and 1967. Only arrogance dares refute the possibility that these holyday-tetrads, by correlating with those events, are signs from God regarding Revelation's Tribulation without having first investigated all of the available *information*. After all, Jesus did say to watch for "...signs in sun and moon and stars," (Luke 21:25). And that's what these holyday-tetrads seemed most likely to be -signs in sun, moon, and stars. Else they were the most incredible coincidences ever, because of their alignment with historical events which even more coincidentally fulfilled ancient Biblical prophecies. Only diligent and humble investigation will discern whether they are mere coincidences or significant communication.

Jesus chided the Jewish elitists for being able to discern the meaning of weather patterns but not the meaning of the signs of the times (Matthew 16:1-3). A heavy cloud cover with thunder and waxing gusts of wind usually means rain. Ground tremors and a

Clear Signs of Trouble and Great Joy

growing dome on the side of a mountain meant the Mt. St. Helen's blast. Volcanologists were wise enough to know that. They issued warnings because they saw *information*. Now they're paying attention to Yellowstone's signs. Empirically observable events are the medium in which signs are painted. And as science brought us enough mental capacity to understand the approach of events by the nature of their causes, a mental framework constructed from The Holy Bible's *information* recognizes the relevance of these special tetrads to God causing the end of this evil age.

Science is the investigation of ordered pattern found amongst random spew. Minor coincidences can be ignored. But the uninvestigated dismissal of enormous coincidences is as intellectually irresponsible as it is emotionally arrogant and spiritually risky. Moses saw a bush perpetually burning on a mountain; he didn't turn away in the same scoffing bias as normal science turns away from the Bible. Nor did he immediately remove his shoes and drop to his knees. He only knew it to be a very unordinary pattern in his randomly normal surroundings. So he attended what he saw to attain *information* he lacked.

The same intellectual responsibility obliged me to turn aside and examine this peculiar sight described by John Hagee rather than arrogantly writing it off as preacher puff. Truly, I expected my search to be short. I wasn't confident of these holyday-tetrads being anything more than glorious coincidences correlating with spiritually significant events for drawing attention to the fact that I AM WHO I AM is the God of Israel -and Yippee!- someday He will return.

The Tribulation followed by Jesus' return is no small matter. It will be the biggest paradigm shift mankind has ever experienced. It will be an even bigger paradigm shift for normal scientists, though maybe not their most significant one. According to the Bible, as much as two thirds of the world's population will perish in those tumultuous seven years. Afterwards, God will be on Earth guiding a remnant of people who loved truth more than deceit and loved righteousness more than evil. Whatever signs of The Tribulation might appear in the sun, moon, and stars, I reasoned, should be more than enormously coincidental. They should thoroughly correlate with Biblical prophecies regarding the end. Their very essence should perspicuously connote the same message of evil's end as does the Bible. Whether the symbols involved are letters forming words, sentences, and chapters, whether they are road signs before potholes, construction zones, and mountain curves, or whether they are somewhat rare astronomical alignments occurring on the special days denoted by an ancient Book, sets of complex symbols correlating in their essence with prophesied events do not happen by chance alone. Their purposeful design should be unmistakable to the honest mind. Nothing less than such a set of symbols in the heavens would convince me.

So I set out to invalidate Hagee's proposition about these signs indicating some Biblically major occurrence in 2015, or maybe portending something bigger soon thereafter. Sound reason demanded that I expect these four holyday-tetrads to be embedded within a much larger system of self communicative signs before I could understand them to be a clear warning of anything. So I first poked around the internet to learn whether all four eclipses of these special tetrads were indeed technically total. God is not a stooge. He's not sloppy. If He's going to make a sign out of a coincidence, He won't fudge around its boundaries or color across its lines like Haekel fudged evolution's embryo charts. I quickly found two websites confirming the totality of all the eclipses. But a third site indicated the April 4th eclipse of 2015 was just barely, by a tiny sliver, partial. Aha! Gotcha, John!

But not so fast, Stevie. That's two against one. Democracy does not determine reality; reality is what it is. Everything of reality is coherent with everything else. I needed to see unanimity if the signs were real. The matter at hand was too important to have been given an ambiguous sign, or to have been left to the risks of democracy. So I consulted NASA's lunar eclipse list, as I should have in the first place. It showed the April 4, 2015 eclipse to be total by just a tiny bit more than a sliver. It was so much barely total that when my precious wife and I sat in our pickup, romantically watching this eclipse, its brief totality escaped us. But then again, we were somewhat distracted.

Understanding that Jewish holydays begin on full moons drains some of the rarity out of these tetrads. Any total lunar eclipse has a four in twelve chance of happening on a major Jewish holyday, since there are four full moon holydays each year and twelve full moons available to be eclipsed. It also happens that Passover and Tabernacles are exactly six full moons apart from each other. So, if the first eclipse of a tetrad occurs on Passover, the following eclipse will definitely occur on Tabernacles.

There are many tetrads with either their first or second year of eclipses occurring on both holydays. Since Passover sometimes occurs in an intercalary month, some tetrads won't have eclipses on all four holydays. Still, I found more than several special tetrads having both years of Passover and Tabernacles adorned with eclipses. And they rather clump together in astronomical cycles of a century or two interspersed by a few or more centuries of just a few to no tetrads, somewhat similar to the cycles of the Christmas star discussed in Chapter 3.

Regardless of these holyday-tetrads being a bit normalized, their occurrence at major events involving fulfillment of the significant, Biblical prophecy about Israel's return to nationhood is nothing short of stunning. And Israel's return to nationhood is not only among the most central prophecies of the Bible, it also is a condition precedent to the seven years of Tribulation described in the Bible as Christ's final defeat of evil for a remnant of mankind. That is beyond stunning. It is even a bit communicative, like a road sign preceding a very sharp curve in a narrow, cliff hugging, mountain road.

Under the pressure of Spanish pogroms[7] begun in 1391, some two-hundred-thousand Jews had converted to Catholicism by 1492, over half of Spain's Jewish population. Most of the Jewish population had arrived there in the seventh century after Muslims had captured Jerusalem. Then, in the ninth century, other Muslims conquered and settled Spain, too. But the Muslims who conquered Spain were better disposed towards the Jews than were those who captured Jerusalem. These "Spanish" Muslims considered Jews to be "People of the Book".[8] This friendliness of the Moors (as they were called) attracted Jews from other pogroms around Europe over the next few centuries.

Then, in the eleventh century, came a Christian conquest in all its holy, religious fervor to free Spain from these "horrible" Muslims. Religious fervor usually does not bode well for those of a less fervent religion, whether Islamic, Judaic, Christian, or what have you. Man's holiness stitches better blindfolds than it does work clothes. Not only were the Jews less fervent, but they had been enjoying a few centuries of rather good life amongst the

Clear Signs of Trouble and Great Joy

kinder Moors. But as usual, within the Jewish plight, anti-Semitism was catching up with them again, and a somewhat Hitlerish anti-Semitism to boot. More shamefully, it was a "Christian" anti-Semitism.

Pogroms percolated around re-Christianized Spain for almost one-hundred years. Although Jewish communities were allowed, the Jews were slowly being bullied into converting to Catholicism. Some defiantly continued being Jews outwardly, but more feigned conversion. A few actually did convert. They were called "conversos". Converso families continued in the social and commercial successes they had enjoyed for the few centuries before all this wonderful religious fervor ruined their peace. While pogroms increased the converso population, most conforming families continued their Jewish culture in private, always teetering, but not entirely abandoning their beliefs. The foot dragging remnant of the braver and more hardened Jewish hold-outs became perceived as a threat; their courage might influence the mushier conversos to take up Judaism again.

Ferdinand and Isabella feared a massive return to Jewish belief. So they issued the Alhambra Decree on March 31, 1492, giving the more stubborn Jewish holdouts until July to either convert or abandon their possessions and flee Spain under the penalty of death. Then occurred the first of four blood moons on Passover, 1493. Moreover, the day before Tabernacles of 1493 produced another interesting event.

While the good King and Queen began booting Jews from their land in 1492, Christopher Columbus had been discovering a new land for them (the good King and Queen, that is, or so they thought) It had become time to explore, settle, and exploit his discovery, which led to the European colonization of the New World, which led to the founding of America, which God has since used to help establish and support the prophesied nation, Israel, after having used it to kick the Holocaust out of Hitler's britches. Indeed, Manifest Destiny was a reality; but its meaning was completely misapplied to the stealing of the poor Indian's homeland, since they weren't as manifestly destined.

Columbus set sail to explore and colonize the New World on September 24, 1493, the day before the second total lunar eclipse of that year. Think about it. Tabernacles is the Jewish holyday celebrating God's residence with His believing peoples. The Jews were entering their final epoch of pogroms, which would end at their national homeland, while God was seeding a nation, in whose hearts He would dwell, for assisting Israel along the way to becoming that reestablished nation for the Jewish Messiah's reign.

The beginning events of the United States of America happened with a religious fervor, too. But this fervor began a bit differently. It was more of an attitudinal fervor than it was a fervor to go do God's work of killing people who don't "do God's work."

> During the period of its founding, an incredible collection of men and events shaped the destiny of America. I believe that words such as *random* or *fortunate* are wholly inadequate to describe this phenomenon.[9]

In his farewell address to the revolutionary armies, Gen. George Washington noted this same phenomena, having been its firsthand witness of the closest kind.

208

Apophis and the Blood Moons

> The singular interpositions of providence, in our feeble condition, were such as could scarcely escape the attention of the most unobserving; while the unparalleled perseverance of the armies of the United States, through almost every possible suffering and discouragement, for the space of eight long years, was little short of a standing miracle.[10]

And Samuel Adams, also known as The Father of the American Revolution, agreed with The General.

> There are instances of, I would say, an almost astonishing Providence in our favor; our success has staggered our enemies, and almost given faith to infidels; so that we may truly say it is not our own arm which has saved us. The hand of Heaven appears to have led us on to be, perhaps, humble instruments and means in the great Providential dispensation which is completing.[11]

Peeled of history revisionism's opaque whitewash, God's hand in America's founding shines of its own accord. But of course, "We can't have such light shining about! Got'ta hide that from the general public!" Therefore, most all of America's founders were Deists according to the detractors of established and ancient histories and to deniers of the testimonies about those astounding events. For example, Wikipedia dutifully explains the Christian religious aspirations of the early settlers. Then it smugly notes:

> Historians debate how influential Christianity was in the era of the American Revolution. Many of the founding fathers were active in a local church; some of them, such as Jefferson, Franklin, and Washington had Deist sentiments.[12]

America's three biggest names are conveniently dropped into the proposition of Deist sentiments in order to convince the general public that America was a Deistic happenstance. How crafty is deceit! America's principle classroom textbook during its most formative one-hundred-fifty years (as many as eight million were sold) was "The Little Bible of New England" *The New England Primer*. This short reading primer employed mostly Biblical text for reading exercises. By carefully chosen mnemonics and exemplars, it instructed children in the ways of Puritan life, ways far displaced from any Deist concept, big or little. Comfortably nestled within this time of The Little Bible's giant popularity was the American Revolution and the founding of the United States of America. Deists were the exception there. We must pay more attention to the *information* of the past than we do to the various revisions of history. Letters, documents, and quotes of the founders disagree with the implied Deism Wikipedia hopes the general public will swallow without investigation.

But in the words of The Father of the American Revolution, the "hand of Heaven" quite obviously led the Founders to being "humble instruments and means in the Great Providential dispensation." Deism does not lead. Deism is not providential. Deism dispenses nothing. Deism is a disengaged tick of a disinterested clock, therefore, Deism humbles

Clear Signs of Trouble and Great Joy

nothing. America's early Deism is intellectual B.S. (biased stretch). God's leading and providence correlate with faithful, abiding interest and action rather than with arrogant aloofness. Read Washington and Adam's words again. Do they describe the arrogant ticking of a clock's disinterest? Of course not! They shine truthful light upon the constant drumming of Wikipedia's bias.

Regardless of how little the historical documents might say about some dedication of America to God at some formal ceremony, the founders' countless prayers seeking the providence of an active and personal God testify to Whom they bowed. Since they bowed, they embedded His principles into the essence of the laws they passed and the freedoms they perceived, however much they patterned America's governmental processes after those of Greece and Rome. America's beginning was Christian in its most fundamental concepts, if not in the preponderance of its population. Invocations of God in the preambles of almost every state constitution, Scriptures carved onto monuments across its land, and even the basic, philosophical nature of limited government testify of the deliberate dedication to I AM WHO I AM made by the hearts and minds that formed the exceptional U.S.A.

The spirit of Deism was not active in America's foundation. But it is very active in America's destruction. Today, most of Christ's churches sit like lukewarm bumps on a log

Figure 41

engulfed in the flames of communist revolution while signs they should know humbly knock on the door for their stiff necked, brass foreheaded attention.

Apophis and the Blood Moons

The New England Primer drew its teaching material from the Bible, not from philosophers of the Bible. From cover to cover, it was designed to persuade children about the authority of God and parents. Biblical education was important to the people shaping the new nation because God was important to them. America's first universities were Christian institutions for training ministers, evangelists, and missionaries to spread the Gospel of Jesus Christ. Christian faith so interwove America's early fiber that almost two-hundred years later we were still singing patriotic, Christian hymns and praying to Jesus' Father in the small country school I attended. How thankful I am for that!

Today's political wall of separation only separates the truth from the general public. The past happened like it happened, while history is the past soaked a little in each bard's personal dye. Look up America's beginnings for yourself in earlier, less revised history books. Or take a ride through the way it was with Rush Revere and his talking horse, Liberty.[13] Then ponder why twisting America's history was called revision, and who's been pulling our leg. God's purpose for America is readily apparent to truth loving minds. The Gospel not only participated in American thought, but American feet participated in carrying the gospel around the world. The strength of American faith crashed across Nazi occupied beaches on D-Day, not in predominance of military strength, as CNN reminds us,[14] but as the indispensable component of fifteen Allied armies.

Detractors of America's contributions to World War II and outright America haters contend that the Soviet Union could have defeated the Axis power without America's help. Both Russia and America mobilized about the same number of soldiers. And the effect of an Eastern Front weakening Hitler's Western Front was the same as a Western Front weakening his Eastern Front. But realistically, if the British Isles would have been left without America to defend against Hitler's unhindered occupation of Western Europe, Eastern Front or not, they would have lacked the resources and industrial capacity to survive. Even if they could have defeated Hitler at their borders, they most likely would have immediately needed to defend against a Soviet communist Europe next. Without American involvement, Russia may have been the winner of war torn Europe. And have we considered how long North Africa might have stood against communism's lust for enslaving humanity? Communism, one ideological degree away from Nazism, would have quickly spread from the Bering Straights in the east through Europe and the British Isles to the land of ice and fire, had the world been without American participation in World War II. And what about the Middle East? Should we think the expansionist mentality of communism would have humbly limited itself to Eastern Europe and East Germany after it had beaten Hitler? To even think the outcome of World War II would have been the same without American involvement is to speculate sixty percent of the Allied strength could have won what one-hundred percent of its strength struggled frightfully to win. Moreover, American war production during World War II was unparalleled, except maybe by Nazi Germany. One Michigan factory turned out a B-24 bomber every hour. One American shipyard produced a merchant ship every week. The US Navy became the largest naval fleet in history by 1944, and

> In just four years, the United States would produce more airplanes than all of the major war powers combined. Germany, Japan, Italy, and the Soviet

Clear Signs of Trouble and Great Joy

Union could not build a successful four engine heavy bomber. America, in contrast, produced 34,000 excellent B-17s, B-24s, and B-29s.[15]

So, if there had been no America to aid the Allied forces, what would have subtracted its enormous contribution from the war? What have we forgotten? Oh! Of Course! Japan!! Pearl Harbor. The loss of any one of several harrowing Pacific battles would have quickly dropped America off WW II's map. Or a liberal, "make love not war" retreat from Pearl Harbor into America-hating self pity would have invited the land of the rising sun through Hollywood right into the heartland of America. If FDR was little Barry Obama begging the world to accept America's apologies for World War I and demanding the Greatest Generation to bow in humble subservience to political correctness, Japan would have swallowed China in one, unhindered bite, and then would have washed it down with a big swig of America.

The Soviet Union would have then been pinched between two fronts: 1) a Hitler plagued Europe to its west, and 2) a Japanese controlled China to its east (and south). Both would have been more than difficult fronts, to say the least. Both would have been more industrious than all the other nations combined, if America had somehow been subtracted from the WW II equation.

Maybe we were a bit premature in writing off Hitler in the absence of America's contribution to WW II. Nazism, communism, totalitarianism, authoritarianism, socialism, and American Progressivism are all left-wing evils of the same collectivist type. America's participation in World War II was indispensable to morality's light still faintly flickering on a world stage of left-leaning goose-steppers.

But even if America would have fought the tyranny of Nazism alone, it would not have won that war alone.

> Unless the LORD watches over the city,
> the watchman stays awake in vain.
>
> Psalm 127:1b

Four-hundred-fifty years earlier, I AM WHO I AM attested to His participation in World War II's history by setting a sign in the heavens at the very beginning of America's developmental path. America became bigly involved in God's work of preserving Israel through the onslaught of murderous, left-wing assaults. Israel survived Hitler's holocaust by God's use of America. The Tribulation event will be another murderous left-wing assault on Israel. But America's hindrance of leftist assaults will have been removed by then.

God restrains evil assaults to whatever degree serves His purpose. America served as a watchman at God's side and a tool in His work. But America is not The Seer, nor is it The Tool Maker. It was neither America nor any other human force doing the crucial work of preserving liberty in the face of tyranny, for instance, raising up an FDR instead of a Barry Obama for those times, or inspiring "Rosie the Riveter" then, instead of "It isn't my bag, man; make love not war, dude." Four blood moons, each one on a Jewish holyday

Apophis and the Blood Moons

celebrating God's wrath having passed over the righteous and His dwelling with the righteous, marked the planting of America's seed into the soil of history the same year Columbus stumbled across America and the Alhambra Decree institutionalized Israel's next path of troubles.

But America's greatest contribution to the world was the independent spirit of autonomous participation as opposed to the collective spirit of servile subjugation, a spirit of shared truths instead of the spirit of intellectual thefts, and a spirit of duty, virtue, and morality instead of a spirit of dependence, vice, and debauchery. America participated with God and fourteen other nations in drop kicking recent history's worst collectivist, Hitler, off the world stage. Then America participated in Germany and Japan's rise to independent, sovereign, dignified nations. America participated with Great Britain, Poland, Pope John Paul II, and I AM WHO I AM in taking down the Iron Curtain. I AM WHO I AM and the America of His 1493-4 road sign pounded their way into Berlin and toppled a political evil bent upon exterminating the Jews and dominating a world of faithless, chilly, fascism.

Every time history places the extermination of the Jews bet, God's hand shows up at the table. He crushed Hitler's Nazis and brought forth His Jews from the death camps. America became the second biggest human glove I AM WHO I AM wore in creating His tiny Jewish state. (The Israelites, themselves, were the biggest.) Thereafter, America has supported Israel's existence and stood by its side. God created America for this purpose of hindering the mystery of lawlessness amongst the nations until the time for revealing the lawless one within the Tribulation has come.

These ideas don't set well with everyone. But even if they only please I AM WHO I AM, dayenu![16] If He only created the motions of the heavenly bodies to blink total lunar eclipses on Jewish holydays at the beginning of America's development and Israel's long trek to nationhood, dayenu! Has any man fudged the motions of the heavenly bodies like men fudge history today? The holyday blood moons of 1493-4 loomed above the beginnings of two historical paths walked by two peoples of enormous significance. Those paths wound through extraordinary events to converge four-hundred-fifty-five years later when I AM WHO I AM honored His ancient promise to Abraham, marking it as a special event by a second set of four, holyday blood moons.

Israel was not even a nation having a land when Moses told them how completely God would uproot them if they abandoned their covenant with Him. Four-hundred-thirty years earlier, God promised Abraham that Palestine would belong to his descendents. (It was called the Land of Canaan in his day.) When Abraham asked how he could believe these things would come to pass, God performed a common-to-that-day and very solemn, covenantal ritual. He directed Abraham to split a heifer, a female goat, and a ram in half and lay the halves out apart from each other in two rows with a turtledove and a young pigeon also torn asunder (Gen 15:1-20). Although we might wonder why such a messy fuss, Abraham knew precisely what it meant. In his day it was the customary manner of establishing a most solemn contract. The two contracting parties walked each other between the gruesome animal halves indicating the same fate to befall the party who failed to perform his contractual duty.

Yet, when the time arrived for the walking to begin, God walked alone. I AM WHO

Clear Signs of Trouble and Great Joy

I AM showed with definite certainty that this contract was entirely unilateral. It was God's most solemn promise to Abraham. Abraham had believed God and left plush Ur for dusty Canaan. This pleased God greatly. So He unilaterally promised, since Abraham was where God led and doing the things God meant him to do. I AM WHO I AM walked alone through the demonstration of what would happen to Himself if He did not give Palestine to Abraham's descendents.

Behold! Israel in Palestine as God promised! Behold! We see holyday-tetrads testifying to His ancient promise today kept! Does this not say a word or two about the truth of God's Holy Bible? Do we have any thanks for I AM WHO I AM's faithfulness? Or do we feel hostile about such profound, heavenly coincidences intertwining the "dastardly trio" of the Bible, Israel, and America?

When God led His Israelites out of Egypt through forty years of desert wandering into the Promised Land, a particular Hebrew term became used to connote their horde. The term extends from a primitive Hebrew root, *ayd*, which is concretely used to mean *a witness*, and abstractly used to mean *testimony,* and specifically used to mean *a recorder*, or *a prince.*[17] *Aydaw* is its feminine form meaning "congregation". Throughout the Old Testament Israel is called "the congregation" by use of this Hebrew word constructed from the idea of *witness* and *testimony*. Throughout the Old Testament, Israel's interactions with God, both good and bad, are the congregation's witnessing of God's goodness, truth, and faithfulness. Today, Israel in their land continues to be testimony to I AM WHO I AM, and a further witnessing of His providence.

> …Let God be true though every man be false, as it is written,
> "That thou mayest be justified in thy words,
> and prevail when thou art judged."
>
> Romans 3:4b

From 1948 forward God has been presenting His case to our current generations of scientifically mesmerized, God-judging people. Scientists scoff. Atheists scorn. Paganism obfuscates. The world's plethora of idolatries all mock Israel, Christ's Church, the Word of The Holy God, and JEHOVAH (I AM WHO I AM). But these bold, blood moons don't scoff. Look folks, especially you scientists, this is your mechanistic material existence clearly displaying the purpose of glorifying God's faithfulness to the message of His Holy Bible, and to Christ's Church, and to JEHOVAH's chosen nation of people. Their testimony is empirically observable. Their message is obvious. It's just not what most people want to know. So people refuse to look. They purposefully forget. They subjectively deny.

This nation whose existence today testifies to God's faithfulness is the witness of His love and power. Israel witnessed God's reality while the Egyptian army drowned in the returning sea. They witnessed His care in the manna they ate, and in their shoes not wearing out while trekking forty years through the desert. They witnessed His control of destiny in His scattering of the Canaanite nations before them, and His faithfulness in the land given to them. The correlation of the February 25, 1362BC total lunar eclipse, archeology, and ancient documents attests to the reality of those acts of I AM WHO I AM empirically observed

Apophis and the Blood Moons

by His congregation. Consequently, this same nation of people experienced His chastisement when they cheated Him. And they experienced His grace when they repented, even though they would cheat again. Israel's testimony is why secular historians refuse to correct their ancient chronologies.

> After all, it should be remembered that history -as constructed by historians- is not the past. The past is what happened, while history is just our best guess as to what happened. [18]

When the biased guess of placing the Exodus event in Ramesses' reign has been corrected by the empirically observed February 25, 1362BC total lunar eclipse (see it for yourself in the NASA lunar eclipse list), the past comes pouring out of archaeological discoveries, flushing away mounds of biased history, and then flowing into accurate correlation with Biblical narratives. Even blood moons testify of that history, as we will see in Chapter 8. The correlations between archeology and the Bible are so remarkable that the evidence of the 1362BC total lunar eclipse should not have been necessary. Still, secular historians kick like blind mules against any implication of the past being what the Bible says it is. Their minds recede into myopic paradigms of every imaginable explanation other than the Biblical one, however ridiculous their fallacious "evidences" might be (e.g., the 1659BC and 1631BC eclipses). Scoffers and skeptics fail to see the simple, inescapable truth of the Bible's most basic and important statement about God, "I AM WHO I AM." Glowering out of mental darkness, they foolhardily proclaim He isn't who He is.

God entrusted His oracles to Israel. Testifying of them to the rest of the world was a purpose He had for His congregation of witnesses. Other people in other places at other times claim to have received God's oracles. But the undeniable difference between God's oracles given to Israel and oracles claimed by others is fulfilled prophecy. God packed Israel's oracles abundantly full of prophecies. He then filled subsequent history with the occurrences of those earlier prophesied events. In Chapter 2 we analyzed the undeniable observation of not just one of these fulfilled prophecies, but of the one which is the essential framework of all Biblical prophecy. We found this framework to be indisputable by empirical evidence and data alike. We see Israel today, the empirical observation of fulfilled Biblical prophecy. God made fulfilled prophecy to be a point for distinguishing between His oracles and everyone else's religious writings.

> [21]And if you say in your heart, "How may we know the word which the LORD has not spoken?"—[22]when a prophet speaks in the name of the LORD, if the word does not come to pass or come true, that is a word which the LORD has not spoken; the prophet has spoken it presumptuously, you need not be afraid of him.
>
> Deuteronomy 18:21-22

Most Biblical prophecies have been fulfilled by now. Many well documented books have been written on the subject. A few remaining prophecies are yet to happen. Most of those are developing in the course of our current history; these special tetrads warn of it.

Clear Signs of Trouble and Great Joy

The impossible enormity of coincidence about these holyday-tetrads extends from the fulfilled (and being fulfilled) prophecies found in the oracles Israel has carried down through its history. It is time for everyone to wake up at this clearly ringing bell.

Before Joshua led the Israelites into Canaan, Moses warned God's congregation of witnesses against worshipping other gods. Paying any respect, honor, or obeisance at all to something not real destroys one's testimony about reality. Moses warned that God would throw them out of the land if they practiced idolatry. But what about the walk between those bloody carcasses I AM WHO I AM took alone? And then He threatens to take the land away from the people He promised to give it to? Doesn't that destroy God's witness?

Isaac and Jacob didn't receive the land either. Nor did their descendants who were enslaved in Egypt. Their children received it. But their children's children lost it. Later generations received it again, but only the Jewish ones of Judah and Benjamin. And still other generations lost it again (AD70). They were all Abraham's descendants, as were the generations of those wandering the world until the ones of our day staggered, barely alive, out of Hitler's death camps into Israel's blessed homeland, this time, prophecy says, forever. To our current generation of Abraham's descendants, the "forever" God promised is being fulfilled. Indeed, Israel will again teeter on the verge of defeat, BUT, their land will never be taken from them again, EVER. Today, they are in Palestine to stay. Their enemies will soon exhibit that fact, having become smoking rubble. Is that an audacious statement? Or does it simply extend into the future what we've learned from the past about God's prophecy always coming true? Ask the congregation. Note these signs.

In spite of witnessing God's hand throughout their history, the Israelites declined their responsibility of testifying to the other nations. They didn't remain faithful. They thought God's covenant was just to bless them alone. So, in their pride, they gave their minds first to the idolatry of other gods, and next to the idolatry of thinking they knew God without observing His *information*, but always by "hearing" what I AM WHO I AM did not speak. God sent them prophets begging for their repentance in the face of looming catastrophes. Biblical narrative, archeology, and even secular historians all agree on what followed.

In 722BC, God used Shalmaneser V and Sargon II of Assyria to remove the ten northern tribes of Israel from their land. Then through Nebuchadnezzar, He tore asunder the lower two tribes and removed them from the Promised Land in 586BC. Judah and Benjamin were restored seventy years later for the purpose of receiving their King, the Messiah, mankind's Christ. (The Desire of Nations we discussed in Chapter 4.) His reception should have been their next step of testimony before the eyes of the God-judging world. Yet their idolatry persisted. Their resolve had refocused sharply and singularly upon God's Law. But they were now observing The Law for their own purposes rather than for God's purposes. Their idolatry merely moved from worshipping the wrong object to wrongly worshipping the correct object. They loved the gift but rejected the Giver. They passionately returned to The Law as a testimony of their own righteousness instead of a testimony to His faithfulness. They were so certain of themselves that they became their own idols. Consequently, they lived for and worshipped HE IS WHO WE ARE instead of I AM WHO I AM. They thought they knew God. But they really knew only themselves. Instead of a relationship with Him they related to their own concepts. So when their King finally did come, they were given to

Apophis and the Blood Moons

The Law rather than to The Law's Author; they were so filled with themselves that their minds had no room left for understanding. They didn't recognize their own Messiah King. They rejected the One of whom they were raised to testify.

Yet God's purpose for Israel would be satisfied in the end. He was still as much in control of destiny as when He cleared the Canaanites from the land. Israel would yet testify to I AM WHO I AM before all mankind. But this testimony would arrive over a much rougher road leading to a more distant time. For their rejection of both Messiah and obedience they were again torn from their land and sent wandering through the nations.

> [14] For I will be like a lion to Ephraim,
> and like a young lion to the house of Judah.
> I, even I, will rend and go away,
> I will carry off, and none shall rescue.
>
> Hosea 5:14

Jesus coming as their King should have been no surprise to them had they known I AM WHO I AM through sincere obedience to His Law and Prophets. The Law and Prophets define God's nature. Obedience discovers the nature prescribed by The Law. Observation of fulfilled prophecies identifies The Law Giver. But His nature is understood and received only through faith in I AM, and His righteousness is donned only through the grace He worked in Jesus Christ. The Law is merely the definitive guide. Without the nature of God's righteousness in the mind's composition and the heart's beating, His oracles are not understood. The oracles God entrusted to Israel gave them a simple formula by which anyone rightly searching for the Messiah King could have known the time of His coming.

> [26] And after the sixty-two weeks, an anointed one shall be cut off, and shall
> have nothing; and the people of the prince who is to come shall destroy the
> city and the sanctuary. Its end shall come with a flood, and to the end there
> shall be war; desolations are decreed.
>
> Daniel 9:26

Who wants to know their king will be a cut-off anointed one? The Israelites wanted to know their promised King would be the victorious hero to free them of Roman rule. Yet *information* said He would be cut off. It seems the first century Jews had the same disease of ignoring *information* which also festers within our twenty-first century science. *Information* calls for sincerity towards reality, not for pride in what we know. Had they accepted the Oracles for what they said, their King would not have been cut off, or at least they would have repented as a congregation and turned to Him for healing after they had cut Him off.

But they gave no attention to the Oracles' *information*. They didn't regard finishing the transgression, ending the sin, atoning for iniquity, and the bringing in of righteousness importantly enough to note the four-hundred-eighty-three prophetic years which had passed since Nehemiah was given the decree to rebuild Jerusalem. This was a solid clue, though it was a tiny clue. If, with that clue in mind, they had listened to the faint stirrings in the backwaters of Bethlehem, if they had been humble enough to regard what a few smelly

Clear Signs of Trouble and Great Joy

sheep herders were saying about some angels they heard bearing tidings of great joy, then their own souls would not have crashed into another wall of unbelief.

Information is not significant because of its discoverers or its deliverers. *Information* is significant because of the realities producing it. The *information* in their Book of Daniel was a critical clue about an existing reality. If they had not had Daniel's prophecies, they could not have been faulted for being ignorant of the time of their Messiah's coming. But they disregarded their Oracles' *information* and watched for their Messiah through eyes made of themselves. Consequently, they turned their King over to Pilate for crucifixion.

God spent centuries inspiring His messages, instructions, and warnings into The Holy Bible. We must watch for what its *information* actually indicates, rather than for what we think it says. This is why His name is I AM WHO I AM, not HE IS WHAT WE THINK. Although Isaiah and the Psalms are filled with very picturesque themes regarding a suffering, mistreated Messiah, connecting them with Jesus requires a sincere desire to know the truth, a desire sufficiently resolute for dissecting and critically examining one's own dreams and nightmares.

Truth has an awareness by which it finds only those who yearn for it.

Had the Israelites desired the truth they would have at least recognized Him after they crucified Him. Reports of His resurrection were beginning to spread everywhere. That was a clue. Jesus' disciples were coherently matching all His characteristics, His life, His death, and His resurrection with prophecies from the ancient Oracles given to them, even though lowly fishermen and a had-been Pharisee were the ones striking up those matches. But the Jewish leaders remained so blinded by their own, self-made light that they even lied to their followers to cover up Jesus' resurrection which they well knew had occurred.

No other nation of people were given oracles proclaiming beforehand things that would happen to them the way those things eventually did happen. Prophecy is the verification of Israel's testimony, and it is the Holy Spirit's testimony to God's plan.

> [7] Surely the Lord GOD does nothing,
> without revealing his secret
> to his servants the prophets.
>
> Amos 3:7

As the Oracles proclaimed, after the anointed one was cut off, Jerusalem was besieged, destroyed, and its survivors were sold around the world into slavery. They were carried off, and none could rescue. It happened just like Hosea wrote it would happen seven centuries before, and like Daniel also prophesied five centuries before it happened. The spurned, crucified, and resurrected Messiah, now the Lion of Judah, tore those faithless people and returned to His place at the right hand of God. Maybe the ability to notice that Jesus was "the anointed one cut off" required more careful thinking in their day. But missing that same point today in light of all this available *information* requires a genuinely insincere, intellectually irresponsible pursuit of "truth".

But in spite of man's faithlessness, God will faithfully achieve His purposes. Israel has always been and is continuing to be the testimony of God's faithfulness and truth.

Apophis and the Blood Moons

Today, before our own empirically observing eyes, those dry bones of Ezekiel 37 have been risen, reconnected, and fleshed out as the modern nation of Israel. The ancient books of the Bible are copiously sprinkled with prophecies of God doing what we are empirically beholding with our very own eyes, today. By simply being a nation again in the land promised to Abraham's descendents. Israel is testifying to God's faithfulness, as well as, to the truth about The Holy Bible being the Oracles given to Jacob by God.

What other nation of today began as a promise to one man who faithfully followed God's voice thirty-nine-hundred years ago? What other nation was gestated as slaves in the belly of a great empire, and then birthed out of that empire's seemingly spontaneous collapse? What other nation took its land while yet a ragtag band of ex-slaves' children, and then developed into a kingdom to eventually be destroyed and removed from the land only to be returned to nationhood again in the same land? And what nation after such incredible turns of tragedy and fortune was again conquered mercilessly, torn again from its land and scattered again throughout the world to again wander pitifully for eighteen-hundred more years through hideous extermination camps before emerging as a sovereign nation in the very land from which it was long ago torn as by a lion, the very land promised to them forever by an entity claiming to be The Creator of All Existence? Let's patiently wait while the skeptics draw up a list of those nations. But our wait is short, for the list is only one nation long: Israel! God's witness.

> [5] He established a testimony in Jacob,
> and appointed a law in Israel,
> which he commanded our fathers
> to teach to their children;
> [6] that the next generation might know them,
> the children yet unborn,
> and arise and tell them to their children,
> [7] so that they should set their hope in God,
> and not forget the works of God,
> but keep his commandments;
> [8] and that they should not be like their fathers,
> a stubborn and rebellious generation,
> a generation whose heart was not steadfast,
> whose spirit was not faithful to God.
>
> Psalm 78:5-8

The numerous correlations of modern Israel with the ancient prophecies of the Bible make it what many recent preachers and eschatologists are calling God's canary in the coal mine. Miners carried caged canaries into their mines to forewarn of poisonous air. Bad air kills a little bird long before it does a big miner, giving the miner a short warning to get out when he has walked into bad air. Similarly, the closer God moves Israel towards the complete conditions prophesied in the Bible, the closer time has moved us towards the horrific end of this world's evil paradigm. We can know the time is drawing near by watching what happens to the canary.

Clear Signs of Trouble and Great Joy

> ¹⁴ I will restore the fortunes of my people Israel,
> and they shall rebuild the ruined cities and inhabit them;
> they shall plant vineyards and drink their wine,
> and they shall make gardens and eat their fruit.
> ¹⁵ I will plant them upon their land,
> and they shall never again be plucked up
> out of the land which I have given them,"
> says the LORD your God.
>
> Amos 9:14-15.

In the latter half of the nineteenth century, Russian pogroms against the Jews turned many Jewish hearts towards their old homeland. Some started moving there. Others engaged new, Enlightenment influenced conversations regarding the Western World's Jewish Problem. Discussion of a Jewish homeland was taken up by the European social elitists and politicians somewhat unwillingly. But if their new, "enlightened" way of considering world problems did not engage the Jewish Problem also, Europe's influential classes would be seen as the smiley faced, bigoted hypocrites they too much really were. They didn't want such genuine exposure, so they discussed the Jewish Problem. But, as usual with men, their hearts were not with their tongues.

In this smug atmosphere of ritualistic compassion, Theodor Herzl, a Jewish, Austro-Hungarian journalist/political activist, crafted the World Zionist Organization from sentiments stirred up by the more ideological Lovers of Zion movement shortly preceding it. Herzl shifted Jewish concerns into a higher gear, displacing the Jewish discourse of mere discussion with a demand for

> immediate political activities to obtain international recognition of the Jewish claim to Palestine.[19]

Still, the world was not facing the Jewish people with hearts quite as smiley as were their "enlightened" faces. The Ottoman-Turk Empire was pivotal in this matter of a Palestinian homeland for the Jews. At least it was momentarily pivotal. Regardless of the gracious shelter it gave the Jews from Spain's 1492 pogrom, and regardless of the rather caring way it occupied the Holy Land from 1517 to 1917, The Empire would allow no Jewish control of any Jewish state on any Palestinian soil whatsoever. We will return to the breathtaking result of this Ottoman-Turkish mistake in a more appropriate context. Suffice it to say here, this mistake is why their pivotal role was only momentary. As the past reveals, the Zionist movement outlasted its Empire problem. In time, the Jewish state emerged from Hitler's death camps. And that state hasn't looked back since. Cities were built; fallow land has sprung to life bearing a wide variety of quality produce; and the lives of Israel's people are bolstered by great scientific discoveries and inventions, all as Biblically prophesied. Numerous signs in the sun, moon, and stars agree with the Oracles regarding Israel's continuing prosperity until their Messiah gives all their promised land over to myrtle trees, gardens, unimaginable beauty, and unavoidable prosperity.

Apophis and the Blood Moons

Amos said God's people will "never again be plucked up." This prophecy was not about their return from Babylon. They were plucked up after that return. The prophecy was about 1948. 1948 was the only time God returned Israel to their land without again plucking them up. Of course, history isn't yet finished with this version of "returned". Nor is God finished protecting them, for their enemies are not yet finished attacking. Israel's current condition and the return of its Messiah, the Savior of all who repent from denying Him, is what these signs in the heavens are all about. Evil's end will come with Israel's second chance to get acknowledging their King and Messiah right. God is not finished protecting His cub, Israel, the apple of His eye. Therefore, heeding the warnings which He wrote into His Book, painted into history, and organized amongst the motions of the sun, moon, and stars will be as wise as stepping out from between a cub and its mama bear.

> [12] Let the nations bestir themselves,
> and come up to the valley of Jehoshaphat;
> for there I will sit to judge
> all the nations round about.
> [13] Put in the sickle,
> for the harvest is ripe.
> Go in, tread,
> for the wine press is full.
> The vats overflow,
> for their wickedness is great.
> [14] Multitudes, multitudes,
> in the valley of decision!
> For the day of the LORD is near
> in the valley of decision.
> [15] The sun and the moon are darkened,
> and the stars withdraw their shining.
> [16] And the LORD roars from Zion,
> and utters his voice from Jerusalem,
> and the heavens and the earth shake.
> But the LORD is a refuge to his people,
> a stronghold to the people of Israel.
>
> Joel 3:12-16

What shall we say about the faithfulness God maintains towards His promises? All we need do is look with our eyes. Archaeological evidence correctly aligns with the Biblical record upon the testimony of just one blood moon beyond man's tinkering reach. Yet God, being The Piler-On, has piled an enormity of testimonial blood moons above the corrected history; we shall examine it in Chapter 8. We stand without excuse before the testimonies of Israel and of Israel's God, as Moses noted before his death:

> [32] For ask now of the days that are past, which were before you, since the day that God created man upon the earth, and ask from one end of heaven to the other, whether such a great thing as this has ever happened or was ever

Clear Signs of Trouble and Great Joy

> heard of. ³³Did any people ever hear the voice of a god speaking out of the midst of the fire, as you have heard, and still live? ³⁴Or has any god ever attempted to go and take a nation for himself from the midst of another nation, by trials, by signs, by wonders, and by war, by a mighty hand and an outstretched arm, and by great terrors, according to all that the LORD your God did for you in Egypt before your eyes? ³⁵To you it was shown, that you might know that the LORD is God; there is no other besides him. ³⁶Out of heaven he let you hear his voice, that he might discipline you; and on earth he let you see his great fire, and you heard his words out of the midst of the fire. ³⁷And because he loved your fathers and chose their descendants after them, and brought you out of Egypt with his own presence, by his great power, ³⁸driving out before you nations greater and mightier than yourselves, to bring you in, to give you their land for an inheritance, as at this day; ³⁹know therefore this day, and lay it to your heart, that the LORD is God in heaven above and on the earth beneath; there is no other.
>
> Deuteronomy 4:32-39

Israel's history is emphatically unique. Has the establishment (reestablishment) of any other nation been adorned with four total lunar eclipses, two each on its two most significant holydays commemorating God's development of and dwelling with that nation? Does any other nation have a written history to back up such commemorative eclipses? And the evidence of astroarcheology and fulfilled prophecies to back up that written history? No. Israel is the one unique nation. The Bible is their unique Oracles from God.

Israel announced its statehood at midnight, May 14, 1948, the precise time the British Mandate protecting them ended. Almost simultaneously, President Truman sent a telegram through the Department of State to six U.S. missions and six U.S. consulates announcing the United States government recognized

> the provisional Jewish govt as the de facto authority of new Jewish state.[20]

Two courses of history begun in the same year at the hands of Ferdinand and Isabella converged at the formation of the state of Israel four-hundred-fifty-six years later. One holyday-tetrad marked the joint beginning of both those historical paths. The very next holyday-tetrad marked the convergence of those paths at the astonishing fulfillment of God's promise to Abraham, the reestablishment of the Israelite nation! May 11, 1949, by a vote of 37 to 12 with 9 abstentions, the United Nations admitted Israel to a seat. Twenty-eight days later a total lunar eclipse occurred on Passover. It was followed by three more total eclipses, each of those on a Passover or a Tabernacles, as well! How testimonial is that!?!

On May 15, 1948 Egypt, Syria, Jordan (then Transjordan), Lebanon, and Iraq invaded Israel to nip this fledging nation in the bud. The invasion began a new phase in Jewish history, the phase of God protecting Israel in their homeland like a Mamma Bear protecting her cub.

> ¹⁰ I will bring them home from the land of Egypt,
> and gather them from Assyria;

Apophis and the Blood Moons

> and I will bring them to the land of Gilead and to Lebanon,
> till there is no room for them.
> [11] They shall pass through the sea of Egypt,
> and the waves of the sea shall be smitten,
> and all the depths of the Nile dried up.
> The pride of Assyria shall be laid low,
> and the scepter of Egypt shall depart.
> [12] I will make them strong in the LORD
> and they shall glory in his name,"
> says the LORD.
>
> Zechariah 10:10-12

By January 7, 1949 the alliance of Arab nations were only able to achieve a cease fire. Israel passed through the "sea"[21] of Egypt and the pride of Assyria was laid low as the alliance of Arab nations could not defeat this fledgling, tiny nation. By God's outstretched arm Israel remained the nation which He founded the previous year. Those four total lunar eclipses celebrated. Israel was tested again in 1956 when Gamal Abdal Nasser of Egypt nationalized the Suez Canal and restricted Israel from its use. After eight days of fighting, Israel won continued access to the Indian Ocean through the canal, and control of the Gaza strip to boot. For a few years there was relative peace.

But in May of 1967, Nasser closed the Gulf of Aqaba to Israel and mobilized military units in the Sinai Peninsula. On June 5, 1967 Israel crippled the Arabic allied air power with a massive air strike. By June 10, Israel had captured Jerusalem's Old City and the Golan Heights. Forty-seven days previously was a total lunar eclipse on Passover. One-hundred-thirty days later occurred another total lunar eclipse on Tabernacles. Then six lunar months later, in 1968, was another eclipse on Passover, and yet a fourth eclipse six months later on Tabernacles. These holyday eclipses of 1967-8 celebrated something even more meaningful than Israel's sovereign achievement:

> [2]Thus says the LORD of hosts: I am jealous for Zion with great jealousy, and I am jealous for her with great wrath. [3]Thus says the LORD: I will return to Zion, and will dwell in the midst of Jerusalem, and Jerusalem shall be called the faithful city, and the mountain of the LORD of hosts, the holy mountain.
>
> Zechariah 8:2-3

No. God didn't come to dwell in Jerusalem on June 10, 1967. At least not like Zechariah meant. We shouldn't be like the first century Jews who thought God to be HE IS WHO WE ARE. We are engaged by His processes; He is not engaged by our imaginations. And as you saw the Jews move from 1492 through 1948, His processes move slower than our thoughts. June 10, 1967 was just one of many steps in the process which leads to God's dwelling in Jerusalem.

It was a very noteworthy flutter of the canary in the cage. After nearly two-thousand years, God gave Jerusalem back into the hands of His chosen people in preparation for His

Clear Signs of Trouble and Great Joy

reign. The very idea of it makes anti-Semitic blood boil with contempt. But the jealousy to pay attention to is the jealousy of I AM WHO I AM, for He is drawing very near to doing something drastic about the scoffing and contempt shown towards His Word.

It seems we might want to pay some attention to these tetrads God set on Israel's holydays.

Mark Biltz discovered the phenomena in 2010.[22] The eclipses themselves were plenty obvious, but only as being four in a row. The orbital mechanics of Earth and the moon have been known for centuries to produce four total lunar eclipses within a span of two years. I can't imagine nobody actually noticed these tetrads occurring on Passover, Tabernacles, and again, Passover, and Tabernacles as they were occurring. Most likely someone did. Maybe reports of them didn't reach very many people. But surely they were somewhat reported. Until Mark Biltz paid attention to them, they were not publicly available information. Thank God for watchmen! Biltz pointed out that it would happen a fourth time, in 2014-5. (Note the cohesiveness of the numbers: four eclipses in each tetrad, and four tetrads of those four, holyday eclipses.)

Eschatologists' most popular thoughts regarding the import of the holyday-tetrad to come focused on the Temple Mount in Jerusalem. The reasoning went like this: Israel was first completely driven from their Temple when it was destroyed in AD70. Then they were driven from the entire city of Jerusalem during the same assault. Seventy years after that siege had begun, and as a result of the Simon Bar Kokhba Revolt of AD136, the Jews were completely driven out of Judea. They lost their promised land after losing their city after losing their Temple.

Since the 1949-50 holyday-tetrad marked Israel's regaining the land, went the popular thinking, and the 1967-8 holyday-tetrad marked Israel's regaining God's Holy City, then surely the 2014-5 holyday-tetrad meant Israel would regain control of the Temple Mount. It was nearly logical. But logic isn't like horseshoes; close doesn't count. Had they carefully considered the 1493-4 holyday-tetrad, they might have reached a very different, less embarrassing conclusion.

These holyday-tetrads do not just coincide with historical events. A particular nature of Jewish history flowed from the 1493-4 holyday-tetrad to the 1967-8 one. It was not entirely different from what they had experienced from AD136 until 1492: pogroms and running and pogroms and running and pogroms and running. However, the new Alhambra Decree barring Jews from Spain sprang up simultaneously with the seeding of a new nation God would use for preserving His chosen people through a Holocaust in order to reestablish His nation in The Promised Land.

The Alhambra Decree traveled through history alongside the germination and growth of the United States of America. But when God gave His Holy City into the hands of His reestablished nation, that epoch of Jewish history ended. Another epoch began. After the final eclipse of the 1967-8 special tetrad, with Israel standing firmly as God's special nation possessing I AM WHO I AM's most treasured city, on December 16, 1968, the Alhambra Decree was officially revoked. Is that somewhat amazing? Is it not revealing? Then, at nearly the same time, America's Christian culture, which had served God in preserving Israel, caught the modern strain of an old, atheist virus. America's cultural cold of the late 1960s developed into a severe case of political pneumonia by 2015, gently settling into a tyrannically Covid panic by 2020.

Israel was on the incline to self sufficiency while Americans went sliding into a state of socialistic dependency. I AM WHO I AM isn't a Deist.

After the final eclipse of 1494, God stitched together a glove He would wear for re-establishing the nation of His people in the land He promised to Abraham. Biblical principles of human interactions and law were stitched into the newly developing America, individual freedom, justice by the rule of law, and goodwill towards all. In 1938 this aspiration came to be articulated by the now old (and unfortunately impugned) adage "Truth, justice, and the American way". Coincidentally, this was the same year Hitler began his plan, on the night of Kristallnacht[23], to exterminate the Jews. From America's planting forward, God was preparing America for a special moment in His story. Within seven years after Kristallnacht, it had led the Allied forces to the restraining of a great evil. No Jews would have remained to fulfill God's promise to Abraham had Hitler won World War II.

2015 has now gone by. Israel has not regained control of the Temple Mount. Eschatologists were wrong. Yet, there remains a better sense to be understood about the 2014-5 special tetrad. A sense which emerges from events concurrent with the first three of Biltz' holyday-tetrads. But before we discuss that sense, we must again digress.

There are four basic ways eschatologists interpret the prophecies of the Bible's Revelation to John. Preterism views most of the book's prophecies as having been fulfilled in the time of Jerusalem's destruction, AD70. Historicism sees the fulfillment of its prophecies in the history of The Roman Empire's fall. Futurism places its prophesied events in some future time of a seven year span of horrors called "The Tribulation", after which Christ returns. Idealism sees Revelation's prophecies as meaning the many kinds of glorious and disastrous events which happen in every person's life. God did not inspire the Bible to agree with any one school of eschatology. But each of the four interpretations of Revelation offers some of the same sense the Word of God, itself, makes of Revelation. It was an extraordinarily Sethite way to present a prophecy. Each school of interpretation insisting that it alone is correct is an naturally human way of being false.

Consider Revelation's seven headed beast which rose out of the sea. The angel who gave this revelation helped John understand it. And although its rising out of the sea was definitely a futuristic statement about the Antichrist to come, the angel presented this beast as embedded in a very real context, thrashing from the past into our present time to become that specific figure in the future, the Antichrist, risen from a vast sea of historic, churning and foaming malevolence against God and His innocents of humanity.

> [9]This calls for a mind with wisdom: the seven heads are seven mountains ... [10]they are also seven kings, five of whom have fallen, one is, the other has not yet come, and when he comes he must remain only a little while. [11]As for the beast that was and is not, it is an eighth but it belongs to the seven, and it goes to perdition.
>
> Revelation 17:9-12

Clear Signs of Trouble and Great Joy

"Mountain" is a Biblical metaphor for "kingdom", therefore, the seven heads are also seven kings. The eighth king yet to come is the Tribulation's Antichrist, the beast and his doomed kingdom. He is generally thought to be the dispatched seventh-head having been returned to life again. The first six, historical kingdoms are identifiable as: Egypt, Assyria, Babylon, Persia, Macedonian Empire, and the Roman Empire. (Historicists, Preterists, Idealists, and even some of the Futurists will object to that observation, but the coherence of God's Word with history supports it.) John knew those six kingdoms, being culturally Jewish and devoutly Christian, though he didn't know the seventh, because that kingdom was yet in his future. The time of its demise and the cause of its fatal wound are well known history, today, although its end has rarely been viewed in Revelation's light, if ever.

The prophecies in God's Word do not foretell only events. They disclose threads of spiritual processes upon which historical events string like beads. Each of the seven dragon-heads attempted to spread its control over the entire world. It is essentially what the beast is about, world domination, elevated selfishness. This beast of Revelation is a bead strung on the lust-for-world-control thread of the "I AM But You're Not" weave. God will grant forty-two months (three-and-a-half lunar years) of success to its attempt to control the world. Then Jesus shows up. He will come responding to Israel's forlorn call to their ancient God in a time of unparalleled crisis when they're hopelessly cornered by this selfish beast. After treading the grapes of wrath, Jesus will lead the world into love, peace, and joy.

> ...and the government will be upon his shoulder.
>
> (Isaiah 9:6)

Jesus' reign is a single historical bead strung upon a unique thread of spiritual process within the righteous-effects-of-truth weave. For although it is not concretely visible, today, His eternal reign is discernible in the abstract of all He has made and its history. His coming and crushing and binding Satan for sequestering from mankind and to throw the Antichrist into the lake of fire will be His final dealing with the opposing spiritual string upon which is strung countless beads of corrosive liars calling themselves politicians, rulers, governors, and religious leaders.

The string to be vanquished came into God's physical creation at the Tree of Knowledge of Good and Evil. There it was tied to the snake, the ancient serpent. It strung first Eve's heart, then Adam's, and every human heart after them, even stitching into Christ's heart as He necessarily became everyone else's sins strung up on the cross. This string's enticing flavors and scents whip into an even greater, modern wonderland of godless fancy fluffed up by wayward pop-stars, deceptive educators, fake-news, lying politicians, and slick religious leaders. The string continues on to the Lake of Fire, wherein the snake will be cast with everyone who failed to restring their hearts at the cross of Jesus.

But don't think the seven-headed dragon is a symbol of toxic government alone. It empowers governmental misbehavior enabled by every individual's willingness to settle for anything less than the truth, which, after all, was Eve's mistake made at Draco's behest.

History is a long running horror of kings slaughtering brothers and families out of power lust. The Antichrist will murder masses of people over its lust for God's crown.

Apophis and the Blood Moons

Power corrupts the lessers of God. Even the one as wondrously made as was Lucifer could not resist corruption. We find both ends of this corrupt, spiritual string tied to Satan, first as the initiating influence of evil embodied in Eve's smiley faced snake, then finally as evil's complete embodiment in the person of the blood drunk Antichrist. Between the beginning and end of this evil string, seven heads of the beast are clearly discernable by their involvement with Israel and relationships to one another. Even the early Sumerians, depicting a slain head of a seven-headed dragon, sensed this spiritual string of evil centuries before Moses began writing The Law.

It is interesting how God drops subtle hints upon men of all times and persuasions. We could, instead, say the author of Revelation merely reflected a cultural knowledge of the Sumerian's seven-headed dragon sporting a slain seventh-head, if it weren't for the actual, historic slaying of the seventh empire to have controlled Israel's Promised Land.

As we noted earlier, the angel informed John that seven kings were represented by the seven heads of a draconian beast. He said five were, one is, and another is yet to come. The angel placed them in time, thus placing them in reality. Although history shows many more than seven kings, it shows only seven who engaged two specific aspects: 1) each ruling for a time over God's land promised to Abraham, the place for I AM's reign, and 2) each defeating the one before it, excepting, of course, the first. Their kingdoms interlink to form an historic chain chronologically stretching from ancient Egypt through 1923. This chain binds the Middle Eastern world to a continuous history of aspiration for world dominion by draconian rulers who each thought himself to be God (or to be Caliph for Allah, in the case of the last one). Each empire opposed I AM WHO I AM's rule and lorded it over God's people. The eighth king, the fatally wounded seventh-head resurrected, actually will achieve world dominion. But only for a comparatively short moment.

So there is in this seven-headed beast a metaphorical representation of lust for God's rule -Serpens stretching forth to seize the Northern Crown. The seven headed dragon represents thirty-six-hundred years of humanity's continuous effort to achieve world dominion. However, a particular Antichrist will come. His kingdom will be a revival of the Ottoman-Turk Empire, or at least in some manner, a redevelopment of the Islamic Caliphate, which was also dissolved with the end of that Empire in 1923. The Antichrist will tyrannize the world from Jerusalem for the final three-and-a-half years of a seven year Tribulation.

Another example of this kind of metaphorical/particular representation seen in Revelation's imagery is the portent of the woman clothed in the sun with the moon at her feet and crowned with twelve stars. Two millennia ago, a Jewish virgin gave birth to a child. Satan pursued the child into the wilderness but could not win His worship there. So Satan assaulted the child by crucifixion, but could not secure His death. The child was resurrected and caught up to God's throne away from the dragon's gluttonous teeth. The child is easily recognizable as the historic Jesus Christ. Metaphorically, the woman was Israel. When Satan could not capture the Christ by either allure in the desert or assault with the cross, he poured out a flood of offense against Israel. Jerusalem was destroyed; the Temple was destroyed; and the Israelites fled into the wilderness of Gentile nations.

Many theologians say the metaphor of the dragon trying to devour the child even stretches back through Israel's history to Cain slaying Abel, Satan's first attempt to avert the

Clear Signs of Trouble and Great Joy

coming of The Promised Seed. Thereafter Satan tries to destroy the woman's offspring: the Hebrews' male babies were thrown into the Nile, Ahasuerus' decree to destroy the Jews from all Persian provinces, Herod's murder of Bethlehem's children. The flood pouring forth from the dragon's mouth flows on from the destruction of Jerusalem, through the shattering of Bar Kocba's rebellion, through all the antiSemitic pogroms thereafter, through the death camps of Hitler, to the final days of the Tribulation when Israel will call upon their God, I AM WHO I AM. Israel's history amongst the nations has been a flight from one pogrom after another, as if it were pummeled about in the eddies of a flood. Having reached a near crescendo at Hitler's Final Solution, the flood will crest at The Antichrist's bloody rampage. God will again take Israel upon His wings to faithfully preserve a remnant for His promise, as He always has.

And yet another eschatological prophecy finds a metaphorical fit in history and a literal fit in the future Tribulation. It's sense doesn't come from The Book of Revelation, though it is closely tied to Revelation's prophecies.

> [7]For the mystery of lawlessness is already at work; only he who now restrains it will do so until he is out of the way. [8]And then the lawless one will be revealed, and the Lord Jesus will slay him with the breath of his mouth and destroy him by his appearing and his coming.
> II Thessalonians 2:7-8

What is this mystery of lawlessness? John writes,

> Every one who commits sin is guilty of lawlessness; sin is lawlessness.
> I John 3:4

The Greek word for "sin" in this verse is *harmatia,* a word well known to mean *"missing the mark",* or simply, as being *"an error",* like just *"a simple mistake"*. It is any imperfection or mistake, whether intentional or not. The Greek for lawlessness is *anomia,* meaning *"illegality,* that is, *violation of law* or (generally) *wickedness."*[24] Lawlessness involves everything from the subtlest of intentional errors to the worst offenses. The mystery of lawlessness is seen in the serpent's enticement of Eve to eat the forbidden fruit. Just eating a fruit seems totally innocent. But eating a fruit was not the point. Ignoring a law was. Before this event, no imperfection had occurred in God's physical creation. No law had been ignored. Creation's physical substance was still perfectly orderly, as was Adam and Eve's spiritual substance. But Eve's teeth breaking the skin of that fruit put the first crack into the orderliness of physical creation. The consequences of her anomic[25] snack planted lawlessness directly into the soul of mankind.

There is now a seed of disdain for God's law in every heart. Eve's mistake tainted the very essence of the human spirit, thus separating our spirits from God's Spirit, for He can not commune with imperfection. So Adam and Eve died an immediate, spiritual death in the day they ate the fruit. Also, in that very moment, the eternal life process of their physical bodies ended and a physical death process began within all the creatures of the earth. Her eating of the fruit did not consummate ultimate physical death; it immediately consummated the death process. Disorder entered their DNA as surely as the second law of thermodynamics spread to all the material of their physical being, becoming the most basic law of physics, everything constantly seeking its lowest state of being, that ever trickling flow of things from order to chaos, i.e., death. Death did not culminate that day. It came. Eve died with Adam on their feet as the nature of their situation completely changed. They now felt naked. They felt the need to hide from God. And from that moment on, every child, every animal, every plant sprout conceived was born to die. Man became a spirit separated from God and attached only to a physical body now living towards death.

    Imperfection's constraints sequestered man from God and His angels and became the taint of the entire physical creation. Man was now a reflective creature cut off from the light of righteousness. Lawlessness received its chance to ruminate within all the reflections of mankind's population. By the process of popular influence and the nature of vengeance, the waxing of lawlessness should have ended mankind long ago. But throughout history, the mystery of lawlessness has been restrained by one means or another of God's doing. He's not a Deist.

> [7]For the mystery of lawlessness is already at work; only he who now
> restrains it will do so until he is out of the way.
> II Thessalonians 2:7

    From its first settling until recently, America has thought of itself as a refuge for lawfully minded people who are concerned about maintaining justice for everyone, without partiality. America wasn't perfect. After all, it was formed by humans which the Bible says are all false, even if its formation was influenced by God who is wholly true. In spite of its failures, America stood up for the individual's right to choose the place and nature of his or her life, to make his own decisions, set his own course, establish his own business or work his own desired job, enjoy the produce of his own hands, love his chosen bride, make his own family, teach them his values, or reject culturally taught values. At least in theory.

    We all know that America's practice wasn't perfect. But the heart and soul of what made America great was the "love for thy neighbor" embedded inside each person, a love made of honor for every person's freedom. The maintenance of orderly lawfulness was the extent to which America's government was legally empowered to rule over its sovereign citizens. Its governance was exceptional in a world of pompous, self serving, fascistic governments.

    But America is not the kingdom of righteousness. God never intended it to be. Nor did it advertise itself to be perfect. In fact, The Bill of Rights is a legal limitation of the government's power because America's founders understood the tyrannical dangers unleashed by empowering any human soul. A citizenry empowered with collective sovereignty might eventually begin voting to eat each other for dinner, as the movie "Soylent Green" imagined and the Swedish Professor, Magnus Soderlund, has now suggested. America advanced itself to defend goodness and a moral sense of individualism from the collective powers of democracy in spite of its own shades of darkness. The point isn't whether or not

Clear Signs of Trouble and Great Joy

America ever was perfect. The point was America's desire and effort to empower its citizenry for each to individually seek righteousness instead of evil.

Certainly America's "righteousness" was not perfectly legislated, administered, and adjudicated. But it lived in the general character of its people rather like a mystery of lawfulness. And as such, John Adams, second president of the United States, expressed a deep truth about reality in his 1798 speech to the military,

> We have no government armed with power capable of contending with human passions unbridled by morality and religion . . . Our Constitution was made only for a moral and religious people. It is wholly inadequate to the government of any other.

Ethics, morality, godliness, righteousness, and such can not be legislated no matter how much truly wonderful people may try. These things can not be pressed upon a man's soul from outside his being any more than squeezing a ball of dirt will produce a stalk of corn. They are character traits which must root within a person's spirit and grow into his soul from a sincere and basic desire to do good in every situation. Seeding and cultivating goodness is the importance of godly family and culture. A good culture inspires ethics, morality, godliness, righteousness, and such into its people like a good field produces healthy wheat. A good family roots these attributes in its children by conveying to them the desirability of righteousness' benefits. It is a principle of psychology that people generally do right to those who do them right. The individual sovereignty of America depended upon godly character being reflected in the behavior of its citizens more than it depended upon the skills, abilities, and social gifts of overbearing leaders.

> Only a virtuous people are capable of freedom. As nations become corrupt and vicious, they have more need of masters.
>
> Benjamin Franklin

Masters love to be needed. They make everything about themselves. The adept con artist knows his most vulnerable target is a desperate, sensual mind ready to grasp for whatever it wants before considering what it really needs. Mastering a population requires ruining the ethics, morality, godliness, and righteousness of its culture. Demoralized souls lack the ability to fend for themselves. They need masters. They are uprooted from the nutritious soils of mutual interaction. Then con artists sew mysteries of lawlessness into fields flattened by the social distancing of cowardice and selfishness. Con artists are history's emperors, kings, politicians, and their advisors. They are today's journalists, entertainers, and educators puffing themselves up to be influential and important on account of their audacity to imagine more than to know by research. They obfuscate the common sense of what is good to put lipstick on vice for cultivating lawlessness. They are draco.

So profoundly active was this mystery of lawlessness early in the twentieth century that even America dipped a toe into its draconian waters. Most Western countries were experimenting with fascism, a new political fad. This tendency was called the fascistic moment of the Western World. The resulting lawlessness of that moment rose to the level of

Apophis and the Blood Moons

Hitler's Holocaust, something the world should never have allowed. American influentials were giddily aflutter with Hitler's optimism for socialistic prosperity supplied by the new tower of technology. France snoozed at Hitler's reentry into the Rhineland. Everyone made excuses for his taking Austria. The Jews paid the price for the stupor of left-wing politics. Today, the world snoozes again as left-wing politicians ascribe peace to Islamic fascism.

> [8.65] O Prophet! urge the believers to war; if there are twenty patient ones of you they shall overcome two hundred, and if there are a hundred of you they shall overcome a thousand of those who disbelieve, because they are a people who do not understand.
>
> Koran: The Ascension 8.65

Granted, America did not single handedly end Hitler's nightmare. But America was an indispensable element restraining his lawlessness, which, if left unchecked, would have annihilated the Jewish people. The gates of the concentration camps swung open, and the helpless innocents wandered into the blessing of their own sovereign homeland three years later.

How big was President Truman's immediate recognition of the new Jewish state?

> As the UN session was drawing to a close, simultaneously with the British mandate in Palestine, without a new decision, Israel, on May 14, took matters into its own hands and proclaimed its existence as a sovereign state. Boldness had its instant reward. Without warning even to the State Department officials who, to the last minute, were seeking to stave off partition and Jewish independence together with all their consequences, the White House immediately announced recognition of the new Jewish commonwealth. The Soviet Union swiftly picked up its cue, recognizing Israel *de jure* three days later.[26]

Nobody can off load his belongings into the middle of a neighborhood street, pitch a tent beside the pile, and call himself "a neighbor in the hood" without recognition as such by the other folks on the block. People are not "neighbors in the hood" until the neighborhood determines they are neighbors, however right or wrong that might be. Neighborhood is a mutual matter. So, the first neighbor to accept the newcomer begins the process of others doing the same, because man is a reflector. President Truman's immediate recognition of Israel's announced statehood broke the international ice. And Israel became a neighbor in the world's hood.

America began helping with Israel's expenses of settlement in 1949, and has thereafter become Israel's biggest support in hindering Arabic opposition.

> The United States has provided Israel with $233.7 billion in aid (after adjustment for inflation) since the state was formed in 1948 through the end of last year, research by TheMarker has found…
> In nominal terms, total American aid was $112 billion over the

Clear Signs of Trouble and Great Joy

years, according to data that appears on the website of the U.S. Congress.[27]

But for God's purposes to be achieved in Israel's time of trouble (The Tribulation, a.k.a, Jacob's Trouble), Israel must face the world's unrestrained hostility alone.

Ever since God led Israel out of Egypt, His message to her has been, "Rely on Me. I am your army. I am your supplier. I am your providence. The other nations are not your providence, I AM." Hosea's prophecy said they would be raised up after they called out to God, not after they turned to America, or to any other nation. Of course, Israel, even now, has abandoned I AM WHO I AM to worship evolution and secularism, as they have historically whored with one idol or another. So, eventually, they must stand naked before the world's savagery, stripped of allied support, left with no call out for simple, human decency, like a wet, cold, animal cornered in the sight of all mankind, angels, and God, to witness I AM WHO I AM's faithful response.

To this day God is Israel's army in a very literal sense. Israel's time of great trouble will end with Jesus Christ coming at their desperate cry for help. The mere breath of His word will end the rabid dog pack of nations gathered in The Valley of Decision thinking to destroy His people. He will purge wickedness from Earth in a word, and then

> [14] The sons of those who oppressed you
> shall come bending low to you;
> and all who despised you
> shall bow down at your feet;
> they shall call you the City of the LORD,
> the Zion of the Holy One of Israel.
> [15] Whereas you have been forsaken and hated,
> with no one passing through,
> I will make you majestic for ever,
> a joy from age to age.
>
> Isaiah 60:14-15

Israel's situation in the Tribulation, alone against the world, will in the end set them at the forefront of all the nations.

So America, which God raised to stand by Israel, began a decline into becoming the same evil it had formerly hindered. The ways America has collapsed from its early quest for righteousness seem innumerable. Fifty years ago our traditional values and conservative ways were put on public trial by socialist inspired children rebelling against what gave them the freedom to rebel, as if they were prophets of a better and newer order with some dawn of Aquarius mesmerizing their better senses. The deception of this youth movement eventually scuttled the public's ability to think rationally. Its intellectual substance re-rooted in the pragmatic, Progressive ambitions of the early twentieth century. Most news outlets, books and journals, movies, sitcoms, musicians, artists, authors, and poets joined the crowd of barely twenty-something infants of the 60s. Together they coerced America into rolling over,

Apophis and the Blood Moons

like a love struck mutt, for nihilistic[28] socialism camouflaged as concern for the environment, the poor, and the minorities. The mesmerized public lapped up the same utopian dream their parents fought against two decades earlier. But this time utopia's pushers wore a friendlier face of a more smiley fascism. America turned from its restraint of lawlessness and entered the race for which draconian government would carry the utopian baton across the anomic finish line first.

To think we still chase man made utopia after observing the results of Hitler's fatal attraction to it makes much of my point. The promise of utopia intoxicates people of minimal thought. But utopia is not for man to create. It is for God to work. The righteousness utopia requires must well up from inside individuals, as America's Founders correctly stated. It can be neither legislated nor bullied into them. The prospect of satisfaction and justice for all can only occur within a people who each sacrifices his or her own selfish ambitions to think and do what affects everyone rightfully. It must come from a sincere, personal desire for everyone's good and a genuine respect for everyone's sovereignty. And that's why John Adams said America's Constitution would only be able to govern a moral and religious people. It is why America's first reading primer was composed of Biblical Scriptures. It is why America's first universities were theological institutions. It is why the Constitution recognized the sovereignty of the individual as a right given to everyone by God Himself, not by man. The Word of God insists that I AM WHO I AM creates utopia, while man only creates messes.

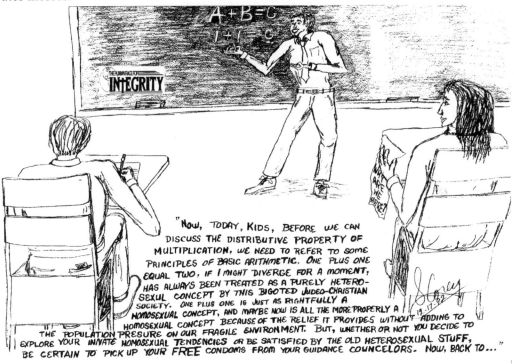

Figure 42

Clear Signs of Trouble and Great Joy

Socialist philosophy mesmerizes the intellectually irresponsible public into willfully engaging in political crime while masquerading as lawful Progressives. God did not make unique individuals to transform their moral characters into nihilistic weaves of socialist excuses. As America began drawing upon hormone charged adolescents for advice in the 1960s, the rules of responsible thought were thrown into the same dumpsters which had caught any *information* that might possibly falsify science's guessing against God. Consequently, we are no longer taught to look down the front of our pants for knowing which sex we are. The main point of school before the 60s was to equip minds for thinking. Today, it is to stoke minds against the traditional foundations of American culture. We now eat America's apple pie without casting so much as a thought onto America's old ways of truth and justice for all.

The deceitfulness of the Obama administration reached its pinnacle under the 2014-5 special tetrad. Once proudly (but not perfectly) standing for truth and justice, America had never sunk to the anomic depths of political lying achieved by Barack Obama's cadre of deep state illusionists propped up by a leftward driving media. Never before was America's hope so audaciously misdirected and its government so fraudulently transformed into the very despotism its Constitution was crafted to prevent. America's resistance to Islamo-fascism all but vanished. In its place arose a disdain for Israel. The "ingenious" Obama gave Iran a friendly veil in the form of a nuclear arms deal behind which they could advance nuclear armament, and several pallets of cold cash for funding terrorism.

The scheduled adoption day for the Iran nuclear deal, July 14, 2015, came eleven weeks (the Biblical number of chaos and disarray) before the final total lunar eclipse of the 2014-5 holyday-tetrad. While withdrawing American troops from Iraq, the Obama administration had been covertly supporting the Muslim Spring. By the end of 2015 it appeared the restraint of radical Islam had turned into an outright jailbreak for jihad. The nation once hindering the mystery of lawlessness was beginning the anomic goose step of the twentieth century draconian fascisms. The nation which fifty years earlier staunchly defended the individual sovereignty of its citizens was now being led by a Marxist President vowing the fundamental transformation of America's essential core of freedom.

America was not the ultimate hindrance of lawlessness; The Holy Spirit is. The eventual end of America's participation in this hindrance was clearly displayed in the nature of what happened under the 2014-5 holyday-tetrad. The beginning of the process of ending the hindrance of lawlessness was the point of that 2014-5 special tetrad, rather than Israel winning control of the Temple mount. Like the 1493-4 special tetrad, it began an epochal process more than it marked any event of great historic significance. Never before did Americans march through their own neighborhoods chanting, "What do we want? Dead Cops! When do we want them? Now!" and "Pigs in a blanket! Fry 'em like bacon!" while receiving blessings from the news media, praise from Hollywood, and defensive cover from the Oval Office. The 2014-5 holyday-tetrad marked the beginning of a process which would soon meld America with the very lawlessness God rose it to restrain, leaving Israel to face the hostility of the Tribulation alone.

Now that Israel has developed into a nation again, they must become isolated and cornered by a hostile world for God to show the special relationship He reserves for their

Apophis and the Blood Moons

remnant in accord with His promise to Abraham. This set of four tetrads with each of their sixteen eclipses occurring on a Jewish holyday marks the development of Israel and Jerusalem for the Tribulation stage. There Islam will momentarily steal the glory God meant for Israel and briefly usurp the crown belonging to Jesus Christ.

We must admit the remarkable nature of Israel's history. It isn't the oldest culture. China has been a culture longer than Israel. But no other nation has twice been expelled from its homeland and twice returned there to become a nation again. No other people has maintained its cultural identity while wandering homelessly through so many hostile nations for eighteen-hundred years to become a nation again where it had been a nation before. These historic realities can not be refuted. Israel's history is remarkable in itself. But its own Oracles written at least twenty-five centuries earlier (according to its accusers, thirty-five centuries earlier according to evidence) predicted both the wandering and the return.

Those circumstances overburden the definition of "remarkable". They exceed all sense of "coincidence". Then what term can we possibly apply to total lunar eclipses happening on that people's major holydays at the occurrence of events developmental to their becoming a nation again in possession of the city I AM WHO I AM long ago proclaimed to be His?

> [10] Foreigners shall build up your walls,
> and their kings shall minister to you;
> for in my wrath I smote you,
> but in my favor I have had mercy on you.
> [11] Your gates shall be open continually;
> day and night they shall not be shut;
> that men may bring to you the wealth of the nations,
> with their kings led in procession.
> [12] For the nation and kingdom
> that will not serve you shall perish;
>
> Isaiah 60:10-13

It goes beyond the concept of "remarkable coincidence". When we consider Jesus' statement that signs would happen in the sun, moon, and stars as His return approaches, what can we rationally call the coherence of these holyday-tetrads with Biblical prophecy and prophecy-fulfilling history? Enormously remarkable coincidences having nothing at all to do with the (yuk) Bible and Jesus' coming? How much coincidence can reality produce without it meaning a thing? Could Darwin have guessed wrong about the universe having no design or purpose?

Then Apophis conveniently comes fourteen years after the last of these four holyday-tetrads to bookend a seven year period made metaphorically similar by celestial signs to the seven years Revelation states will be for evil's end. It participates coherently in this system of orderly, quite perspicuous signs, prophecies, and historical events. Is it possible that Apophis' two brushes past Earth might actually indicate the time of the Tribulation's beginning and end when Israel will be raised to the forefront of the nations? Holyday-tetrads stood like road signs at the reestablishment of Israel possessing God's Holy

Clear Signs of Trouble and Great Joy

City. maybe Apophis is also a road sign. Have we seen sufficient *information* to regard Apophis as a warning to stop equating Jesus with the tooth fairy and start spiritually preparing for His physical return? Or should we look for more evidence?

Footnotes

1. McCoy, Dan. http://egyptianmythology.org/gods-and-goddesses/apophis/ 2014-2016
2. McCoy, Dan. http://egyptianmythology.org/gods-and-goddesses/ra/ 2014-2016
3. https://en.wikipedia.org/wiki/Ape. Wikimedia Foundation, Inc.
4. https://en.wikipedia.org/wiki/99942_Apophis. Wikimedia Foundation, Inc.
5. Kovacs, Joe. "Is 2nd coming of Jesus etched in night sky?" April 18, 2010. WND.com. 1997-2013
6. Ancient Book of Jasher: Biblefacts Annotated Edition. 2008. Ken Johnson, TH.D. Biblefacts Ministries. biblefacts.org. Ch 1, Sec 5, Vs 9. (Pg. 13)
7. A pogrom is more than persecution. Although a persecution can include a massacre, it can also merely be the unreasonable limitation of a certain group, class, or race of people by crafty regulations. Its severity can rise to imprisonment and confiscation of property. But when the severity rises to actual massacre, as at Wounded Knee, Cripple Creek, Auschwitz, and Planned Parenthood, they are pogroms, being organized massacres of certain, targeted categories of people.
8. The Muslims are not as detached from Judaism and Christianity as most folks think. They trace their own heritage to Abraham through Ishmael. However, their religious beliefs were not articulated until AD610. In fact, before AD610, there was no concept of "Islam". Islam came from a powerful spiritual entity supposedly sitting on Muhammad until he feared for his life and relented to receiving "Allah's Revelations" for correcting the corruption the Israelites had allowed to accumulate in "The Book" (the Old Testament). And so, the Jews were considered by these Muslims of Medieval Spain as "People of the Book".
9. Spivey, Larkin. Miracles of the American Revolution: Divine Intervention and the Birth of the Republic. God and Country Press (an imprint of AMG Publishers) 6815 Shallowford Road, Chattanooga, Tennessee 37421. 2010. Pg ix.
10. http://www.valuesvotersnews.com/2008/12/george-washingtons-view-of-gods-role-in.html
11. Federer, William J. Samuel Adams knew what would overthrow America's liberties. 2016. Americanminute.com.
12. https://en.wikipedia.org/wiki/History_of_Religion_in_the_United_States. Wikimedia Foundation, Inc.
13. The main characters of Rush Limbaugh's enlightening American history set for children.
14. http://www.cnn.com/2014/06/05/opinion/opinion-d-day-myth-reality/index.html.
15. Hanson, Victor Davis. Why America Was Indispensable to the Allies' Winning World War II. National Review. May 14, 2015. http://www.nationalreview.com/article/418329/why-america-was-indispensable-allies-winning-world-war-ii-victor-davis-hanson.
16. Dayenu is a Hebrew expression meaning, "It is sufficient for us." The traditional Passover seder includes a song titled *Dayenu*.
17. Strong, James. Strong's Hebrew and Greek Dictionaries. Quickverse Electronic Addition STEP Files. Findex.com, Inc. 2003. At H5712 and H5707.
18. Rohl, David. Exodus: Myth and History. Thinking Man Media, 6900 West Lake Street, St. Louis Park, MN 55426. 2015. Pg 263.
19. Halpern, Ben. The Idea of the Jewish State. Second Edition. 1969. Harvard University Press, Cambridge M, Massachusetts. Pg 15.
20. https://www.archives.gov/files/education/lessons/us-israel/images/recognition-telegram-1.jpg

Apophis and the Blood Moons

21. The "sea" is an often used Biblical metaphor of people acting without harmony, showing no correspondence with the actual truth of matters, such that the "roaring seas" and "raging nations" are similar (though not identical) concepts of selfish, God-ignoring, chaotic activities of mankind which usually turn violent and murderous. Israel has passed through the "sea of Egypt" to exist as a nation again. But the smiting of the seas' waves, the drying up of the Nile, the departing of Egypt's scepter, and Israel's eventual glory in the Lord's name are among other culminating events of this same prophetic process begun at Israel's passing through Egypt's "sea". We are yet in the middle of the prophecy's fulfillment.

22. The popular book on the blood moons phenomena written by John Hagee, as well as the video he produced on the same topic imply Mr. Hagee discovered the phenomena. But indeed, Mark Biltz discovered it and showed it to Mr. Hagee.

23. November 9-10, 1938, Hitler's paramilitary Brownshirts stirred death and destruction into Jewish lives throughout Nazi Germany. Businesses were raided and windows were broken throughout Jewish communities, whence the name: Kristallnacht. Sledge hammers were taken to what wouldn't break so easily; buildings were burned to the ground. No restraint for health and life was shown. Estimates of Jewish fatalities vary. Considering those who died from maltreatment while incarcerated, and more who committed suicide, besides those who were murdered in their streets, businesses, synagogues, and in their own homes during those two days, death tolled in the hundreds. Thirty thousand were incarcerated in concentration camps. Live, worldwide reports of the event shocked the world as much as 9-11 did sixty-three years later. And like 9-11 changed our world, once Kristallnacht launched Hitler's Final Solution -The Holocaust- onto the world stage, the world then was never the same again. Mankind's shame-filled disdain for God's chosen people came to a festering head over the following seven years, like an ugly zit on the face of human history. But zits return, usually on prom night, bigger, and uglier, when we think ourselves to be the most beautiful. The Tribulation will begin with the look of prom night. Everyone will be enamored with the character who will become Antichrist; everyone will be saying, "Peace! Peace! Peace!" The next holocaust will be around the corner.

24. Strong's Hebrew and Greek Dictionaries. James Strong. Quickverse Electronic Addition STEP Files. Findex.com, Inc. 2003. At G458.

25. Merriam-Webster's Eleventh Collegiate Dictionary defines "anomic" as: "social instability resulting from a breakdown of standards and values; *also* : personal unrest, alienation, and uncertainty that comes from a lack of purpose or ideals."

26. Halpern, Ben. 1969. Pgs 375 and 376

27. http://www.haaretz.com/israel-news/business/u-s-aid-to-israel-totals-233-7b-over-six-decades

28. Merriam-Webster's 11th Collegiate Dictionary defines nihilism as: "1 a : a viewpoint that traditional values and beliefs are unfounded and that existence is senseless and useless b : a doctrine that denies any objective ground of truth and especially of moral truths 2 a : a doctrine or belief that conditions in the social organization are so bad as to make destruction desirable for its own sake independent of any constructive program or possibility b *capitalized* : the program of a 19th century Russian party advocating revolutionary reform and using terrorism and assassination." "Nihilistic" is the adjectival expression of "nihilism".

Clear Signs of Trouble and Great Joy

Dayenu.

Chapter 6
The Rest of the Blood Moons

^{25}And there will be signs in sun and moon and stars, and upon the earth distress of nations in perplexity at the roaring of the sea and the waves, ^{26}men fainting with fear and with foreboding of what is coming on the world; for the powers of the heavens will be shaken.
Luke 21:25-26

    These tetrads of total lunar eclipses occurring on major Jewish holydays at formative events of Israel's statehood and repossession of Jerusalem are empirically observable. Use NASA's lunar eclipse list to find the tetrads, and then use any of several websites to determine the historical Passover dates for seeing the holyday-tetrads. Therefore, since you can see them, they are undeniably real. The only remaining question about this phenomena is whether it happened by mere coincidence, as if there is no God, or whether they are signs designed into the physical universe specifically for the purpose of a Holy God's warning to mankind. Common sense says these four special tetrads show more concurrence than mere happenstance can rationally explain. Some philosophers of science say the best explanatory theory is the one able to explain the most phenomena. There is just too much informative order between these four special tetrads, the significant events at which they occurred, and the Biblical prophecies of those events written thirty-five hundred years earlier. Coincidence explains little. A loving, warning, righteous God determined to deal evil out of humanity's game is the only reasonable theory for explaining this ancient system of dots on history's map.

    If the mechanistic material universe could produce one most amazing of all coincidences, then it would be this peculiar timing of eclipsed moons with holydays adorning the key historical events of today's Biblically prophesied Israel. And we yet have three more chapters of extraordinarily converging realities to examine. Maybe the material universe is not as mechanistic as much as it is designed; maybe it isn't as purposeless as much it is communicative. Then again, maybe coincidences can be bigger than anyone imagined.

    In the semi-arid desert between Delta and Grand Junction, if you look closely at the hilltops to the east of the highway, you can see a few, distant piles of rocks generally set up on the western edges of a few hills. I've been told the Indians piled them there to use as blinds when driving game across the desert floor and over the edge of the cliff to the west. That narrative possibly explains those rock piles. But I also suspect the presence of the valley floor, the nearby canyon, and the known practice of ancient hunters to drive their game over cliffs might have ladened those piles of stones with that way of game hunting. They could just as well have only been markers, maybe even astronomical markers. Ancient

Clear Signs of Trouble and Great Joy

cultures around the world have left behind sky observatories ranging from the most rudimentary combinations of a few carefully situated stones to the complexity of Stonehenge and the glory and mystery of Gobekli Tepe. But those monuments also laden these stone heaps with "observatory" sense. At the end of my pondering those stone piles, the only thing I know about them for certain is that nature did not heap up the stones; beasts did not stack them; people did. To know why people stacked them, we need to know their stackers.

Sometimes, when hiking Colorado's trails through the brush and timber, we will find a pile of stones at an intersection of trails. We automatically suspect someone arranged them to be a marker, since everyone knows nature doesn't neatly stack piles of rocks at trail intersections. The context of the trail and people and intention and intersection naturally ladens the rock stack with a sense of some communication regarding that particular intersection, like maybe, "Turn here." Man marks places, times, and things of relevance. Recall the markers made by Indians centuries ago where they had evidently seen dinosaurs. (Or, where ancient Indians had discovered the marvels of paleontology and correctly assumed the shape of dinosaurs from a few bits of fossils they found.) Man is a marker maker. So he is also a marker reader. God knows this. So maybe God has made some markers for man to read? The Bible says He did (Genesis 1:14). And behold, we see within the chaotic backdrop of five-thousand years of lunar eclipses some total ones stacked upon Jewish holydays beside times of very significant events relative to Israelite history prophesied in the very holy Bible.

Understanding exactly what the marker pile means requires knowing the marker maker. If your friend told you to follow a trail to an intersection of paths at a pile of stones and snap a picture of the magnificent peak in the background, everyone else would think the rock stack might indicate where to turn, not knowing what your friend said. But if everyone fleeing a campground under attack by sasquatch and grizzlies sees a rock pile beside an arrow with the word "hurry" scratched in the dirt at an intersection of paths, the whole situation entails enough informative elements of its own to achieve a perspicuous message without the stack observer knowing the rock stacker. If there were no attacking sasquatch and grizzlies when the observer found the rock stack, he might assume somebody left directions for his girlfriend. But the sasquatch and grizzly raid, the fear, the flight, the stack, the arrow, and the word, naturally laden the marker with the meaning, "direction to safety," although a morbid prankster could have been the marker maker, or a soon to be embarrassed couple.

The point is that eventually enough *information* can be contained in a system of circumstances for it to be understood as communication. And when the circumstances of such a system also correspond to the symbolism of an ancient narrative, then the communicated meaning becomes even more discernable.

Maybe the prolific attacks of radical Islamists and insatiable Progressives are hardly sasquatch and grizzly raids (yet). But a lot of people are beginning to get an ominous feeling that a bad moon is on the rise, and that it might be nice to know where the tall grass is.

Is it only happenstance that the Bible foresaw the wandering Jews becoming a nation again under the bad moon of Islamic and Progressive raids on mankind's homesteads? These concerns raise the coherence of those holyday-tetrads, other astronomical events (e.g.,

The Rest of the Blood Moons

the Jupiter/Regulus "Christmas star"), Biblically prophesied history, and the undeniable reality that no one except the Holy Creator can stack motions of the heavenly bodies over prophesied events. Biblically prophesied events coordinated with meaningful alignments of the celestial bodies should verify the Bible's propositions about the identity of The Creator and the need to learn from the Bible where to find the tall grass and when to run for its cover.

These sixteen lunar eclipses of Jewish, holyday full-moons concurring with Biblically important events of a Biblically important people being reestablished as a Biblically prophesied nation are inherently ladened with a sense of important communication. Their convergence with the historical events of I AM WHO I AM keeping His promise to Abraham adds clarity to what they might be meant to communicate. But even that coherence does not verify the proposition that they specifically communicate when the Tribulation arrives, as opposed to just communicating, "We win someday, que serah."

However, if these eclipses were found to be part of a much greater system of vastly more correlations, their message would become more discernable the same way a puzzle is clearer when more pieces have been set into their places. Although the signs we've discussed so far do not inherently indicate a "when" for the Tribulation, we have arrived at enough coherence in them to understand they clearly indicate a "nearness" about that prophesied end of this age. The holyday-tetrads have been discernable for many decades. They should be old news to the entire world by now, along with their speculative meaning. But they are yet barely known. Even many preachers refuse to compare realities in the sky, history, and their Bibles for understanding the signs of the times, like the Pharisees refused to see the signs of their times.

Why do we struggle so hard against sixteen "stacked" eclipses? Is it simply because we do not want to believe the Bible is God's communication to man? Or do we believe it is His communication except that part about the sun, moon, and stars having been created for making signs? Like Darwin only guessed about nature being purposeless and non-designed, subsequent science only guesses about I AM WHO I AM not being who the Bible says He is and cold-nor-hot preachers only guess about the sun, moon, and stars not making God's talkative signs. It isn't that either of them discovered any *information* suggesting God's non-existence, or proving He has never written messages in heavenly patterns. It is only that they do not want to believe God is almighty, lest their teaching and preaching be the lamer, so they misconstrue *information* into data and facts they think will hide God from their ideas. Yet, these eclipses coherently intersect Biblical themes: the Israelites, Passover, Tabernacles, the land of Israel, God's Holy City, the one who hinders lawlessness, seven years of tribulation, and even the concept of the sun, moon, and stars' purpose for making communicative signs.

> [14]And God said, "Let there be lights in the firmament of the heavens…and let them be for signs…"
>
> Genesis 1:14a

> [25]And there will be signs in sun and moon and stars…look up and raise your

Clear Signs of Trouble and Great Joy

>heads, because your redemption is drawing near.
>
>Luke 21:25a, 28b

What light brightly shines from the timing of these eclipses! Coincidence explains far less about them than the Bible directly reveals. But does the light of their timing shine only to emphasize God's involvement in Israel's events? Or does it shine to warn of the impending approach of Christ's decisive blow to evil and it's workers? When Jesus said there would be signs, the context of His discussion was the end of our evil age. As such, do the aspects of these blood moons entail the end He was discussing? The holyday-tetrads definitely entail the key prophecies of the end, Israel being a nation again with Jerusalem for its capitol. But their core meaning is the Tribulation trouble passing over Israel (Passover) for Christ's dwelling there with mankind (Tabernacles).

These four special tetrads stacked at those prophesied events along history's pathway inform us of God's existence and His communicative nature just like rocks stacked at an intersection of trails informs us of another human's existence and communicative nature. Moreover, they inform of the universe's real design for a specific purpose. Darwin guessed wrong. Stacked rocks require design. Stacking them at certain places involves purpose. Designed purpose makes communication. The Bible states that, I AM WHO I AM made the heavenly bodies to communicate. And with our own eyes, we can empirically observe these holyday-tetrads from NASA's own records.

What shall we think!?

Maybe we should begin thinking about the significance of those tetrad-tagged events having, millennia ago, been foretold to happen when and where they did happen. Maybe we should be led to ponder the reason the Bible foretold they would happen. According to The Holy Bible, Israel became a nation in possession of Jerusalem for Christ to reign from Jerusalem, raising Israel to the head of the nations, and hoisting it up as an ensign to God's reality. And in that communication is found God's providential warning to not ignore the truth of His Holy Bible.

Rhinnie's Principle notes that misassembled *information* distorts the holes in a puzzle to being what only assumptions can fill. Distorted holes no longer match their appropriate puzzle pieces. Therefore, every assumption filling a deformed hole in a misassembled puzzle displaces the appropriate puzzle piece, rendering that piece into a useless anomaly headed for one of normal science's dumpsters. Every trashed reality is another rejected stone upon which God meant to found a bit of sanity. The assumption of evolution can not explain the Bible's ancient predictions of Israel's present condition, and even of its being adorned with meaningful eclipses, because a scrambled puzzle can not imply the shapes or colors of the real *information* meant for its assemblage. It only begs for more guessing to fill more holes made by the rejection of undesirable puzzle pieces.

But the puzzle as assembled around The Holy Bible has a space for every piece of *information* which normal science discards, as well as for every bit of *information* normal science exhibits. It isn't that all science is deceitful; it is that normal science will not correct its obvious errors and deliberate biases by any empirically observable *information* involving the Bible. But ALL *information*, especially the Bible, has a proper place in reality's puzzle

The Rest of the Blood Moons

since every piece of *information* is smoke of reality. And being *information*, the Bible is one of reality's loudest, smoking guns, the Central Piece of reality's own puzzle.

Without evidence, even against available yet ignored evidence, Charles Darwin insisted the universe wasn't designed for any purpose, such that all things forever were and would forever be nothing more than happenstance. But the correlation of ALL *information* shines brightly, conspicuously, and authoritatively upon The Holy Bible as being reality's centerpiece of knowledge purposefully designed into the universe. Of course, normal scientists will deny that this empirically observable mass of *information* in the heavens is attesting to God's design of and purpose for our universe. But that's what normal science does; it throws away *information* in order to safeguard pareidolic guesses and assumptive theories. But, regardless of what normal science says, the informative stars still circle the globe, night in and night out, day by day telling the glory of God and proclaiming His handiwork in the terms of ancient concepts and symbols.

Most people encountering the *information* formed by these special tetrads will choose either the irrationality of thinking it's purely coincidental, the irresponsibility of denying its existence without investigation, or the mindlessness of admitting its existence but refusing to think about how or why such clearly informative order exists. Modern man's mental habit is to deny The Holy Bible without honest consideration while castigating its propositions without investigation. Therefore, the deniers of God's truths truly overflow with opinions and assumptions, having given their minds a bill of divorce from evidence led research. Without sufficient knowledge to correctly discern the implications of this mass of very real signs, they will reject its message, deny its relevance, and meet the end the signs forewarn everyone to avoid.

> [6] My people are destroyed for lack of knowledge;
> because you have rejected knowledge,
> I reject you from being a priest to me.
> And since you have forgotten the law of your God,
> I also will forget your children.
>
> Hosea 4:6

The science-priests of the nineteenth century have led the twenty-first century science worshippers into teaching children the rejection of God's Word without study, consideration, or even simple respect for curiosity. A century-and-a-half ago, they proclaimed the Bible to be just another dusty old book well before they had found any evidence for swapping ape-man for The Creator. Even as these signs in the heavens were testifying to the historical fulfillments of key Biblical prophecies, the science-priests were fecklessly digging in dirt for some hint, any hint, of evolution. Although a little knowledge of the Bible is necessary to recognize the signs in the sky, the persistent lack of evidence for evolution in the fossil record doesn't need much knowledge at all to decipher. No irrefutable missing link between any "kinds" has been discovered to verify Darwin's imagined processes of evolution. On the other hand, the hidden link between the science-priests' theories and their biased ambitions is clearly discernable in their dumpsters being packed full of real evidences

Clear Signs of Trouble and Great Joy

contradicting their long held theories. Thomas Kuhn discovered that "missing" link.

Numerous evolution contradicting discoveries were silenced by the early geologists and paleontologists so their imaginations could propose evolution from a few scattered fossils of manlike monkeys. Michael Cremo and Richard Thompson discovered that missing link. The enormous complexities of the living world, DNA, the incredibly intricate bio-machinery of the cell, the unimaginably vast networks of the brain, and the irreducible complexity of interdependent organs, which simply could not have evolved one without the other, conclusively attest to the impossibility of evolution. Through the eons required for a fish fin to develop into a leg, how efficiently would the fin have worked while being half leg, or the leg have worked while being half fin? How well would the dinosaur's arm have picked at its prey while being half wing? How well would its scales have shielded its skin while being half feather? Just a little bit of clear thinking unlinks the entire theory of evolution from reality. Still, evolutionists preach the inescapable certainty of ape becoming man based only upon faith in the favorite fairytales of the science-priesthood.

Against all rational assessment of available evidences, which certainly conclude that monkey never become man, evolution continues to be taught in our schools while the morality and reality of God continues to be denied, impugned, and even punished there. Then Progressives accuse guns for the same murdering of human life they commit enormously more in abortion clinics. They demand a guilty conscience for every school shooting and a hearty blessing for every abortion. That ape-man for God swap has driven us insane.

The present world mess is mostly due to the socialized disregard for any meaningful consideration of The Holy Bible. God's Word commands what man's mind flees. It exhorts making mutual peace and benefit, even if their nature gets in the way of manifest destinies. Human nature is to kill whatever gets in its way. Therefore, from the day Nimrod seized control of our ancestors to this day, governments have wielded power to murder far more than they have to protect. The twentieth century's scientifically blessed, socialist governments murdered more people than all of history's previous governments combined, including Nimrod's. The American government blesses the murder of its own children, in staggering numbers. The science-priests pronounced man to be no more significant than nematodes. By that pronouncement, political-priests bully everybody into mindless submission to the performance of draconian evils. Thus the modern world rushes headlong towards the ultimate level of rationalized killing. What began with the eighteenth century science-priests matured through twentieth century draconian governance and will peak with the vile slaughter of God's elect in the Tribulation.

Darwin's insanity replaced God's wisdom in the current world puzzle by swapping the acknowledgment of God at school for the blessing of guessing. Progressive liberals blame the resultant mayhem upon inanimate objects, like guns, and upon the traditional understanding of God's Word. But the Bible teaches the sanctity of every human life and the Holiness of the God who intimately knits life in the womb. The consideration of life's sanctity and God's Holiness inspires sanity into civilization. The fear of God leads to wisdom while the guessing of man leads only to strife and struggle. Attention to His Word builds understanding. The Holy Bible proposed that God would return Israel to its homeland in the process of destroying evil so righteousness can be restored to a remnant of wiser

The Rest of the Blood Moons

people. It presents Revelation's Tribulation as His time to steer the world from insanity back to righteousness. Therefore, beholding Israel in the land God promised to Abraham's children verifies the foolishness of teaching the insanity of Darwin's nihilism to our children.

But a loving God would not wreak havoc upon a world of misguided people without giving clear and sufficient warning to everyone sincere enough to pay attention. The Bible proposes God will warn mankind of its impending doom simply because I AM WHO I AM is a loving, giving God. So maybe His showing us the pathway leading to safety from a coming grizzly/sasquatch raid is as simple as stacking eclipses along The Way pointed out by the Word. These special tetrads come stacked in complete correlation with God's Bible pointing like an arrow to eternal safety, if not necessarily temporal safety. It's what love does. Jesus said His gospel will have been preached to every nation before He returns. Therefore, the Bible will be everywhere available for man's understanding when God's heavenly phenomena begins to light up the end of evil's path.

How many coincidences must pile upon one another for scientific imagination to recognize the simple message made by a meaningfully ordered stack? That isn't to mean every coincidence has a message. Most are meaningless happenstance. But when coincidences intertwine through multiple layers of inherent meaning, happenstance becomes the irrational explanation, and purposeful communication becomes the rational one.

The Lincoln and Kennedy assassinations, for example, were surrounded by a mass of coincidences. Even Snopes.com fails to explain away them all. Yet those events share only one layer of coincidence: interrelationships between the many trivial circumstances of people, places, and times. But their coincidences had no book written two-thousand years earlier predicting those two assassinations would happen as they eventually did. That's a missing layer. Since there was no such book of prophecy, then the assassinations had no advance narratives or themes cultivated into a church-like society awaiting the prophesied assassinations for two-thousand intervening years, another missing layer. The assassinations were never made universally relevant to every individual of the entire world by some redemptive purpose they offered to mankind globally. That layer missing from the Lincoln/Kennedy coincidences is the enormous layer these clear signs of trouble and great joy have in abundance. The Lincoln/Kennedy assassination "mystery" exists only at the first layer of interrelationships, that curious, entertaining mixture of loosely correlating trivia.

But these sixteen total lunar eclipses not only share the first layer of trivial coincidence with the Lincoln/Kennedy "mystery", they engage every layer it doesn't. Their overarching layers rise out of an ancient Book predicting the first layer of "coincidental events" (layer #2), ladening all of the "coincidences" with the theme of an improved manner of life on Earth (layer # 3), and coloring them in shades of disastrous trouble coming to everyone loving God-less life, but of great joy coming to those who love life with The Warning God (layer #4). The layered interrelationships of these clear signs demands their explanation extend beyond mere coincidence. Communication better fit's their nature of layered, systemic order.

Maybe the "trivial coincidences" of these holyday eclipses can be imagined as the product of simple chance. But the multi-layering of their correlations is unattainable by

Clear Signs of Trouble and Great Joy

chance alone, otherwise, monkeys with typewriters did Chaucer's work. These signs can only happen by design for a purpose, like the readability of any book happens only by its purposeful design in accord with overarching layers of language and cultural expressions.

But must the communication of these signs necessarily portend the time of the Tribulation and approximate the time of Jesus Christ's return? We know from Biblical prophecies throughout the Old Testament and the Book of Revelation that the nation of Israel would exist again after its destruction by the Romans. They indicate the nation's role in the occurrence of evil's end. This can be understood by anyone who honestly investigates the Bible's prophecies, narratives, and themes instead of blindly believing the incredulous coincidences of evolution. Accordingly, the current existence of Israel as a nation advances our considerations of these signs another step closer to understanding them as warnings of the impending Tribulation. The holyday-tetrads of 1493-4, 1949-50, and 1967-8 are indeed confirmational evidences of God's providence in Israel's current nationhood with Jerusalem as its capitol, by design, for a purpose. That in itself is a giant message to the world. Even by the unbiased analyses which science should follow, those special tetrads which happened at Biblically prophesied events confirm the truth of the Bible regarding the definite reality of the God it describes. Nothing else could cause such deep layering of circumstantial correlations amongst precisely ordered eclipses, ancient prophecies, and current history.

But do they communicate *when* the Tribulation event might occur? Were these blood moons intended to indicate more than a mere confirmation of God's reality and His plan to save a remnant of mankind? If yes, there should be layers of correlation beyond the ones we've noted, and their approximation of evil's end should be quite clear, considering the significance of such a matter. Therefore, I needed to find more correlations before I was willing to understand these holyday-tetrads as approximating when the Tribulation will strike before Jesus' return. To dismiss that possibility without full consideration would have been as intellectually irresponsible as accepting it without investigation.

Information verifies or falsifies possibilities; people don't. Our responsibility is to seek *information*, or the lack of it, rather than to be issuing unverified proclamations. Only the lack of any corresponding *information*, or the existence of contradicting *information*, can dismiss a possibility. So I was now captured by the need to verify or falsify a possibility of great import in these eclipses. The only way I could break free of this investigation was to search until evidence itself provided a clear answer.

It was rational to expect additional *information* to come in the form of more special tetrads beyond the four we've been discussing. Since the development of Christ's Millennial reign entails the Tribulation, which entails the nation of Israel, which entails its history as a people, which entails their first Passover, then the signs Jesus said would indicate the immediacy of His return might need to entail more than just the eclipses Mark Biltz saw happening from five centuries ago until now. So I combed through NASA's lunar eclipse list logging every tetrad and the dates of their individual eclipses from the approximate time of the first Passover (ca 1500BC) through what would be the end of Christ's Millennial reign (ca AD3000). NASA's list ends fifty years short of what I needed for a thorough study. But it was close enough to either verify or falsify the possibility that Mark Biltz' holyday-tetrads were tied into a system of alarming communication. My list grew to one-hundred-twenty-six

The Rest of the Blood Moons

tetrads.

Logging these tetrads and their dates was the easy part. Determining which total lunar eclipses occurred on Passover was the difficult part. Identifying Passover is not as simple as noting the first full moon occurring after a particular Spring equinox. The Israelites used a lunar calendar instead of a solar one. Lunar months are 29.53 days long. Twelve lunar months are 354.36 days long, instead of the 365.24 days of a solar year. The lunar calendar must account for those 10.93 additional days, otherwise the beginning and end dates of the lunar year, all of its months, and every Jewish holyday would wander through the seasons as the years passed.

To prevent such wandering, the Israelites added a second first-month to their calendar approximately every third year. The general effect is that, whenever a full moon occurs within a week or so after the Spring equinox, it may or may not represent Passover. If the year happens to be a "leap year" of the Jewish calendar, then Passover will occur on the second full moon after the Spring equinox, which would be later in our April.

So we need to know what determines a "leap year" for the Jewish calendar.

Before the disbandment of the Sanhedrin in the fourth century AD, the first month of the year, Nissan, was delayed if Spring conditions had not prevailed by the time of the Spring equinox. They weren't very scientific about this. The barley had to be ripening and the trees blossoming,

> Spring should be felt; it should be bright and green.[1]

But even if things were greening up by the time of the Spring equinox, if the roads and bridges were still in disrepair from a hard winter, then Passover might be delayed a month anyway, because it is a big-travel holyday. Trying to analyze which full moon was Passover using the Sanhedrin's system was easy as running pylons on a Harley with flat tires. Fortunately, with the fourth century decline of the Sanhedrin the Jewish sage, Hillel II, and his rabbinical court developed a Jewish calendar with mathematically fixed intercalations (leap years).[2] His system employed a nineteen year cycle, a significantly meaningful number we soon shall discuss and will see many times in our analysis. Astronomers recognize nineteen years as a cyclical aspect of the earth and moon's motions.

There are a few calculator programs and a couple intimidating formulas available on the internet for determining all of the Passover dates since the beginning of Hillel II's intercalated calendar. They show whether Passover would be on the full moon of the first several days after the Spring equinox, or on the full moon occurring later in April. But any full moon between those dates is definitely a Passover full moon. I tested all of the soon-after-equinox and later-in-April full moon dates of my tetrad list by using the calculator at *http://www.covert.org/paschaldate.html*. It indicated that all full moons occurring between March 27 and April 19 were definitely Passover full moons.

But for the soon-after-equinox and later-in-April full moons before the fourth century AD, the determination remained ambiguous. Were the plants green? Could you smell the flowers? Could you get across the bridge to smell the flowers? That's been a while back, and since nobody wrote home about it, we don't know. Therefore, in order to reasonably

Clear Signs of Trouble and Great Joy

estimate how many holyday-tetrads occurred from Israel's Exodus until AD3000, I considered every March 28 through April 18 full moon as definitely being a Passover full moon. I asterisked as mere possibilities those tetrads occurring before the fourth century AD with full moons on our ambiguous dates.

Why bother with how many of these tetrads were holyday-tetrads? Maybe I was being too critical, but if this God who does everything perfectly is attempting to indicate to us the soon coming of evil's messy end, a period of doom beyond our imagination, then the indicator He used should be inherently informational. It should not need me or any one else to ascribe meaning to it. Its message should rise of itself from the commonly known meanings of the elements involved, the implications of their interrelationships, and even the meaning of their measurements. The meaning should be inherent.

Numbers in the Bible often supply *information* about the subjects they enumerate beyond just the enumeration. For example, The Revelation of John is full of sevens: seven churches, seven lamp stands, seven torches around God's throne, seven-headed dragon, the seven-headed beast, seven seals on a scroll, seven trumpets, seven bowls of wrath, including the seven years from the Tribulation's beginning to its end, etc. The Hebrew word for seven, *sheba*, extends from the primitive root, *shaba,* meaning an oath,

> ...to *seven* oneself, that is, *swear* (as if by repeating a declaration seven times).[3]

Seven is made of four, which represents the physical creation, and three, which represents God. These are the two realms of existence proposed by the Bible and attested by many witnesses. God has taken on an oath that all things of both heaven and earth will be done right forever, without error, without sin. Seven, therefore, represents the spiritual completeness of what is being numbered, that is, an essential completeness extending from God's involvement in the enumerated circumstances. The "seven" number of lamp stands are the spiritual fullness of the Holy Spirit's guiding light for Christ's Church. The "seven" number of heads of the beast represent the spiritual fullness of human tyranny expressed in this chain of empires. Three represents the fullness of totally divine involvement. The three foul spirits released by God from the mouths of the dragon, the beast, and the false prophet go out to draw the entire world into Armageddon (Revelation 16:13), the place where God will defeat evil. The "three" of them indicates God's involvement in their release for assembling the world's evil tyrants to receive a thorough pounding. In fact, the dragon, beast, and false prophet are a complex of the Antichrist playing God by faking His trinity. Revelation is also filled with tens, twelves, and even a twenty-four, which is a two-times-twelve. Ten represents perfect, ordinal completion, like a bucket perfectly full of water; one more molecule would overflow it; one less would leave an empty space. The Bible lists ten generations from Adam to Noah and ten generations from Noah to Abraham. God poured upon the Egyptians ten plagues and laid upon the Israelites ten commandments. The ten kings of the Antichrist's empire (Revelation 17:12) are the fullness of its reign over the entire earth. Twelve represents perfect order.[4] There were twelve tribes of Israel. These tribes show up in Revelation to define the one-hundred-forty-four thousand ($12 \times 12 \times 10^3$) who follow the

The Rest of the Blood Moons

Lamb wherever He goes. And we who follow Christ will forever join Him in His reign (Revelation 22:5). Maybe that is the sense in there being twenty-four elders (2X12) surrounding God's throne (Revelation 4:4). Even Ishmael, of whom God said He would make a great nation (the Arabs, who later became Islamic), had twelve sons (Genesis 17:20). Numbers bring additional meaning to the Biblical text. They also add meaning to this system of signs.

I had been reviewing Old Testament history when John Hagee's book brought Mark Biltz' holyday-tetrads to my attention. For several months I kept a chart of the post-Solomon kings and prophets of both Israel and Judah slipped under the glass desk topper at my office to help me keep their names and dates in order (a "twelve" kind of thing to do). One morning curiosity inspired me to count the kings on the chart. Maybe it was because I understood the significance of numbers regarding Biblically relevant matters, for the ends of both kingdoms were very relevant to Biblical prophecy. Israel, the northern kingdom, enjoyed nineteen kings[5] from the dividing of the kingdom until its demise in 722BC. Judah had twenty kings before it was swept away by Nebuchadnezzar in 586BC. But only nineteen of Judah's kings were actual kings. The twentieth was a usurper. She was Athaliah, the grandmother of Joash, who was the rightful king. She was not a king. Sorry, ladies. Therefore, Judah also had nineteen, rightful kings during the time of the divided kingdom. Then it ended, too.

This same count of nineteen, rightful kings for both Israel and Judah piqued my interest. We already know ten is the number of ordinal completeness. So what concept might the number nine add to ten?

> [Nine] is the last of the digits, and thus marks the end; and is significant of the conclusion of a matter...and is thus significant of the end of man, and the summation of all man's works. Nine is therefore the number of finality or judgment...It marks the completeness, the end and issue of all things as to man -the judgment of man and all his works.[6]

Being the combination of ten and nine, nineteen represents the fullness of judgment upon man's affairs. Is it only coincidental that Israel and Judah each had nineteen kings before God ended their idolatries by disastrous invasions? Or is it that God is in such ultimate control of history that He even works communication into the numbers measuring history? We will analyze that possibility in the next chapter.

Once I had listed the tetrads and highlighted those I could reasonably discern to have occurred on the Passover and Tabernacle holydays, I counted them. There were nineteen. What happens to mankind's evil ways when Christ returns? They are summed up in immediate destruction for His righteous reign to begin. What happens after Christ's Millennial reign on earth? Judgment Day. Both are definitely nineteen kinds of things! That caught my attention. From Moses' receiving the Law by which man's works will be judged until the time for judgment, the number of these special tetrads is the number representing complete and final judgment. I found that to be rather communicative.

But maybe not totally communicative. We must remember that my count of nineteen enumerates only those tetrads reasonably certain to have involved two Passover full moons

Clear Signs of Trouble and Great Joy

and two Tabernacle full moons. Indeed, amongst the one-hundred-twenty-six tetrads, there are two more which may or may not have been special tetrads. Those were the 1057-1056BC and 377-376BC ones, should anybody wish to analyze them more closely. Moreover, one Passover date given by the online calculator I used completely did not make sense to me. It is the third eclipse of the AD842-3 tetrad, the one on March 19. (Just a coincidence, maybe?) The date is a couple days before the Spring equinox. Oops! But I counted this tetrad to be consistent with the use of the calculator as a criteria for a tetrad's inclusion in the count. Furthermore, we need to remember that NASA's eclipse list ends about forty years short of what would be the end of Jesus' Millennial reign, if this reign were to begin shortly after a Tribulation period indicated by Apophis' two flybys. As impressive a message as these nineteen special tetrads presented, winking one at a time at Earth during the span of history God's Law was given until the coming of His judgment, we must keep this caveat in mind: the number of them is uncertain. There might be eighteen. There might be twenty, or maybe, twenty-one.

Indeed, upon first analysis, I thought there were twenty. There is a tetrad yet to occur 2032-3. Its first eclipse happens April, 25. Its third one happens April 14. April 14 is definitely Passover, therefore, its fourth eclipse will be on Tabernacles. But I mistook April 25 to be Passover. I was so impressed to find this tetrad snuggled within Apophis' seven year cycle that I forgot full moons after April 19 are only possible Passover moons. Alas! The Passover calculator at *www.covert.org* showed Passover 2032 was March 29, not April 25. I lost my holyday-tetrad number twenty.

From experience over the years I've developed a rule of thumb which says whenever I lose one thing, then, before I can find it, I will lose something else. Maybe it's like a redemption: something paid to the god of lost stuff for the return of something from the land of the lost. Or maybe there must be a contribution to bad fortune to balance good fortune.

The relevant meaning of final judgment in the possible number of these special tetrads was coherent with the possibility they were collectively a message of warning. What I noticed next significantly added to this sense. I was about to find this tetrad to be more than special. I was about to find it to be completely unique. That's when I lost my sense of humor about Apophis marking the Tribulation years. Apophis' seven year cycle was about to take on a very ominous sense from this 2032-3 tetrad.

But first we must digress. Consider the Revelation to John. Don't try to sort, identify, or set its events into chronological order. Just try to understand the nature of what the angel showed him.

A world full of people give worshipful approval to a deceptively slick leader. Three-and-a-half years later, this leader becomes evil beyond all the leaders who preceded him; he becomes possessed by Satan. Then he sits on the throne of God's Holy City, Israel's capitol, and proclaims himself to actually be God. His faithful followers descend upon God's elect and anyone else refusing obsessively worship him. Those who continue refusing will lose their heads. (Any resemblance to Islamic practice is purely thought provoking.) It is that "find one thing-lose another" rule of thumb again. But, as an O^2 witness, I assure you, finding I AM WHO I AM is worth far more than just the loss of your head. While all this wicked bullyhood is mounting, the entire earth is being racked by disaster. Life is rushing out of mankind like a torrent through a breached dam. Then, with all the world assembled against them, a remnant of Israelites realize they are cornered. Their utter extermination is looming once again. Holding Israel's back against the wall, the world is playing the extermination card, again. So Israel repents from denying their God's power and authority to cry out to Him for help.

> [15]I will return again to my place,
> until they acknowledge their guilt and seek my face,
> and in their distress they seek me, saying,
> [6:1]"Come, let us return to the LORD;
> for he has torn, that he may heal us;
> he has stricken, and he will bind us up.
> [2]After two days he will revive us;
> on the third day he will raise us up,
> that we may live before him."
>
> Hosea 5:15-6:3

> [11]Then I saw heaven opened, and behold, a white horse! He who sat upon it is called Faithful and True, and in righteousness he judges and makes war....[16]On his robe and on his thigh he has a name inscribed, King of kings and Lord of lords.
>
> Revelation 19:11, 16

This same King of Kings came to Israel once before. The Old Testament is filled with direct prophecy, allegory, and symbolism regarding the first time He came. There was no excuse for the Jews' having missed Him. Numerous details of Jesus' life and death perfectly matched the Scriptural projections of the Messiah. Yet the people did not look for I AM through the template of those details. They viewed everything through the template of what the social movers and shakers, the priests, elders, Pharisees, and scribes, were teaching. Their understanding was constructed of biased group think, human traditions, and misinterpretations of Scripture. Therefore, they did not recognize Jesus when He came. They regarded their paradigm more highly than the actual *information* of the Bible.

The prophecies about His coming were not beyond the understanding of an ordinary man. For example, when Jesus called Phillip to follow Him, Phillip went to Nathaniel and said, "We have found him of whom Moses in the law and also the prophets wrote, Jesus of Nazareth, the son of Joseph." (John 1:45) This simple fisherman regarded *information* above the group think and cultural bias of the religious leaders. Later Jesus said to those elitists refusing to believe Him,

> [44]How can you believe, who receive glory from one another and do not seek the glory that comes from the only God? [45]Do not think that I shall accuse you to the Father; it is Moses who accuses you, on whom you set your hope. [46]If you believed Moses, you would believe me, for he wrote of me. [47]But if you do not believe his writings, how will you believe my words?

Clear Signs of Trouble and Great Joy

John 5:44-47

Daniel even gave a formula for calculating when the Messiah would come. We discussed it thoroughly in Chapter 2. Any first century Jew with enough interest in knowing the Messiah's advent could have deciphered it. And God also sent a star signaling and a forerunner proclaiming Jesus' first coming. The last sentences of Malachi, the last Scriptures God inspired before Jesus was born, warned the people to heed Scripture and watch for the forerunner...

> [4]Remember the law of my servant Moses, the statutes and ordinances that I commanded him at Horeb for all Israel.
> [5]Behold, I will send you Elijah the prophet before the great and terrible day of the LORD comes. [6]And he will turn the hearts of fathers to their children and the hearts of children to their fathers, lest I come and smite the land with a curse.
>
> Malachi 4:4-6

There had been numerous charlatans proclaiming to be the Messiah in the previous few decades. But John the Baptist was not difficult to miss. He was born amongst a flurry of signs, as was Jesus. He boldly spoke the message Malachi predicted. And Jesus did come performing wonders attesting to whom He was and to what John the Baptist had preached.

Israel was given advanced notice of its King's coming. They did not miss Him by lack of *information*. But instead of attending the available *information*, they busied themselves with imagination, presumptions, and ultimately, with scoffing. They were sure the coming of their Messiah would be for ending their subservience to Rome. They imagined His coming was to restore their national glory. But instead, He came to end their subservience to the penalty of eternal death. Consequently, the religious leaders didn't recognize Him. They perceived Jesus as a threat to their own power and personal glory. Self-interest obstructed their ability to know truth by reality's details because they considered their own perceptions more than they considered the evidences of reality and their Oracles. In short, they didn't get it right the first time Christ came. They were spiritually and conceptually unclean. That error was the terrible part of the great day of His first coming.

> [39]You search the scriptures, because you think that in them you have eternal life; and it is they that bear witness to me; [40]yet you refuse to come to me that you may have life.
>
> John 5:39-40

They rejected Him. They became more tightly bound to even greater sin. They eventually met terrible deaths at the destruction of Jerusalem.

But the Tribulation is Israel's second chance to get their King's coming right. The prophetic timeline of Daniel 9:24-27, The Christmas Star of 3BC, the visit of the wise men, the visits of angels to Zechariah at the table of incense, and to Mary, Joseph, and the shepherds, and John the Baptist's preaching of repentance were among many signs having

The Rest of the Blood Moons

sprouted up like witnesses around Jesus' birth and ministry. Still the religious leaders wouldn't see. They chose cushy circumstances over Jesus' austere truths. They led millions of people into eternal doom. It is why Jesus likened the end to gathering thorns for the fire.

As there was the first time Christ came, there will be advance notice (Luke 21:25) of Jesus' second coming. It will be Israel's second chance to get it right. Could we be so bold as to say God gave a greatly organized system of celestial signs to be an "Elijah" for announcing the King's second coming as well? Maybe the statement is a bit daring and somewhat exaggerated. But the Western World's discovery of Enoch's book also fit's this "Elijah principle". Enoch addressed his book to the generation of the end as a warning to it. But his book went lost to the world in the first few centuries of our era. It couldn't very well warn while it was lost. It was finally discovered again in Ethiopia and made available to the entire world at the very time infantile geology was conspiring an ape-man theory for deluding billions of people into rejecting the truth of God's Holy Bible. It was like a warning coming into the world against the insanity of brewing a bunch of BS (bad science). Yet it went ignored for another century before it was translated into the international language of English. There is just something prescient about that synchronicity.

But eventually Christians began learning that scientific exploration was not entirely the enemy of God, which the hiding of *information* and denial of observations and testimonies are. They began scientifically exploring the world without the lens of normal science's pareidolic ape-man paradigms. They deliberated outside normal science's paradigm guarding clubhouse of peer reviewers. The evidences they've discovered of The Holy Bible's truth about creation, the Global Flood, the promise, the scrawny nation (that promise kept), and the resurrected, returning King have eroded the foundations of evolution like a second coming of Elijah. The Pharisee-led Jews paid no regard to Elijah (John the Baptist) the first time Christ offered Himself to be their king. Is anyone paying attention to this "Elijah" today?

Israel will have no choice but to pay attention. They will be cornered by a morbidly nasty reality. Unlike the first time their King came, Israel will be facing its end at the hands of a savage world gathered against them in the plains of Armageddon. Their entrapment will be seemingly inescapable. A repentant return to I AM WHO I AM will be their last resort. So, using Teddy's Theory, they will try it. They will like it!

As Hosea 6:1-2 prophesies, He will come and revive them and raise them up again. As important as it is to form paradigms and theories for investigating reality, it is even more important to prune them, correct them, and build upon them in accord with the ongoing discovery of *information.*

> [3] Let us know, let us press on to know the LORD;
> his going forth is sure as the dawn;
> he will come to us as the showers,
> as the spring rains that water the earth.
>
> Hosea 6:3

In August of 1977, I sat down on the edge of my bed and hung my head over the barrel of an 8mm Mauser. I pulled the trigger on an empty chamber four times to get the feel

Clear Signs of Trouble and Great Joy

for what I planned to be my last voluntary motion. Then I chambered the round. With my thumb on the trigger and the barrel between my eyes, I searched one last time for a reason not to sum up my affairs with one twitch of my thumb. Obviously, I found one. My brother had died in an auto accident three years earlier, and Mom wasn't handling that. Pulling the trigger would drag her over the edge with me. I couldn't; I loved her. But neither was I willing to live with my manic-depression. So I decided to die anyway. But just not physically.

I rejected everything I had come to believe, except for my faith in Jesus Christ. It wasn't like I had been convinced by evidence of His reality. It was more like Teddy's Theory: the Bible's story of Him was by far the best thing I knew to believe. But in order to continue in this life, which I had grown to abhor, I needed to shove everything I believed off my mind's table and scrape all decorum of thought off its walls before I could put my Humpty-Dumpty-like life back together again, correctly, so to speak.

My chosen "therapy" was to examine every thought, idea, inkling, feeling, emotion, mood, or attitude welling up from within me for its correspondence with reality, i.e., for the measure of its truth by it's relationship to evidences. I would accept what of it I was able to find true, correct what of it I found to be twisted, reject what I discovered to be irreparably false, while placing on hold for further consideration whatever I could not conclude for sure, since in all reality, I was still exploring to find truth at the same time. Although it sounds subjectively circular, discovering truth is actually a perpetual procedure of enormous importance. It is an adventure into reality by the building of verified truth upon verified truth. I did the same for all *information,* data, and suggestions that I encountered from the world around me, moment by moment. As such, I committed to realigning myself with what was truly discoverable, matter not what its reality turned out to be. For it was truth that I needed, since in the final analysis, I've always known that I am overarched by a reality I can not fundamentally transform, because after all, Reality Is What It Is. Consequently, I ventured into a perpetual quest for correcting my soul to match discernable evidence, no *information* discarded, none discounted, but ALL of it considered together.

I am convinced God's Spirit answered my ambition to know things like they are. Over the next forty-three years He steered my search for reality right to these signs purposefully designed into the sun, moon, and stars for warning a deceitful world of Truth's return.

I know second chances personally. So I took somber notice of the Tribulation being Israel's second chance to accept the coming of her King. Until I saw this correlation, Apophis had not meant much more to me than good entertainment. But April 25, 2032 is the next full moon after Passover, 2032. So what?

So, when Israel observed its first Passover after exiting Egypt, a few distressed men brought a vexing problem to Moses' attention.

> [6]And there were certain men who were unclean through touching the dead body of a man, so that they could not keep the passover on that day; and they came before Moses and Aaron on that day; [7]and those men said to him, "We are unclean through touching the dead body of a man; why are we kept from offering the LORD'S offering at its appointed time among the

The Rest of the Blood Moons

people of Israel?" ⁸And Moses said to them, "Wait, that I may hear what the LORD will command concerning you."

⁹The LORD said to Moses, ¹⁰"Say to the people of Israel, If any man of you or of your descendants is unclean through touching a dead body, or is afar off on a journey, he shall still keep the passover to the LORD. ¹¹In the second month on the fourteenth day in the evening they shall keep it;"

Numbers 9:6-11

The full moon after Passover is the full moon of second-chance Passover, Pesach Sheni. Second chance Passover is for anyone who could not keep Passover because he was unclean. As a nation, Israel was unclean with the idolatry of subjectivity when Immanu-el first came. To them, I AM WHO I AM was HE IS WHO WE ARE. To them, God's Scriptures explained everything they thought. It becomes kind of a myopia wherein you don't know the Scriptures which don't fit your perceptions the same way science doesn't know evidence contradicting its theories. So, their King, Jesus, instead of taking up His reign and killing everyone who disagreed with Him (as is the practice of purely human kings,) sacrificed Himself as The Passover Lamb establishing the way for everyone seeking truth to live. He was the anointed one cut off. His sacrifice ended Daniel's sixty-ninth week of years.

The Tribulation is the seventieth week of years at the end of which Israel will be given a second chance to get their relationship with I AM WHO I AM right. (The almost precisely two-thousand years between those two weeks were also prophesied in the Bible. We will discuss them in the next chapter.) The Jews' will be spiritually and psychologically cleansed as an instantaneous result of their repentant call upon God. Jesus will rush to their defense like an enraged mamma bear protecting her cub. After the last lover of evil has been flushed into death, Jesus begins His reign in Jerusalem for a thousand years as The Lion of Judah, The King of Kings, the righteous Lord of everyone who loves righteousness instead of lies. Israel will be elevated to the head of the nations. What a giant mulligan!

The prophecy correlates with the April 25, 2032 eclipse of Pesach Sheni's full moon within the seven years marked by an asteroid named after the Egyptian concept of chaotic evil. This tetrad now appears to be extra special, unique, actually. And it becomes unique tetrad number twenty, even though it is not a holyday-tetrad, just as Athaliah's uniqueness made her "king of Judah" number twenty, though she was not a king. They both involve an usurpation. If God intended this coincidence to be informative, then surely I had lost all of Apophis' entertainment value to a newly found intellectual sobriety. I had found an amazing statement being made by these special tetrads seemingly in the number of mankind's affairs summed up, plus an usurper.

Messages are made of meaningful patterns. When both the message sender and the message recipient share the same frame of reference, then even simple patterns begin developing the characteristics of language. This tetrad set between Apophis' two passages now seemed ominously linguistic.

Until the Pesach Sheni correlation surfaced, I had been expecting the Biblical and historical correlations of the 1493-4, 1949-50, 1967-8, and 2014-5 special tetrads to end

Clear Signs of Trouble and Great Joy

short of indicating any "when" of the Tribulation. I had been looking for their meaning to end at the mere verification of God's involvement in the reestablishment of Israel with Jerusalem as its capitol, and then whatever else the 2014-5 tetrad would be discovered to mean. It is the way we've been trained to think: cynically, skeptically, and accusatorily. We're supposed to safeguard the normalcy of our previously assembled thoughts. We're supposed to maintain faith in group-think and herd-mentality. We're supposed to conform to cultural and political paradigms modeled by people we audaciously trust to inform us about things we've never researched ourselves. But as the first century Jews discovered, this is an eternally dangerous way to think. Just because we have the privacy to think our own thoughts does not mean our own thoughts are well put together. Only their correlation with reality means the correctness of their assemblage and the sanity of our conclusions.

Many Christians panic at any possibility of knowing even generally when Christ's return will be. And knowing anything about the Tribulation's occurrence would posit the knowledge of when He will return. They shelter misperception within the "not knowing the day or the hour" statement Jesus made.[7] For today's Christians, it is similar to the old Jewish "the Messiah will free us from Rome" error. Of course, Jesus would have freed them from Rome had the whole nation accepted Him. And today, some preachers of Christ's gospel, in the secret places of their hearts, rue the day they must lay their enormous wealth and popularity aside for Christ's returning presence, so consumed are they by their own greatness. The Pharisees rued the same prospect two-thousand years ago for the same reason. Both mistakes entail knowing the way things are while ignoring the implications made by why things are that way. Certainly we won't know the day or the hour of Christ's return. But we can and must know the flow of life's implications regarding the time in which that day and hour will be embedded. History is a contextual thing because man's affairs are the accumulations of his events. Therefore, history's flow into the future can be approximated by the historical content of the way the recent past has flowed into the present.

Not knowing the day or the hour has only one definite meaning: neither the actual day nor the precise hour of that day will be known. All other meanings are mere possibilities. Indeed, there can be useful speculation that Jesus metaphorically meant we can't know the decade, the century, or as some Christians mesmerized by evolution think, we can't even know which millennium it might be, or which geological epoch. But Paul wrote to the Corinthians

> [6]...learn by us not to go beyond what is written, that none of you may be puffed up in favor of one against another.
>
> I Corinthians 4:6

The first century Christians at Corinth engaged in numerous, judgmental squabbles between each other and their people-groups. St. Paul's first letter written to them explained Christ as being the foundation of every believers' endeavors, therefore judgment of each other based on popular ideas of popular people and their people-groups is not right, since any ideas beyond what the Bible itself says may or may not be right. Only what the Bible specifically says can be known as being specifically correct, to which this colossal system of

The Rest of the Blood Moons

signs purposefully designed into the motions of the heavenly bodies attests. Everything else are just possibilities. Jesus didn't say you won't know the century or decade. He didn't say you can't know the year, month, or even week He comes. Jesus said you won't know the day or the hour. Therefore, the idea that you can know the approximate time of His return, and the idea that you can not, are both beyond what is written in the Bible. But one or the other must be true. That you can know seems to be written in the stars, where the Bible says His glory is told, His work is proclaimed (Psalm 19:1), and His signs are posted (Genesis 1:14 and Luke 21:25). Certainly Christ's return to be earth's King is His work. Certainly the stars do not proclaim the day or the hour. But they appear to be shouting the time. Jesus did say to watch (Luke 21:25). Watching entails seeing, if you are honestly looking.

If Apophis' two excursions by earth approximate the time of the Tribulation, then they also approximate the time of Christ's return. Many Christians, regardless of going beyond what is written, will find this point problematic enough to discredit the entire proposition. My proposition joins theirs at not knowing the day or hour. But it accords more with The Holy Bible's proclamations about God's talkative stars. Indeed, maybe the stars are spilling the beans, and this CPA has been counting!

The concept of not knowing the hour of His return was mostly why I had been searching to discredit the idea that these four holyday-tetrads were telling of it. I, too, began my research as a skeptic. But intellectual responsibility obligated me to consider the proposition that Apophis does mark the Tribulation, at least until I could find no more *information* to indicate such a conclusion. So I kept searching.

Those four special tetrads occurred at historical events clearly forming the central elements of the Tribulation: a national Israel in possession of Jerusalem facing the unhindered man of lawlessness, the Antichrist. Those events will divert the flow of world history into the Tribulation course, according to Biblical prophecy. The four events establishing those Tribulation elements having been accompanied each by four lunar eclipses on Israel's most important Biblical holydays is no mere coincidence. But it is the unique tetrad of 2032-3 which kicks the last consideration of "mere coincidence" off the road of reason into the weeds and brambles of unwarranted guessing, where also Darwin found that "non-designed, purposelessness of the universe" wreck.

Instead of failing in quantity and cohesiveness of message, as I expected these tetrads would do as my search continued, the more I researched the more correlations I found. Eventually, the correlations became so extensive, consistent, and coherent that I turned from searching to dismiss any Tribulation connotations to allowing their correlations to speak for themselves. I simply followed their lead.

Apophis arrives on a Friday 13th, a day culturally ladened with the sense of doom. Opposite of this meaning is Easter Sunday, a day culturally ladened with the sense of mankind's opportunity to blissfully live forever. April 13, 2036, the day of Apophis' next return, is Easter Sunday. Moreover, the number of spiritual fullness, seven, is the number of years between Apophis' two near Earth experiences. Seven is also the Biblically prophesied number of years between the Tribulation's beginning of doom and its end at the return of the resurrected Christ. The Tribulation gives Israel a second chance to receive its King. The Tribulation begins with mankind's widespread penchant for utopian inspired deceit. It peaks at

Clear Signs of Trouble and Great Joy

the Antichrist's proclamation of himself being God. Then it ends with the unimaginable glory of Christ's return, instantly destroying evil as only truth can. And winking at us from between those two days of April 13 are four total lunar eclipses beginning on Second-chance Passover and ending on Tabernacles, the Jewish holydays representing God's passing over His elect to destroy evil (Passover) and then to dwell with those whom He passed over (Tabernacles).

What more does such a set of picturesque circumstances need in order to be a sign? Does it need an ancient book of prophecies saying there will be signs? Well. It has that. It also has congruity with that Book's characters, holydays, and historical themes. And it has prophecies about the remnant of mankind whom I AM WHO I AM will rescue. It even has the Biblical meaning of numbers measuring its system of signs.

I could now see message.

But one last bother remained.

The April 25, 2032 total lunar eclipse occurring on Second-chance Passover is indeed a very impressive "coincidence" in light of the Tribulation being Israel's second chance to receive its King. And the last two eclipses occurring on Passover and Tabernacles are very convincing about the possibility of this tetrad being a prophetic event marker in the same sense as the 1949-50 and 1967-8 special tetrads were. But the October 18, 2032 lunar eclipse falls on no Jewish holyday. To me, this was a major problem. God does not deal in loose ends. He ties all ends to their purposes, even if the tie is not made by the measure of the last tick of a Bulova watch. One, detached eclipse was left to fend off the end of my lingering doubt.

As I pondered this loose end, the eclipses of 2014-5 were happening. My beautiful wife and I were making it a point to observe every one of them. Our hearts were filling with wonder and joy (and romance.) We thanked I AM WHO I AM for all four eclipses being visible in Colorado. As we watched each one, God's control over everything of this physical universe stirred awe into my soul. He controls more than the clockwork motions of the heavenly bodies; He influences direction into the courses of human history. And He has steered motions and courses of history to coincide with prophecies He inspired in the Oracles He gave to His chosen people.

> [4] The LORD has made everything for its purpose,
> even the wicked for the day of trouble.
>
> Proverbs 16:4

The purposes He made for everything are the deepest undercurrent of existence, and the most pervasive one. I AM WHO I AM is not a Deist, as verified by Israel's formative events adorned with eclipsed moons upon Judaism's important holydays.

But the October 18, 2032 eclipse hung amongst the other three like a fly in the ointment. Those others occur on Jewish holydays, but this one doesn't. If this 2032-3 tetrad was really meant to portend the Tribulation by its correlation with Apophis, I didn't think any of its eclipses should fail to correlate with the central theme of the Tribulation, at least in some substantial sense. It seems that God would have tied up all loose ends in any sign He

The Rest of the Blood Moons

meant for warning of such enormous devastation. Giant propositions need giant validation. But this dangling eclipse rather spoiled the validation of Apophis marking the beginning and end of the Tribulation. So I tried to put the idea of these signs portending the Tribulation out of my mind.

A few weeks later my curiosity rose to the eclipse again. I hadn't measured this tetrad's chronological orientation within the seven years marked by Apophis. As a CPA, I often must tally days between dates. So it was a simple task for me to locate the midpoint between April 13, 2029 and April 13, 2036.

Why date the midpoint? Revelation places an importance upon the halfway point of the Tribulation. Its first three-and-a-half years build up to a paradigm shift occurring at the beginning of its last three-and-a-half years. The paradigm shift is the nearly universal popularity of the Antichrist in the first half of the Tribulation accelerating into the inevitable product of populism, a ruthless fascism of tyranny conquering people such as the world has never before seen. Hitler would be proud of it. It is the time when the Antichrist proclaims himself to be God and demands worship at the penalty of death. It is the time of the false king, the usurper, the twentieth tetrad. It is Serpens' penultimate struggle to seize the Northern Crown.

The midpoint between April 13, 2029 and April 13, 2036 is October 13, 2032, only five days before the second, dangling eclipse of the 2032-3 tetrad. It would seem these five days are a matter of inaccuracy. But five is the Biblical number meaning grace. God's grace will become a priceless commodity after the Antichrist's seizure of world dominion. In the next chapter we will see this number measure grace into the events of some dastardly, historical developments.

All of the four holyday-tetrads mark their historic events not by precision, but by correlation. The first tetrad sensibly followed the Alhambra Decree's issuance and the discovery of America by slightly over a year. The Alhambra Decree was revoked seventy-one days after the end of the third special tetrad. Precision aside, that's correlative. The second special tetrad followed Israel's declaration of sovereign nationhood by somewhat less than a year, although it began in the first year Israel took its seat in the United Nations, but not on the very day. Nor did the third of these tetrads begin or end at the beginning or ending dates of the Six Day War, which effected Israel's control of Jerusalem. Nor did any eclipse of the fourth special tetrad fall upon the specific date of any major event, although the tetrad itself portentously winked over a period of events correlative with the worldwide outbreak of a Tribulation like radicalism. I was finally impressed. The October 18, 2032 eclipse correlates meaningfully with the Antichrist's proclamation of himself to be God, only delayed by the number of grace. Those entering the darkest door of the Tribulation will need grace. Those exiting the horrors of the Tribulation will do so by God's grace. Now is the time to seek His grace. Don't wait. Follow the evidence, written in the sky, to The Holy Bible and know Jesus.

The seven years of tribulation are commonly taken to be what's called prophetic years. Prophetic years are three-hundred-sixty day years. The Book of Revelation refers to the "…half of the week…" of Daniel 9:27 as one-thousand-two-hundred-sixty days (7X360/2). But Apophis passes Earth on two occasions seven solar years apart (7X365).

Clear Signs of Trouble and Great Joy

That's thirty-five days more than the seven prophetic years of Revelation's Tribulation. So, some people might be interested in thinking October 18, 2032 is the exact midpoint of a prophesied Tribulation period beginning April 27, 2029 and ending with Christ's return on April 15, 2036 (which would be a real bad day for the IRS). In such a case, Jesus would not only return to tread the grapes of wrath on tax day, 2036, but we would know the day of His return before it arrives. That isn't as much "beyond what's written" as it is counter to what's written. Those tempted to think October 18 is the middle day of the Tribulation must acknowledge that October 13 only might be the middle day. We don't know. It could even be any day thereabouts. This system of signs doesn't indicate the middle day. It is about message, not precision. Therefore, the five day discrepancy, by failing to help calculate the day and hour of Jesus' return, does more to verify the prophetic nature of this unique tetrad than it does to invalidate it.

A number of video presentations on YouTube attempt to debunk the "four blood-moon" theory. The most rational of them is Chris White's "The Blood Moon Theory Debunked". It is rather well thought, although not so well researched. Mr. White bases his skepticism largely on the failure of the holyday-tetrads to perfectly align with dates of the events Hagee and Biltz claim they portend.

> Another point is that the dates of the historical events for which these tetrads supposedly correlate do not seem to correlate very well at all to the dates of the tetrads themselves. For example, the Spanish Inquisition actually started some fifteen years before the 1493-94 tetrad and ended roughly three-hundred-and-fifty years later. They try to give this some credibility by saying that what the tetrad is really connected with is the so called Alhambra Decree issued on the 31st of March, 1492, which officially expelled the Jews from Spain. But even then the first eclipse didn't occur until over a year later, and the last eclipse over two years later. So, unless you call being off by a year God's way of predicting something, then this isn't a match.
>
> The next so called match is supposed to be when Israel declared its independence in 1948 and won the war for independence the same year. The dates of the 1949-50 tetrads again didn't occur until over a year later and didn't fall on the dates of any of Israel's victories, or on the day that the UN recognized them as a state, or on any other significant dates.
>
> Trust me, if there was any significance to the actual dates of these tetrads, you would have heard about it. But the best they can do is, as you will see, is the next one coming within ten months of an event.
>
> So, yah, the last one they say occurred within a conjunction with the Six Day War. But in reality it didn't start until ten months after the war ended. And the last eclipse didn't occur until a full year after that.[8]

Mr. White is quite right about these four special tetrads missing perfect alignments with the dates of key events. If perfection of alignment is necessary for identifying the significant events in God's historical process of finishing transgression, ending sin, atoning for iniquity, bringing in everlasting righteousness, sealing vision and prophet, and anointing a

The Rest of the Blood Moons

most holy place, then these special tetrads are indeed feckless. But, we already know from the Bible that Israel will need to become a nation again, Jerusalem will need to be Israel's capitol while sporting a temple, the one hindering the mystery of lawlessness will need to have been removed, an Antichrist will need to arise in the midst of a very messy seven years, after which a wrathful King of Kings will come. What we did not know before Pastors Biltz and Hagee shared their priceless *information* was the language I AM WHO I AM would use to indicate the most significant warning mankind has been given since Noah warned the people of The Great Flood in his days.

The 1949-50 and 1967-8 holyday-tetrads didn't inform us of the nation's reestablishment and capture of Jerusalem. We could see as much. And from the Bible we knew those events were prophetically important. But the special tetrads empirically verified those events to be the product of the designing, purposing Creator of Everything by being a coordination between His Written Book, His book of Nature, and His book of human history. Consequently, they empirically verified that He is a communicator. Indispensable to our study of His signs in the sun, moon, and stars, Israel's reestablishment and repossession of Old Jerusalem empirically verified the communicative nature of holyday-tetrads. All of them since the inception of the Alhambra Decree against God's Spanish Jews have marked prophetically significant events.

We need to remember, every human alive today is a Sethite. The Sethite line extended from Adam through Enoch and Noah to all three of his sons, and through those sons to all of us. No humans other than Noah's family survived the Flood. Therefore, we are all genetically Sethites. Josephus says of the pre-flood Sethites:

> Now this Seth, when he was brought up, and came to those years in which he could discern what was good, became a virtuous man...so did he leave children behind him who imitated his virtues...They also inhabited the same country without dissensions...upon Adam's prediction that the world would be destroyed at one time by the force of fire, and at another time by the violence and quantity of water, they made two pillars; the one of brick, the other of stone; they inscribed their discoveries on them both that in case the pillar of brick should be destroyed by the flood, the pillar of stone might remain, and exhibit those discoveries to mankind.[9]

God's obedient Sethites were preaching and co-operating with His message. They worked together to be informative of God's plans as they were learning them. They didn't shoot each other in the feet over what the actual length of a gnat's whisker might mean to God's communication. Being Sethites in our blood and God's children in Christ's blood, we need to build our informing pillars upon the good sense of each other's efforts. We should not be technocratic, little nerds. We are all false (Romans 3:4). None of our favorite hobby gnats' whiskers measure up to Christ's perfect knowledge. Our rightness is in the propriety of our quest, not in the microscopic precision of our knowledge. Otherwise, we would destroy each other to the very last drop of blood.

So I consider it wholly appropriate that, by approximation rather than by technicality, God informs of these final few significant events marching towards evil's demise.

Clear Signs of Trouble and Great Joy

Nobody is precisely right about everything. Every call for measurement has its degree of tolerance for error. It's what man must do because his ability to be precise is limited in every way. Therefore, *information* needs a little slack for fitting its affects into these sloppy, imprecise human heads. So rather than tearing each other down over the length of our favorite hobby gnat's whisker, let's have the common sense it takes to build upon and square more accurately what others get approximately right. For which of us lives to himself and dies to himself? Which one of us was God's gift to mankind? I am sure God's gift to mankind is coming again. I see His signs; I've counted the stars' spilt beans.

What eventually impressed me about these tetrads approximating the dates of events instead of exactly specifying them is that approximation involves coherence while exactitude invokes equation. Equation must find a precise match of identities, while approximation avails metaphor. Metaphor makes communication out of essential similarities while equation produces only technical lexicons.

Those special tetrads beginning in the year following Biblically relevant events are not predictive of their events, just as Mr. White lamented. Signs are not always prophetically predictive. Obviously, the occurrence of these signs after their correlating events is meant for verification rather than for prediction. Who in his right mind would think to predict an event which has already happened? The sign reading, "Grand Junction 60," just north of Montrose, predicts the result of a sixty mile drive North because it indeed precedes the drive. The sign just outside Grand Junction which reads, "Entering Grand Junction City Limit," verifies that you drove sixty miles in the right direction. They are both signs. They are both meant to be communicative of Grand Junction's relationship to the drive. But the meaning of each communication is affected by the context of its sign. Context is an export of approximation. Specificity is an import of equation. We don't need exactitude to discover good sense; we only need contextual coherence. God is trying to communicate. He's not interested in showing off.

Israel's historical development is rationally discernable as having been Biblically prophesied. Throughout the Old Testament Israel's presence again in Palestine, after a long absence, is prophesied to be part of God's endgame. Behold! Israel is present in Palestine after a long absence! Prophecy is that simple. Unless one wants to be skeptical. Then behold! Twelve blood moons just happened on twelve Jewish holydays, some when Israel was run out of Spain, some when they resettled Palestine, and some when they recaptured Old Jerusalem from Jordan, and well, so what!?

The contexts of the blood moons argue the prophecy's point against the skeptics' lack of good sense. Seeing those three special tetrads appear in contextual relationships with the three developmental events of Israel's Biblically predicted reestablishment suggests a coherent understanding of the 2014-5 special tetrad. It's as if the signs are developing their own, useful language. And having now seen the fourth special tetrad wink over what was arguably America's most deceitful, destructive, antithetical presidential administration, which essentially loosed radical Islam into the world and was hell bent upon fundamentally transforming the once Christian America into a people conquering, atheistic socialism, adds even more sense to the development of that language. The man of perdition will not be revealed until the one hindering lawlessness is out of the way (II Thessalonians 2:1-8).

The Rest of the Blood Moons

Israel's national existence was the world's first step towards the Tribulation. The unleashing of lawlessness pushes the world another step towards God's endgame objective. Technical precision can't speak it. But sensible coherence paints its undeniable picture.

Indeed, this series of four special tetrads develops a language for the entire system of signs. The thinking observer is given enough *information* to see these holyday-tetrads like markers of particular nouns: Israel, pogroms, Holy City, and America (the one which has recently been restraining lawlessness). These nouns are moved by verbs of historic, prophecy-fulfilling action: process of pogroms, leading to nation formation, acquisition of most holy place, and a jailbreak of perdition's pernicious horde. The Holy Scripture supplies more nouns and actions yet to come. We analyzed those for correlations with the special tetrad of 2014-5 and the unique one of 2032-3. Very close correlations can be found between the 2014-5 and 2032-3 tetrads, Apophis, and the prophesied time of trouble for ending unholiness to begin holiness.

Is this not the same treatment scientists give *information* -analyze, categorize, discern patterns, correlate discernments, interpret correlations, formulate theory, and then predict? It is the scientific method. What is not scientific is denying evidences leading towards conclusions the researcher does not desire to accept. Rejecting Biblical *information* just because Biblical *information* is...well...Biblical is biased, prejudiced, and ignorant. Ignorance is not supposed to be scientific. Nor is prejudice. Nor is bias. If we can not understand the communicative elements of signs, then indeed, do we have the intelligence to understand the communicative elements of an alphabet? Evidently you do, because you're still reading. So let's apply our ability to understand communicative signs and continue.

I AM WHO I AM, in fact, added a prophetic angle to the verification aspect of these three special tetrads. The pogroms following the Jews everywhere, pressing them towards the dream of a homeland, were actions of the Alhambra Decree's nature. We must acknowledge this astronomically marked period of time as being more than just another ordinary spell. For eighteen-hundred-and-seventy-eight years God's chosen people wandered amongst the nations after having been stripped from the land God promised to their forebear, Abraham. (Abraham was promised that land in 1877BC.) 1492 through 1968 was an era of pogroms unique in itself. It was the era in which God used these pogroms to press His chosen people on to possessing the land He promised, complete with His most prized Holy City. The Alhambra Decree covered a period of epochal, prophesied history. And as Mr. White revealed for us, the three special tetrads followed the events they marked within that epoch, except one. The official end of the Alhambra Decree followed the last total lunar eclipse of 1968 by less than one-hundred days. That eclipse had a prophetic nature, if it was meant to have any relationship at all with the rescinding of the Alhambra Decree. They are too proximately situated not to have been designed to correlate.

Using this template of understanding, we can suggest that the time extending from the 2014-5 special tetrad through the 2032-3 unique tetrad also marks an epoch. Having learned the language of the 1493-4 through 1967-8 special tetrads, we can hear the 2014-5 special tetrad whisper, with an accent of God's Holy Word, a prophetic indication of the Tribulation's approach. Now that God's epoch for establishing the nation of His chosen people with Jerusalem taken into its hands has culminated, what epochal development of

Clear Signs of Trouble and Great Joy

momentous proportion does the Bible say remains?

> ⁷ I will tell of the decree of the LORD:
> He said to me, "You are my son,
> today I have begotten you.
> ⁸ Ask of me, and I will make the nations your heritage,
> and the ends of the earth your possession.
> ⁹ You shall break them with a rod of iron,
> and dash them in pieces like a potter's vessel."
> ¹⁰ Now therefore, O kings, be wise;
> be warned, O rulers of the earth.
> ¹¹ Serve the LORD with fear,
> with trembling ¹²kiss his feet,
> lest he be angry, and you perish in the way;
> for his wrath is quickly kindled.
> Blessed are all who take refuge in him.
>
> Psalm 2:7-11

Do you think the rulers of the earth will tremble at the sight of these warning signs enough to kiss the Lord's feet? Let's not be ridiculous. These signs will be no less ignored by the nations than is God's Holy Bible. So this epoch begins as inconspicuously to us as did the Alhambra epoch. But it will end with the painfully apparent dashing of the nations into pieces, God's endgame, the Tribulation occasioning Christ's return. The Bible speaks of it. Our current history smacks of it. And these concurring eclipses really do seem to point right at it. It's a coherent thing.

But the advantage in this endgame is reached through a gambit. The short epoch between the last holyday-tetrad and the coming unique tetrad is the time for removing the one who has been restraining lawlessness.

> ¹Now concerning the coming of our Lord Jesus Christ…that day will not come, unless the rebellion comes first, and the man of lawlessness is revealed…so that he takes his seat in the temple of God, proclaiming himself to be God… ⁷For the mystery of lawlessness is already at work; only he who now restrains it will do so until he is out of the way. ⁸And then the lawless one will be revealed, and the Lord Jesus will slay him with the breath of his mouth and destroy him by his appearing and his coming.
>
> II Thessalonians 2:1,3-4,7-8)

We've discussed many Biblical attributes involved in these "tremendous coincidences". Let's ponder some more patterns converging with this apparent language of four special tetrads and a unique one.

> Behold, I stand at the door and knock; if any one hears my voice and opens the door, I will come in to him and eat with him, and he with me.
>
> Revelation 3:20

The Rest of the Blood Moons

Minds of ordinary intelligence get that way by comparing and categorizing experiences. We see things and events; we compare and sort them by similar attributes; we form understandings from noted similarities; we develop knowledge by repeating the process wherever we find interrelating similarities. Allowing correlation to suggest knowledge is not only the essence of the scientific method, it is part of another process of building understanding and wisdom -the propensity to believe what keeps presenting itself as believable. Science calls it paradigm. The Bible calls it faith.

But opposite of faith is the propensity to keep denying what continues to present itself as believable. It is skepticism and doubt. Left unrestrained, it is the bold process of assured foolishness. The fool will deny correlating similarities just because he doesn't like the paradigm their coherence suggests, or the faith towards which it points. He will use whatever excuses he can imagine to support his denial. Usually those excuses are irrational, since his denial is irrational. He will even close The Book, set it aside, and stop thinking about it, as if that will make God and all His pervasive rationality go away.

But most often the fool makes the excuse that similarities are not identical. He will demand absolute identity in a world understood only through metaphor and approximation (even the closest scientific and engineering measurements entail a specified tolerance for error, i.e. approximation). He will proclaim every empirical interrelationship countering his ambitions and beliefs to be only circumstantial and anomalous when they do not precisely and exactly correspond to the minutest level of detail. Supporting his state of denial, the fool demands a labeled identity for every attribute shown to him before he will acknowledge any of its interrelationships with anything else, if he will even do so then. Oh! But of course! The correlations constructing the his preferred paradigm need not fit precisely! In fact, his correlations don't need to fit at all. They only need to sound good and satisfy the natural lusts of the human heart while appearing to be rather concurrent.

Real similarities and comparisons are for wisdom to find and for foolishness to hide. We sort ourselves between wisdom and foolishness by how we respond to our desires long before we consider any correlating patterns.

Knock, knock, knock went the three convergent tetrads. KNOCK, KNOCK plead the final two. (Notice the number of them all: five?) Opening the door to the obvious will not only make you gloriously alive and properly fed by God's grace, it will also shed Jesus' light upon Apophis, this way coming.

The interrelationships of these three special tetrads spanning the epoch of the Alhambra Decree and concurring with events significantly relative to Biblical prophecy form a demonstrative statement. God is faithful to the promise He contracted with Abraham. He is also faithful to where He next calls the world. The grace He extends to everyone who opens his door to Jesus Christ's knocking was a contract signed by God in blood on the cross. The Bible concludes that Jesus Christ's physical return to the earth is part of His contract kept with those who opened their doors to eat truth with Him. Since the previous, three special tetrads were about contract keeping, the 2014-5 and 2032-3 tetrads might also be about contract keeping. Now that the contract with Abraham has been kept, which contract is left to keep? That's right. I hope you're dining with Jesus, because He's bringing

Clear Signs of Trouble and Great Joy

a rod for smashing the insolent pots who refuse to open at His knocking.

Even the Bible asks us to test and verify its own statements by studying how they correlate with history, by watching the effects they produce in the lives of whoever gives them an honest, believing try, and by experiencing their effects by trying them yourself (Teddy's Theory). So Jesus implied that we might want to be watching for signs in the sun, moon, and stars, another Biblical statement. If ever there was a good reason for fundamental transformation, for something worthy of our hope and change, would it not be the eternal life promised by God's Holy Bible? Has anything else enjoyed such an amazing display of meaningful interrelationships than the promise God made to those who let Christ speak truth into their lives at the dinner table after His knocking?

Yet we gullibly swallow massive bunkum about ape-man, secularism, and "walls of separation" served up by soapy-faced, stuffed-shirt politicians, so-called professors, and hot-nor-cold preachers, ideologies which are all based merely upon scientific imagination, guesses, and presumptions, not one of which has so much as a single sign in the sky. Their signs are in the destruction of human life produced by all of the nihilistic political movements preaching the wonders of scientific progress built upon God's supposed grave.

Effectively testing for the truth of what we've been taught and commanded demands, first, that we take off the horse-blinds strapped onto us by social illusionists and slick deceivers. It next demands we rationally consider everything our unhindered eyes will thusly be seeing more clearly, including the empirically observable Holy Bible. We should then use the same audacity to mix some Bible into our thinking as we used to stir copious influences of Hollywood and head fakes of fake-news into our hearts. But since the rejection of God's Word is irrationally clutched tightly by the world's majority of people, consider the possibility that the 2014-5 special tetrad indicates the beginning of an epoch in which the man of lawlessness lures a gullible world into a Progressively anti-Semitic fascism meant to corner, kill, and bury Israel beside its murdered God.

[4] He who sits in the heavens laughs;
the LORD has them in derision.
[5] Then he will speak to them in his wrath,
and terrify them in his fury, saying,
[6] "I have set my king
on Zion, my holy hill."
[7] I will tell of the decree of the LORD:
He said to me, "You are my son,
today I have begotten you.
[8] Ask of me, and I will make the nations your heritage,
and the ends of the earth your possession.
[9] You shall break them with a rod of iron,
and dash them in pieces like a potter's vessel."
[10] Now therefore, O kings, be wise;
be warned, O rulers of the earth.
[11] Serve the LORD with fear,
with trembling [12]kiss his feet,

The Rest of the Blood Moons

lest he be angry, and you perish in the way;
for his wrath is quickly kindled.
Blessed are all who take refuge in him.

Psalm 2:4-11

Paul reminds the Thessalonians that the mystery of lawlessness was being restrained even in their day. It is yet being restrained at the time I write this, as Paul said it would be until the one who restrains is removed from its way. Nobody has lived from Paul's time to now. Therefore, no person is the one restraining. No nation existing then exists today (except Israel), so the one working the ultimate restraint isn't a nation. God Himself, angels, and the church existed then and today. It could be one of those, or a mix of them.

My thoughts on the one restraining turn to Ophiuchus, the serpent bearer. He is restraining the serpent from seizing the crown while His right foot crushes Scorpio's head. The church, angels, and even men and nations have down through history taken part in restraining the mystery of lawlessness at the influence of God's Spirit, like various pieces engaged in a very long chess game. In the past century the United States was moved to play a big role in restraining lawlessness. But it has been the Holy Spirit, God's living motion, involvement, and relationship with everyone who loves Him, which has been working into the place of restraint whatever piece will next be effective. The Holy Spirit is the one who restrains in the same sense Hasan Abbasifar[10] is the one who chesses, even though his chess pieces wield the game's action. The Holy Spirit plays truth into the movement of His pieces; the unholy Antichrist plays deceit. America has been jerked around by the affects of their gaming, and was eventually thrown captured by endgame deceit.

At the end of the seven Tribulation years, at Israel's acknowledgment of guilt, Israel calls out in distress to God. And Christ comes. The evil of all the world's *information* denial will end with the nations gathering at Armageddon to destroy the Jews. But, at the sight of Christ descending to Israel's aid the hostiles will foolishly train their weapons on Him. One of only two responses will occur to the mind behind every eye seeing Christ's return: 1) unrepentant opposition against Him, or 2) repentant anguish over not having followed Him sooner or more closely. Only the anguishing will survive. The opposing seed of the serpent will be crushed by the word baited on Christ's breath. The serpent will then be bound and sequestered, locked away from mankind, just as the serpent sequestered the evidences of truth, righteousness, and the glory of I AM WHO I AM from the general public and their children's education. For a thousand years Jesus will dwell in Jerusalem, reigning with righteousness in the empirical sight of everyone, making justice, causing peace, spreading truth, and teaching righteousness. And yes, my scientific friends, His presence will be as empirically observable then as His otherwise inexplicable burial Shroud of Turin is empirically observable today, but no denier will be left to deny simple observations.

But none of this happens until the ultimate restrainer of the mystery of lawlessness is out of the way. The restraining one isn't entirely America. It is entirely the Holy Spirit. America has only been the last piece the Holy Spirit has played in His endgame gambit. The Barack Hussein Obama administration withdrew America from restraining the spread of

Clear Signs of Trouble and Great Joy

Islamic radicals, infused an unprecedented level of lawlessness into the American government, and transformed the American stance upon freedom and justice for all into a worldwide apology tour. By the time Apophis arrives, Iran will have had opportunity to add nuclear weapons to its lust for annihilating Israel, and sleeper cells of Islamic radicals around the world will have had fourteen years to draw and perfect plans for Islamic global dominance.

Consider the number NASA assigned to Apophis in its log of asteroids: 99942. Recall the Biblical meaning of 9, the conclusion of a matter, and of 3, divine fullness, God's doing it. Might three nines represent the final conclusion of a matter by God's hand? E.W. Bullinger suggests that 42

> ...is a number connected with Antichrist. An important part of his career is to last for 42 months (Revelation 11:2, 13:5), and thus this number is fixed upon him...Its factors are six and seven (6X7=42), and this shows a connection between man and the Spirit of God, and between Christ and Antichrist...Being a multiple of seven, it might be supposed that it would be connected with spiritual perfection. But it is the product of six times seven. Six, therefore, being the number of Man, and man's opposition to God, forty-two becomes significant of the working out of man's opposition to God.[11]

Might the number 99942 express God concluding the matter of man's opposing Him? That is the final effect of the Tribulation. And we find that numbered rock marking out a seven year period during which four total lunar eclipses each express a significant aspect of the Tribulation by the nature of the particular days on which they occur. Or maybe it is just another coincidence in this mounting avalanche of happenstances. How interesting is it that this particular asteroid was reported at just the right time for NASA to have assigned it that prescient number? How interesting are the four eclipses winking Tribulation attributes from between Apophis' two approaching rendezvous with Earth? How much more interesting is it that Mr. Tucker and Mr. Tholen gave it the name of the Egyptian destroyer-serpent? Does all this convergence of imagery and representations not form an expression? Whether coincidentally or by design, it presciently portrays the Tribulation aspects described in the Book of Revelation. To the ends of the world SETI would shout the irrefutable discovery of intelligent, extraterrestrial life communicating with us, had it observed even a hundredth this much pattern in the universe's background radio noise; news stories throughout the world would focus on such a discovery far more than they focused on Covid-19. But since the immensity of this discovered pattern entails the Bible instead of E.T.'s mom, you won't find a journalist in sight of its *information*.

Still, it is an eerie reality. These are not products of mere imagination like are the theories of ape-man, socialism, and pollution-caused global warming. You can discover the *information* of these signs for yourself, if you are willing to invest the time.

Now, consider this. When the second angel of Revelation blew his trumpet, John saw something like a great, flaming mountain thrown into the sea, and it killed a third of all living things in the sea. Then, when the third angel blew his trumpet, John saw a great star fall from heaven, ablaze like a torch. It poisoned a third of the springs and rivers (Revelation 8:8-10). Apophis can't be either of those two events. Those impacts occur during the Tribulation. But it can be metaphorical of the time of those events. Apophis' 2036 appearance approximates the Tribulation's end. The 2029 brush with Earth approximates its beginning, if indeed Apophis really does mark the Tribulation years at all. If Apophis and these seeming asteroid strikes must have a concrete connection, then it could only be a speculative connection. And there is something fascinating to speculate about that possibility.

Catastrophism is a geologic model attributing much of the earth's geological formations to the momentary chaos of catastrophes rather than to epochs of gradual change. Many geologists of the early nineteenth century fought hard against the catastrophist model; it smacked too much of the Biblical Flood. They were gloriously basking in the victory of having bullied the Neptunists[12] into silence when catastrophism sprouted out of the mud and sedimentary layers of Earth's crust. Eventually the bully-hold Vulcanism[13] attempted to maintain over geology through the nineteenth and well into the twentieth century was beginning to be washed up by a flood of evidence requiring enormous movements of astronomical quantities of water. By the late twentieth century most geologists and paleontologists were agreeing about some catastrophic asteroid impact at the northeast corner of South America having put the dinosaurs out of business. Then various theories of a comet impact emerged by the turn of the twenty-first century to re-explain the cause of the Younger-Dryas cooling event at the end of "the last great ice age". By the close of the twentieth century, catastrophism had impacted scientism, and Vulcanism solidified into a buried layer of thought.

Today, a few astronomers are hypothesizing the possibility of a small, brown dwarf star having an extremely eccentric orbit around our sun. Others hypothesize an additional planet with a highly eccentric orbit. Even ancient Sumerian mythology invoked an additional planet called Nibiru which was thought to orbit the sun once every thirty-six hundred years or so. D.S. Allen, a science historian, and J.B. Delair, a geologist and anthropologist co-authored *Cataclysm!*, a comprehensive analysis of coherent astronomical, geological, paleontological, archaeological, and anthropological evidences of a possible, massive object having passed through the solar system approximately 9500BC, causing a global deluge to wash over Earth, yanking its axis into a wobble.

The Bible details a global deluge, but it implies a 2350BC date. Calculations of the energy involved in raising a mass of water deep enough to flood the entire globe almost demands a cosmic event. Allen and Delair present numerous examples of world mythologies entailing either an impact event or some other form of a cosmic snafu. Theories attribute the Younger-Dryas cooling to an impact event. In fact, it is the event which Andrew Collins and Graham Hancock theorize to be the inspiration and subject matter of Gobekli Tepe's construction and burial, the most ancient of known temples. *Information* abounds for the possibility that Earth was set awash in a flood resulting from some astronomical cataclysm. Did Apophis come out of nowhere? Or could Apophis be a piece of debris left over from whatever cosmic assault God may have launched upon the earth? Could a left over piece of debris from that snafu have been orbiting the sun in an approximate seven year cycle throughout subsequent history until we find it book ending this uniquely picturesque tetrad?

Clear Signs of Trouble and Great Joy

Having seen how this imaginative God correlates motions in the heavens with meaningful events on Earth, should we be surprised to find that so? Could its gravitationally massive parent accompany it in 2036, causing the enormous earthquakes prophesied in Revelation? Whatever will be will be seen.

    We have analyzed several cosmic alignments of the sun and moon shedding Biblical light on Earth's history. We will analyze many of their correlations in the next chapter. In Chapter 8 we will notice an astonishing testimony of blood moons. And, in Chapter 9 we will discover the Great Flood, the Gospel of Christ, and the wrath at His second coming undeniably chiseled in ancient stone. Scientists go all aflutter over any trace of water found in space. Only one of those molecules proclaims to them the possibility of life way out in that vast, vacant cold. If a mere water molecule can vouch for the presumption of life in an otherwise deadly vacuum, then surely this mass of "coincidences" interrelated with the Bible just might suggest Apophis' association with both The Great Flood and the Tribulation.

    Any speculation relating Apophis to the Flood is simply an entertaining aside. Getting back to descriptive business, with the October 18, 2032 eclipse being meaningfully near the midpoint of Apophis' seven year cycle, all four of the 2032-3 total lunar eclipses make clearly relevant associations with the fundamental elements of the Holy Bible's Tribulation period: 1) Israel's second chance, 2) Satan proclaiming himself to be God, 3) God passing over His remnant to end evil, and 4) Christ thereafter tabernacling[14] with man. I certainly am not suggesting every eclipse portends the date of a prophesied event. Their communicative nature is too big to serve such trivia. I suggest the date of each eclipse portends a fundamental attribute of Revelation's Tribulation event for casting Apophis' seven year period in Revelation's hues. I'm proposing that God has painted for us a clear sign over the time for trouble and great joy.

    Now, let's measure that theory.

Footnotes

1. Posner, Menachem. "How Does the Spring Equinox Relate to the Timing of Passover? About the Jewish Leap Year". http://www.chabad.org/holidays/passover/pesach_cdo/aid/495531/jewish/How-Does-the-Spriing-Equinox-Relate-to-the-Timing-of-Passover?-About-the-Jewish-Leap-Year.html
2. Ibid.
3. Strong's Hebrew and Greek Dictionaries. James Strong. Quickverse Electronic Addition STEP Files. Findex.com, Inc. 2003. At H7650.
4. Bullinger, E.W. Number in Scripture: Its Supernatural Design and Spiritual Significance. Alacrity Press, 2014. (Originally published by Eyre & Spottiswoode Ltd., 1921.) Pg 75.
5. Some people think Tibni became king of Israel from 885-881BC, between Zimri and Omri, raising the number of Northern Kingdom kings to twenty. But I Kings 16:16 makes it clear that "...all of Israel made Omri, the commander of the army, king over Israel..." It was afterward that "...half of the people followed Tibni...to make him king, and half followed Omri. [22]But the people who followed Omri overcame the people who followed Tibni...so Tibni died and Omri became king." (I Kings 16:21-22). Tibni's four year "reign" was at most a rebellion. Scripture states that people attempted "to make him king". It doesn't say he became the king.

The Rest of the Blood Moons

6. Bullinger, E.W. Pg 167.
7. Matthew 24:36,40,50; 25:13; Mark 13:32; Luke 12:40,46.
8. White, Chris. Four Blood Moons Debunked. 00:6:54. January 1, 2014. YouTube video.
9. Josephus, Flavius. The Antiquities of the Jews. Book I, Chapter II, sec. 3. Josephus: Complete Works. Translated by William Whiston AM. 1960. Kregel Publications, Grand Rapids, Michigan, 49501. Pg. 27
10. The 2013 chess grandmaster.
11. Bullinger, E.W. Pg. 191
12. The Neptunists were a product of the intersection of natural history and the Word of God. They were men of faith seeing geological layering as evidence of the earth's having been formless, void, and submerged in water before God called forth dry land which He then flooded again with a totally submerging but brief expanse of water. Neptunism theorized that one geologic process had created the rock layers, a related process carved the valleys, and then post-flood formations developed by much less catastrophic processes.
13.   Dr. James Hutton advanced the idea that geological formations offered no information of a beginning and no clues to an end, but only the evidence of slow, gradual change. The earth's crust was produced by myriads of epochs of the same gradual processes we observe happening around us today. Over countless eons vulcanism shaped and reshaped the landscape many times and will continue to do so for eons to come.
14. Merriam Webster's 11th Collegiate Dictionary. At "tabernacle"; ": to take up temporary residence; *especially* : to inhabit a physical body"

Clear Signs of Trouble and Great Joy

Dayenu.

Chapter 7
Measurements

...behold, there was a man, whose appearance was like bronze, with a line of flax and a measuring reed in his hand...
Ezekiel 40:3

¹Then I was given a measuring rod like a staff, and I was told: "Rise and measure the temple of God and the altar and those who worship there,"
Revelation 11:1

Twice the Jewish Temple was measured at times when it didn't exist. In 586BC, Nebuchadnezzar destroyed the Temple. Twenty-five years later, God sent a messenger to measure the Temple for Ezekiel, although it was still rubble. The measurements were for Ezekiel's use in stirring the exiled Israelites to begin practicing and performing God's ordinances and laws again.

> ¹⁰And you, son of man, describe to the house of Israel the temple and its appearance and plan, that they may be ashamed of their iniquities. ¹¹And if they are ashamed of all that they have done, portray the temple, its arrangement, its exits and its entrances, and its whole form; and make known to them all its ordinances and all its laws; and write it down in their sight, so that they may observe and perform all its laws and all its ordinances. ¹²This is the law of the temple: the whole territory round about upon the top of the mountain shall be most holy. Behold, this is the law of the temple.
> Ezekiel 43:10-12

In a vision regarding Jerusalem after the exiles will have returned, Jeremiah was told the measuring line "...shall go out farther..." (Jeremiah 31:39). Nearly a century later, Zechariah saw in a vision a man with a measuring line who said he was going "to measure Jerusalem, to see what is its breadth and what is its length." (Zechariah 2:2). Within decades of these visions the Second Temple was rebuilt; Jerusalem was rebuilt; and its wall was extended. Measurement is taken of realities, not of unrealities. Ezekiel, Jeremiah, and Zechariah measured propositions to express the certainty of their future culmination. Measurement expressed the prophetic aspect of the propositions. History now shows those prophecies were indeed fulfilled.

Around twenty-five years after Titus destroyed Herod's Temple in AD70, an angel of the Lord told the Apostle John to take a rod and measure it, too, which also no longer

Clear Signs of Trouble and Great Joy

existed. Some eschatologists consider this to mean a rebuilt Temple is part of God's plan for the Tribulation, the seventieth week of Daniel. I agree. The expectation is consistent with the realities which followed Ezekiel, Jeremiah, and Zechariah's measurements.

The oracles given to Israel also measure God's plans embedded in mankind's history. The measurements are made with numbers of special meaning. Threes, fours, fives, sixes, sevens, eights, nines, tens, twelves, nineteens, forties, fifties, and seventies are some of the more common numbers which deliver a Biblical meaning beyond simple enumeration. Elevens and thirteens add ominous sense; fourteens and twenty-ones add emphatic sense. The importance of Ezekiel and John's measurements was their assurance of God's basic plan for Israel and mankind to receive their Savior, Priest, and King who will replace mankind's evil with His righteousness. Should we expect the historical events actualizing God's promise to abol-ish evil and establish righteousness to be any less measured than the restored Temple and city were? So when we notice history consistently happening in correlating patterns measured by numbers of particular Biblical meaning, should we not pay attention? Isn't it commonly said that science's aim is to search for order amongst chaos? If events can be discovered amongst history's chaos correlating with Biblical prophecy and themes in patterns measurable by numbers of Biblical meaning, then we can be sure we are observing a medium in which I AM WHO I AM has written a coherent message. And why should science not pay attention to the nature of that reality?

The number "four" expresses God's Creation, since it is additional to God (3+1). Being additional to God, rather than a part of Him, Creation requires His gracious regards in order to exist, the meaning of five. Man subjectively takes advantage of grace to sidestep God's objective laws, coincidentally (or not), the number of man is six. Man's own mind and emotions add to God's concerns whatever man desires to think in a world so accommodative towards subjectivity. Man's tastes are subjective. His proposals are subjective. His religions are subjective. (Biblically prescribed Christianity is God's religion, not man's; man's religion is imagination; God's religion is revelation, as the orderliness of the universe testifies.) Even man's science is subjective; ask Thomas Kuhn.

But God is objective. He created an objective world in an objective universe. And as science knows, objectivity is guided by measurements. Unlike subjective relationships, objective relationships have a measurable quality. Objective meaning is reflected in the measurements of reality. God calls man to be objective with Him. So He calls him to see in accord with measurements. Subjective eyes see in accord with whatever they want to see. But measurement hinders subjectivity. Objective eyes see in accord with reality, whether physical or spiritual, so they see by careful measurement. Challenger's icicles were an objectively discernable measure of O-ring inflexibility. Other historical interconnections are also objectively discernable by measurements of spiritual significance. Joe Biden's Freudian slip ten days before the election (the number of perfect completion) spiritually measured the vast number of polling anomalies, inconsistencies, and outright malfeasance to be the cheating of the 2020 vote. Measurements form a coherent expression of Biblical themes wherever man's history is found to be measurable in Biblically meaningful terms.

I AM WHO I AM measures the interrelationships of all deeds woven into the fabric of history. We do not have either the sensory or the mental capabilities to know those

Measurements

interrelationships as intricately as He does. Yet, within history's overall weave, the purposes of some events are so critical for us to know that God used both prophecy and measurements to make them apparent to our limited discernment. Biblical prophecy is somewhat like a dye applied to the fabric of history, making the interrelationships of fundamentally relevant events stand apart from history's normal clutter like printed patterns on an apron. Meaningful measurements of time and/or quantity in patterns emphasize their importance. Sometimes the measurements themselves are given as a prophetic template for identifying prophesied events and characters against the background noise of history.

But the Bible often gives a measurement in undefined units, or by using a defined unit so absurd, when taken literally, that it has to be metaphorical. Daniel was told the end of all the prophetic wonders shown to him would come after a time, two times, and half a time. I doubt Daniel found it to be entirely informative, not knowing how long "a time" was. The angel said there would be one-thousand-two-hundred-ninety days from the cutting off of the continual burnt offering and the setting up of the abomination which makes desolate. No significant corner of life was turned on February 21, AD74, one thousand-two-hundred-ninety days after Titus destroyed the Temple, cutting off the Jews' ability to make daily burnt offerings. Obviously, the unit of this measurement is not actual days, unless the measurement might extend from a different "cutting off" than what occurred in AD70. Some eschatologists find that measurement to be calibrated in literal days for measuring some occurrence to happen shortly after the end of the Tribulation, calling upon a different "cutting off of the burnt offering" for the measurement's starting point. Others treat the days as representing years, and then go looking for correlating dates of relevant, historical events. Both interpretations might be right in the manner of a Sethite-like dual prophecy.

If we peer into the weave of known history while bearing in mind Biblical prophecy and numbers, we can observe numerous measurements between interconnected events revealing meaningful patterns, such as Daniel's "seventy weeks of years", or the one-hundred-twenty years God told Noah would be the limit of man's days. They express important parts of God's purpose designed into the universe which would otherwise be lost amidst the spew of history. Other Biblically relevant events beat through history in measures of time like the notes of a melody meted out by numbers of Biblical meaning. Prophetic Scriptures add lyrics to the measured melodies as the creation raises its voice in melodious praise of the work God is performing.

Even non-prophesied events join in the tune. They can be the most insignificant of events, on a global scale, but sufficiently significant on an individual scale to deliver spiritual expressions. For example, it occurred to me, one weekend, that a pile of splintery, jagged edged wood scraps that had been accumulating around the end of my outdoor workbench was becoming hazardous. I cleaned it up before that weekend ended. The next weekend, a compressor hose lassoed my ankle as I rushed out my shed door, sending me airborne, completely parallel to the ground, before I slammed face first into the soft, accommodating dirt where that pile of jagged, splintery wood had been. I laid there, laughing, as I thanked God for giving me the inclination to clean up the weekend before instead of cuts and bruises giving me that inclination the weekend after.

Carl Jung called meaningful coincidences like this "synchronicity".

Clear Signs of Trouble and Great Joy

Synchronicity can seem even enchanted. I needed to verify a report of seven towns named "Salem" having been within the August 21, 2017 total solar eclipse's path of totality before I could use that information later in this chapter. Finding Salem, OR and Salem, SC was a snap. I verified four of the other five Salems with a little more difficulty. But I wasn't finding any Salem in the southwest corner of Wyoming, where the report said it was. I searched roadmaps; I Googled; I Binged, and I pinged "Salem, Wyoming" for nearly a month-and-a-half until I was ready to write this interesting tidbit out of my book. Then one morning I got a phone call. I don't remember the name she asked for, but there was no one at my office by that name. She had dialed a wrong number. But she seemed a bit chatty, so we talked a couple minutes. Then I asked from what part of the country she was calling.

"Wyoming."

"Oh really!? What part of Wyoming?" I asked.

"South," she replied

You don't say!

"Maybe you can help me with a problem. For weeks I've been looking to verify the existence of a town named Salem in the southwest corner of Wyoming."

"Oh yes. There was a Salem there; I mean, it's still there, but they changed its name years ago. It isn't much more than a wide spot on the road," she offered.

And so, by synchronicity, you will read about the seven Salem's within that eclipse's path of totality later in this chapter.

Einstein, Wolfgang Pauli, and Jung spent many occasions discussing synchronicity, for they all noticed it in their lives. I'm sure you've noticed it in yours, too. Of course, like Sadie, I know Who to thank for the synchronicity I often experience, the woodpile, the Salem, the beautiful lady to whom I am married, and even the occasion of my writing this book to you. He created all things for a purpose (Proverbs 16:4).

So, we should at least be curious about measurable phenomena, if not soberly attentive. When events revolve meaningfully around a core purpose, their import might be more than simply "coincidence". When they consistently involve a particular, Biblically described people and nation central to Biblical themes, "coincidence" might not adequately describe their measured interrelationships. "Synchronicity" might describe them somewhat better. But when many other thusly measured events also correlate with an amassing set of Biblical prophecies, they are no less than patterns printed onto history by a very real, communicative Creator bearing a stark warning we should hear. Ears to hear can clearly hear The Creator's call for a return to belief in the objective reality of The Holy Word of God, and for giving its communications the attention they are due.

Is it too much to think I AM WHO I AM has marked the road our history now travels? Is it too much for us to understand that "...[man's] days shall be a hundred-twenty years...", "...seventy weeks of years are decreed...", "...let them be for signs...", and "...there will be signs in the sun and moon and stars...", etc. (Genesis 6:3; Daniel 9:26; Genesis 1:14; Luke 21:25) might mean the map of history is marked for easy understanding? Jesus told the Pharisees and Sadducees,

You know how to interpret the appearance of the sky, but you can not

Measurements

interpret the signs of the times.

(Matthew 16:3)

The Pharisees and Sadducees had better eyes to see their own pomp and glory than they had to see God's Oracles. They viewed the prophecies of the Scriptures by the subjectivity of their own lives, just like today we disregard The Bible by the subjectivity of our own science. We must not refuse to objectively consider the prophecies of the Bible when experiencing current history. Religious leaders crucified Christ when He came the first time. When He comes again, the warring nations will turn their weapons on Him. It will be their horrific mistake. Ignorance caused by misdirected attention is dangerous.

Any route marked on a roadmap necessarily identifies a starting point as well as a destination. When Israel arrived in the Promised Land, they were to remember their covenant with God, their starting point forty years earlier. Artaxerxes' decree to restore and rebuild Jerusalem started the count of four-hundred-eighty-three lunar years to the destination of the cutting off of an anointed one (Daniel 9:26). The year of Artexerxes' decree was known in the early, first century AD. Anyone paying attention to its date in light of Daniel's prophecy would have been expecting an enormously significant death of an anointed one approaching soon.

Jesus' ministry was not worked in a corner. By the time of His execution He was known throughout the land. Speculation of His being The Messiah was everywhere. His crucifixion in the four-hundred-eighty-third lunar year after Artaxerxes gave Nehemiah the decree to rebuild the city was an exclamation point for anyone paying attention. It was a mark on history's map. It was a confirmational prophecy of Christ's identity. The Bible gave His people a sufficiently marked map of their time. The gospel writers also noted numerous prophecies in the Jewish scriptures that came to happen in Jesus' life, death, and resurrection, further marking Him on the prophetic map as the Jewish Messiah come to save everyone who would objectively believe. Objective reason would have easily identified this extraordinary fellow. Some people were attentive enough to recognize Him before He was "cut off". A few of those were at His crucifixion. One of them brought a linen for His burial.

The angel told Daniel to "…shut up the words, and seal the book" (Daniel 12:4). It would be truly sad if this was all the angel said about his prophecies. The measurements given Daniel by the angel would have been as useless to us as "…a time, two times, and half a time…" was to Daniel. But that wasn't all the angel said. His charge to Daniel was not to shut up the words and seal the book forever, but UNTIL. "Until" acknowledges a destination point in time. The very meaning of the word reaches out for an end for attachment to its beginning. In this case, it means there is a time when the words and the book will be objectively unsealed. Then understanding will flow forth. The twenty-five-hundred years following the angel's revelation to Daniel has brought sufficient measurements of history into our view for unsealing Daniel's book.

We can even see in our current time a remarkable measurement confirming the arrival of Daniel's "until". It places the modern state of Israel at the center of an entire pattern of signs, not a chronological or spatial centrality, but a centrality of Scriptural meaning and prophetic purpose. Seeing it only requires some focused attention, a reasonably

Clear Signs of Trouble and Great Joy

honest memory, and a little intellectual responsibility, the same skills required to aim a telescope or to align a slide under a microscope.

We would be remiss to think the Old Testament is the sum total of ancient Jewish history. Recall that the term selected by the Holy Spirit for expressing "the congregation of Hebrews" in the Old Testament is a form of the primitive Hebrew root meaning "to witness and testify". Those testifying, Hebrew witnesses would write history books carefully, since history has always been to them an important part of knowing their God. So, it is quite natural to find the Bible also referring to other ancient Jewish books of history: the Book of the Wars of the Lord (Numbers 21:14), the Book of the Acts of Solomon (I Kings 11:41), the history of Nathan the prophet, the prophecy of Ahijah the Shilonite, the visions of Iddo the seer (II Chronicles 9:29), and the Book of Jasher (Joshua 10:13 and II Samuel 1:18). Fortunately, The Book of Jasher was not lost, as were most of the others.

Jasher was enormously popular amongst the Jews down through the first century AD. Then it went lost to most of the world. Its recent rediscovery makes a colorful addition of information to many Biblical narratives. And its information might be forming part of the understanding now unsealing Daniel's "until". Although the debate over just how fanciful the color of Jasher's information might be is beyond the scope of this book, Jasher presents one extraordinarily interesting, somewhat amusing, suspiciously prophetic tidbit of information, fanciful or not. It is found in The Book of Jasher's propensity to date its stories by the year after Creation in which they occurred.

Jasher doesn't fail to date Abraham's birth: 1948AM (Anno Mundi, AM, is Latin for "in the year of the world"). Whether God directed this to be the date Israel's venerated Forefather was born, or not, whether the date is accurate or fanciful, it is greatly coincidental, and it is even a bit beyond synchronistic, that the year of Israel's reestablished statehood is expressed by the same number that is the year Israel's Forefather was born. It was Abraham to whom God promised Israel's nationhood in Palestine forever. Abraham's stated birth: 1948AM. Israel actually reborn: AD1948. Go find another nation thusly measured for bearing testimony to both I AM WHO I AM and the prescience of the Oracles He gave them.

Haven't found any yet? OK. Let's continue discussing this nation. Even more is added to this marvel from no less than The Holy Bible, those prescient Oracles given to Jacob (a poetic term the Bible uses for Israel). Galatians 3:17 says the Law was given four-hundred-thirty years after God promised to make a great nation of Abraham. Abraham was seventy years old at the time.[1] 1447BC is held by traditionally minded, Biblical scholars to be the date of the Exodus as determined by combining the length of the king's reigns given in the Bible, the number of years the Bible says the Temple was completed after the year of the Exodus, and events of otherwise known dates, for instance, the date Sennacherib lost his army at Jerusalem, 701BC. God gave Israel His Law a few months after their Exodus from Egypt. 1447BC is not an exact date; it could be off by a year or two, or a few. But it's close enough to see this other incredibly "coincidental" measurement.

Four-hundred-thirty years before the Exodus, at approximately 1877BC, God promised to Abraham a great nation. (In Chapter 9 we will see 1877BC mirror the 1948AM/AD1948 phenomena.) As we noted above, seventy years earlier Abraham was

Measurements

born, about 1947BC. And as I AM WHO I AM's propensity is to pile on: 1947 also measures the year of the most significant discovery concerning the Oracles God gave Israel: The Dead Sea Scrolls. Getting curious?

It isn't so much that these three events roughly measure the six-thousand years of Biblical narrative and prophecy into thirds as it is that the Biblical narrative runs from Creation 1,948 years to the birth of The Man of Faith, and then another 1,948 years to the time when God faithfully brought forth from a virgin The Man of Salvation, and then yet another 1,948 years to when I AM WHO I AM faithfully returned Israel to nationhood in their Promised Land for receiving The Man of Salvation's return. Like a line of praise sung for God's victory over evil to reestablish righteousness, the history of Salvation from Creation to the preparation for the resurrected Messiah's return is measured into three bars of equal count. 1948 is a testimony fascinating enough to get its own chapter in our little adventure through God's enormous wonders.

If Apophis and the total lunar eclipses of 2032-3 are signals of the end of evil's age, then we are the people of God's endgame to whom that line of prophecy has been sung. By it we have received enough perspective to finally understand what the angel told Daniel.

> [6]And I said to the man clothed in linen, who was above the waters of the stream, "How long shall it be till the end of these wonders?" [7]The man clothed in linen, who was above the waters of the stream, raised his right hand and his left hand toward heaven; and I heard him swear by him who lives for ever that it would be for a time, two times, and half a time.
>
> Daniel 12:6-7a

If Apophis' return on Easter Sunday, 2036, does foretell Christ's second coming, then the ambiguity of "a time, two times, and half a time" resolves into crystal clarity. The Bible states that after Christ returns He will reign for a thousand years (Isaiah 2:2-4, Revelation 20:4), then Judgment Day occurs (Revelation 20:11). Judgment Day is the end of all these wonders about which Daniel asked. On that day the heavens and the earth will flee from the face of The Judge, beginning the new day of eternal bliss forever free of sin and error. If we consider the "time" of "a time, two times, and half a time" to be a thousand year unit, we will see this prophecy remarkably fit our proposition that Apophis' two rendezvous around that unique tetrad do approximate the time of the Tribulation. Moreover, we will see a confirmation of that fit by how well Biblical sense correlates with history.

A thousand year unit of measure for each "time" places Judgment Day at about three-thousand-five-hundred years from when Daniel was talking to the angel (approximately 500BC). This unit would place "The end of these wonders" at about AD3000. If Christ returns around 2036 in accord with our proposition, then His Millennial reign would end at approximately AD3036, bringing on that day in which the heavens and the earth will flee from God's face. The angel's measurement for Daniel seems to meaningfully verify our proposition, even though its fit is not to the precise tick of Mr. White's Bulova watch.

The correlation of the measurement given Daniel with that 1,948 year pattern of

Clear Signs of Trouble and Great Joy

human history is even more apparent when considered from the perspective of the Day of Judgment. If we were to look back at history from Judgment Day, we would see the last epoch first, the thousand year reign of Christ from approximately 2036 to 3036: A TIME. Next we would see its preceding epoch, two-thousand years spanning from the death, resurrection, and ascension of Christ to the time of His return, from AD33 to around 2036: TWO TIMES. This epoch is commonly known today as the Church Age, or The Age of Grace. It is also called The Time of The Gentiles, distinguishing it from the epoch prior to it, the age of Law and Sacrifice given to God's chosen witnesses, Israel. The HALF A TIME is the last five-hundred years of that Law and Sacrifice epoch. It spans from when the angel was presenting this prophecy to Daniel to the resurrection of Jesus in AD33, a little more than five-hundred years.

The angel gave Daniel a measure of time which closely corresponds with what we know today about salvation worked into history. The measurement presciently correlates with the unfulfilled remainder of Biblical prophecy. It coherently measures the spiritual epochs of the working of the end of all these wonders from the day Daniel and the angel conversed to the final Day of Judgment, if our proposition about 2036 holds true.

Even the unit of measure employed astoundingly testifies to this measurement's coherence with Biblical themes: 10x10x10, the number of perfect fullness (or in this case, perfect fulfillment) raised by the number of God's involvement (3), I AM WHO I AM fully fulfilling His promises. What other unit of measure could be more Biblically meaningful for such a portentous measurement? What of more significance could this measurement point to beyond that of Christ's glorious return to defeat all liars and every scrap of subjectivity? How well it correlates with history! Does the proposition need any further verification? But of course, by this point in our adventure, you should be suspecting there will be more.

Nobody is better at piling on than is Jesus' Daddy. He has piled up more verification whether we need it or not (or want it or not.) It comes through His prophet, Hosea. In the late eighth century BC, Hosea warned Ephraim and Judah (The Northern and Southern Kingdoms of David's Israel) that, like a young lion, God would rip them to shreds and return to His place because of their unfaithfulness. Then, after two days, when they returned to Him, He would raise them up, restoring them on the third day.

> [14] For I will be like a lion to Ephraim,
> and like a young lion to the house of Judah.
> I, even I, will rend and go away,
> I will carry off, and none shall rescue.
> [15] I will return again to my place,
> until they acknowledge their guilt and seek my face,
> and in their distress they seek me, saying,
> [1] "Come, let us return to the LORD;
> for he has torn, that he may heal us;
> he has stricken, and he will bind us up.
> [2] After two days he will revive us;
> on the third day he will raise us up,
> that we may live before him."

Measurements

Hosea 5:14-6:2

If Ephraim and Judah's restoration occurred a couple days after they had been torn apart and destroyed in AD70, certainly we would have found that in the annals of history. But we don't see any occurrence like it. The term "days" obviously does not express a literal meaning for this prophecy. Unless the accusative pronoun of verse 6:1 is switched to "Himself" so the Lord's resurrection can be seen in the "third day". In that case, it would go like this: He tore Himself (which He did on the cross) and went away (from Scripture we know He went into Hades) for two days, and on the third day He raised Himself up from Hades, bringing with Him the keys of death by which He will loose us to be raised up as well, including Ephraim and Judah. But this sense only parallels the text, since the pronoun of 6:1-2 articulating who was raised up on the third day is not "Himself"; it is "us". Interpreting it as "Himself" must also overlook the inconsistency of Hades, to where Christ went while in the grave, certainly not being "His place". His place is at the right hand of His Father in Heaven, to where He indeed went after His resurrection, and from where He has not yet returned. Moreover, Hosea specified that Ephraim and Judah would be raised AFTER calling out to their Lord, which they definitely did not do three days after crucifying their King. Refusal to repent and call on their God after crucifying Christ is why Jerusalem was destroyed in AD70. Besides, misinterpreting the pronoun to mean "Himself" in order to find Christ's resurrection in the prophecy is a subjective action. Granted, it paints an artistically useful picture. But it is not a picture faithful to the text. Interpreting the Scripture in accord with the pronoun the text supplies, "us", is the objective treatment.

Holding true to the rule of antecedents for pronouns, it is Ephraim and Judah who will be utterly torn, and then, after they repent and turn to the Lord, will be raised on the third day. Therefore, the "days" in Hosea 6:2 can not be twenty-four hour periods; if they were, the passage would have no fulfillment in history. The less literal, more artistic, parallel of Christ raising up Ephraim and Judah on the third day by raising Himself from the grave is possibly a dual meaning, full of grace for its grammatical errors, in the great Sethite tradition of dual prophecies. But applying the wholly meaningful 10x10x10 unit of measure to the "days" of this prophecy, we find it identifying the third millennium AD as the one in which Christ returns from His place in heaven to raise up Ephraim and Judah from their apparent defeat at the end of the Tribulation.

As a result of Israel rejecting and crucifying its King in AD33, Ephraim and Judah were torn asunder and removed from the Promised Land again, in AD70, when Jerusalem was destroyed and the Temple was leveled, cutting off, even to our day, the continual burnt offerings of the Jews. Soon after, a rebellion arose amongst the Jewish remnant allowed to stay in the land. Their rebellion was brutally crushed. The Jewish remnant was then thoroughly scattered throughout the nations by AD135. Ephraim and Judah had been torn as by a lion. The Lion had returned to His place. There He awaits their call on Him. Ephraim and Judah spent the rest of the first millennium AD, the first day, and all of the next millennium, the second day, in continuous rejection of their Priest and King, Jesus Christ. Then on the third millennium (day), our current millennium, Israel will become cornered by a world full of anti-Semitic bigots. They will have only one option for survival: the ancient

Clear Signs of Trouble and Great Joy

I AM WHO I AM, the butt of this modern world's jokes, the God perceptually murdered by Darwin's imaginary ape-man. Ephraim and Judah will repent of their ignorance and realize the truth of I AM WHO I AM. So they call upon Him for help with their backs shoved against yet another extermination wall.

Jesus Christ, the one they crucified, picks up the phone. Do you think He will be vindictive? Of course not! He is faithfully merciful to the truly repentant. At their desperate call on God, He will raise Ephraim and Judah above all nations to live before Him in glory and great joy. The end will indeed seem entirely coincidental to everyone of the end who refuse to recognize the truth of I AM WHO I AM, refusing to pay attention to the signs of the times and the measurements marked on history's roadmap. But for us who are of the beginning, who have come to know the Lord God by testing, learning, and living His Holy Word, none of this seems coincidental in the least. It's precisely what we would expect from the communicative Author of All Things.

The angel intimated to Daniel that the prophecies about the end of evil's age would be unsealed for the people of the end. Apophis' fit with those two prophecies show us to be the people of the end. Signs are showing that Daniel's prophecy has been unsealed. Today, more people than ever before are presenting consistent, cohesive, coherent *information* they've discovered (been shown by God) about this coming end. What they say is nothing like the old "THE END IS NOW" prophecies tossed around on the basis of Jesus' face having been seen on a potato chip, or a piece of burnt toast, or on some such other thing. The angel gave Daniel a time frame for the end of the world: a time, two times, and half a time. Although it was ambiguous to Daniel, enough *information* has accumulated over the intervening twenty-five-hundred years to illuminate its clarity for us. Hosea wrote the confirmational prophecy. And signs in the sun, moon, and stars correlating with prophetically relevant, historical events and conditions on Earth, measurable in Biblically meaningful numbers, define all of this illuminated clarity as the approaching return of Jesus Christ. I can't help saying it again: a time of Christ's Millennial reign, two times of the Age of Grace, and half a time of Law and Sacrifice! Is it just that those sneaky ole priests who made up the Bible in Babylon were better guessers than was Darwin? Or has I AM WHO I AM authored both the Bible and human history within the universe He designed to sing the glory of Jesus Christ, your Savior, to reality's tune of empirical observation?

Discovering the Biblically implied six-thousand year narrative of man's history being roughly expressed in three 1,948 year periods, finding total lunar eclipses on Israel's major holidays at Israel's becoming a nation again in possession of Jerusalem, knowing that four more total lunar eclipses of very ominous dates will neatly split in half the seven year cycle of an asteroid named for chaos at the one, epochal moment precisely fitting Daniel and Hosea's prophecies is far beyond any rational consideration for coincidence to explain, or even for synchronicity to suggest. As the passage of time draws even more evidences of the Bible's truth into open sight, skepticism against God's Holy Word must stand upon ever more irrationality to maintain its boasting.

> [3]Let no one deceive you in any way; for that day will not come, unless the rebellion comes first,

Measurements

II Thessalonians 2:3a)

And we've only begun to discuss the Biblical measurements of man's time on this earth. Let's observe another prophecy measure out of history a very key player in not only the Tribulation, but also in the last few centuries which have been setting a stage for the Tribulation. This prophecy was also given by the angel speaking to Daniel.

> [11] And from the time that the continual burnt offering is taken away, and the abomination that makes desolate is set up, there shall be a thousand two hundred and ninety days.
>
> Daniel 12:11

Some "cutting off of the continual burnt offering" is a very significant event for prophecy. God commanded His people to sacrifice an unblemished, one year old male lamb twice every day, one in the morning, and another in the evening. They were continually being offered. They kept Israel's sins rolling forward during the epoch of Law and Sacrifice until God would sacrifice His own, unblemished Son to begin the epoch of His grace and mercy, the "two times" we discussed above. Antiochus IV Epiphanies, a Seleucid king who ruled from 175 to 164BC, briefly ended the continual burnt offering. The Maccabee brothers quickly ended that hiatus. Maybe that event was the beginning of this prophecy, as some Biblical scholars suggest. But it could have only been metaphorical at most, and dual at best. Daniel's prophecy did not imply a brief cutting off, as Antiochus' "cutting off" was. But in AD70 the Romans ended the continual burnt offering so resoundingly that it has not been restored since.

The one-thousand-two-hundred-ninety day part of the prophecy didn't tell Daniel much more than did the "time, two times, and half-a-time" of the preceding prophecy; it was sealed up until the end, too. For Daniel, it was only a vague hint. But history has now dropped his vague hint into the context of much more enlightening *information*.

The angel told Daniel that the measurement was from the cutting off of the continual burnt offering and the setting up of the abomination which makes desolate. By "abomination that makes desolate", was the angel referring to an abominable event which makes desolate, or to an abominable entity who makes desolate? In the great Sethite tradition of dual prophecy, as the "abominable event", the days more clearly apply to the Tribulation by representing actual days. But as an "abominable entity", the prophecy identifies a spiritual line of idolatrous people who will work the abominable event into history.

We are here concerned with measuring patterns in history which are relevant to the conditions necessary for the beginning of the Tribulation event. Therefore, we are concerned with the idolatrous line of people. Their idolatry is the driving force of the seventh and eighth heads of Revelation's dragon/beast (the seventh kingdom having died, and the eighth to be the seventh revived), since Revelation places those heads in its story regarding idolatry's end. It is the Antichrist, the emperor of the eighth kingdom, who cuts off the continual burnt offering within the Tribulation week of years (Daniel 9:27; II Thessalonians 2:3-4). Therefore, to fully appreciate this one-thousand-two-hundred-ninety day measurement, we

Clear Signs of Trouble and Great Joy

need to identify the particular spiritual line of idolatrous people who will take away the continual burnt offering during the Tribulation.

In the last chapter we discussed the seven heads of the dragon/beast shown to the Apostle John. Each was an empire which had defeated the one preceding it (except the first), and each controlled the land promised to Abraham. We found the first six embedded in well known history: Egypt, Assyria, Babylon, the Medo/Persians, the Macedonians, and the Roman Empire. Now we must concern ourselves with the identity of the seventh.

Any message requires consistency in the patterns delivering it, whether those patterns are made by light rays, sound waves, ink on paper, or the orientation of objects, such as a pile of rocks at an intersection of paths, or total lunar eclipses on Jewish holydays neatly stacked at the time of events significant to Jewish history. Abandoning the established pattern of the first six kingdoms chain-linked through history obscures the identity of the seventh kingdom and subordinates the search for it to subjectivity. It opens the possibilities of the dragon/beast's identity to a jumbled chaos of historical trivia through which misled eschatologists have rummaged for centuries to find the identity of this seventh kingdom. Subjectivity isn't The Bible's way. God is objective.

Regardless, most eschatologists have made trivia into much ado about some revived Roman Empire being the seventh-head, say, the Holy Roman Empire, or maybe the European Union, or something of those sorts. Those guesses read well in books. But they jostle this seventh-head of God's Book out of coherence with the Bible's basic theme. A revived Roman Empire not only garbles the very real historical and Biblically coherent pattern made by those first six empires, it twists the Scriptures regarding the seventh-head. Attributing a revived Roman Empire to the seventh kingdom mixes a "came back to life" aspect into the sixth-head which the Bible never attributed to it. The angel never informed John that the sixth-head would be revived as the seventh-head, like the Holy Roman Empire or the European Union are perceived to be the old Roman Empire revived. So the seventh-head, being some later version of the Roman Empire, would have been nothing more than an extension of the sixth-head, convoluting their nature of being two different heads. But if the Roman Empire, as the sixth-head, permanently died, such that the revived Roman Empire wouldn't have to be a mere extension of the sixth to be the seventh-head, then the seventh couldn't die because the angel said one head was mortally wounded, not two. Therefore, any form of the Roman Empire being the seventh-head could only be the sixth-head extended, which indeed makes it the sixth-head still, making the eighth actually to be the seventh. That entire interpretation rolls into another barrow pit of twisted up ideas. Let's step back from the wreckage to gain a better perspective.

The angel specifically stated the seventh kingdom was yet to come. As an empire unrelated to the sixth, which indeed defeated the sixth, the seventh would continue history's chain of empires through time according to the pattern established by those first six, no rollover, no barrow pit, no wreck, no spinning hubcap left on the highway of ideas.

Can we find in history such an empire which defeated the Roman Empire and also controlled the Holy Land?

The collapse of the Roman Empire is commonly taught to have happened in AD475 at the hands of barbaric hordes. But that was only the beginning of the collapse. The

Measurements

administration of the Roman Empire had been split into western and eastern territories by the beginning of the fifth century. The western half collapsed in AD475.[2] But the eastern half continued unimpeded until 1453, when the Ottoman-Turks breached Constantinople's walls. From the beginning of the sixth century to its defeat by the Ottoman-Turks, the eastern half was yet the remains of the old Roman Empire.

> Byzantium...was an ancient Greek colony in early antiquity that later became Constantinople, and then Istanbul. The Greek term Byzantium (or Byzantion) continued to be used as a name of Constantinople during the Byzantine Empire, even though it only referred to the empire's capital...The name "Byzantine Empire" was introduced by the historian Hieronymus Wolf only in 1555, a century after the empire had ceased to exist.[3]

> ...the citizens of Constantinople and the rest of the Eastern Roman Empire identified strongly as Romans and Christians...and...Byzantium was ruled by Roman law and Roman political institutions, and its official language was Latin.[4]

Although its culture was Greek, its formality was Roman. The Ottoman-Turks, who defeated Byzantium, the remnant of the Roman Empire, satisfy the first criteria. Then they seized Jerusalem in 1517,[5] fulfilling the second criteria. 1517 also began such an astounding measurement of history that it certainly verifies the Ottoman-Turk Empire as the seventh head, which we will discuss shortly. The Ottoman-Turk Empire, having defeated the sixth-head, controlled the Holy Land until it was dealt a mortal wound by World War I. During its life, Constantinople had become the seat of The Caliphate. That Caliphate died with the Empire, leaving the Islamic world headless. The historical criteria established by the first six heads identifies the Ottoman-Turk Empire as being the one and only candidate for the seventh-head of Revelation's dragon/beast, if you want to be objective about Scripture. And indeed, it was mortally wounded, just as Scripture prophesied, by a wound as carefully measured into history (see Figure 4, pg. 153) as was The Empire's conception.

Not only does this seventh empire fit those two criteria, it also reflects the wound the ancient Sumerians depicted of a seven-headed dragon, the same wound which Revelation predicted would happen to the dragon/beast. We could think of this as mere pareidolia, or simply as a tidbit of ancient history coincidentally known and applied by John to the Revelation story. But the reality of seven kingdoms chain-linked through four-thousand years of history, each controlling the Promised Land for a time, converging with our time as fulfilled Biblical prophecy in concurrence with Revelation's theme of the dragon/beast, agreeing with the Sumerian depiction, rises far above the level of a merely interesting anomaly.

The thousand-two-hundred-ninety day measurement in Daniel Chapter 12 verifies the accuracy of this identification. Let's return to the "cutting off of the continual burnt offering". The Roman General, Titus, destroyed Jerusalem and the Temple with a regiment assembled mostly of Syrian stock. Several centuries later these Syrians would become Muslim. Considering all of the available *information*, "the prince who is to come" of Daniel 9:26 is Revelation's Antichrist making the strong covenant of 9:27, giving rise to the

Clear Signs of Trouble and Great Joy

Tribulation, causing the sacrifice and offering to be ended, having come on "the wing of abominations". The Antichrist is "the prince" of Titus' Syrian regiment by his being Muslim. Jonathan Cahn noted this. The one-thousand-two-hundred-ninety day measurement of Daniel 12:11 confirms that proposition as well.

Before this measurement will become apparent, we first need to note an interesting Hebraism at verse 11. The above quote is taken from the RSV "...from the time... the...offering is taken away, and the abomination...is set up, there shall be...days." From the first time I read this verse, some fifty years ago, I have been entirely perplexed about what time period is being addressed. We are told the time spans twelve-hundred-ninety days. Ok, fine. But twelve-hundred-ninety days from what to when? Although the verse indicates *from what* (the cutting off), it doesn't seem to indicate *until when*. At least it doesn't in most English translations.

But the original Hebrew is markedly different.

The Hebrew text supplies no word for "and" at verse 11. Neither does it supply a word meaning "until". In the Hebrew text, the concept of a cut off sacrifice merely precedes the concept of a desolating abomination without any adjoining conjunction or adverb. Of the four-hundred-twenty-five times the RSV uses "until" in its translation of the Hebrew Old Testament, forty-five times there is no word in the Hebrew text expressing "until". Every one of those forty-five Hebrew sentences are structured to imply "until" contextually. Hebrew does this especially when the word "from" is present earlier in the sentence, just as it is present in Daniel 12:11. In fact, the entire sentence structure of 12:11 is typical of those forty-five times "until" is implied, and so is the surrounding context typical. Yet, of the seventeen different translations I consulted at this verse, only the International Children's Bible and The New Century Version translate the verse to say twelve-hundred-ninety days from the cutting off of the continual burnt offering *until* the abomination which makes desolate is set up. Now we have a period of time represented by a beginning "what" to an ending "when", and surprise, surprise, surprise, it comes from dealing with the Hebrew text objectively instead of subjectively.

Why do so many educated people translate such an often used and obvious literary device completely wrong? Eschatologists have commonly seen these twelve-hundred-ninety days as somehow belonging only to the Tribulation. It's a tempting trap. The Antichrist's proclamation of his being God is the abomination making desolate being "set up" during the Tribulation. But the Antichrist also ends the continual burnt offerings having been reinstated shortly before he rises to this offense. The ending of the burnt offering is generally perceived to happen when the Antichrist has seized the throne and demanded worship, being the event which is "the abomination that makes desolate" in the context of the Tribulation. But the setting up of the abomination can not happen twelve-hundred-ninety days later because it is part of cutting off the continual burnt offering by the Antichrist, unless we dump a consistent Hebraism in order to translate "and" in the place of the more proper "until", thus combining the two events so that the twelve-hundred-ninety days brings us to some completely unknown event thirty days after the Tribulation. Alas, we're back at the site of another wreck in the barrow pit of twisted up ideas.

"Set up" has an interesting consistency amongst the various translations of this

Measurements

verse. The Hebrew word, *nawthan*, translated as "set up", conveys a wide range of concepts. It denotes ideas ranging from "...add, apply, appoint, ascribe, assign..." to "...cast, cause, charge, come..." to "...direct, distribute, do...give (forth, over, up), grant, hang (up)...leave, lend, let...perform, place, pour..." and I think we get the drift; it is a Hebrew word "...used with great latitude of application."[5] The choice of "set up" for its almost universal translation at this verse appears to be another popular leaning towards applying the verse only to the Tribulation years. "Set up" and "cut off" both deliver a useful serving of eschatological sensationalism.

But if this verse is a dual prophecy, it might be making a pinpoint on history's roadmap to verify the identity of the dragon's seventh-head, which from John's perspective, was yet to come, while from our perspective, is dead but faunching in the grave to return as the eighth. If by "abomination", the verse is referring to the Antichrist's idolatrous people who make desolate, then, at some point in time, their idolatrous empire would have to be "set up", "made", or "brought into being", consistent with the meaning of the Hebrew term, *nawthan*. That conception would be the setting up of the abomination, the seventh-head, which eventually was killed to soon be resurrected as the abominable eighth-head, making desolation by demanding worship as God.

The twelfth century AD deterioration of Seljuk power left Anatolia controlled by regional leaders of independent principalities called beyliks. Most of them peacefully abided within the bounds of their territories. But Ossman I controlled his beylik in central Anatolia with an expansionist mentality. He began capturing border towns, and eventually entire beyliks. After him, his son, Orhan, won control of all northwestern Anatolia and advanced into the Balkans.[6] When Orhan had reached his eighties, he planned to leave this expanded beylik to his eldest son, Suleyman Pasha. But in 1357, the favored Suleyman Pasha died after falling from a horse. Orhan's despair exacted a heavy toll on his own health. He died five years later, in 1362, leaving control of the expanded beylik to his second son, Murad.[7]

It was Murad who established the former <u>Osmanli</u> tribe into a sultanate.[8]

The American Heritage Dictionary defines "sultinate" as the reign of a sultan. It defines "sultan" as the ruler of a Muslim country, especially in regard to the former Ottoman-Turk Empire. Merriam-Webster agrees. History verifies. Less than a century later this sultanate collapsed the Byzantine Empire.

Allowing the "days" of Daniel 12:11 to metaphorically represent years, one-thousand-two-hundred-ninety years after the destruction of the Temple cut off the continual burnt offering comes AD1360, three years (the number of God's doing it) after Suleyman's death. There we find Orhan in the midst of grieving himself into the grave, steering the beylik into the hands of Murad, who would set it up as a sultanate -the Ottoman-Turk Empire. Orhan died five years after his eldest son's death, the number of grace. That five-year stretch of history gave control of this beylik to the one who would set it up as an empire. We will see this five-year pattern of events again in relation to the end of the dragon/beast's lawless seventh-head, and once again in relation to America, the nation God commissioned to restrain the mystery of lawlessness amongst nations.

Clear Signs of Trouble and Great Joy

Certainly Ossman began the expansionist mentality Orhan built upon. But Murad added the audacity of a Caliphate over the entire Islamic world to that expansionist mentality, an abomination eventually ending the Roman Empire within the next ninety-one years and taking control of the Holy Land within one-hundred-fifty-five years. Soon it would rule the vast expanse of the Islamic lands until its final end at the deposition of Caliph Abdul Mejid II on March 3, 1924, only a few years before a handful of Muslims established the organization aimed at reviving that lost Caliphate. That organization is today known as the Muslim Brotherhood. The angel measured for Daniel the heart of a five-year historical process setting Murad up as Orhan's successor, who developed the Ossman beylik into a Caliphate. This measurement has now been unsealed for us to see.

God's plan for the final days of evil is not complicated; it is complex. Many peoples and nations will interplay amongst natural disasters, hysterias, mass deceptions, demonic activities, persecutions, and fortunately, angelic respites. This seven-year situation does not just drop out of the sky all wrapped up in birthday-paper to be opened at the time God scheduled for an eventual Tribulation party. Instead, the Tribulation happens like all other history does: it develops out of many years of man's foolish, interconnecting deeds. So, what can be seen in the Bible's various prophetic passages about the end of evil are historical processes spanning through time, places, and peoples. Those processes eventually develop into the peoples, places, and chronology of the Tribulation situation, as we just saw the seventh-head of the dragon historically develop out of a prophecy evidently given as early as that Sumerian depiction prior to even the Mesopotamian cultures. History is not complicated either; it is complex.

Jerusalem is a major scene in the Tribulation act of God's plan. The development of its set is highlighted by one of two perspicuously measured, paths of history. Jerusalem is the place I AM WHO I AM picked to be His one and only Holy City. It was the place of God's ancient priest/king, Melchizedek (Genesis 14:18). It will be the place of the eternal reign of the King and Priest in Melchizedek's order, Jesus Christ (Psalm 110:4, Hebrews 5:10). It is Mount Zion in the far north (Psalm 48:2b). But its historical emergence has been the epitome of ebb and flow. It was long ago overseen by God's priest/king, Melchizedek. It eventually fell into Canaanite idolatry. Then it came under control of God's chosen people. They stirred into it a Jewish imitation of Canaanite idolatry. God chastened that nation and restored Jerusalem to worshipping Him. But then it fell into the idolatry of religious arrogance, equating the objective I AM WHO I AM with the subjectivity of HE IS WHO WE ARE. It was then turned over to the Gentiles after the Jewish leaders crucified Jesus and failed to repent. Populations of Jews lived in Jerusalem on and off, returning to be driven out to return to be driven out, until the Ottoman-Turk Empire captured it in 1517. That event begins our next measurement of history regarding Jerusalem, The Jubilee Path.

God is hotly jealous over His Holy City. He will have His city over evil's dead body. Of this He forewarns. So, He will preserve the City through its desolations and wars until the time He draws it to Himself. Every empire of the dragon which captured God's people and His city also sheltered them. Egypt sheltered Jacob's family until the Exodus. Assyria did not slaughter all the Israelites, but moved most of them to new homes throughout its empire. Babylon and Persia gave the Jews the most respect of all those empires as the

Measurements

Jews were scattered throughout their regions. Macedonia and Rome brought their rule to the Jewish homeland approximately four-hundred years later. The Ottoman-Turks opened their cities to the Jews fleeing Spain's Alhambra Decree. And they later opened Palestine for the Jews fleeing Russian pogroms near the end of the nineteenth century. Jerusalem's history was turbidly war torn until 1517 began a somewhat more prosperous and peaceful period of Ottoman rule over Jerusalem.

The history of Jerusalem's development from 1517 into its Tribulation role is The Jubilee Path, since it is measurable in fifty-year periods widely known as Jubilees. Four key events on this path prepare Jerusalem for receiving The King of Kings and Lord of Lords at the end of the Tribulation. The chronology of those events is measured in units of the Jubilee.

The Jubilee was Israel's most economically, civilly, and spiritually meaningful measure of time. Its fiftieth year stood for the restoration of all things. The King of Kings will restore all things when He returns to tabernacle with mankind throughout The Millennium, the "TIME" of Daniel 12:7. In the Jubilee year, all previously purchased land was returned to its sellers; slaves were set free; debts were forgiven. It expressed the process of preparing Jerusalem for its King's reign to restore the righteousness, peace, and joy Adam and Eve's liberalism destroyed. Such preparation is precisely what this measured path of four events is about. Therefore, the unit of this process's measure is appropriately the fifty years of a Jubilee.

> [10]And you shall hallow the fiftieth year, and proclaim liberty throughout the land to all its inhabitants; it shall be a jubilee for you, when each of you shall return to his property and each of you shall return to his family...in it you shall neither sow, nor reap what grows of itself, nor gather the grapes from the undressed vines. [12]For it is a jubilee; it shall be holy to you; you shall eat what it yields out of the field.
>
> Leviticus 25:10-12

Being about return, "Jubilee" is also about relationship with God. Trusting Him is especially important in regard to the Jubilee year. Imagine the whole of your last forty-nine years of affairs being turned upside down and shaken like an Etch a Sketch. You are forbidden to even plant and cultivate crops in this year long process of reformation. You must completely rely upon God's having blessed your preparation. The Jubilee involves return, trust in God, and wise preparation.

The Ottoman rule over Jerusalem began a new process for the Jews in 1517. It would bring Jerusalem from having been the Middle Eastern soccer ball of pre-sixteenth century nations to being the cup of trembling for all the world's nations at the time Israel calls upon I AM WHO I AM (Hosea 5:15-6:1). The number "eight" represents new beginning, Biblically speaking, whether it is the beginning of a new list, series, process, or paradigm, and with a sense of abundance in that new beginning. The new heaven and new earth occur on the eighth thousand-year day after Creation, the day after the seventh Millennial day of rest, Christ's thousand year reign on Earth. The Eighth Day is abundantly

Clear Signs of Trouble and Great Joy

different from all those before it. The eighth head of the dragon is also a new type of head, indwelt by Satan personally, and abundantly abominable.

The Jubilee Path begins at Jerusalem's fall into Ottoman-Turk control in 1517, its first key event. The first count of Jubilees on the path spans the four-hundred years when Jerusalem was controlled by the Ottoman-Turk Empire, eight Jubilees, from 1517 to its release from control in 1917, its second key event. This count of eight Jubilees indeed developed a very different time for Jerusalem. By 1917, Jews were pouring into the Promised Land from around the world, as those "dry bones of Ezekiel" had begun rattling together. In 1917, Jerusalem and The Promised Land were given their first step towards true, long awaited independence.

An amazingly clear timeline measured in seven year counts crosses this Jubilee Path at 1917, further defining the importance of that year. Jonathan Cahn discovered and communicated a prophetic process, measured by a seven year count, delivering a special warning to America. Further research into Cahn's idea of Shemitah years (every seventh year), as he discussed them in his book, *The Harbinger,* reveals an amazing path of events developing the world situation towards the Tribulation as much as the Jubilee Path prepared Jerusalem for receiving her King. We will explore the Shemitah Path soon.

But concerning the Jubilee Path, the patterns of Jerusalem's key events being measurable in periods representative of God's restoration of all things stretches "coincidence" beyond its rational meaning. Jerusalem is God's Holy City where mankind's restoration begins. The seventh-head of the dragon coughed up Jerusalem after eight Jubilees of indigestion.

1917 led to the British Mandate (administrative British control, but not British possession, of The Promised Land). Britain's relationship with Jerusalem was a conservatorship. During this time of the mandate, Jerusalem's population grew from some twenty-thousand to over one-hundred-fifty thousand. Eventually, the population became two thirds Jewish. Then Israel captured the western part of the city in its 1948 war of independence, while Jordan took possession of the eastern part, The Old City of Zion, where all of the Christian and Jewish holy sites are.

Those sites fell into disrepair, and some were even desecrated. About half the synagogues were either razed to the ground or converted to animal shelters. Even the ancient Jewish graveyard at the Mount of Olives was desecrated. This is what Islam does. It desecrates everything not of itself, just as the Apostle Paul wrote about the man of lawlessness, the son of perdition

> ...who opposes and exalts himself against every so-called god or object of worship...
>
> (II Thessalonians 2:4a)

with which the Koran agrees,

> And fight with them until there is no more persecution and religion should be only for Allah.

Measurements

(Ascension 8.39)

Old Jerusalem remained in this condition for nineteen years. Recall the spiritual significance of the number nineteen, the full summation of human affairs. At the end of those nineteen years, in 1967, Jordan's rule over the city was summed up by only six days of war.

1967 became the ninth Jubilee of The Jubilee Path, (the number of summation). Israel won control of Old Jerusalem, the Path's third key event. Four total lunar eclipses of Passover and Tabernacle full moons celebrated the glorious return of the Old City, Zion, with all of its ravaged holy sites, to the witnesses of God's faithfulness. At Israel's repossession of Jerusalem in 1967 (the ninth Jubilee), the Jubilee Path intersects the series of four, crystal clear, holyday-tetrad signs. (The Shemitah Path intersected the next holyday-tetrad.) The Six-Day War fell only an inch or two and a couple nuances short of being a miracle, by secular estimates. It is known to Christ's redeemed and I AM's witnesses as a completely obvious miracle, by all objective estimates. The Six-Day War enjoyed God's providential interaction even more than did America's war for independence.

In 1967 Jerusalem entered the final Jubilee of its preparation to be the ruling seat of The King of Kings and Lord of Lords.

The fifty years following 1967 completes the tenth and final Jubilee of this path (the number of perfect completion). This final, key event occurred in 2017 when the amazing President Donald J. Trump pronounced to the world America's recognition of Jerusalem as being Israel's capitol. His pronouncement was a major event, considering the United States was still recognized as the leader of the free world. By 2017, Jerusalem had taken its position as Israel's capitol historically prepared to host the Tribulation play of God's end-game gambit, written in the ancient playbook of I AM WHO I AM. All the world's militaries will meet on the plains of Megiddo, sixty miles north of Jerusalem, the world's most international city, claimed to be a most holy site by all three of the world's Abrahamic religions. By moving the American Embassy to Jerusalem, President Trump signaled to the world an abiding American/Israeli allied relationship, a relationship every nation will have with Jerusalem after the Tribulation ends. Twelve years after Trump's announcement (the number of perfect order), the Tribulation will begin, according to Apophis and the unique tetrad. Establishing God's order is the purpose of the Tribulation, the process which sums up all of mankind's chaos nineteen years after 2017, as indicated by Apophis' 2036 return on Easter Sunday of that year, the day after Passover. Man can't make this stuff up.

If Robert Mueller was one-tenth the investigator he was purported to be, he would truly have been humbled by what there really was to discover about President Donald J. Trump. If Democrats were anywhere near the Biblical followers Nancy Pelosi insinuates herself to be, they and Trump would seamlessly effect good and benefit for all of America. Biblical prophecy, itself, called upon the full character of this President's audacious integrity to announce to an objecting, denying, anti-Semitic world, that in the eyes of I AM WHO I AM and The United States of America, Jerusalem is Israel's capitol! Amen! and amen. It took fifty years to drag the Gentile nations to the recognition of this truth. Eight American Presidents lied about giving Jerusalem its rightful recognition (the number of new

Clear Signs of Trouble and Great Joy

beginning). President Trump's courageous character may not have spoken a very long sentence into the annals of history, but his short sentence prophesied an immense truth.

> [21] My son, fear the LORD and the king,
> and do not disobey either of them;
> [22] for disaster from them will rise suddenly,
> and who knows the ruin that will come from them both?
>
> Proverbs 24:21-22

> [1] The king's heart is a stream of water in the hand of the LORD;
> he turns it wherever he will.
>
> Proverbs 21:1

> [1] Let every person be subject to the governing authorities. For there is no authority except from God, and those that exist have been instituted by God.
>
> Romans 13:1

> [15] By me kings reign,
> and rulers decree what is just;
> [16] by me princes rule,
> and nobles govern the earth.
>
> Proverbs 8:15-16

> ...the Most High rules the kingdom of men, and gives it to whom he will.
>
> Daniel 4:25b

Thank God for The Don! They'll reap the dusk who've impeached The Don.

Then 2018 begins the process of God's leading Israel with Jerusalem as its acknowledged capitol to face alone, without America's help, a hostile world, demanding, pursuing, and warring for Israel's destruction and the extermination of I AM WHO I AM's witnesses.

> [13] And I saw, issuing from the mouth of the dragon and from the mouth of the beast and from the mouth of the false prophet, three foul spirits like frogs; [14] for they are demonic spirits, performing signs, who go abroad to the kings of the whole world, to assemble them for battle on the great day of God the Almighty. [15] ("Lo, I am coming like a thief! Blessed is he who is awake, keeping his garments that he may not go naked and be seen exposed!") [16] And they assembled them at the place which is called in Hebrew Armageddon.
>
> Revelation 16:13-16

Even America will eventually turn against Israel. However America's rebellion is effected, at Christ's return, the whole of the world will be assembled for battle against The Lord God and His people (Revelation 16:14). We discussed how the import of the 2014-5 special tetrad was naturally ladened with the concept of America turning away from restraining the mystery of lawlessness. The Obama administration abandoned Israel by sweet

Measurements

dealing a capability for amassing nuclear weapons to Israel's arch enemy, Iran. He set America's military might aside so Jihad could blossom in the refreshing showers of the Muslim Spring. Obama had, before, proudly announced his goal to fundamentally transform America from individual sovereignty into nihilistic socialism.

Note how interestingly Israel and America are tied together by these signs and measurements. Jerusalem developed from 1968 towards the truth of its being God's Holy City, while, in 1968, America began falling from its Christian faith into a deceit led by dope smoking children, pop-stars, obfuscating entertainers, bigoted professors, fake-news journalists, lying politicians, and women biting into the forbidden fruit once again, spitting its seeds into

Figure 43

God's face, even choosing the horrific deaths of their own children rather than a few months of personal discomfort. The same special tetrad (1493-4) announcing Israel's final trek through pogroms unto nationhood announced also the beginning of America's rise through Christianity to its descent into blaspheming the very God who fashioned it. Truman recognized and stood by Israel's nationhood first. President Trump recognized and stood by Jerusalem as Israel's capitol next. Then, immediately began an amazing set of astronomical coincidences specifically and demonstratively concerning America. But to fully appreciate the reality and perspicuous meaning of those signs, we must first discuss the Shemitah Path, measured in the same seven year units Jonathan Cahn discussed in *The Harbinger* .

Clear Signs of Trouble and Great Joy

The Bible is full of sevens. The number expresses spiritual purpose regarding the topic it enumerates. Even when it enumerates evil, as it does in the seven heads of the dragon/beast, it signifies an interconnection with God's purposes,

> [4] The LORD has made everything for its purpose,
> even the wicked for the day of trouble.
>
> Proverbs 16:4

Rabbi Jonathan Cahn theorized that significant events occurring in the final year of particular seven year periods (Shemitahs) were warning calls for America to repent of its secular mindset and once again acknowledge and serve the God of its founding.[9] This immense system of signs closely correlates with Rabbi Cahn's proposal.

As we earlier considered, even though there is no document in any government archive formally and legally dedicating America to the service of the Ultimate Sovereign, far more than such a formality was in the hearts of the men who risked and sacrificed their futures to establish the United States of America. More of The Founding Fathers served the Ultimate Sovereign than not, although today's expertly subjective snipers beg you to think otherwise.

One expert sniper, Simon Brown, shoots at George Washington's inaugural address of April 30, 1789 from the weeds of the Americans United for the Separation of Church and State,

> On April 30, Washington was, indeed, sworn in as the first president of the United States. He gave an address of just seven paragraphs and made no mention of God or Jesus.[10]

Another expert on religion, David L. Holmes, who was the Walter G. Mason Professor of Religious Studies at the prestigious College of William and Mary, and who is the noted historian of American religion, determined that most of The Founding Fathers were loyal to Christian practices but were influenced by Deism.[11] He even stuffs George Washington into this Deist pigeonhole. Modern American culture obliges us to forget what the founders of our own nation expressed. George Washington's inaugural address was only seven paragraphs long, as Mr. Brown notes, but among those few paragraphs George Washington expressed from his heart the following not so small contemplations:

> ...it would be peculiarly improper to omit in this first official Act, my fervent supplications to that Almighty Being who rules over the Universe, who presides in the Councils of Nations, and whose providential aids can supply every human defect, that his benediction may consecrate to the liberties and happiness of the People of the United States...and may enable every instrument employed in its administration to execute with success, the functions allotted to his charge.[12]

Deism is defined by dictionaries in the following manner:

Measurements

> The belief, claiming foundation solely upon the evidence of reason, in the existence of God as the creator of the universe who after setting it in motion abandoned it, assumed no control over life, exerted no influence on natural phenomena, and gave no supernatural revelation.[13]

Compare "abandoned it", "no control", "no influence", and "no supernatural revelation" to "rules over", "presides in", "providential aids can supply", "benediction may consecrate", and "enable". What do you notice? So do I. George Washington's concepts are neither consistent nor cohesive with a universe abandoned by god to its own mechanistic horrors. Any Deist who makes "fervent supplications to the Almighty Being" has annulled his own belief. Why ask a god to be closely involved with your problems when you believe it abandoned everything? So how is the claim that George Washington was influenced by Deism even coherent? Wouldn't you expect a university professor to think more capably than incoherently? And how does belief in an interactive God even mix with any concept at all of some damnable, hermit-god? Let's hear President Washington continue expressing from his heart.

> In tendering this homage to the Great Author of every public and private good I assure myself that it expresses your sentiments not less than my own; nor those of my fellow citizens at large, less than either.
> George W. Washington

Washington continues from a non-Deist frame of mind, referring to God as "the Great Author of every public and private good". He attributes to God the effecting of not only a quantity of good, but also of an intimate quality of good. And more than just a good quantity of quality sprinkled about publicly, he attributes God's effects as being delivered into the private depths of individual lives, too. He not only recognizes God's close attachment to our affairs, but he acknowledges the very personal nature of that attachment. Those are not Deist concepts. From this frame of mind he expresses assurance that, also, the people of the American nation understand the Almighty as being providential and deeply engaged in the affairs of man. Washington could state this assurance because he was an eye witness to it, while the best Brown, Holmes, and their company of snipers can do is raise accusations, for they weren't alive then to witness any of the intimate details to which Washington was deeply and personally privy. Truly they know not what they talk about.

Washington was savvy enough to understand how greatly damaged his own social and political capital would be if he raised fake assurances about such a matter. American literature of the period bears out his assurance. Private expressions of deep faith in God's providential interactions with their lives and circumstances abound in their diaries, books, poetry, music, etc. Regardless of the Deists and atheists who also existed then, the American communities grew into Christian cultures. Those who were committed to Christ, His Father, and the Holy Word of God were more influential in shaping this nation than were the obfuscating deniers of God who have deconstructed, today, what the Founders created back then. History bears it out. God grew America into a non-Deist, Christian nation. So Christian it

Clear Signs of Trouble and Great Joy

was that we still began every school day with prayer and sang religious, patriotic songs when I was in grade school. That's not Deist.

Brown accused Washington of making no mention of God or Jesus in his inaugural address. But Washington's references to "Almighty Being" and "Great Author" are in no way abdications of I AM WHO I AM and Jesus Christ's identities. To the contrary, if Mr. Brown were more honest with his knowledge of literary devices he would have pointed out for us Washington's use of synecdoche in referring to God and Christ by the aspects of their being "Almighty" and "Author". These aspects were well known Christian concepts for "*the deity*" in Washington's day. And they still are today. Synecdoche is not useful where one or both of the referenced entity and that entity's trait are not culturally known. Therefore, Washington's use of synecdoche verifies the Biblical nature of the public attitude towards I AM WHO I AM and Jesus Christ in Washington's day. If the public did not perceive of God by those attributes, Washington's synecdoche would have been received as gibberish.

Let's see if you can spot more synecdoche, and maybe even a bit of metonymy…

> No people can be bound to acknowledge and adore the invisible hand, which conducts the affairs of men more than the People of the United States. Every step, by which they have advanced to the character of an independent nation, seems to have been distinguished by some token of providential agency. And in the important revolution just accomplished in the system of their United Government, the tranquil deliberations and voluntary consent of so many distinct communities, from which the event has resulted, cannot be compared with the means by which most Governments have been established, without some return of pious gratitude along with an humble anticipation of the future blessings which the past seem to presage. These reflections, arising out of the present crisis, have forced themselves too strongly on my mind to be suppressed. You will join with me I trust in thinking, that there are none under the influence of which, the proceedings of a new and free Government can more auspiciously commence.
>
> …Since we ought to be no less persuaded that the propitious smiles of Heaven, can never be expected on a nation that disregards the eternal rules of order and right, which Heaven itself has ordained. And since the preservation of the sacred fire of liberty, and the destiny of the Republican model of Government, are justly considered as deeply, perhaps finally staked, on the experiment entrusted to the hands of the American people…
>
> Having thus imparted to you my sentiments, as they have been awakened by the occasion which brings us together, I shall take my present leave; but not without resorting once more to the benign parent of the human race, in humble supplication that since he has been pleased to favour the American people, with opportunities for deliberating in perfect tranquility, and dispositions for deciding with unparalleled unanimity on a form of Government, for the security of their Union, and the advancement of their happiness; so his divine blessing may be equally conspicuous in the enlarged views, the temperate consultations, and the wise measures on which the success of this Government must depend.
>
> George W Washington

Measurements

Washington pronounced no words of America's official dedication to God's purposes. But Washington's heart clearly pronounced that dedication through the coherence of his speech. And he was confidently sure that his sentiments were common. Why was he sure? Because Christianity was the American culture of his day, not Deism, nor atheism, or Islam, Hindu, and not Buddhism. Christianity and hospitality were the expected attitudes in his day, the same way we are expected to think evolution, abortion, and socialism today.

Although Washington's statement does not fit well in our day, we must not project our disbelief onto the belief of his times. No official dedication was necessary then. Today, it would be very necessary. The dedication was in their hearts like ape-man and selfishness are the dedication of America's heart today. So America's dedication to The Almighty God is left for those who have eyes to see to see it, and for those who don't to disagree, which after all, is the American way of allowing fools to stumble into their own garbage heaps.

The dedication to God of "the experiment entrusted to the hands of the American people" reached through individual hearts into the deliberations and decisions of individuals in Washington's day, as it still does in a remnant of Americans today. The student of religion is oblivious to such concepts. But the students of Christ understand them. As a student of Christ, Rabbi Cahn noted the purposes of God to which America was guided in its beginning, the same purposes away from which it stumbles today. It is useless for the Browns and Holmeses of the world to obfuscate and hide *information* in the confusing mess of their theory ladened data and facts. God's warning for America is truly seen in Rabbi Cahn's Shemitahs by any eye willing to see reality.

Yet how does Rabbi Cahn know these particular years strung in sequences of seven are the actual continuation of the same sequence begun by God for Moses at Mt. Sinai? Mr. White probably thinks it is very important. But does God? God Himself informs us by revelation that all men are false. God's perfection is able to keep track of such things. But our faultiness can not. It makes sense that a God empathetic enough to lead broken, frail creatures out of destruction's way will not warn them by technicalities beyond their ability to perceive. This principle can be seen in the words God inspired St. Paul to write at Romans 14:5-6:

> ⁵One man esteems one day as better than another, while another man esteems all days alike. Let every one be fully convinced in his own mind.
> ⁶He who observes the day, observes it in honor of the Lord. He also who eats, eats in honor of the Lord, since he gives thanks to God; while he who abstains, abstains in honor of the Lord and gives thanks to God.

We know two opposites can not both be true at the same time in the same sense. Either one day is better than the rest, or not. Either God expects us to abstain from eating certain foods, or not. But His truth about some issues exceeds our ability to certainly know. Not knowing is part of our existing falsehood for which we enjoy His grace. Yet we still must relate to God. Therefore, in accord with His empathy, for the sake of His participation in our lives, He relates to each God-fearing individual regarding debatable issues in the way

Clear Signs of Trouble and Great Joy

that individual relates those issues to Him. We do the same thing with our children and friends. For some situations, precision is less important than essence. We should never sacrifice the essence of relationship for the greater precision of a lesser point. These Shemitahs seem to be one of those situations.

Each seventh year being an actual Shemitah year was important to the message of Rabbi Cahn's *The Harbinger*. Mark Biltz shows compelling evidence that these are indeed the continuing sequence of Moses' Shemitahs. He proposes that each year, as measured by the Hebrew calendar, being divisible by seven, is a Shemitah year.[14] But that depends upon the accuracy of the Hebrew calendar. Regardless, a precise correlation with Moses' Shemitahs is not the main point of the Shemitah Path. The main point is in the Biblical meaning of the number seven and the series of prescient events occurring within these seven year cycles. Understanding draws more from implication than it does from precision. The implications of the numerous "coincidences" Rabbi Cahn discovered within these seven year cycles rise to a communication from General Washington's God about what America has, today, abandoned, whether Cahn's Shemitahs are Shemitahs exactly, or Shemitahs metaphorically.

Seven is a special number regarding God's non-Deist relationship with the physical universe for our sakes. It is the combination of the Divine number three, and the number of the physical Creation, four. This combination expresses the basic, spiritual significance of their relationship. All things are about that relationship. Throughout eternity the people of that relationship, Christ's beloved church, will reign with Him. Seven represents I AM WHO I AM's fundamental spiritual purpose in what it measures. In addition to the economic warnings Rabbi Cahn saw, The Shemitah Path measures interactions between Israel, America, Antichrist, and Tribulation into a boldly highlighted roadmap of God's final purpose as it develops in our generation: the defeat and sequestering of evil to establish divine order.

Twenty-five years after The Alhambra Decree was instituted in the year America was discovered, the seventh-head of the dragon, the Ottoman-Turk Empire, seized control of Palestine and God's Holy City. The Empire gave the Jews fleeing Spain refuge. It even allowed them to settle in Palestine. But it restrained Jewish immigration. They could move there in single family units only. Group migration was not tolerated. The Empire chose, instead, to preserve that land for the Palestinian population.

But Russia initiated new pogroms against the Jews in 1881. To the people of the Empire, the Jews were the people of The Book (the Bible) even though they believed the Jews introduced errors and misrepresentations into those Oracles God gave them. So, the Empire eased this restriction in honor of the persecuted people of The Book. It dropped the single family" rule, allowing Russian Jews to migrate in groups (a "seven" kind of thing to do).

Thus, The First Aliyah began in 1882,[15] the first, massive migration of Russian Jews into Palestine. The First Aliyah occasioned the development of the Zionist Movement, the drive for a national Israel which eventually developed into the Israelites taking up arms May 15, 1948 after boldly declaring independence the midnight before. The nation of Israel is fundamental to the reality of God's faithfulness. For He promised Abraham that He would make a nation through Isaac, forever, in the land which was once Palestine, known formerly as Canaan. If indeed He were God, if surely He were faithful, then certainly He would keep

Measurements

His promise to Abraham, having walked between those bloody carcass halves alone. The First Aliyah began the process of gathering Israelites to become the promised nation for all Earth to receive its King of kings, certainly a seven kind of purpose.

In Chapter 9 we will discover a year measured by signs which occurred five years before the Shemitah Path began in 1882. AD1877 bore no event relevant to the Path. Instead, its relevance is in "5" meaning grace, and in its having the same enumeration as the year in which God promised the national heritage to Abraham, 1877BC.

If we follow the seven year cycle of Shemitahs forward from 1882, we will find ourselves walking through the reestablishment of the nation promised Abraham right into the middle of Rabbi Cahn's Shemitah years of 2001, 2008, 2015 and on to Apophis' Shemitah years of 2029 and 2036. Divine purpose is written all over this Path, from the grace which must precede its process to the relevance of its significant events within the prophetic theme of evil's defeat at that final and glorious Shemitah year, 2036.

God's number, 3, indicates His involvement in what 3 measures. The First Aliyah is commonly noted to have lasted from 1882 through 1903, twenty-one years, which entails three Shemitah cycles, 1889, 1896, and 1903, indicating the non-Deist involvement of I AM WHO I AM in gathering His people of promise back into the land promised to their forbear, Abraham. Moreover, from the beginning of the Shemitah Path in 1882 until the ominous approach of Apophis in 2029, the portion of The Path purposed for preparing the Tribulation stage, is also twenty-one Shemitahs, three sets of seven, the same number of years as was the First Aliyah, the number expressing complete spiritual significance worked by God. Certainly it is God's non-Deist, providential hand leading today's nations into the Tribulation, whether we call Him Almighty Being, Great Author, I AM WHO I AM, or Jesus' Daddy. Then the last Shemitah, the final year of the Tribulation, is The Path's twenty-second Shemitah, the Biblical number of doubly intense chaos and mayhem. Within this framework of measured meaning, the Shemitah Path becomes incredibly fascinating, on the one hand, and perspicuously prophetic, on the other.

Two Shemitahs after the end of the First Aliyah (fourteen years being twice the number of spiritual significance), in 1917, the seventh-head of the dragon, the Ottoman-Turk Empire, disgorged Palestine and I AM WHO I AM's Holy City, Jerusalem. 1917 ends the fifth Shemitah cycle of the Shemitah Path, the Biblical number of grace. That count of five Shemitahs doubly emphasizes the grace this Shemitah process requires as indicated by the five years preceding 1882, even as 1917 is the eighth Jubilee of the Jubilee Path, the Biblical number of new beginning, that which grace is meant to enable. The release of the Holy Land from the dragon/beast's control is the beginning process towards reestablishing the Israelites' sovereignty to become a nation forever, a truly gracious event God worked for them.

Let's consider the additional meaning the number "1917" piles onto the significance of "grace".

> Seventeen stands out very prominently as a significant number. It is not a multiple of any other number, and therefore it has no factors. Hence it is called one of the prime (or indivisible) numbers. What is more, it is the

Clear Signs of Trouble and Great Joy

> seventh in the list of prime numbers...being the seventh of the series [of primes], it partakes of and intensifies the significance of the number seven. Indeed, it is the combination or sum of two perfect numbers -seven and ten- seven being the number of spiritual perfection, and ten of ordinal perfection.
>
> Contrasted together the significance of these two numbers is clear; and when united in the number seventeen we have a union of their respective meanings, viz., spiritual perfection, plus ordinal perfection, or the perfection of spiritual order.[16]

Mankind's materialistic evil will be entirely summed up (19) for the complete spirituality (17) of Christ's righteousness.

As the Jews flowed into the Promised Land, having nationhood on their minds, the Ottoman-Turk Empire's denial of their independence was becoming its own major faux pas. It wasn't merely that these immigrants were Jews. They were the genealogical offspring The Most High God promised to The Father of Faith. Any denier standing between this Bear and that cub was destined to become quick and certain carnage, a.k.a. bird food. The year the Empire disgorged the Promised Land began the Jubilee within which I AM WHO I AM started raising up "those dry bones" gathered there during the First Aliyah to become a nation for taking control of God's Holy City on the ninth Jubilee of the path, 1967, the summation of Jerusalem's years spent under the control of others.

In 1917, the pathway of historical signs leading into the Tribulation crossed Jerusalem's pathway of preparation for becoming the throne upon which The King of Kings and Lord of Lords will reign. 1917 was a very significant year.

While the Jubilee Path extends past 1917 by a count of two more Jubilees to a set of signs apparently warning America specifically, the Shemitah Path continues for seventeen Shemitahs to a year portrayed by this system of signs as the coming of the King of kings, The Desire of Nations. With that end of the Shemitah Path begins a thousand years of peace, the "time" of Daniel's "time, two times, and half a time". The coherence of these signs in sun, moon, and stars with prophesied events of Israel's reestablishment, and with measurements finding numbers of Biblical meaning between events significantly relevant to those prophecies, confirm the thousand years of peace as being the work of I AM WHO I AM, not of Hitler, nor of Islam, nor of American Progressivism, nor of Covid-cowardice, nor of any other human effort to fashion utopia by fascistic means.

The end of the eighth Jubilee, at 1917, begins a new type of history for the Jewish people. Their land was released to become an independent and sovereign nation. Jerusalem was released to become its capitol. Twenty-five-hundred years had passed since Judah last acted as an independent and sovereign kingdom (586BC). Israel was first established to demonstrate to the world God's faithfulness to a people humbling themselves before Him. The Israelites found every way possible to turn from Him. Yet, God did not scatter them throughout the world to destroy them. He scattered them to demonstrate to the world His faithfulness by returning to them. He would gather them together as a nation, again, for their second chance to receive the righteous reign of their King of Kings. The beginning of His restoration work will indeed be a new kind of history. The events along this pathway are not

Measurements

just developmental to God's goal of a national Israel. They more relevantly develop into evil's defeat for righteousness to begin at the end of the Shemitah Path. Considered together, the events carefully measured into the Shemitah Path clearly warn of the trouble this way coming for the lovers of evil, and the great joy coming to those crushed by evil's lovers.

As three Shemitahs were the measure of God's working the First Aliyah, three Shemitahs from 1917 measures His doing of Kristallnacht at 1938. It seems the blessings of His beloved always come to them trouble first. Abraham's receipt of the promise came through the abandonment of his homeland. Israel's rise to nationhood came through their slavery to Egypt. Israel's honor at the forefront of all the nations will come through their horrors experienced during the Tribulation. Israel's becoming a nation again had to trek through Hitler's death camps. The blessings of those who love righteousness come in that sequence. For no matter how long it must take, what comes first ends so that what comes next will never end. God gives Israel's antagonists their turn to enjoy glory first so the turn He has planned for Israel will be eternal.

Thus Kristallnacht came luring the world's worst cockroach to his demise. Hitler set his Final Solution, the extermination of the Jews, into motion on the night of broken glass in 1938, the third Shemitah year after 1917. Hitler's Final Solution was of the same mentality and behavior towards God's people which the eighth beast will repeat in the Tribulation. All the world's nations today shout proudly about the Holocaust, "Never again!" They are so assured in their self-righteousness! But the nations are all made of the same spiritual substance: 1) accuse God, 2) claim to be God; and 3) destroy righteousness: accuse/deceive/destroy, those same, prominent Biblical concepts ascribed to Satan. But the Antichrist's Final Solution will work to his own demise as surely as Kristallnacht began Hitler's end. Ever since God's chosen people entered Egypt to gestate into His chosen nation, every attempt to exterminate them has ended in the annihilation of that endeavor's mastermind.

The Biblical meaning of the number of Shemitahs from 1917 to the 2036 end of the Shemitah Path characterizes both the blessing to which God's people come and the trouble into which their antagonists are thrown. And confirming God's orderly purpose for this stretch of time, the number of years from 1917 to 2036 is the same number as the Psalm praising the wonders of God's holy ordinances, 119. Kristallnacht came for the Jews three Shemitahs after 1917, the number of God's doing it. Kristallnacht triggered the destruction of the destroyers before the blessing of the patient remnant of God's tortured people at the end of Hitler's Holocaust trap.

The seventeen Shemitahs from 1917 to 2036 are the Biblical number of complete spiritual significance. In it we find another significant Sethite-like dual prophecy. Seventeen can be viewed as the combination of three and fourteen: God working doubly significant spiritual purpose for His people, or it can be viewed as the combination of four and thirteen: the lovers of the physical creation careening into evil's doom.

1938 was not a good year for the Jews. It was three Shemitahs into Jerusalem's ninth Jubilee. And the following seven years waxed even worse for them. But those Jews who patiently waited through Hitler's Holocaust came to a blessed 1945! Some came there barely alive, like kicked and flicked bone bags. Those who avoided or survived Hitler's slaughterhouses surely lamented over beloved children, parents, siblings, husbands, wives,

Clear Signs of Trouble and Great Joy

aunts, uncles, cousins, friends, and the general mass of Jewish humanity crushed in Hitler's camps. Their patience was profound. Stepping through the thrown-open gates of Auschwitz, Treblinka, Sobibor, etc. into an only somewhat freer world, where their homes had been taken from them or destroyed, and where most of their loved ones were dead, could not have entirely been the relief for which they had patiently awaited. But the Holocaust set the Jews up for what they had patiently awaited through all the previous twenty-five centuries: their national homeland, sovereignty, free at last!. Free at last.

    Hitler's Holocaust shocked the world conscience. Especially shocked was the Western World. It had been descending into fascistic socialism since the beginning of the century. Hitler was surprisingly admired throughout the Western World before the war. But after it, the entire world found him repulsive, not for what he had done to the Jews, but even more, it had become public knowledge. World resistance to the idea of a Jewish state momentarily lowered. A path to the nationhood for which the Jews had patiently awaited was the blessing in those open gates of 1945. 1948 began the blessing of God's promise to Abraham kept. From 1938 to the 2036 raising of Israel above all nations is fourteen Shemitahs, the number of spiritual significance doubled. The establishment of Israel in 1948 followed Kristallnacht by ten years, the number of perfect completion. Israel's new nationhood will not end. Israel will be lifted above every nation after the Antichrist's holocaust is crushed.

    The context is now ready for explaining the second measurement of days the angel gave Daniel to be sealed up until the time of the end. This measurement directly expresses the glory of God's triumph over the nature of man's rebellion. It contrasts the ungodly, fascistic nature of the seventh dragonhead with the godly blessing of Israel's nationhood. The death of the obstinate dragonhead released the land to become a nation; the horrors of genocide catalyzed the nation; and, in spite of, maybe even, because of the entire world's hostility towards the Jews, Israel will soon be elevated to the place of righteousness from where the King of Righteousness will reign.

    The prophetically relevant nature of the seventh-head wasn't its having been Ottoman-Turkish. That was merely its human vessel. Hostility towards I AM WHO I AM expressed through warring against His people is the nature of the dragon's spirituality, and therefore, of the seventh-head, as it was also Hitler's spiritual nature. They share it alike.

    Islam, the religion and soul of the Ottoman-Turk Empire, is a rather recent creation compared to the historical existence of the seven-headed dragon. Recall from Chapter 4 that the depiction of this dragon has been amongst mankind's symbols since very early times, at least from the beginning of Sumer in the third millennium BC. Moreover, the ancient Sumerian depictions of it portray the seventh-head as slain at least forty-five-hundred years before it was literally slain in 1917. But Islam wasn't created until AD610.

    Islam is not the fiber of the dragon's nature. Rebellion is. Rebellion is the nature of Islam's fiber. Therefore, Islam is dragonish. It is the nature of which many accuse Cain: a determination to slay all God's people by destroying the Sethite line of descent through the Jews to the woman's Seed, who crushes the dragon's head. The attack on His lineage has flooded through world history: the Egyptians killing Israel's male babies, Haman bidding the Persian king to annihilate the Jews, the massacre of Bethlehem's infants, the crucifixion

Measurements

of Jesus Christ, the Alhambra Decree, the Final Solution, and eventually, the Tribulation. This spiritual nature has been a flood of rampage against the Jews escalated in AD70, tagging them through history with many pogroms. The intent to exterminate God's people entails the lusts of both Adolf Hitler and the religion of the Ottoman-Turk Empire, Islam.

And so we should not be surprised to see the measurement given to Daniel express a correlation between these two antagonists of God's saints and the destruction God promises to anyone who curses the Jews. But that is not all of its expression. A patient wait through many horrors to be raised as a surviving remnant into joyful and secure nationhood, promised and delivered by God Himself, is the main import of this beautiful and prophetic expression. The measurement begins with the conception of Israel's final antagonist. It ends at the demise of the Jewish people's penultimate antagonist. The correlation ties these two antagonists together with the cords of a common nature: accuse, deceive, and destroy God's chosen people, then catch hell for having done so.

During the heat of World War II, the Islamic leader at Jerusalem, Amin al-Husseini met with Adolf Hitler to request Nazi support in resisting the establishment of a Jewish state in Palestine.[17] He offered to stir up a revolt in the Arab territories, intending the strife to develop into a war against the Jews in the Promised Land similar to what Hitler was waging against them in Europe.

> ...prior to Kristallnacht, these Nazi policies had been primarily nonviolent. After Kristallnacht, conditions for German Jews grew increasingly worse. During World War II (1939-45), Hitler and the Nazis implemented their so-called "Final Solution" to what they referred to as the "Jewish problem," and carried out the systematic murder of some 6 million European Jews in what came to be known as the Holocaust.[18]

Ideologically, Hitler perceived World War II as a battle between National Socialism and the Jews.[19] Hitler and al-Husseini were both certain the Nazis would be victorious. But Hitler's hands were full of the war in Europe. The Nazis could not service another theater of conflict. So Hitler took a rain check on Husseini's offer.

He hid his death camps from public view, since his Final Solution crossed the line of social acceptability, although it was personally acceptable in the darkness of most individual hearts. So the world's perception of the Jewish problem turned a corner when the Russians and Americans discovered those camps and crematoriums during the Nazis' final days. The New York Times knew of Hitler's extermination effort all along, but they chose to back-page the story while millions of innocent civilians were being incinerated. Eventually, the inevitable exposure of Hitler's massive cruelty raised so much world shock and horror that...

> ...leftist intellectuals redefined fascism as 'right-wing' and projected their own sins onto conservatives, even as they continued to borrow heavily from fascist and pre-fascist thought.[20]

Hitler's misdeeds indelibly altered world mentality towards genocide, complicated world sympathies towards fascism, and temporarily softened international aversions towards the

Clear Signs of Trouble and Great Joy

Jews.

The gates of the concentration camps swung open in 1945, freeing a fortunate few Israelites to enjoy a short lived bubble of world sympathy almost deep enough for creating a nation of Israel, but certainly big enough to accept the Jewish state (which arose without the world's assistance) when the British Mandate ended. David Ben-Gurion, the chairman of the Jewish Agency, declared the statehood of Israel at midnight, May 14, 1948.

Four total lunar eclipses on the Passovers and Tabernacles of 1949 and 1950 attested to God's hand producing this order of business, as did the three years by which 1948 followed 1945. Moreover, the prophecy given Daniel two-and-a-half millennia earlier measured together the demonic fibers of the Holocaust, patiently endured by the Jews, and Islam's genocidal lust which the Israelites will again need to patiently endure.

> [12]Blessed is he who waits and comes to the thousand three hundred and thirty-five days.
>
> Daniel 12:12

What "thousand three hundred and thirty-five days"? The angel seemingly hung this number in mid air, extending from an unknown beginning to culminate at an untold end. But the verse's immediate context gives a clue. And the genocidal lust for Jewish blood on the part of both Hitler and the soon to be revived seventh-head of the dragon gives a clear sign.

By prophetic measurement, the immediately preceding verse verified the setting up of the Ottoman-Turk Empire to become the seventh-head of the dragon, the abomination which makes desolate. The abomination entails the hunting of Jewish heads, e.g., Hitler's Holocaust and the Antichrist's beheadings. God led the Jews to wait through this head-hunting for their ultimate blessing. As we noted above, disaster happens to God's people first, then blessing. Although Hitler's head-hunting was merely personal, the Antichrist's is Islamic. Islam was the religion of the slain seventh-head which will return to the world as the Islamic kingdom of the Antichrist, again, head-hunting the Jews. Could the inception of Islam have been the beginning of this measure of days to the blessing which the Jews patiently awaited? Most sources date Muhammad's reception of the Islamic revelation at AD610. From then to the swinging open of Hitler's Holocaust gates in the Shemitah year, 1945, is one-thousand-three-hundred-thirty-five years.

God draws attention to the anti-Semitic genocides in the time of evil's end by connecting the beginning of the Antichrist's anti-Semitic Islam to the end of Hitler's anti-Semitic Nazism with a Biblically given count of days representing years. But the blessing to which those gaunt figures toddling out of Hitler's death camps came was less about the end of their tortures and more about the establishment of their nation the number of God's doing (three years) later. They were barely alive, their bones showed through their skin. The world met them with a short outpouring of sympathy. The terror at Kristalnacht, the horrors of the Holocaust, a blessing of survival, and a short lived, worldwide sympathy emboldened these naked, starved Israelites to step forward and declare nationhood, that for which God's chosen people have awaited since Nebuchadnezzar captured Judah twenty-five-hundred years ago.

Measurements

And it wasn't just that they now had a nation. That was the least of their blessing. They now had the beginning of the forever part of God's promise to give Abraham a nation raised above all of the world's nations. Israel' reestablishment even had a prophetic mark on the world's most ancient, mysterious pillar. No wonder the beast is full of the dragon for making war on God's saints! The end of the Tribulation's anti-Semitism, fourteen Shemitahs after 1938, begins Jesus Christ's reign from Jerusalem, elevating this hated little nation to the head of all the nations, all of the world's traps set for them having sprung upon the trappers. Jewish patience with being perpetually head-hunted indeed brings them to a great blessing of spiritual significance doubled, prophesied even from Neolithic times. It's coming is not only soon, it is certain.

The torturers met their doom while the patient victims received wondrous blessing. It's just another of I AM WHO I AM's many ways of snaring fools by their own devices. Those who drag flaming bags of crap onto other people's door steps, glorying in their ability to conquer and control through fear, overpower by accusation, enslave with deception, and destroy a people's ability to socially know truth from falsehood, right from wrong, and good from evil, scrubbing every semblance of faith and perseverance out of culture, there remains the other way of understanding the seventeen Shemitah's from 1917 to the end of the Shemitah Path.

Four Shemitahs after 1917 the Jew torturer met his end in his own trap: 1945. Hitler reveled for seven years in the power God allowed him to wield over the Jewish people. Then he was horrifically crushed. Berlin and much of Germany lay in heaps of smoldering rubble. Nooses were preparing for the necks of those who aided the perverse horrors he piled upon God's chosen people. The godless endeavored to own the world first. They gloried first. Then all hell came collecting the eternal bill for their self-elevation above the ones chosen by I AM WHO I AM to lead the nations. Four is the number of what the godless first love and worship: the world, the physical universe, the materialism of Darwin's non-designed purposelessness. Thirteen Shemitahs after 1945 ominously speaks the eternal doom every God denier will receive in the bloody winepress of 2036. Of this I AM WHO I AM lovingly warns, and I write. Pay to it your attention.

The measurement of seventeen Shemitahs from the 1917 intersection of pathways to 2036 is a clear sign of great joy and horrific trouble, an intimate enumeration of God's great orderliness (119) in achieving His traditional goals. Make sure your sensibilities are set for joy. Three your life. Love I AM WHO I AM first by approaching Him through His gate - Jesus Christ- and patiently persevere the abuse poured upon godly peoples by liberals, the accusers/deceivers/destroyers of decency and godliness. Endure the public assassination of your good character by the godless media. Persevere the Progressive assault on your liberties. The truth is your best resistance. Stand on it. In the end, truth will trump evil.

So, walk the way of three and fourteen, patiently wait, and soon Christ will come for you, bringing to you eternal joy after your brief time of trouble.

The most destructive lie ever told about Hitler was told also by American Progressivism, although it may have originated with Stalin. Left-wing pundits of socialist governments could not abide Hitler's spectacle having been conceived on their side of the political fence. It shined too much light of truth on the very same political paths they were following

Clear Signs of Trouble and Great Joy

through collectivist meadows. Hitler's system was without a doubt socialist. "NAZI" is merely a German acronym for The National <u>SOCIALIST</u> German Workers' Party (as expressed in German, of course). Taking advantage of the common man's impatience for detailed *information*, the American Progressive socialists, in order to build a similar collectivism on the backs of an unsuspecting working class, convinced the world that Hitler's fascism came from the right side of the political fence and that fascism was ultra-conservative rather than commonly Progressive, even though fascism is the fiery heat in collectivism's rude blood.

> Before the war, fascism was widely viewed as a progressive social movement with many liberal and left wing adherents in Europe and the United States; the horror of the Holocaust completely changed our view of fascism as something uniquely evil and ineluctably bound up with extreme nationalism, paranoia, and genocidal racism. After the war, the American progressives who had praised Mussolini and even looked sympathetically at Hitler in the 1920's and 1930's had to distance themselves from the horrors of Nazism.[21]

Truly Hitler was on the very same, left side of the political fence, in the very same fascistically mildewed meadow where American Progressives feed on the same grass full of the same diet of accusing, deceiving, and destroying everything in the way of whatever they desire to do.

The real difference between Nazism and American Progressivism is a matter of degree rather than of left or right. The political left is collectivism, synonymous with fascism. All other political philosophies form the right, according to the brilliant Thomas Sowell. American Progressive socialism and Hitler's National socialism are the same form of bullying into servility everyone cowardly enough to surrender in the face of fear and abuse. Both "disappear" anybody sufficiently courageous to never surrender, as was done to the great Donald Trump. But Hitler's socialism was nationalistic, the elevation of its own nation towards world rule. On the other hand, American Progressive socialists are internationalists. They strive to crush all of the world, including America, under one, tyrannical foot. International and national socialists are both draconian governments born of that late eighteenth century flight from God.

The first horrors of the fascistic gauntlet through which the Jewish people must run to reach their Millennial righteousness was occasioned by the great, left-wing liberal, Hitler. Since his demise, we are seeing liberals prop up and excuse the anti-Semitic head-hunting of the Islamic form of draconian governance. Hitler's phase of the gauntlet was called the Holocaust. Islam will create the next phase of the gauntlet, called the Tribulation. Hitler's liberalism took its shot at the Jews first. Today's liberalism is performing CPR on the anti-Semitic dragonhead left for dead two Jubilee's ago, eloquently speaking like the false prophet of Revelation 13:11-17, praising Islam, castigating the Bible.

Eight Shemitahs after the blessing of those who came to the one-thousand-three-hundred-thirty-five days was the 2001 Shemitah of September 11th. Eight was also the number of Jubilees the Ottoman-Turk Empire occupied Jerusalem until 1917, representing

Measurements

its new beginning. Just as 1917 began the rise towards nationhood of Israel's accumulating dry-bones, 2001 began the world's fast track towards the Tribulation, after which evil will be fully restrained and righteousness will be given full reign.

It is often said everyone remembers where he was and what he was doing when he first heard of the 9-11 attack. Most of us remember how many times we heard it said the world changed on September 11th. Although situations irreversibly change with each deed of every day, the change on September 11 was a universal new beginning for the entire world, a day which affected us all the same way. The change wasn't so much the affect of a great shock. Nor was it merely the change of new and increasing security measures. Or even the great escalation of Islamic terrorism. As we will shortly discuss, beginning with 2001, the signs of an approaching Tribulation started occurring each and every Shemitah rather than upon intermittent Shemitahs separated by numbers expressing Biblical meaning. Every Shemitah after 2001 presents an amazingly perspicuous sign of Jesus' return!

Previously, spiritual meaning was found along the Shemitah Path in the number of counts between Biblically significant Shemitahs. For example, The Path's first five Shemitahs led to 1917, then three more to Kristallnacht, fourteen more to the return of Jesus Christ at the end of the Tribulation, and of course, these eight Shemitahs from 1945 to 2001. But after 2001, each successive Shemitah brings forth an astonishing event prophetically relevant to the world's fast track into the Tribulation, where it will meet the King of kings face to face in 2036. There are five Shemitahs from 2001 to 2036, an expression of God's grace that will be needed to endure the Tribulation for greeting the Lord.

In 2004 (not a Shemitah year), Apophis was discovered. Then, in 2008, a colossal gamma-ray burst was observed in Bootes, the constellation portraying Christ's quick return. The Christian anticipation of evil's nearing end and righteousness' beginning was escalated in 2010 by Hubble's photograph of a debris field made by two colliding asteroids. From Hubble's vantage point, that debris field vaguely displayed the likeness of a cross circumscribed by The Star of David, seemingly in an onward rush leaving a dusty streak behind (see Figure 50, pg. 349). Then the attentive Mark Biltz discovered God's signposts above the events of Israel's reestablishment as a nation in possession of God's Holy City.

If we could determine when the seal came off Daniel's final prophecies, the time, two times, and half-a-time, the 1,290 and the 1,335 days, I think September 11, 2001 would be a very rational, if not entirely accurate guess. For those with eyes to see, 9-11 was also the new beginning of an abundantly prescient perspective by which to view signs and recognize them as portents of evil's end. I do not believe it too audacious to state that the world entered a time of austere warning on September 11, 2001, a call for everyone to quickly climb aboard the Argo, Christ ferrying us through a flood of deceit to enjoy life eternally. And was it synchronistic, or was it inspired, that its date dials 9-1-1?

But for those who deny what God reveals about Himself in The Holy Bible, this new beginning started an overflowing abundance of deceit and destructive horrors cascading from the world's denial of the Self revealed knowledge of I AM WHO I AM, Jesus Christ. For although Christ's will be the only boat to dock its precious cargo in the blessed harbor on the other side of the Tribulation's mayhem, it is not a boat for the "geniuses" who focus on evolution's pareidolia seen in the dirt. They won't believe Christ's story as verified by

Clear Signs of Trouble and Great Joy

the stars. They can't believe simply because they won't look.

The stark difference between the eternal destiny you can receive, or Hell otherwise, begins with something as simple as to where you direct your attention.

In the mid late 1990s, Fouad Hussein, a Jordanian journalist, found himself locked in a Syrian jail with Abu-Musad al-Zarqawi, an intently focused terrorist. They did interviews. They established friendship. As a result, when released, Hussein was able to interview almost every member within the inner circle of al-Qaeda's top leadership. He wrote <u>Al-Zarqawi: The Second Generation of Al-Qaeda</u>, revealing al-Qaeda's plan to establish global dominance and re-institute the Caliphate by 2020. His book explains how September 11, 2001 was the first step of that plan.[22]

Al-Qaeda's plan stepped onto I AM WHO I AM's Shemitah Path in 2001. Happening in the eighth Shemitah after the blessing came to those who waited the thirteen-hundred-thirty-five days seems an appropriate communication regarding the overwhelmingly abundant good to which this path leads for those who abide in the Lord, and the catastrophic horror into which it leads Jesus' deniers. Blessing the patiently waiting was certainly not al-Qaeda's purpose in this first step of their plan. Awakening war was.

> [The First Phase Known as "the awakening"] has already been carried out and was supposed to have lasted from 2000 to 2003, or more precisely from the terrorist attacks of September 11, 2001 in New York and Washington to the fall of Baghdad in 2003. The aim of the attacks of 9/11 was to provoke the US into declaring war on the Islamic world and thereby "awakening" Muslims. "The first phase was judged by the strategists and masterminds behind al-Qaeda as very successful," writes Hussein. "The battle field was opened up and the Americans and their allies became a closer and easier target." The terrorist network is also reported as being satisfied that its message can now be heard "everywhere."[23]

For all those who love to wear out God's elect, September 11th was the ninth Shemitah after Kristallnacht in 1938, Hitler's first step of The Final Solution.

> [Nine] is the last of the digits, and thus marks the end; and is significant of the conclusion of a matter.[24]

September 11th began al-Qaeda's plan. But al-Qaeda's plan began their end, like Kristallnacht began Hitler's end, as all accusers/deceivers/destroyers are eventually trapped by their own bloodlust.

> ³I will bless those who bless [the Israelites,] and him who curses [them] I will curse.
>
> Genesis 12:3

God has so ordered history. These signs along His path to the end of evil read "beginning" for His saints and "end" for His foes.

Measurements

Embedded within these Shemitah measurements was an event Jesus invoked to deliver an awakening of His own. His RSVP warning to al-Qaeda's call to war was so spiritually marked that it did not need to ring nearly as loud as al-Qaeda's rang on September 11th. It whispered, in fact, since its timing sounded most of that alarm: September 11, 2015, 5:23PM AST, 2X7 years and precisely two hours and thirty-seven minutes beyond the precise moment al-Qaeda rang their call to war, 8:46AM EST, September 11, 2001. Those two-hours-and-thirty-seven minutes round up to three hours, the number of God's doing it. 2X7 years after al-Qaeda's call to war, I AM WHO I AM RSVP'd to Mecca 2X7 years before 2029, the year Apophis' arrival portends the beginning of the Tribulation war to and evil. That is double spiritual significance doubled maybe even more than it is trouble doubled! It is also quite symmetrical. Then, thirteen days after God's RSVP, twenty-four hundred hajj pilgrims died in a stampede at Mecca. Note the 13. Note the 3X8X10X10. Four days after that stampede, the fourth total lunar eclipse of the 2014-5 holyday-tetrad occurred on Tabernacles, seventeen days after God's RSVP, the Biblical number of complete spiritual meaning. God only needed to whisper, for reality is known by its measurements.

Ok! Ok, already! What happened September 11, 2015?!

Mecca is Islam's most important city. Every Muslim male is required to march around the Kaaba at Mecca once in his lifetime. Al-Qaeda's twenty year plan was well underway and being considered a burgeoning success by 2015. Therefore, the Grand Mosque at Mecca would need to be sufficiently expanded and glorified to handle the resulting flood of pilgrims an Islamic world domination would bring.

Several giant cranes were at the mosque renovation site. Friday afternoon, September 11, 2015, the site was engulfed in a violent thunderstorm. Amidst frequent lightning strikes (of which a photograph of one was used to embellish the story as having been caused by lightning) a powerful gust of wind toppled a lightning-struck crane onto the Grand Mosque, killing one-hundred-eleven worshippers inside, and injuring a couple hundred more, as I AM WHO I AM lowered the boom on Allah's roof.

A French news source quoted the Saudi Binladin Group engineer (the contracted renovator) as saying,

> "It was not a technical issue at all," said the engineer, who asked not to be identified, "I can only say that what happened was beyond the power of humans. It was an act of God and, to my knowledge, there was no human fault in it at all."[25]

We would expect to hear any construction company's engineer blame God after such a tragedy, even if it was definitely and obviously the company's fault. Then again, isn't it interesting that after several years of no-incident construction, no less or more than fourteen years to the exact day, only two-and-a-half hours beyond the precise minute Islam's September 11, 2001 attack struck New York City, the wind went churlish in Mecca? And note another tidbit of trivial "coincidence", a faintly whispered signature, maybe. The Muslims sounded their alarm on a Tuesday. The day of I AM WHO I AM's RSVP was the third day after Tuesday, a Friday, again, that familiar number of God's doing it, the same number

Clear Signs of Trouble and Great Joy

of hours beyond fourteen years God's RSVP rung al-Qaeda's bell after 9-11.

The Shemitahs immediately before and after God's RSVP to Islam's call to war also bear great curiosities, prophecies, to be more rhetorical, if not completely exact, Mr. White. Mark these next few pages in your mind and recall them when you read Chapter 9.

In Chapter 4 we discussed how the Mazzaroth constellations, Cygnus, the ever circling swan, and Bootes, the farmer rushing to harvest, represent the aspects of Jesus' ever abiding presence in the Spirit and return in His Celestial body. Each of these constellations shout another piece of meaning from a Shemitah year on each side of God's 2015 RSVP.

> At 2:12a.m. EDT, Swift detected an explosion from deep space that was so powerful that its afterglow was briefly visible to the naked eye. Even more astonishing, the explosion itself took place halfway across the visible universe!
>
> "No other known object or type of explosion could be seen by the naked eye at such an immense distance," says Swift science team member Stephen Holland of NASA's Goddard Space Flight Center in Greenbelt, Md. "We don't know yet if anyone was looking at the afterglow at the time it brightened to peak visibility. But if someone just happened to be looking at the right place at the right time, they saw the most distant object ever seen by human eyes without optical aid."
>
> "Even by the standards of gamma-ray bursts, this burst was a whopper," says Swift lead scientist Neil Gehrels, "It blows away every gamma-ray burst we've seen so far."
>
> Swift normally detects about two gamma ray bursts per week. But March 19 was a special day. The satellite detected four bursts on that day, which is a Swift record for one day.[26]

Note, again, the number of bursts detected on that day -four- the same as the number of eclipses in a tetrad and the number of special tetrads calling attention to the rise and development of Israel and America for God's Tribulation endgame (1493-4, 1949-50, 1967-8, and 2014-5). Four is also the number of stars comprising the anciently perceived King Star, Regulus. Science forbids any call upon God for explaining a physical reality, yet it calls upon coincidence to explain more than coincidence can rationally imagine. How meaningfully can enormous accumulations of occurrences be interconnected by Biblically descriptive numbers and still be considered nothing more than a heap of mere happenstance? Four is the Biblical number representing our physical realm, as if all of this were a message addressing the worshippers of science, who, after all, first saw that informative, gamma-ray burst in Bootes.

According to the nineteenth century description of the Mazzaroth, Bootes represents Christ rushing back to His physical creation for harvesting His saints into the storehouses of eternal glory before treading the grapes of wrath, the deniers of His righteousness and stealers of His reality. This record outburst in Bootes occurred during the Shemitah year between al-Qaeda's 2001 call to war and I AM WHO I AM's 2015 RSVP. March is the third month of the year, the number of God's doing. Nineteen represents the complete

Measurements

summation of man's affairs, one of the central purposes of the Tribulation. Christ harvesting His followers into eternal bliss and treading the deniers in the great winepress will certainly sum up man's affairs. And eight is the Biblical number representing a new beginning, the start of the epoch of God's righteous reign on Earth after Jesus has struck the sum-key with His feet in that winepress of The Lamb's wrath. 03-19-08 was a very interestingly numbered day for sighting the greatest gamma-ray burst ever observed by mankind happening in the constellation representing Christ's quick return to this physical place (four bursts were sighted that day). Considered together, the numbers of this date read, "God sums up man's affairs for a new beginning in this old universe." The very purpose for which Christ returns to Earth was inscribed with the medium of a stellar outburst, by God's hand, alone, upon this constellation representing Christ's quick return to the harvest.

    Maybe it seems like pareidolia. But remember, pareidolia is seeing patterns in reality's spew where there really are no patterns. Seeing dates in Timex rocks and Bulova bones is pareidolia. But these historic events and astronomical alignments correlating with prophecy and Biblically relevant history are actual patterns made by reality, orderly imprints measured onto the fabric of time by numbers bearing Biblical meaning. They are not imagined patterns, like Nebraska Man was a pattern imagined from a pig's tooth. These patterns in the celestial bodies are real; Appendices 1 and 2 describe how you can see them for yourself. They express a concept coherent with that of the world's events building towards tribulation.

    What word expresses the refusal to see patterns where reality has indeed formed patterns? Foolishness?

    These seemingly endless patterns are consistent with Biblical measurements. They are cohesive with Biblical prophecies and coherent in every respect with the Biblical message of salvation and restoration. Apparently, God is trying to get mankind's attention. Or at least He's trying to get yours. Isn't that what you would expect of a God described as the Bible describes Him? Horrific, nearly Hellish events are soon to break out on Earth. Then Christ will return to eliminate evil along with its doers so that everyone loving righteousness, peace, and joy can have those in abundance, forever. You need to quickly climb aboard the Argo, His lifeboat of grace, and be carried through this world's mess to dock in eternal joy. For that purpose His kindness and love shine brighter than GRB 080319B shined from Bootes. I beg you, especially you Muslims and Progressives, don't be fools. Note the patterns. Study The Holy Bible. Or at least apply Teddy's Theory: Try it. You'll like it!

    Revelation describes the Tribulation as a time when over half the world's population will die. It will be far more frightening than Covid-19. But masking mouths with fibs won't relieve those troubles like it eased the faux fear of Covid. Hopefully the Shemitah following 2015 might shine brightly enough to penetrate any remaining cataracts of anti-Biblical bias from eyes trained to only see mechanistic material. The 2022 Shemitah might be the final warning God displays in the sun, moon, and stars regarding the horrific cleansing He is about to work, other than Apophis' 2029 rendezvous, possibly too late to be much of a warning.

    This upcoming event of 2022 gave me an opportunity to test my hypothesis about these clear signs of trouble and great joy being so purposefully stacked with meaning that they might be forming a language of their own. I read a headline on WND.com regarding some "new star" to appear in 2022. The headline itself didn't reveal where in the sky this

Clear Signs of Trouble and Great Joy

"new star" would appear. But I immediately recognized 2022 as being the Shemitah year between God's 2015 RSVP and the presaged beginning of the Tribulation in 2029. I noted that both the "new star" and the year of its expected appearance paralleled the GRB 080319B outburst from Bootes within the Shemitah year nestled half-way between 2001 and God's 2015 RSVP. Extending the parallel further, 2008 and 2022 are both Shemitah years. Both involve astronomically energetic outbursts; each occurs seven years on either side of God's RSVP to Islam's war; and each occurs either seven years after or before a Shemitah year of hellacious outbreak (9-11 in 2001, and the Tribulation at 2029). Science enjoys this kind of symmetry. So I clicked the headline while confidently expecting to read that this "new star" will occur in the constellation with a symmetrical meaning to that of Bootes'.

Cygnus, represents Christ's truth ever present with us, circling above us on outstretched wings while Bootes represents Christ returning His resurrected, physical presence to us. Bootes is the depiction of a land dwelling man. Christ came as a man and will return as one, too. But Christ is also the eternal Lord of the Spirits, who was, and is, and is to come, The Truth about EVERYTHING. Man is both physical and spiritual; he has a body and a spirit. So Christ is also represented in the constellations as a free flying spirit: Cygnus, the swan, and also, as a physical man -Bootes, the farmer.

Symmetry would be maintained by a "new star" shining in Cygnus, I concluded. Indeed, I expected this article to locate that star there. It did. My prediction was confirmed. These signs were forming a communicative "language" from their coherence with the Biblical narrative. This "new, temporary star" would appear in Cygnus.[27] But I didn't realize how spiritual its communication would actually be.

Astronomers have been watching for a binary star system to collapse close enough to Earth to produce an outburst visible to the unaided eye. Cygnus' new star was to be an explosion. Larry Molnar, an astronomy professor at Calvin College in Grand Rapids, Michigan, compiled data recorded from 1999 through 2013 on a pair of binary stars in Cygnus designated as KIC 9832227.[28] Adding a piece of 1999 data to the data his own team collected on this star system, he calculated the collision of these two stars to have occurred approximately eighteen-hundred years ago, meaning the light of their grand explosion would reach earth in 2022, making it the second brightest star in the night sky for a bit less than a year. His prediction went viral throughout the world of astronomy enthusiasts. Being on the fringe of astronomy enthusiasm myself, I also noticed the story. The symmetry between KIC 9832227 and GRB 080319B, as well as between Cygnus and Bootes, and 2008 and 2015, added coherence to the message within these five signs of the 2001 through 2029 Shemitahs.

Of course, the astute astronomy enthusiasts among my readers have probably objected to the light of its collapse reaching Earth in 2022. The 1999 report Molnar's team used in its calculation contained a tiny error. A 1999 eclipse in these binary stars was reported twelve hours opposite from when it actually occurred. Somebody mixed up their AM and PM. This error was enough to destroy Molnar's calculation of when the system collapsed, and therefore, of when its outburst of light would become visible. 2022 will not be the year of a grand, new lightshow in Cygnus.

My detractors are rejoicing! My fans are lamenting. This single fly in all the grand ointment of these signs' consistency, symmetry, and coherence has produced a very foul,

Measurements

intellectual odor. Skeptics will claim the odor nullifies my entire proposition.

But does it? Maybe the message wasn't intended to be expressed by the actual shining of a real, new star. Maybe its not shining was meant to subtly deliver an importantly, specific warning. Would my lose-one-thing-find-another rule of thumb hold true?

It did.

The "not a new star" shines brighter from Cygnus even while I write about it. But, like much of the Bible's prescience and prophecy, its light isn't seen by physical eyesight. It shines from real patterns existing within reality's own *information*, and it is seen by eyes willing to acknowledge realities observable only by the patterns they make. The pattern of Biblical correlation with the concepts of Bootes/GRB 080319B, Cygnus/KIC9832227, Apophis, 09-11-15 at Mecca, 09-11-01 in New York, and their seven year cycle carries the explanation of this phenomena beyond coincidence all the way to the purposeful design made by a communicative Creator. That explanation presents to us realty's call for rational minds to accept The Holy Bible as their guiding truth. It shines forth the spiritual nature of the coming Tribulation, silently shouting one final warning, "Right your thinking by the evidences The Lord of Spirits gave for us in what can be seen!"

> [20]Ever since the creation of the world his invisible nature, namely, his etern-
> -al power and deity, has been clearly perceived in the things that have been made. So they are without excuse; [21]for although they knew God they did not honor him as God or give thanks to him, but they became futile in their thinking and their senseless minds were darkened.
>
> Romans 1:20-21

The expected place of this "new star" being the constellation, Cygnus, the ever presence of Christ's Holy Spirit bringing Truth to our lives, is highly meaningful. And even more meaningful was its expected time of appearance being the last Shemitah year before Apophis marks the Tribulation's beginning, just when a warning needs to be the most austerely received by attention astutely paid to very real patterns. How perspicuous it is that its shining is seen through attention paid to patterns made only of *information*. It could have been expected to shine from any other constellation during some year other than that particular Shemitah year nestled between God's RSVP to Islam's war call and the Apophis presaged beginning of the final war to end evil's grip on man's heart.

Jesus Christ audaciously proclaimed Himself to be The Way, The Life, and The Truth. Jesus knows all things. He's as much I AM WHO I AM as is His Daddy. But the rest of us don't have enough mental capacity to know everything. Comparatively speaking, we don't know diddly-squat. The best we can do is collect observations and note coherent patterns amongst them for concluding a little more knowledge than we had. David Hume noticed our diddly-squat condition, too. Hume's Principal about "too much to observe" inspired Karl Popper to propose that falsification method for testing theories. Yet cosmology has gone viral with Hume's Principle while applying Popper's Theory by the most wily of biases, unlike Molnar employed Popper's Theory without bias. Thus, cosmology suborns its judgment to Charles Darwin's faulty guess about the imaginary, non-designed, purposeless-

Clear Signs of Trouble and Great Joy

ness of the universe, rather than allowing patterns of evidence to lead it to the truth found even within Creation's intricate design. The theoretical science of origins has followed its own imagination into the barrow pit of wrecked ideas instead of simply following reality's road signs without bias, that is, following Jesus without sassing. He knows everything about us and I AM WHO I AM, both, while we know little more than diddly-squat about either.

Molnar was unwilling to subordinate his judgment even to his own desire to find a highly anticipated new star. Instead, he gave up his falsified proposition out of intellectual respect for *information*, a.k.a. evidence, however slight that evidence might have seemed, even so slight as twelve, tiny hours amongst all of science's theorized billions of years.

> "Good science makes testable predictions," Molnar said. "There have been a few other papers that have tried to poke at our project, and we've been able to poke back criticisms that just don't fly. But this one does fly, and I think they have a good point. This illustrates how science can be self correcting."[29]

It also illustrates how science *has not been* self correcting. Oh! I assure you, the science of nitroglycerin has been self correcting. Same with the science of airplanes and bridges, although the science of shuttle launches had its Challenger. But the science of ape-man hasn't touched self correction with a ten-million year pole! Nor has political science self corrected for the observed horrors of Progressivism/socialism/communism/Nazism, Stalin's starved Ukrainians, Hitler's murdered Jews, Chairman Mao's dead landlords, America's aborted babies, not to mention hundreds of millions of good folks depleted of dignity inside collectivism's impoverished covid corrals. Molnar's response to the AM/PM gaff only illustrates that science *can be* self correcting. It does not illustrate that it is employing Popper's Principle, except for the honorable Molnar's exemplar.

David Rohl shined the 02-25-1362BC total lunar eclipse's light upon the errors of Egypt's Old Chronology. But the correction called forth by that *information* will never happen in the "great" halls of secular "education". Heavens no! Rohl's correction aligns a vast heap of archaeological evidence with the Biblical narratives, which secular historians refuse to acknowledge, love to hate, and laugh out of their classrooms. They can't allow Bible stuff to be accepted realities! So, instead of self correction, archaeological science extends to the great David Rohl, the Einstein of historians, their middle finger of fiendship. How much correction of the "dinosaurs went extinct millions of years ago" theory did the sauropods engraved on Bishop Bell's fifteenth century grave make? Or the soft tissues and proteins now being discovered in dinosaur fossils? No correction has been made to the "resurrection of Jesus is a myth" theory as a result of The Shroud of Turin's having been scientifically authenticated by careful, controlled, and well measured analysis of its every aspect.

Molnar didn't spit on the AM/PM error, proclaiming his theory in spite of it. You won't find that AM/PM error in science's dumpster underneath all those polystrate whatnots. Instead, Molnar treasured it, for it corrected an error in his theory.

But archeology spit's on the Anasazi sauropod etched upon Utah's sandstone. How informative of scientific subjectivity is Molnar's not-a-new-star! Paleontology spits on

Measurements

Cremo and Thompson's meticulously researched and documented demonstration of the slight of hand played by early geologists while animating evolution mythology. How informative of scientific arrogance is Molnar's not-a-new-star! Political science hasn't corrected its socialist guess in light of Hitler's incinerated Jews, Stalin's starved Ukranians, Mao's shot landlords, or the thousands of prenatal boys and girls American Progressives are yet sucking through surgical tubes every day. How revealing of political horror is Molnar's not-a-new-star! And the science of origins has convinced the entire world to spit in Jesus' resurrected face while totally disregarding the otherwise inexplicable Shroud of Turin and Oviedo Cloth. How informative of scientific foolishness is Molnar's not-a-new-star.

Molnar allowed one, tiny tidbit of *information* to realign his theory with reality while the rest of science continues impugning the bearers of *information* in order to shelter their not-a-God guess and cover up their "let's all be communists, now" errors. They have no intention of employing Karl Popper's Principle! Surely, everyone else doesn't know what they're talking about! These scoffers just keep hanging around Hume, making things up in spite of *information* revealing reality to be other than what they first imagined. Reality will soon cast them out its end.

But a true remnant of humanity still does acknowledge what eyes to see do see. It intellectually responds in accord with evidence, just as Molner did. This not-a-new-star in Cygnus warns everyone to get honest about *information*, a.k.a. evidences, observations, and common sense, which The Holy Bible is, which this enormous system of signs is, and what we will soon find even Gobekli Tepe to be.

The KIC 9832227 not-a-new-star proclaims, "Yes science can! Yes politics can! Yes academics and Hollywood can! Yes you can! YES WE CAN get it right!" Just like Molnar did. But science's full dumpsters show it hasn't. D.C.'s impeachment charades show politics won't. Good time rock-and-roll shows Hollywood isn't interested. Uncorrected textbooks show academia's hostility towards the truth. Fifty-million destroyed, prenatal lives show cold-nor-hot churches lie about loving their fellow man. Everyone needs to be interested in KIC 983227, because the end of denial is swiftly coming on the wings of Jesus Christ.

Biblical prophecies of the horrific Tribulation intimate that most people will continue denying the Lord of Salvation right up to the moment they are destroyed by the sight of His coming. But you mustn't. He is giving full and clear warning to everyone. 2022 without a KIC 9832227 visibly shining demonstrates how you must correct your theories about not-a-God to accord with the complete order of coherent, perspicuous *information* shining truth down from the skies above, up from the dirt of the epochs, and out from the known events of history, all purposefully designed by The Lord of Spirits to correlate with the message of His Holy Bible. If Rhinnie can understand the necessity of completing a puzzle, then we adults should be able to as well. EVERY piece counts!!! What shines out of Cygnus, the constellation representing the ever presence of Jesus Christ, the way, the life, and THE TRUTH is the importance of responding in rational accord with every observation, i.e., with ALL *INFORMATION*, nothing ascribed to anomaly, no dumpsters employed, lest you be found a liar by The One who uses stars for ink and not-a-new-stars for juries.

"This is arguably the most important part of the scientific process.

Clear Signs of Trouble and Great Joy

> Knowledge advances the most when bold predictions are made, and people question and test those predictions," [the researcher who discovered the KIC 9832227 error] said. "Often the most exciting discoveries happen when our expectations are not met. This is a good example of how scientists from different parts of the world can work together to better understand how our universe works, bringing with them new pieces to the puzzle."[30]

Modern science's definition of Natural is animated by Darwin's wild guess about the universe being purposeless, without design. But we have seen more than enough purposefully designed *information* in this monumental order of celestial patterns correlating with historically fulfilled Biblical prophecy to know Darwin's guess animates deceit, not nature. Yes, we can now acknowledge a precisely designed purpose for the universe! The design and purpose clearly seen in these pesky signs form a Bible shaped hole at the center of reality's jigsaw puzzle. It doesn't matter if you are willing to place the Bible in its space or not. It fits there Naturally. The Creator specifically shaped that hole for The Holy Bible. And at Christ's coming, The Bible will be placed there over the dead bodies of its deniers. Of that these signs warn! Like a key in a lock, the Bible in that central hole will release Christ's truth into science, academia, and public life for a thousand years.

The Bible doesn't need you to place it in its space. Your place is to acknowledge the fit, like Molnar's was to acknowledge the correction. All of this phenomenal order in the signs of Christ's return verifies the fit. We don't have to guess anymore. Normal science's theories need revised to conform to reality's display of that Bible shaped hole at its center. Christ's return to sum up man's affairs for a new beginning will be the ultimate, scientific proof of The Bible's perfect fit into that central space of Nature's puzzle. It will be the griddle frying Darwin's goose-egg guess over the heat of empirical observations.

Today, you can indeed see the whole Bible fit there by faith placed in the Creator of everything. I will leave the Biblical message made by the not-a-new-star's number, 9832227, for you to interpret, should you engage faith in the Lord of Spirits. Then use Teddy's Theory; try fitting the Bible in the central space of your life! You'll like the fit. No. You won't just like it. You'll love it! And Jesus will make it fit eternally.

Apophis will scorch past Earth seven years after Cygnus' not-a-new-star calls mankind to return its attention to finding the truth. Apophis has been passing through the neighborhood every seven years or so, for how long we don't know. But its arrival this time sets a road-sign at the beginning of God's final process to end destruction, deceit, and accusation. The signs are clear. The essence of these five successive events on five sequential Shemitah years, beginning with 9-11-2001, reads: Islam declares war; Christ shall come quickly to its war; He RSVPs Islam's invitation; so take cover in the Truth; deceit shall rise up to meet its doom. Then the last Shemitah, the year of Apophis' return on Easter Sunday, 2036, portrays the glorious arrival of Jesus Christ, The Truth, upon the Mount of Olives to destroy all deceit and evil with their workers by His holy presence, before instituting righteous order. How abundant was that new beginning made on September 11, 2001!

[19] He declares his word to Jacob,

Measurements

his statutes and ordinances to Israel.
[20] He has not dealt thus with any other nation;
they do not know his ordinances.

Psalm 147:19-20

The Bible chose to address these Israelite people by that Hebrew term derived from the root meaning "testimony and witness". Throughout history the Israelites witnessed God's interactions with them. Today Israel's existence testifies to His faithfulness. The Shemitah Path began the Israelite's First Aliyah, 1882, which continued through three

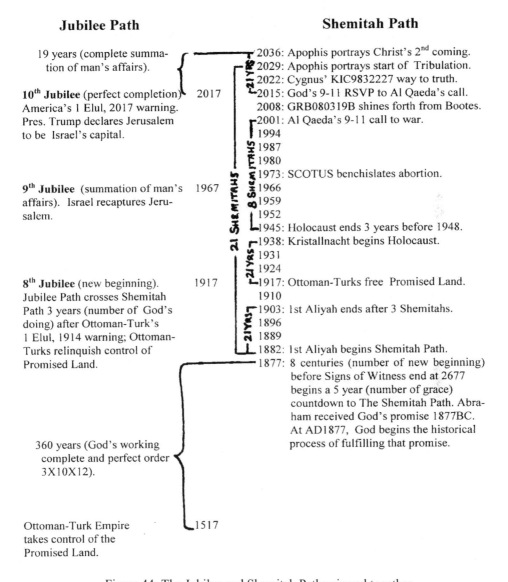

Jubilee Path		Shemitah Path
19 years (complete summation of man's affairs).		2036: Apophis portrays Christ's 2nd coming.
		2029: Apophis portrays start of Tribulation.
		2022: Cygnus' KIC9832227 way to truth.
10th Jubilee (perfect completion) America's 1 Elul, 2017 warning. Pres. Trump declares Jerusalem to be Israel's capital.	2017	2015: God's 9-11 RSVP to Al Qaeda's call.
		2008: GRB080319B shines forth from Bootes.
		2001: Al Qaeda's 9-11 call to war.
		1994
		1987
		1980
		1973: SCOTUS benchislates abortion.
9th Jubilee (summation of man's affairs). Israel recaptures Jerusalem.	1967	1966
		1959
		1952
		1945: Holocaust ends 3 years before 1948.
		1938: Kristallnacht begins Holocaust.
		1931
		1924
8th Jubilee (new beginning). Jubilee Path crosses Shemitah Path 3 years (number of God's doing) after Ottoman-Turk's 1 Elul, 1914 warning; Ottoman-Turks relinquish control of Promised Land.	1917	1917: Ottoman-Turks free Promised Land.
		1910
		1903: 1st Aliyah ends after 3 Shemitahs.
		1896
		1889
		1882: 1st Aliyah begins Shemitah Path.
		1877: 8 centuries (number of new beginning) before Signs of Witness end at 2677 begins a 5 year (number of grace) countdown to The Shemitah Path. Abraham received God's promise 1877BC. At AD1877, God begins the historical process of fulfilling that promise.
360 years (God's working complete and perfect order 3X10X12).		
Ottoman-Turk Empire takes control of the Promised Land.	1517	

Figure 44: The Jubilee and Shemitah Paths viewed together.

Clear Signs of Trouble and Great Joy

Shemitahs, the number of God's doing it. From 1882 to Apophis' presage of the Tribulation's arrival near the 2029 Shemitah year is twenty-one Shemitahs, the number of God's doing times the number of spiritual significance (3X7). Then, the catastrophic end, the Tribulation period's final year, is the twenty-second and final Shemitah of The Path.

> Twenty-two being the double of eleven, has the significance of that number in an intensified form, -disorganization and disintegration, especially in connection with the Word of God. For the number two is associated with the second person of the Godhead, the Living Word.
> [Twenty-two] is associated with the worst of Israel's kings, -Jeroboam (1 Kings 14:20), and Ahab (1 Kings 16:29), each reigning 22 years.
> Eleven, we have seen, derives its significance by being an addition to Divine order (10), and a subtraction from Divine rule (12).[31]

That the Tribulation will arrive is a certainty. How you arrive at the Tribulation is your choice. May I suggest you arrive there, aboard the Argo, hand in hand with Christ?

The context is now appropriate for discussing a fundamental trigger of the world's final, epochal trek into the Tribulation, as it is displayed by a unique, seven year measurement. This seven year measurement is neither a part of the Shemitah Path, nor of the Jubilee Path. Yet it is appended to the end of The Jubilee Path like an exclamation point indicating a particular relevance to Jerusalem. This measurement engages the Shemitah Path at the year of Cygnus' not-a-new-star, demonstrating the importance of following evidence to wherever it leads. Its peculiar combination of threes, fives, and sevens testifies to it's being a message from God. Its observable correlations with history, places, and astronomical events imply that these seven years deliver God's warning aimed specifically at America, essentially similar to America's warning explained in Cahn's *The Harbinge.* But it reads more like a finality than a second warning.

People have long wondered what role America will play in the end times. For many years they've searched to find some hint of it in Biblical prophecy. Their endeavor has turned up a few interesting concepts, but the answer is finally discernable from these signs.

Total solar eclipses are not entirely rare. One is observed from somewhere on the earth about once every eighteen months. Observers at any particular location will see only one total solar eclipse about every four hundred years, on the average. August 21, 2017 was America's turn. The shadow of its totality moved across America from the northwest to the southeast in the final year of Jerusalem's Jubilee Path. Then, seven years later, another total solar eclipse will draw its shadow of totality northeast across America from the southwest. The one lucky location on Earth which gets to enjoy two total solar eclipses seven years apart is a seventy mile diameter circle where the eclipses' two paths of totality cross over the New Madrid Fault Zone (see Figure 50, page 349).

These two eclipses evoke an ominous sense of doom by that giant, seven-year "X" crossing out America's heartland over a major fault line. Their seven-year separation is the same seven years of Hitler's Holocaust and the quickly coming Tribulation, the Biblical number of complete, divine spirituality. God will bring the Tribulation upon mankind full of horrors similar to those plagues He brought upon Egypt. He gave Pharaoh fair warning of the

Measurements

doom any refusal to free Israel would cause. He is giving the world fair warning today. Pharaoh pied, lied, and dallied to the end of God's patience, then the death angel came. God got what He came for, and Egypt smoldered. The prescient Mark Biltz informs us that this same area of the New Madrid Fault Zone enjoying two total solar eclipses seven years apart is coincidentally (or not) that part of Illinois nicknamed America's Little Egypt.[32]

God speaks to the world in terms drawn from the Hebrew culture, as Rabbi Jonathan Cahn demonstrated for us regarding the 9-11 disaster. Let's consider a Hebrew perspective on this situation, too.

Elul is the Hebrew month correlating with August/September. Cultures which developed through centuries of deeply religious aspirations tend to emerge from the past with one religious meaning or another attached to almost everything. The month of Elul also represents a meaning drawn from the Hebrew past. To the Jews, it is the month of repentance.[33] By our faulty nature, we humans enter the month of Elul in denial, with our backs to God. But God expects us to finish Elul facing Him in acceptance. August 21, 2017, the day of this "X's" first total solar eclipse, was 1 Elul on the Hebrew calendar,[34] Rosh Chodesh Elul, as it is called. Should that mean anything to us?

Only if we're superstitious? That might be a "scientific" guess. But scientific solutions are found among correlations, not guesses (except cosmological solutions, for those, guessing is imperative because those "scientists" have thrown all their *information* into dumpsters). One correlation, such as this eclipse occurring on Rosh Chodesh Elul, usually means nothing more than the presence of a coincidence. But if a past 1 Elul total solar eclipse were found to have occurred in the historical context of events developmental to the fulfilling of Biblical prophecy, then a 1 Elul eclipse could rationally be regarded as a warning given from God. The more the outcome of the corresponding situation could be found to have been consequential to a decision facing whatever nation had been shaded by the eclipse, the more a 1 Elul eclipse could rationally be regarded as God giving a warning. And the more relevant to the fulfillment of Biblical prophecy was the decision faced, the more obviously from God the warning could be regarded. Finding such an eclipse would establish a "principle" by which America's 1 Elul eclipse could also be rationally considered a warning from God. If a prophetic relationship between the two "warned" nations was also observed, the possibility that both 1 Elul eclipses were warnings from God rises at least to the level of a probability. Then a chronological pattern between the two eclipses measured in Biblically meaningful numbers would hoist that probability to a near certainty. The seven year span between America's two, convergent eclipses, then, would make that certainty definite.

Maybe it all resembles superstition. But remember, the Bible attests to the making of signs as being one purpose for the celestial bodies. We've been analyzing an enormous quantity of signs in those bodies correlating with earthly events of Biblical relevance. If it is only superstition, then it is the most coincidental of "only superstition" ever known. And not just in coincidence of occurrences, but also in coincidence of the coherence between occurrences and details of salvation's roadmap described by The Holy Bible. That's a whole lot of coincidence going on out there! Maybe more than what can really be "only coincidental". Most likely, it is purposeful. Not all coincidences are the superstition they might at first

Clear Signs of Trouble and Great Joy

seem. These extensive signs appear to be the heavens attesting to the truth of God's Holy Word (Psalm 19:1-4).

Does America's 1 Elul, 2017 eclipse have a correlate in history? To launch the challenge even higher, can we find a total solar eclipse which drew its warning across a nation that has some important, historical relevance embedded in the Shemitah/Jubilee Paths of Biblically prophesied events, a nation bearing some Biblical relationship with America?

We saw the twentieth century open with Jewish refugees fleeing to Palestine from the resurgence of pogroms in Russia while carrying in their hearts an Enlightenment inspired hope for becoming a nation again. The Jewish people called it The First Aliyah. The Bible called it "dry bones coming together" (Ezekiel 37:7). The Old Testament is sprinkled with prophecies about the Jewish people returning to the land promised to them and becoming a nation again, a nation raised above all nations. History confirmed The First Aliyah to be God's work by those three Shemitahs measuring its twenty-one year duration.

Israel's ancient history was never lost to humanity. At least from the beginning of Christianity, three has been recognized as God's number. Seven has been recognized as the number of spiritual significance. Amongst those three points alone the Ottoman-Turk Empire had sufficient *information* to recognize a pattern of truth forming around this Jewish migration into Palestine, the land I AM WHO I AM promised to Jacob's children. Anyone allocating sufficient attention to the things of The Bible could have seen the Author of The Bible at work in The First Aliyah.

But The Empire had been too busy attending the messages of its own book to notice what current events were developing in accord with the messages of that old Hebrew Book. As a result, it failed to recognize I AM WHO I AM's prophesied dry bones coming together in this Jewish migration of twenty-one years. It failed to understand that I AM WHO I AM, like a charioteer, was carrying His people home to reestablish them as the nation of Israel, and that He had even used The Empire to make their migration possible. The Empire was being given the opportunity to join God's work of restoring national Israel by relinquishing the Promised Land to become that national place of God's chosen people.

The world hates the concept of God choosing the Jews and raising their nation above every nation. Therefore, with all attention focused upon its own aspirations and its own Koran, The Empire planted itself squarely between this little, Hebrew cub and its great Mamma Bear, stamping its Ottoman-Turkish foot onto the Promised Land dirt, insisting that a Jewish state in Palestine would happen over its dead body.

I AM WHO I AM is willing to accept that kind of offer. On August 21, 1914, Rosh Chodesh Elul, The Lord God drew a line in the Ottoman-Turkish dirt with a total solar eclipse. (Look it up in NASA's solar eclipse canon.) The Empire had not attended all of its available *information*. Five times every day it faced Mecca and stuck its butt in the air against everything not of the Koran, just as Americans have become educated to stick their butts in the air against everything not "scientific". If either the Koran or "science" were everything there were to know about reality, then subordination to one or the other would not have been a mistake. But there's far more to know about reality than just the Koran, or just "science", as Cygnus' not-a-new-star warns. Neither The Empire nor "science" opened The Holy Bible even once a day. The Empire paid no attention to I AM WHO I AM's

Measurements

warning drawn across its dirt, because it failed to note the Hebrew accent of that eclipse.

I AM WHO I AM speaks in terms of the Hebrew culture because it was to them He gave His Oracles. Therefore, to understand His warnings, enough attention must be paid to Hebrew concepts in order to interpret what I AM WHO I AM writes upon Earth's wall, or in the cases of the Ottoman-Turks and America, what He has drawn across their dirts.

Rosh Chodesh Elul, 1914 drew a solar eclipse path of totality across the Ottoman-Turk Empire. Rosh Chodesh Elul, 2017 drew one across America. More impressively, the Lord God designed both of these eclipses to occur in August, when the sun aligns with Regulus, calling out to us from The King Star, as if to warn us to align our work in accord with The King's plans. Were the Ottoman-Turk Empire, then, and America, today, being specifically called to repentance by the Natural language God made of these two signs?

The Biblical correlation between these two peoples can not be missed. Murad's elevation of the Ossman beylik into becoming the Ottoman-Turk Empire accommodates the Daniel 12:11 prophecy about the setting up of the abomination which makes desolate, the eventual empire belonging to the lawless man of perdition. America's history clearly displays a struggle to restrain a mystery of lawlessness at work amongst the nations until it is time for the man of perdition to be revealed (II Thessalonians 2:6-7). That man's empire will be the eighth, having risen out of the mortally wounded seventh empire, i.e., the Ottoman-Turk Empire having been dispatched by World War I. These two eclipses are found tightly correlating with each other through the Biblical relationship of the two peoples blessed (or cursed) by their warnings.

The Ottoman-Turk foot of refusal remained planted across I AM WHO I AM's line. It didn't repent. The Empire insisted Palestine would belong only to the Palestinians (the same bait on today's anti-Semitic hook). Maybe The Empire wasn't aware of what it was holding against Jacob's God, or even of its beautiful eclipse being a Rosh Chodesh Elul warning to get its foot off the land where I AM WHO I AM intended to plant those dry bones rattling down from Russia. The Empire ignored the signs of its times. The year of their Rosh Chodesh Elul eclipse, 1914, began World War I. World War I ended the unrepentant Ottoman-Turk Empire. Coincidence? Or writing on the wall?

Would the question be easier to answer if we noted the year in which God drove The Empire's foot off Israel's land? 1917 was the third year after The Empire's Rosh Chodesh Elul eclipse, the number of God's doing it, just as three was also the number of Shemitahs in the First Aliyah, and the number of Shemitahs to Kristallnacht from the Jubilee/Shemitah Path intersection at 1917. Three was the number of years Israel became a nation after America and its allies kicked Hitler's Holocaust gates open. This highly significant year, 1917, occurred two Jubilees before America's 2017 Rosh Chodesh Elul warning, and one-hundred-nineteen years before the end of the Shemitah Path at Christ's presaged return. (Psalm 119 is a very long praise for God's perfect ordinances and orderliness, the same kind of orderliness expressing His ordinances that we've been observing in these meaningful patterns intertwining those celestial bodies, history, and The Holy Bible.) And America's Rosh Chodesh Elul eclipse occurred nineteen years before Apophis will presage Christ's return in 2036, the number of man's affairs fully summed up. Humans can't make this stuff up!

Clear Signs of Trouble and Great Joy

God makes it happen. We're only observing His actions patterned in accord with Biblically meaningful numbers regarding Israel, Jerusalem, America, the dragon, and the world's trundle towards a time of trouble like mankind has never before experienced. Right there in the midst of all this pattern was the seventh empire given an opportunity to affect its own future by a choice of repentance, or of arrogant, brass foreheadedness.

So the question is easy to answer. The two Rosh Chodesh Elul eclipses were messages sent from the Lord of Spirits. They were warnings. Only godless, "scientific" superstition can see them as anything otherwise. The reality of their nature of being warnings is empirically evidenced in the orderliness of their patterns.

Patterns are enormously important to human intelligence. Only by patterns of neurons firing together do you perceive anything consciously. Nobody disregards their stream of consciousness just because it exists by neural patterns. Only by patterns of letters do you read this book. Only by visual patterns do you recognize anything around you. Only by patterns of air compression and rarefaction do you enjoy music. Only by studying patterns of material interactions did we come to know physics, chemistry, and the vast stretches of the universe. So why must patterns suddenly become inconsequential and meaningless when they are found intrinsically, intricately, obviously, and extensively correlating with the core theme of God's Holy Bible, historical events relevant to those, and to the fulfilling of Biblical prophecies? Is a pattern of denying such import of patterns not itself indicative of brass-foreheaded, "scientifically" applied superstition?

Being the most basic element of knowledge and communication, pattern casts its vote for, "PAY ATTENTION TO THE WARNING!" How will your forehead vote? From what could the wonderful, Progressive, land of "liberty" possibly have needed to turn repent? Has Woke replaced Jesus as America's new truth? Has murder in the womb really become America's new justice? Has Cancel Culture become the new American way? Are we still tearing God's monuments out of our public places and canceling His reality from our children's education? America was made for and by individual sovereignty, the freedom to think, express, and even discern in accord with following Christ Jesus to wherever He leads. American Progressivism corrals individual sovereignty so the herd can be rustled, harnessed, and put to collective labor.

To fully understand America's predicament, we must digress.

Jesus said anyone coming to His Father needs to come as a child. He didn't mean childishness. Childishness denies what it does not desire and regards only its own interests; childishness is the product of brass foreheads. The workers of evil are consumed in childishness. If making chaos helps reach their goals, then the more chaos the better. But childlikeness desires guidance, instruction, and the security of orderliness. Children naturally understand that parents know (or at least they believe parents know) what's good and right for them. Children know their own place is that of learning. Childlikeness is humility.

> [6]And these words which I command you this day shall be upon your heart; [7]and you shall teach them diligently to your children, and shall talk of them when you sit in your house, and when you walk by the way, and when you lie down, and when you rise.

Measurements

Deuteronomy 6:6-7

Recall from Chapter 2 that spiritual processes happen by Biblical prescriptions. The Bible's prescriptions are just as objective and specific as are the instructions for scientific lab experiments. Deuteronomy 6:6-7 is a specific set of instructions for maintaining a healthy culture in which trust can be placed securely in upright parents and in peaceful communities formed of God fearing neighborhoods. But trust must be validated by truly beneficial actions and choices, otherwise, it will be lost to deception. A culture deceitfully guiding children into perils and fecklessness, matter not how superstitiously deceit is believed to be truth, fundamentally transforms childlikeness into childishness. Thusly transformed children grow up to teach their children the same deceptive ways they learned. Most children raised in such a culture won't see destruction coming, because both the deceit and its resultant destruction are disguised by the cultural biases of misguiding entertainment, deceptive education, and

Figure 45

incredibly faked news. Hillary was right about the village razing a child.

But wiser children think about what they see enough to eventually recognize and abandon cultural misguidance.

In 1960, I entered grade school with great joy and hope from the trust I held in my parents, church, and country. I was Gabriella/Corey, Christian, and American! And I didn't think there was anything wrong with being any of those. The first two years of my education reflected ideas I knew from home. My school day began each morning with a prayer, and then we pledged allegiance to the flag. We usually sang a patriotic hymn or two. Maybe this

Clear Signs of Trouble and Great Joy

was only a withered version of Deuteronomy 6:6-7. But at least it was a version of it. I felt very secure in being a Christian American raised by a good-hearted, justice-minded family.

I was in High School before I fully realized how completely the intervening years had blindsided me. The prayers and patriotic hymns were taken out of our education before my third grade began. My parents divorced by the end of my sixth grade. Our once respected policemen were being called "pigs" before I had finished my sophomore year of High School. Cities were on fire. American cities! Fires set by Americans! After the Supreme Court banned school prayer in 1962, Illinois and Connecticut quickly became the first states to repeal sodomy laws. My home state, Colorado, repealed its sodomy laws when I was a senior in High School. Eventually it became illegal to call a man a man when (s)he wants to be known as a woman. Truthfully speaking, that is a legislated obligation to lie. You can surgically reshape your body into a plumb, should you wish, but it will yet be a plumb of the same sex into which it was born. Feeling male or female is attitudinal. Reshaping the body is merely sculptural. Sexual reality is genetic. Sexual identity is subjective. DNA doesn't lie. Attitudes do.

America was a Christian nation when I entered High School, but it was a rebelling nation as I finished. The year after I graduated, the Shemitah year of 1973, the United States Supreme Court benchislated women's rights to choose death for their prenatal girls and boys. After deceit had cut education's last stitch to God, America's Christian fabric unraveled quickly. Before my High School years, Christian ways were regarded to be the cultural norm. But after, they were considered abnormal, almost taboo.

I walked away from High School with no more trust in my culture, complete bewilderment about my parents, a high degree of doubt for the neighborhood churches, and a deep mourning for the God-fearing America of my childhood. Even my own duplicity was freed from the social constraints of a traditionally careful culture, entrapping me in such despair that I loathed the very feeling of being alive. That's when I learned the psychological value of repentance.

I arrived at my head hung over the barrel of a Mauser 98 rifle by believing that my character, personality, and general tendencies had emerged from the cumulative affects of all the events, misfortunes, and trespasses that ever happened to me. In other words, everything was somebody else's fault. But pulling the trigger would greatly wrong good people. I couldn't. When I laid the rifle aside, I realized nobody was to blame for my condition other than myself. The truth is, all of us are the product of *how we respond* to the events, misfortunes, and trespasses happening to us and by us. Have we continually wept over lemons? Or do we enjoy an ever abundant supply of lemonade? It's simply an attitudinal thing. Your inner plumb is shaped by your own decisions.

Accepting the sole responsibility for my poor mental condition placed the opportunity for its correction firmly in the hands of the only person who would make any effort to correct it…me. I drew a reassured breathe; I faced the necessary correction with confidence, because the Bible promised Christ's Holy Spirit to us, leading always to truth.

Concepts are like bricks. Vast quantities of *information* are processed into ideas like many bricks stack into walls. Concepts are composed of numerous ideas processed into multiple layers of reasoning through both troublesome and joyful situations. But thoughts

Measurements

and ideas do not accurately reflect reality just because "they are my reasoning", or "they are my favorite entertainer's reasoning", or Madalyn O'Hair's, Barack Obama's, Billy Graham's, or Donald Trump's. They reflect reality only when careful research for and honest consideration of *information* has squared them with what really exists.

Any good bricklayer will tell you the results of laying poorly shaped bricks into a wall, or of laying well shaped bricks sloppily. The structure will become unsightly, maybe even unstable and tottery. And every good carpenter will assure you that a quarter-inch of error in the foundation will multiply into an inch of error at the roofline. Similarly, information, memories, and rational processes must be squared with reality before the concepts they construct are useful for discerning accuracy from error, truth from deceit, right from wrong. Since we must deal with reality, an accurate reflection of it might come in handy.

So, each brick of thought is sufficiently important for us to true and straighten its rational fit with other thoughts, squaring its reflection with the way things actually are. The proper shaping and squaring of thoughts and feelings are critical for believing discovered truth. Every concept inevitably becomes further information for developing additional concepts. Error in any one concept multiplies through others, increasing error in the conclusion. Either the deceit or the truth of your character will increase in accord with how well you square each brick of thought and how accurately you embed it within proper attitudes.

> It doesn't matter how smart you are unless you stop and think.
> Thomas Sowell

Accurately shaped bricks stacked in proper accord construct fortress walls of true knowledge against trouble and deceit. But distorted bricks, laid pragmatically, pile into prison walls of fantasies, depriving you the safety and joys of a mental paradigm closely reflecting both your physical and spiritual situations, and possibly depriving you the prospect of an enjoyable eternity. It is imperative to confirm every component of your thinking, squaring each with reality, verifying the accuracy of its information, correcting your memories and harmonizing your emotions with what you've unbiasedly found to be true, and engaging it only by valid processes of reasoning.

For reality always arrives like a Mack truck, fully loaded and ominously heavy, in its own time, on its own terms, throttle wide open and the wind at its back, giving not a thought to braking just because it is sweet, little you lollygagging in the road. Your responsibility is to properly assess what is real before that Mack comes blasting around the corner ("when" is too late). It's better to think carefully and ride the Mack, even alone, if necessary, than it is to party with those brass foreheads crowding up the middle of reality's highways. Jesus Christ is coming, and He is the Truth (John 14:6). The better you align yourself with truth, the farther you will be out of His way of ending deceit.

I called this entire process of analysis, truing, and squaring "the human equation", and wrote it as follows: memory + reason = intelligence; $intelligence^{humility}$ = wisdom; $intelligence^{arrogance}$ = foolishness. Humility, in essence, is the analysis of your own concepts and mental processes for correcting and refining their reflections of reality. Intelligence raised by the power of humility produces wisdom. The most verifiable reality, The Holy

Clear Signs of Trouble and Great Joy

Bible, says fearing God is the beginning of wisdom's process.

> ³ [Wisdom] has sent out her maids to call
> from the highest places in the town,
> ⁴ "Whoever is simple, let him turn in here!"
> To him who is without sense she says,
> ⁵ "Come, eat of my bread
> and drink of the wine I have mixed.
> ⁶ Leave simpleness, and live,
> and walk in the way of insight."
> ⁷ He who corrects a scoffer gets himself abuse,
> and he who reproves a wicked man incurs injury.
> ⁸ Do not reprove a scoffer, or he will hate you;
> reprove a wise man, and he will love you.
> ⁹ Give instruction to a wise man, and he will be still wiser;
> teach a righteous man and he will increase in learning.
> ¹⁰ The fear of the LORD is the beginning of wisdom,
> and the knowledge of the Holy One is insight.
>
> Proverbs 9:3-10

I AM WHO I AM is the essence of reality, like the Identity Principle is the essence of knowledge. The fear of missing I AM causes the wisdom of truing your thoughts, feelings, and ways with what can been seen (eyes to see and ears to hear). Employing concepts, thoughts, emotions, and reasoning to imagine whatever you want to believe, or in order to gullibly swallow the consensus of scientists, or to mimic the majority of the population, or harmonize with a yoga master, flee with the panicky, riot with mobs, cheat with the voter fraud organization, lie with politicians and news fakers, or to just lip-sync your favorite acquaintances, teachers, preachers, pro-athletes, movie stars, singers, tripping along with Timothy Leary in hallucinogenic bliss, heaps up great walls of foolishness, a devastating departure from reality. Beware of the Mack! Neither intelligence nor education are the determinative factors of wisdom or foolishness. Humility is the operative element of wisdom. Arrogance turns every Einstein into an abject fool.

> ³ Seek the LORD, all you humble of the land,
> who do his commands;
> seek righteousness, seek humility;
> perhaps you may be hidden
> on the day of the wrath of the LORD.
>
> Zephaniah 2:3

The human equation also operates on the social level. People are social entities. Culture is a population's accumulating ocean of experiences, reflections, thoughts, and feelings formed into a popularly held paradigm for imitating in order to form acceptable behaviors in the individual's search for social acceptance.[35] Culture enables interaction. So a society of wise people will enjoy realistic paradigms able to effectively deal with difficulties,

Measurements

since the truths discovered in their past are honored and carried forward as traditions for imitating into their futures. But a society of foolish people will construct social paradigms out of mutual daydreams, denying the existence of reality and the concept of truth so they can trip along with Timothy Leary, lip-syncing their favorite people, harmonizing with mystics, lying, cheating, and rioting with the disillusioned, just because they rely on the ease of consensus instead of any effort to verify. Evidence, research, discovery, learning, and practicing are foreign concepts to arrogance. Arrogance cultivates, fertilizes, and waters great forests of dark excuses into a culture of imitated daydreams and dismissed warnings.

Man was made to reflect God, not each other. His Holy Bible is primary evidence for research and observation, since He is as true as we faulty ones have need to learn truth. Research and observation of this incredible system of clear signs confirms The Holy Bible's status as primary evidence: The Holy Bible was the first to tell us the sun, moon, and stars were made for signs. And behold! We are engaged in a lengthy discussion of the most enormous mass of coherently meaningful alignments ever observed amongst the sun, moon, stars, Biblical prophecy, and human history, occurring from 2000BC all the way to AD3000, every aspect of it being an empirically observable reality or a rationally valid construct.

Therefore, influencing culture with the truth is more important than being influenced by a culture's daydreams and nightmares. It is imperative to make or accept only influences that are properly squared with the truth about reality. The accumulating interactions of

Figure 46

Clear Signs of Trouble and Great Joy

a social group make and stack bricks into being either community fortresses or prisons, just like accumulating perceptions stack into personal fortress or prison walls.

Nothing of temporal life experiences the evidences of reality more than does living close to the land. Across the farmlands of any culture, reality lumbers down the roadways daily. Livestock are like peaches. They bruise and die, seemingly in a moment (especially sheep). To keep the crops and livestock healthy, the farmer/rancher must be more acquainted with reality than with some liberalized daydream about it. By the nature of their livelihood, country folk learn the ways of abidance, the orderliness of humbling themselves to the truths of their livestock's situation.

And the governance of a farm village happens in a fishbowl.

Therefore, country folk reason deeper into their politics, too, attending evidence more than biases and whims, seeing how biases and whims never kept the sheep alive. But in the big city, the inner workings of the governing class are a distant, hidden phenomena. None of them get their hands dirtied with the soils of everyday life. Moreover, nature's worst tragedy caused by a city dweller's moment of forgetfulness is, maybe, doggie poo on the carpet. Consequently, restraint in the city becomes a thing of the past as the city's bright lights and alluring night life liberally dissolve every rational precaution into a lustrous pool of enticing imagination. "To infinity and beyond!" becomes the social norm and political philosophy of crowded cities as much as careful discernment of and obedience to reality are practiced on the farm. This is why big cities full of well educated, brilliant people tend to be more foolish, while a countryside full of less educated, mundane commoners will most often be much wiser.

Wisdom is milled from attention paid to realities, not from attention paid to popular personalities. Only the humble, intellectual responsibility of researching ideas to their fundamental, root concepts and evaluating rational processes to the ends of their effects will develop wisdom. It doesn't come by chanting "Jesus, Jesus, Jesus," "Om...Om...Om," or any other group-think slogan. It doesn't come from daydreaming, nightmaring, or dishonest (biased) searching. It comes from fearing that Mack-load-of-reality enough to commit to tediously discovering, verifying, refining, and practicing what shows itself true by its consistency with what observably exists. Rejecting commitment to the independent discovery and practice of truth increases the risk of becoming entrapped inside your own prison walls of deceit, soon to be demolished by Jesus' Mack truck-full of return this way coming with throttle wide open.

Most of America's religious, cultural, and political debates, happening in the "bright" minds and night life of the big cities, have abandoned the intellectually responsible processes of research, verification, and deliberation. America's education, entertainment, and news institutions have abandoned the fear of God. They might praise the Bible with their lips, but their decisions and actions are far away from the heart of the Bible. Consequently, they have lost the way to truth for not having practiced it. Indeed, "truth" in today's public discourse is as much a rejected concept as Jesus' return is a dismissed one. From the late 1960s to this day, a deliberately liberal distortion of education, religion, culture, and politics has piled America's cultural bricks into a national prison wall of nightmares, each bad dream fanned into a convenient crisis inciting another call for authoritarian rules.

Measurements

People seeking truth and wisdom together, or at least seeking empirically discernable law and order amongst them, are the exceptions to a world full of ruling authoritarians. The world grew up with kings and kingdoms. From ancient times the king was held responsible for the orderliness of cultural affairs and for the goodness or doom of natural happenstances. The spread of Christianity throughout the Western World did not change authoritarianism. At least, not immediately. The philosophical resources of authoritarianism only switched from pagan misperceptions to Christian misperceptions.

As the deceptions of human kings failed to stack into fortress walls over the centuries, impoverishment and despair imprisoned their people. So, the people turned to a newly rising political star -democracy. They thought it would end the murder, theft, and deceit left in the wakes of governmental malfeasance. But government by democracy continued where those old monarchies left off. Except the general public now gets to elect its most popular criminals to be their dictators.

Eighteenth century philosophers were so ambitious to destroy the foundations of monarchy that they felt a dire need to jettison that "dastardly" Bible from Western culture because it favorably refers to kingship as being a Divine institution, especially Jesus' Kingship. Haste to socially distance their philosophy from God's Word led a biased (dishonest) search into Earth's dirt to find a scientific replacement for God's Creativity. They struck a vein of useful deceit in evolution, milling it into socially fashionable bricks, pragmatically piling them into walls of separation between the hope of eternal life and a newborn despair growing throughout Western culture. That culture went sobbing into a nightmare of "science" having "proven" there's nothing beyond death, whining for the government to create utopia. Man's endeavor to create utopia necessitates a unified population pulling only in one direction. At least, that's what everyone imagines. The idea led to democracy electing authoritarians. Those authoritarians then herded populations into the same impoverished, demoralized, corrals of despair folks had mourned under the old monarchies. Indeed, reality is consistent.

Progressivism, socialism, communism, and Nazism are all political beasts born and fed in the same philosophical pasture on the left side of the authoritarian/laissez-faire fence - collectivism, the obedient servitude of every individual to leaders empowered by the force of masses pulling in one common direction towards one idea, coercing people to think within one paradigm, essentially, a neighborhood run by bullies. Every democracy eventually crosses the fence from laissez-faire liberty into the authoritarian pasture, because daydreams feel good while verification of concepts feels too hard to do. And behold! They are always the minimally thinking people who are first lured by authoritarians into unverifiable, sweet tasting daydreams. Once a population has voted to feed on the mildewed grass of the collectivist pasture, democracy will never cross back into liberty's meadows, since authoritarians rule the left pasture by violence while only feigning elections. Study the world's various Progressivisms, socialisms, communisms, and other authoritarian governments. They all exist by first deceiving the masses at the ballot box, and then entrapping them with censorship, more deceit, and overbearing laws, until the people's hearts, minds, and souls are given to the leader, whether sincerely or not.

But the founding of America's democratic-republic was an exception. While "de-

Clear Signs of Trouble and Great Joy

mocratic" referred to how its representatives would be selected, "republic" referred to its Constitutional agreement to feed on the laissez-faire side of the fence where the right pasture grows much greener grass (check out the unadulterated history of politics; capitalism has always supplied the world, while collectivism always consumes like mildew). America shaped and laid its first conceptual brick in accord with the Biblical reality that all men are faulty, and only God is true. Therefore, it could not place big political power into only a few human hands. Power needs to be spread thin between multitudes of hands, frustrating t authoritarian plots. America's democratic-republic formation constructed a wall of lawfulness around its own powers so deceitful men could not fundamentally transform governmenttal power into authoritarian lawlessness. The citizens' rights were spelled out and protected from all democratic actions that might infringe upon the individual citizen's sovereign possession of his own self, for example, those three wolves and a sheep voting on what's for dinner. American exceptionalism knows that sovereignty was given to man by God in the Biblical concept of free will, and that taking their sovereignty away is an act of war against The Mamma Bear of very precious cubs. The American Constitution was granted authority by God through the prayers and personal sacrifices of His more faithful cubs.

In a faulty population, American exceptionalism develops the goodness of each citizen into each restraining his or her own lawlessness, each striving for his own moral, social, and situational improvement in accord with proper consideration given to his and her neighbors, including all prenatal neighbors.

"America" wasn't just a place on the map. Its land was old, and had been on the map since the creation of Earth; still, its weeds, trees, and rocks were as ordinary as were the rest of the world's. America was not just whoever lived on that unexceptional land. For many centuries families of people lived there in the discord of cultural differences, skirmishing with one another since shortly after God terminated the Tower of Babel (the Democrat slave-masters of the South were so unAmerican they chose war rather than acknowledging the humanity of black folks.) "America" was a particular way of achieving law, order, and social unity amongst a population of individuals improving their own moral and physical situations, minding their own businesses, venturing towards justice for all, no adjectives necessary, no authoritarians allowed, as prescribed in a legal document written to extremely limit governmental powers from jamming misshapen bricks into prison walls.

Historically, America did not feed on the authoritarian, left side of the fence. But it increasingly began poking its head through the fence for an occasional bite of collectively mildewed grass (or for a little toke of it). By observing both true history and indicative signs in the sun, moon, and stars, God's commission for America to restrain the mystery of lawlessness grown of the leftist meadows is clearly discernable. America's role in Biblical prophecy was the maintenance of a well squared and stacked fence protecting lawful governance on its right side from the lawless authoritarians of the left side.

> [2] A wise man's heart inclines him toward the right,
> but a fool's heart toward the left.
>   Ecclesiastes 10:2

Measurements

For nearly two-hundred years after America's founding, Christian ideas in American culture restrained the mystery of evolution's philosophical lawlessness, while truth, justice, and the American way restrained the mystery of political lawlessness. This condition was particularly noticeable near the middle of the twentieth century when America set its own leftward drift aside for a few decades of war and political struggle against collectivist regimes plotting authoritarian thefts of the world (World War II and the Cold War).

America's transformation into what God commissioned it to restrain did not occur all at once, one November 4th, through the wee hours of that morning. Indeed, American history began with slavery's lawlessness flowing through America's first few centuries. Then America's ways ebbed to the right, freeing those slaves into citizenship. But the emerging popularity of Karl Marx' communism enticed a new leftward flow before World War II ebbed America back onto the Bill of Rights and the God who influenced its laws. Then America flowed through a youth rebellion leftward into Carter's Administration, ebbed right to Reagan's, strongly flowed left through the Clinton years, slowing a bit during Bush's term, before it cascaded into Obama's collective chaos under that perspicuous holyday tetrad of 2014-5. America teetered there, its end of restraint portended, riding the authoritarian/laissez-faire fence like a has-been rodeo star, leaning ominously to the left.

"The mystery of lawlessness" doesn't refer to the lack of law. That is anarchy. Hitler, Stalin, Mussolini, Mao Tse Tung, Barack Obama, and other authoritarians make no shortage of laws. But theirs are laws authorizing control of the citizens, rather than laws establishing and safeguarding the independence and security of the individual. Authoritarian laws differentiate the left from the right. The ulterior purpose of the left is to entrap the masses into servitude. God's laws are for provisioning the people with ability, responsibility, morality, and ethical standards, enabling the freedom of each individual to possess his own self, to make his own decisions, and to practice consideration and self-control from his own heart.

So, in addition to the evasion of good laws, "lawlessness" also, and maybe even more, refers to the failure of man's laws to reflect God's laws, especially their failure to reflect His laws of truth discovery (The First Amendment guarantees the right to lie, the great American Achilles' heel). "Mystery" refers to how a little lying grows into "lost in the woods", and to the way tolerated inconsistencies develop into massive chaos, like disgruntled emotions drive wayward thoughts into rebellion against valid observations.

Lawful Americans have been horrified by America's fifty year, Forward march into 2020's lawless rebellion, which, to no surprise, was blamed upon lawful people by political authoritarians and news fakers of leftist inclined TV (Ecclesiastes 10:2). Mob insanity put a rifle to America's head that year, demanding its own way, while calling its demand "love" and naming the resultant mayhem "The Summer of Love" ("Love does not insist on its own way, nor is it irritable, rude, or arrogant." I Corinthians 13:5). Police were murdered in that "Summer of Love", murders inspired by chants in the streets from Obama-blessed mobs, chants which Trump abhorred.

What do we want? Dead cops. When do we want it? Now!
Manhattan mob, 2014

Clear Signs of Trouble and Great Joy

> Pigs in a blanket, fry 'em like bacon!
>
> St. Paul mob, 2015

In the same two years that God hung the 1967-8 holyday tetrad in the world's sky to celebrate Israel's recapture of Old Jerusalem, America's news and entertainment media went sneaking across the political fence into that pragmatic pasture of collectively mildewed grass. There it began constructing an authoritarian paradigm for suckering American citizens leftward through the same gate it used. Academia philosophically deconstructed, reshaped, and pragmatically piled America's laissez-faire bricks of mutual respect for one another into little balls of politico/economic deceit, heaping them into prison walls confining justice from reaching all. The American population, grown whimsical in its burgeoning prosperity, had lost its patience for research, discovery, debate, and deliberation by the time Joe Biden assured them he had the greatest voter fraud organization in American politics. Through those previous five decades, from the ignored holyday-tetrad sign of 1967-8 to Joe Biden's ignored plot, the news and entertainment mill presented far less observation and rationale for public reflection than it interjected misperception and sophistry, patiently

Pigs in a blanket of deceit frying their minds like bacon
Figure 47

Measurements

spiking America's social paradigms with dopey, authoritarian attitudes. From pieing the purpose of the Viet-Nam War to exaggerating Nixon's Watergate, from embattling Reagan over everything to covering up all of Clinton's malfeasance and crimes, from the constant belittling of Bush to the constant construction of that timeless, Obama icon, the talking heads of LeftistTV "news", the nasty mouths of Hollywood stars and entertainers, the arrogant antics of pro-athletes, and the herd-mentality they all cultivated into the population at large danced in unison upon America's main-street stage, turning the people towards a leftward chase of the impossible, utopian daydream. America was almost awakened from this wet dream by the "troublesome" new dawn of American greatness rising again as Donald J. Trump took a solid stand upon attention paid to and respect maintained for the truth.

None of wisdom's processes work at the public level anymore. Even though a reasonable majority of Americans process wisdom at the individual level, the clamor of America's leftist bandwagon drowns the good sense made by well reasoned voices and disturbs the peace of mind necessary for careful, unbiased reflection upon reality. The general public hasn't had a proper dance with sound reason and common sense since the early 1960s.

Like a room full of lying brats, every LeftistTV talking-head, every potty-mouthed media star, and the rest of the influential liberals reciprocally support each others' deceptions, erecting a nearly impenetrable wall of bullyhood against all testimonies from the more honest children in the classroom. Destroying public support for the truth is a key strategy of authoritarians when wresting control of free people. Too many well intending individuals lack courage and audacity for standing solidly against the memes, messages, and demands made by America's Progressive misperception mills of twisted truths and cooked-up lies. These leftward drifting culture mills project nightmares onto tiny molehills continuously, fanning them into major, public crises. They pig up all public discourse and take over the public stages, leaving no place for

Figure 48

truth to be told by the more silent majority of wiser people. Looped over and over again, sheltered from any of truth's challenges by a cascade of constant hysteria, these fanned-up crises ignite frets over exaggerated possibilities of disaster, panicking mindless masses of people to lunge for whatever solution Progressive authoritarians next conspire for entrapping an unwary, misinformed, get-along-go-along populace into corrals for oxen-people.. America's public mindset is aflame with leftist delusions burning up freedom's greener meadows.

Clear Signs of Trouble and Great Joy

> Some things are believed because they are demonstrably true, but many other things are believed simply because they have been asserted repeatedly.
>
> Thomas Sowell

Dr. Molnar demonstrated what to do with *information* when it falsifies selfish theories, plans, desires, and cultural memes. He wasn't piggy about his own theory. He listened to reason from the other side, honestly deliberating the evidences they presented. He humbly accepted relevant evidence, adjusting his desired theory to more closely match reality. Like a truth teller in a class-room full of deceptive brats, he wasn't afraid to lose a theory that would have made him seem great. He didn't haul controverting *information* off to normal science's usual dumpster just because it threatened the sense of his favorite idea or political ambition.

Engaging the reality of the Bible is difficult for liberal America. Biblical wisdom flies into the face of America's burgeoning, Progressive culture from its every page. For man does not make utopia, the Bible intimates; God does. Man is not in control of reality; God is. Man is not to be feared; God is. Substantiating the Bible and its intimations requires honest research and thought, which most people choose not to do. Making American greatness requires observing and analyzing those American experiments which worked in the past, comparing them against today's failing ones, then choosing, without bias, to continue doing what once worked while abandoning our newfangled failures. Working the Bible's prescribed experiments into the American experiment requires following Biblical lab instructions, just like science's experiments require following scientific lab instructions.

But the 1962 Supreme Court blocked any such possibility. Thereafter, America has not responded well to the flow of real *information* falsifying liberal theories, plans, desires, and deceptions. It has not responsibly corrected the shaping and laying of its social bricks in accord with reality's testimonies. Therefore, its social paradigms have filled up with deceptive daydreams. The more we allow deception to accumulate, the uglier the truth seems when told. People don't like ugly. So they've become imprisoned by deceit.

America's early reading textbook, The New England Primer, was filled with Scriptures and Biblical memes and messages. Even the liberally jaded Wikipedia admits,

> Many of its selections were drawn from the King James Bible and others were original. It embodied the dominant Puritan attitude and worldview of the day. [Definitely not Deist, my I add.] Among the topics discussed are respect to parental figures, sin, and salvation. Some versions contained the Westminster Shorter Catechism [instructions on becoming Christian;] others contained John Cotton's shorter catechism.

But, without regard for the fact that the Primer had been resoundingly welcomed into America's education system throughout most all of colonial history and warmly welcomed while the Constitution and Bill of Rights were being written and ratified, without regard for the fact that the Primer continued to be America's predominant reading textbook over the next century-and-a-half when the Constitution and Bill of Rights were being obeyed, the liberal,

Measurements

1962 Supreme Court ruled that prayer and Biblical concepts in education were unconstitutional! The gullible, minimally-thinking public grunted Forward to the educational trough now full of evolution swill, being fully assured that none of their public institutions should reflect God's Word, let alone participate in America's great experiments. The 1962 ruling essentially contradicted the very Constitution those Supreme Court justices swore to uphold. No justices were impeached for breaching their oath. Consequently, those breached oaths piled up worse lawlessness...

Although your prenatal neighbors are dehumanizingly called "blastocysts" and "fetuses", they are people, by the undeniable truths of all observations. These prenatal people meet every scientific description of "life" while in their mothers' wombs, and moreover, they are precisely defined as human by their own, unique DNA. Yet the 1973 Supreme Court ruled that the right to pursue the same life, liberty, and happiness enjoyed by our postnatal neighbors was not to be extended to our prenatal boys and girls. That ruling acutely contradicts the "for all" part of justice. Consequently, human life can now be "legally" sucked out of the womb, piece by torn apart piece, through surgical tubes even while life outside the womb is protected (for the time being). Their ruling turned mommy's life-giving womb into a morbid place of brutally murderous discrimination.

Those two bricks, badly misshaped by the Supreme Court, were deliberately piled on their ends in the American wall of justice, creating judicial chaos ever since (the latter ruling occurred eleven years after the first, the Biblical number of chaos). Those two bricks made the "quarter-inch off at the foundation" which has ultimately piled up to the removal of Lady Justice's blindfold at the top, gagging every critical truth with it since.

Justice is at the very heart of God's laws. Not the "this" justice, or the "that" justice, or the "whatever other" justice of America's liberal illusionists, adjectively aiming justice towards the service of only authoritarian ambitions. America's traditional law of freedom was simply justice FOR ALL, without adjectival limitation, because it was meant for protecting every individual citizen's sovereignty to own his or her self, postnatal or prenatal. Adjectival justice is for sheltering a few, while turning the rest into harnessed oxen-people.

God's law expects protection of the vulnerable and helpless, even by man's faulty governance. It expects the education of children in God's Word, whether at school or at home. God expects those who rule to rule with eyes seeing and ears hearing the evidences of reality (every evidence is a testimony to God's truth) the way Dr. Molnar did it: humbly.

A wise man considers his neighbors' autonomy. But a fool harnesses those neighbors to plow his own fields. On the left side of the political fence, human governance is placed above even The Almighty God, and it is hoisted far above muzzled oxen-people. Liberal governance hammers Utopia into a graven image for public worship, while vainly calling itself "Christian". It forgets the Sabbaths and other special days meant for honoring God - Sunday, Thanksgiving, Christmas, Easter, and the 4th of July. It dishonors father and mother, despises their traditions, and throws parental knowledge into science's dumpster in order to raze their children with Aderall, Ritalin, and history revision, dismissing all of Mom and Dad's boorish, old ways. Liberals kill prenatal boys and girls by the thousands, daily, and they assassinate the character of anyone coming to the aid of their prenatal victims. They laud adultery, applaud fornication (for making more prenates to kill), praise homosexuality,

Clear Signs of Trouble and Great Joy

and glorify transexuality, encouraging the denial of physical gender. They legalize all theft under safe-harbor amounts, while their governments freely steal the wealth of one individual's productivity to buy votes from another's envy. Liberals give false testimony throughout all branches of government, media, and academic institutions. Above all, they covet control over your soul. They covet you being harnessed to their schemes, with blinders over your eyes, plowing their fields of dreams like oxen-people muzzled in "oat-bags".

"Hey! Who was that masked oxen-person!?"

Not to worry! Nobody can save old America now.

For centuries the Ten Commandments served a rational and sane construction of American law against the indignities of liberal lawlessness. Liberalism accepts no such guidelines or guardrails around its "laws". They have no consideration for their fellow man. Liberals marched lawfulness and traditional wisdom out of our culture one crisis at a time, each crisis working towards the end of individual research, analysis, comparison, and careful deliberation in order to stampede a minimally thinking population deeper into herd-mentality, here a crisis, there a crisis, everywhere a crisis, one crisis after another, each and every crisis demanding the immediate action of passing another overbearing law or declaring another fascistic mandate, no deliberation allowed, no dissent tolerated.

> Crises is routinely identified as a core mechanism of fascism because it short-circuits debate and democratic deliberation. Hence all fascistic movements commit considerable energy to prolonging a heightened state of emergency.[36]

The French Revolution escalated into a blood-bath over the crisis of starvation. Hitler pulled the Holocaust out of a hat full of panic over hyperinflation. Stalin starved the Ukrainians; Chairman Mao murdered China's landlords; etc., etc., etc., until finally, Biden cheated America's vote, all by panicking the undereducated, highly indoctrinated masses over natural problems magnified into useful crises. Even ancient history acknowledges the great utility of a crisis for driving mindless masses to serving authoritarian goals. Nimrod herded the people of his day into building the Tower of Babel for alleviating panic over the possibility of another flood. Joseph bought the Egyptian people in a slave market he created by taxing their excess grain in good times to sell back to them during a crisis. Two-thousand years ago, Josephus testified about this human penchant for cultivating and harvesting herd-panic. Let's hear him again:

> 2...Now a man that is in adversity does easily comply with such promises; for when such a seducer makes him believe that he shall be delivered from those miseries which oppress him, then it is that the patient is full of hopes of such deliverance.
>
> 3. Thus were the miserable people persuaded by these deceivers...while they did not attend, nor give credit, to the signs that were so evident, and did so plainly foretell their future desolation; but like men infatuated, without either eyes to see or minds to consider, did not regard the denunciations that God made to them.[37]

Measurements

The political cold America caught in the days Israel recaptured Jerusalem developed into a bad case of Obama's Marxist pneumonia, just the co-morbidity Covid-19's crisis needed to squash truth and justice out of the American Way. Nothing of our deceptive bus-ride illustrates the Progressive ambition to seize authoritarian control over America (and the world) more than this handy-dandy, Covid crisis. Nothing distorted public perception more destructively than the subtle twisting liberals made of Covid truths. Oxygen-restricting facemasks still blue American lips with medical lies and numeric fibs. Five-hundred-thousand "Covid" deaths, having been denied the life-saving treatment of Hydroxychloroquine in order to prolong the heightened sense of Covid's emergency, were incessantly bugled across the nation to panic three-hundred-forty-nine-million Covid survivors into laying aside their Constitutional sovereignty in order to plow fields of lies on the Obama/ Biden plantation.

Democrats craftily harnessed America's oxen-people to the Covid plow for tilling up as many states' election laws as possible in order to grow a bumper crop of fraudulent votes. Then, right on que, ten days before America's election (the Biblical number of perfect completeness), Joe Biden confirmed what rationally deliberating Americans already knew. He announced how completely plowed, planted, and ready for harvest was the crop of big-cheat for the 2020 election…

> Secondly, we're in a situation where we have put together, and you… d…guys di…did it for our admini…for President Obama's administration before this; we have put together, I think, the most extensive and inclusive voter fraud organization in…in the history of American politics.
>
> Joe Biden, October 24, 2020

Naturally, all good liberals spun Biden's admission as an innocent slip of the tongue. They were doubly cautious to neither use the enlightening adjective "Freudian", or to allow its use by anyone else, when addressing his admiss…uh…pardon me…slip of the tongue. When his massive field of Democrat voter fraud sprouted, late that election night, and matured unto great fruition throughout the following week, eyes to see clearly saw, while eyes to dream blissfully slept through the truth of enormous election fraud.

The same liberal eyes which, over the previous four years, neurotically investigated every nonevidenced accusation imaginable against Trump, insisting he worked some sort of 2016 vote fraud, refused to take nary a peek at extensively evidenced charges of massive 2020 election fraud, regardless of Biden's confession (sorry, I'm not deaf, blind, irrational, nor deceitful). Every reality of Biden's enormous fraud was carefully washed off all public lips by the consensus keepers of social media, like Bishop Bell's sauropods thrown into science's dumpsters, until America's public mentality was sufficiently sedated to swallow giant troughs full of "valid election" swill, no questions asked, no honest investigations made, no evidence attended, no Democrat debate, no humble deliberation, just irrational assurances liberally mixed in to cover up the fraudulent flavor of the swill they served to a minimally-thinking public. Donald Trump wasn't cheated out of his second term. America was cheated out of its last restraint of the mystery of this very brand of lawlessness.

Clear Signs of Trouble and Great Joy

> ^{2} O LORD, how long shall I cry for help,
> and thou wilt not hear?
> Or cry to thee "Violence!"
> and thou wilt not save?
> ^{3} Why dost thou make me see wrongs
> and look upon trouble?
> Destruction and violence are before me;
> strife and contention arise.
> ^{4} So the law is slacked
> and justice never goes forth.
> For the wicked surround the righteous,
> so justice goes forth perverted.
>
> Habakkuk 1:2-4

America's 2017 1 Elul call to repentance was not about voting for the Republican party. God does not truck with cavemen any more than He reclines with liars. He chooses courageous standers, e.g., Trumpsters telling truths and never caving to lies. America's 1 Elul call was for our return to a courageous stand upon reality's evidences against the deceptions of the panicking masses, even alone, if necessary. It was about recognizing the value of truth over the wealth of silver, gold, and popularity. It was about America reflecting the lawfulness of God's Word one more time, returning to intellectual responsibility.

Donald Trump and his seventy-five-million supporters live in their own world where realities are known by many reflections shining up from pools of their crystal clear evidences. Evidence is easily found, clearly seen, and plainly understood by anyone honestly desiring to know the truth. Ambition to verify only what supports a desire is arrogance. The desire to accept as truth only what accords with reality's evidences lifts the human equation into wisdom. The fear of God pays attention to reality rather than to desires.

The most rational hope for healing human misery and the bringing of eternally blissful life comes through the well evidenced truth of Jesus Christ, as presented by God's Holy Bible, the ultimate nemesis of authoritarian governance. For Jesus instructs His students to die to themselves first, draining from them all fear of anything except His Daddy. Fearlessness unleashes their minds to search for evidences of reality for deliberating into solutions. Deliberation of evidences renders frightful situations into realistic understandings. Those understandings burst the harnesses off oxen-people pulling Trojan horses built from crises' wooden nickels, each Trojan horse secreting hellacious authoritarians in its belly. Jesus raises His fearless believers to courageously stand upon truth and serve what is right, even unto death. Therefore, His followers shall reign with Him forever, which beats a few short years of authoritarian collectivism all to Hell. The Holy Spirit leads their lives through sufficient evidences of blissful promises kept by the God of faithfulness, as affirmed by the many testimonies of the late, great, 1948, and, as will be proven by the archeological discoveries discussed in Chapter 9.

We observed America's true President, Donald J. Trump, standing alone in front of the liberally burned out St. John's church, boldly holding up God's Holy Word as The Standard for American Propriety, the very point of America's 2017 Rosh Chodesh Elul

Measurements

warning.

> [23] The fear of the LORD leads to life;
> and he who has it rests satisfied;
> he will not be visited by harm.
>
> Proverbs 19:23

Trump boldly opposed America's culture of deceit, exposing every leftist accusation by standing firmly upon truth, sometimes alone (to all our shame). Never before has an American President survived such a manic swarm of hysterically charged, political "colonoscopies". But the liberals' nasty search for any Trump misbehavior at all served only to further evidence his great integrity. They found nothing, to their own, unnoticed shame.

His many accomplishments revived America's liberally trashed greatness, fundamentally transforming little Barry's "new normal" into the strongest wave of economic prosperity since the Reagan Revolution. Trump re-inspired America's hope in the God of its founding. He pulled great American truths out of normal science's dumpster while throwing government interference, red-tape, and over-regulation there instead. Trump's foundation upon reality restored a fresh, relieving, new hope for America's future, throwing water on all those liberal crises. And he made no little call upon the Lord of The Holy Bible while doing it.

The Ottoman-Turk Empire opposed God, and lost. Why did God kick America into abysmal lawlessness even while Trump, his seventy-five-million supporters, and their millions and millions of children were drawing back the victory arrow carefully aimed to pierce Progressive oppression and return truthfulness to American governance?

God's will, regarding the Ottoman-Turk Empire, was the freeing of The Promised Land so Israel could again become a nation. Whether the Empire survived or died afterwards depended upon its joining the Lord's will, or fighting it. It fought. It died. The perspicuity of the 2014-5 holyday tetrad portrayed the world's first step towards the Tribulation by removal of the one restraining the mystery of lawlessness, America. That restraint had to end before the man of perdition can be revealed in the Tribulation. But it did not need to end in 2020; clear signs portray the Tribulation's beginning to be 2029. America chose, by popular vote, to continue restraining the mystery of lawlessness with Donald Trump's mystery of lawfulness. But America's tolerance for deception ended the restraint of lawlessness in the third year after the 2017 Rosh Chodesh Elul warning, although not entirely ending America.

Two years after God allowed lawless hands to steal American governance, the perspicuous not-a-new-star "shines" forth from the constellation of Christ's ever present Truth, challenging authoritarian Democrats to govern with eyes to see, humbling themselves in the presence of evidences observable by anyone willing to look, deliberating and adjusting their governance to reality by following all observable evidence to wherever it leads, just like Dr. Molnar did, just like all honest people do. But will these authoritarians continue hiding the evidence of their past frauds, lies, and vote-thefts in order to gaslight the population, censoring truth out of the American experiment? America can yet repent of its lawlessness and lies

Clear Signs of Trouble and Great Joy

AN INTERESTING FIVE YEAR PATTERN

	Five years	
Three years		Two years
1357	**1360**	**1362**
Orhan's son, favored to be his successor, dies, sending Orhan into a deep grief.	AD70 plus the 1290 years to the setting up of the abomination that makes desolate prophesied to Daniel by the angel is AD1360.	Orhan dies of his grief, leaving the Ossman beylik in the control of Murad. Murad's mind is set on expanding the beylik into an empire and Caliphate, which he does, setting up the abomination making desolate, the seventh head of the dragon/beast.
1914	**1917**	**1919**
A total solar eclipse on 1 Elul, 1914 draws a path of totality across the Ottoman-Turk Empire, the seventh dragonhead, calling it to relinquish control of the Proised Land.	The Empire doesn't repent; 1914 begins WW I. By 1917 the Empire is so defeated that it must relinquish control.	Five years after its 1 Elul warning, the Empire effectively dies with the end of WW I.
2017	**2020**	**2022**
A total solar eclipse on 1 Elul, 2017 draws a path of totality across the United States of America, warning it to repent of deceit based lawlessness. Pres. Trump stood upon the truth to proclaim America's recogni- of Jerusalem to be Israel's capitol.	The party of lawlessness swamped America's elections by the most massive, concerted program of election fraud in American history. Unless the American people go to the Lord God in repentant obedience, America has now become the lawlessness God called it to hinder.	KIC9832227 of Cygnus, the constellation of ever-present Truth, demonstrates how truth is discovered only by following ALL *information* to wherever it leads. Will the liberals now steering the American bus-ride turn it back to the avenues of truth?

Figure 49

Measurements

to enjoy a final few years of greatness before the Tribulation's hell breaks out, if those who stole America's reins will join that repentance. But this choice is for those liberals to make. Truth lovers have already made their choice to repent and go speaking truth into America's streets.

The search for truth is often no more than a simple use of your memory. Deliberating truth often happens by merely comparing memories of past news to more current news. Obama used a health care plan structured for failure to severely crash our economy, pronouncing economic despair to be a "new normal", while he accused and impugned America for every crisis around the world. Trump swiftly recovered our economy and put America first in foreign affairs, holding high America's contributions to world stability, and working as much peace into Middle-East affairs as Obama had released chaos there. Obama choked off America's energy resources, requiring us to buy ever more energy from our own enemies. Trump tapped into those resources, turning America into a net energy exporter for the first time since the 1960s, freeing us from dependency upon our own enemies. Yet, nearly all mainstream news about Obama was positive, glowing, and glorifying, while nearly all mainstream news about Trump was negative, glowering, and impugning. Unless failures, crashed economies, and economic despair are good things, if last place is America's place, if mayhem in the Middle-East is peace, if energy shortage is productive and enriching our enemies for our life's necessities is success then the mainstream news got it right. A truly massive crowd of Trump supporters attended his inauguration. But Joe Biden's inauguration was a quiet day in Washington. Instead of a giant crowd celebrating their successful vote, or even a happy gaggle, Biden was sworn in before a shallow show of flags spread over a lifeless facility. That Trump was not in the Whitehouse became more important than whether or not the election was legitimate. All around D.C., rather than surging throngs of rejoicing citizens were many thousands of armed troops, as if the Covidly plowed field had become more important than all of those oxen-people who had mindlessly plowed it. The liberals chased Trump through four years of false accusations about the illegitimacy of Trump's Presidency. But the same liberals censor every evidence of Biden's massive election fraud and assassinate the character of anyone who dares even question the legitimacy of Biden's "presidency", or who dares to question the liberal denial of that greatest voter fraud organization in the history of American politics.

Reality is consistent. Deceit isn't. Consistency leads to truth. Hypocrisy extends from deceit. Memory and comparison form eyes to see the difference. Censorship and intimidation form blinders for turning people into useful oxen. What is true? Who's to say? The answer to those questions is discovered by listening to and considering both sides of any conflict. One side consistently presents evidence and good reasoning laced into a coherent network of further meaning. The other side engages itself mostly in theories, contemplations, and accusations, maybe even thrashing around a stick or too of evidence like a weapon, defending their position by the emotive powers of social media rather than by well reasoned evidence. Wisdom is known by her children, reality is known by its evidence. Listening to both sides of a controversy always clarifies which is true, and which is faking.

Cygnus' not-a-new-star calls us all to throw off our blinders and open our eyes to again see by evidences. It challenges all politicians, potty-mouthed entertainers, and fake-

Clear Signs of Trouble and Great Joy

news talking-heads to fly through 2021 into the prescient year of Cygnus' not-a-new-star while observing and deliberating the very evidence they've been hiding. If we don't respond correctly to this 2022 sign, we might end all shaken up at that sign of 2024's ominous "X".

Trump's stand upon the truth, regardless of consequences, was a prophetic message given us by The Lord of Spirits. In Trump's first year, the tenth and final Jubilee of Jerusalem's Jubilee path, in accord with God's Word, Trump acknowledged Jerusalem to be Israel's capitol. Then, at the beginning of the prophetically strange year of 2020, he honored America's Truth Detector, Rush Limbaugh, The Mayor of Realville, for the role he played in prolonging America's restraint of lawlessness unto that day, awarding Rush the Presidential Medal of Honor for thirty years of exposing deceit and broadcasting truth. When both men are raised into the Great Hall of Eternal Glory there'll be an enormous welcome raised by angels and eternal people, alike, for their modeling of honest stands upon truth, even unto the death of their public images, a Jesus sort of thing to do.

Trump's "prophecy", the 2017 Rosh Chodesh Elul warning, and the 2022 not-a-new-star direct a call for the current American governance, and all Americans, to FOLLOW ALL EVIDENCE TO WHEREVER IT LEADS, no dumpsters employed, no censorship allowed, no gaslighting, pieing, or lying to elevate human authority above what's true. Trump's sovereign stand upon the truth as known by its evidences demonstrated his love for every citizen and his desire for their right to possess their own thoughts and direct their own lives, owning their own souls as given to them by I AM WHO I AM.

Trump's for all Americans is sincere. God loves man truly. Love is a commitment to working actual benefits for the ones loved. Thus, before 2016, Donald J. Trump noticed America's continuing decline into Progressive chaos and the suffering it was causing good Americans. God noted in the Bible mankind's abysmal decline towards eternal despair and the suffering it was causing all peoples. Donald Trump descended from his tower into the Presidency to make America great again for his fellow Americans. Jesus descended from Heaven into the form of a man to make eternal joy available again for the truly meek. Trump was a billionaire; he needed no personal gain from the Presidency. Jesus was God; He needed no personal gain from sharing in mankind's sufferings. Yet, throughout Trump's mission in office, the rulers of Progressive culture deceived the mobs, falsely accused him, and sullied his message. Throughout Jesus' mission on Earth the Jewish rulers deceived the mobs, falsely accused Him, and tried to sully His message. At the end of his mission, the enemies of truth crucified Trump's character, stole his office, and tried to scatter his followers, just like the enemies of Truth crucified Jesus, scattered His followers, and tried to steal His office. But in the end, truth will overtake its enemies.

Donald J. Trump's Presidency was an enormously spiritual part of the clear signs warning the world to prepare for Jesus' return. King David was a metaphor of King Jesus to a nation God was constructing for greatness. President Trump was a metaphor of crucified Jesus to a once great nation now deconstructed by Progressives. As such, one particular Psalm of David's metaphor lends some essence to Trump's metaphor.

[1] Save me, O God, by thy name,
and vindicate me by thy might.

Measurements

² Hear my prayer, O God;
give ear to the words of my mouth.
³ For insolent men have risen against me,
ruthless men seek my life;
they do not set God before them.
Selah
⁴ Behold, God is my helper;
the Lord is the upholder of my life.
⁵ He will requite my enemies with evil;
in thy faithfulness put an end to them.

Psalm 54:1-5

The new dawn of Trump will not return; it was merely a prophetic metaphor. But Jesus will return, bringing destruction to all enemies of the truth, even to those who call themselves by His name while working against His nature, bullying people into thinking like they do, and crucifying the characters of people who think carefully upon the truth. Of Truth's return Cygnus' not-a-new-star warns from the great Shemitah year of America's ominous "X", the final Shemitah year before the Tribulation-portending Apophis arrives in 2029.

God marked Trump's term in office with five signs of His doing it (five being the number of His grace). Three years (the number of God's doing) after the 2014-2015 holyday-tetrad sign began, Donald Trump was inaugurated. Three years after the end of that holyday-tetrad began three total lunar eclipses in a row, extending through the middle of Trump's term. Trump was given three Supreme Court seats to fill with Constitutional originalists (who rarely continue standing against the unConstitutional head-winds of liberal deceit, witness the once conservative, now lying, pieing Chief Justice John Roberts[38]). And Trump's Presidency was ended by the third of three Democrat coups: 1) a Russia Hoax, 2) the Impeachment Pie, and 3) The Greatest Voter Fraud Organization in the History of American Politics. (The 2021 impeachment was merely an arrogant slamming of the door on his backside as the vote-defrauders threw Trump out of America's classroom.)

If there is anything Biden, Pelosi, every Democrat, every Never-Trumper, and all the pieing Rinos and caveman Republicans should fear, it isn't the kicking toes and stomping soles of jack-booted, global fascists. It isn't America's hot-nor-cold, feckless preachers. They should all fear The One who hung in our skies these signs of His graciously working Trump's Administration. That One is The Spirit and The Authority, coming soon to a Jerusalem near you! Lawlessness stole the Presidency, an act of war against God, an act worthy of great Tribulation. If repentance should run through America's cheating heart, America might enjoy a brief respite from trouble before falling into the Tribulation's hell.

God even measured Trump's Administration for us to understand its spiritual significance. Trump was President number 45, the number of the summation of man's affairs (9) times the number of God's grace (those five threes showing God's doing it graciously). God could not have been more graciously honoring of America's most transparent and truthful President, whom He scripted to be run over by lawlessness. 2018, the

Clear Signs of Trouble and Great Joy

beginning of those three tell-tale eclipses, followed Jerusalem's AD70 destruction by 1,948 years, the number testimonial of God's faithfulness to His prophecies and promises, in this case, the promise to remove the one restraining the mystery of lawlessness so evil can be led to its slaughter. And let's be sure to note that the year lawlessness seized control of American governance aboard Joe Biden's voter fraud organization was precisely four-hundred years, to the month, after the Mayflower Compact was signed aboard the Mayflower. Those eight Jubilees of lawful governance matched, to a year, Jerusalem's eight Jubilees under the control of the Ottoman-Turk Empire (1517-1917 and 1620-2020). Eight is the number of new beginning, Jerusalem's new beginning for receiving its King, and America's new beginning of lawless tyranny. Moreover, the end of America's lawfulness occurred one-hundred-three years after Jerusalem was released from Ottoman-Turk control, the same one-hundred-three years by which America's 2017 Rosh Chodesh Elul warning followed the Ottoman-Turk Empire's 1914 Rosh Chodesh Elul warning. The Mayflower Compact was signed one-hundred-three years after the Ottoman-Turk Empire won control of The Promised Land (1517-1620). One-hundred-three is two Jubilees (the number of deliverance[39] doubled for emphasis) plus the number of God's doing it, three. And as an interesting aside, America's lawfulness was overcome by lawlessness in the eleventh month of 2020, the Biblical number of chaos (and coincidentally, the same month the Mayflower Compact was signed). Even Covid-19 measures spiritual significance into the prophetic year of the fraudulent end put to America's new dawn, 2020. 2020 followed the year of the 9-11, 2001 warning (as discussed by the prophetic Jonathan Cahn) by nineteen years, the number of the complete summation of human affairs. And, as I am sure you have had eyes to see, nineteen is Covid's number.

If I were these pieing, lying, lawless usurpers populating almost the entirety of American politics and culture, censoring evidence and blacklisting truth-tellers, I would die of fright in my boots, to Hell with just trembling!!! The Lord God's fingerprints were all over Trump's Administration as much as they were all over America's history, the same as His footprints will soon be all over those who destroyed America's great, new dawn. Jesus is coming to tread the grapes of wrath in a very bloody winepress. Then, He will make the entire world great again, not just America. And the great Donald J. Trump will be there to participate, having stood for truth in the midst of deceit.

> [2] The kings of the earth set themselves,
> and the rulers take counsel together,
> against the LORD and his anointed, saying,
> [3] "Let us burst their bonds asunder,
> and cast their cords from us."
> [4] He who sits in the heavens laughs;
> the LORD has them in derision.
> [5] Then he will speak to them in his wrath,
> and terrify them in his fury, saying,
> [6] "I have set my king
> on Zion, my holy hill."
>
> Psalm 2:2-6

Measurements

> [11] ...the king shall rejoice in God;
> all who swear by him shall glory;
> for the mouths of liars will be stopped.
>
> Psalm 63:11

Please allow me to write this paragraph specifically to every purveyor of America's culture of deceit (if any are still reading, if any even began to read). There is no excuse for having eyes that do not see. The choice to see, or not to see, has always been yours. Make it to see! Humble yourselves quickly! These signs speak with 20/20 clarity. The sight of them requires no corrective lenses. They are unmistakable! Who cares about controlling this silly country you've stolen!? Save your souls from eternal despair! Humbly correct your wrongs; repent of your lies and frauds; admit the truths you've pied; and follow Jesus onto the Argo for docking safely in the blissful harbor across the coming Tribulation. Next year, guys, comes the final Shemitah warning before that Tribulation, Dr. Molnar's not-a-new-star of the constellation of the Ever Present Truth, warning everyone, especially you, to follow all evidence into the truth, no dumpsters employed, no populism regarded, no censorship, no blacklisting, no bullying. God has now placed every branch of American governance into your hands to observe whether YOU will govern by truth or by deceit. Beware! This same Witness of your every thought and governmental action is also the Highest Judge of all existence presiding over the very Highest Court to where you have been summoned. There He will decide between your eternal bliss or eternal doom based upon which direction you chose to lean: to the right into Jesus' reality, or to the left into insane foolishness (Ecclesiastes 10:2).

> [10] Now therefore, O kings, be wise;
> be warned, O rulers of the earth.
> [11] Serve the LORD with fear,
> with trembling [12] kiss his feet,
> lest he be angry, and you perish in the way;
> for his wrath is quickly kindled.
> Blessed are all who take refuge in him.
>
> Psalm 2:10-12

To those who have been pierced to their heart's core by this lawless swallowing up of the exceptionally great America within the deep, Covid darkness of 2020, may I suggest for you a different Word from The Lord of Spirits?

> [1] Fret not yourself because of the wicked,
> be not envious of wrongdoers!
> [2] For they will soon fade like the grass,
> and wither like the green herb.
> [3] Trust in the LORD, and do good...
> so you will dwell in the land, and enjoy security...
> [6] He will bring forth your vindication as the light,

Clear Signs of Trouble and Great Joy

> and your right as the noonday…
> [28] For the LORD loves justice;
> he will not forsake his saints.
> The righteous shall be preserved for ever,
> but the children of the wicked shall be cut off.
>
> Psalm 37:1-3,6,27-28

As we move on through 2021 into the future, remember two things, you purveyors of deceit as well as those who've been struck in the heart, remember, both of you, that, 1) the righteous get their glory doom first, for what comes first must end so that what comes next will continue forever. But the wicked reap their doom glory first, for their doom will never end like their glory will. And, 2) the presiding Judge over The Highest Court of All is also the Eyewitness of every last deed to be tried in His court. His eye sees into and His ear hears the depths of every soul. His memory is not only photographic, it is phonographic. So be assured that it will testify in HD clarity and hi-fidelity about every last one of your hidden deeds, thoughts, and emotions, right down to cheating America out of the truth unto which Jesus had raised it.

Would the April 8, 2024 total solar eclipse' completion of that giant "X" crossing out America's map at the New Madrid Fault Zone speak any more clearly had God hung within its seven years, in America's high noon skies, a clearly portentous illustration from The Book of Revelation? Would that ominous "X" be less telling had any number other than seven towns named Salem ("peace" in Hebrew) experienced the totality of America's Rosh Chodesh Elul eclipse?[40] Or if the number of years between those two eclipses drawing this menacing "X" were not the same number of years comprising Revelation's Tribulation, 7? Would this curious "X" drawn across the American heartland seem less relevant if the KIC 9832227 not-a-new-star weren't attesting within its seven years to the importance of following evidence to wherever evidence leads? Would Dr. Molnar's not-a-new-star be as revealing had it been anticipated to occur in any constellation other than Cygnus, the Ever Present Truth, at any time other than five years (the number of grace) after America's Rosh Chodesh Elul call to repentance? Would the imminence of this "X's" warning have been as austere without KIC 9832227 "shining forth" from the "X's" only Shemitah year, in fact, from the last Shemitah year before 2029's Tribulation presaging arrival of Apophis?

Thirty-three days (the number of chaos times the number of God's doing) after the 2017 Rosh Chodesh Elul eclipse summoned America's lying ways before The Almighty Witness and Judge of Truth and lies, the Revelation 12 Portent shined across America's high-noon skies.

> [1]And a great portent appeared in heaven, a woman clothed with the sun, with the moon under her feet, and on her head a crown of twelve stars; [2]she was with child and she cried out in her pangs of birth, in anguish for delivery. [3]And another portent appeared in heaven; behold, a great red dragon, with seven heads and ten horns, and seven diadems upon his heads.

Measurements

> ⁴His tail swept down a third of the stars of heaven, and cast them to the earth. And the dragon stood before the woman who was about to bear a child, that he might devour her child when she brought it forth; ⁵she brought forth a male child, one who is to rule all the nations with a rod of iron, but her child was caught up to God and to his throne, ⁶and the woman fled into the wilderness, where she has a place prepared by God, in which to be nourished for one thousand two hundred and sixty days.
>
> ¹³And when the dragon saw that he had been thrown down to the earth, he pursued the woman who had borne the male child. ¹⁴But the woman was given the two wings of the great eagle that she might fly from the serpent into the wilderness, to the place where she is to be nourished for a time, and times, and half a time. ¹⁵The serpent poured water like a river out of his mouth after the woman, to sweep her away with the flood. ¹⁶But the earth came to the help of the woman, and the earth opened its mouth and swallowed the river which the dragon had poured from his mouth. ¹⁷Then the dragon was angry with the woman, and went off to make war on the rest of her offspring, on those who keep the commandments of God and bear testimony to Jesus.
>
> Revelation 12:1-6;13-17

On September 23, as high-noon rolled across America, the sun stood clothing Virgo in her shoulder. The moon was at Virgo's feet. Mercury, Mars, and Venus were in the constellation, Leo, joining Leo's nine prominent stars to crown Virgo's head with twelve stars. And behold! Jupiter, the king planet, had proceeded through her birth canal after having entered her womb nine months earlier. Check your astronomy soft-ware; it is there to see! This was clearly that blood chillingly eerie portent from The Holy Bible's Book of Revelation being empirically illustrated in the high-noon skies over an America summoned to an austere choice between returning to truth, or ending in deceit.

Revelation's portent of the woman clothed in the sun introduces Satan into the Tribulation prophecy. He flies at the woman in the form of a seven-headed dragon cast down from heaven to make war on God's elect. We've discussed the seven empires stretching through history, from ancient Egypt through the Ottoman-Turk Empire, the warring they made against God's elect, and the death of its seventh-head as struck by World War I. Those seven empires represent counter-authority to God, struggling to rule mankind and destroy I AM WHO I AM's chosen nation of people, for that nation possesses the throne of The King of Kings in Jerusalem. The seven-headed dragon, as a player in the time of Tribulation, will soon give power to the eighth-head in the last three-and-a-half years of the Tribulation, when the Antichrist proclaims himself to be God, portended by the second total lunar eclipse of the 2032-3 unique tetrad.

In the end, America will turn against Israel with the rest of the nations before Christ returns to rub evil and its doers out of our age. America's Rosh Chedosh Elul warning wasn't about whether or not America would arrive at Armageddon as another of Jerusalem's antagonists. It was about when and how America will turn onto the path leading there before arriving as the most fickle of them all.

Clear Signs of Trouble and Great Joy

> ²Lo, I am about to make Jerusalem a cup of reeling to all the peoples round about; it will be against Judah also in the siege against Jerusalem. ³On that day I will make Jerusalem a heavy stone for all the peoples; all who lift it shall grievously hurt themselves. And all the nations of the earth will come together against it.
>
> Jeremiah 12:2-3

> ¹³And I saw, issuing from the mouth of the dragon and from the mouth of the beast and from the mouth of the false prophet, three foul spirits like frogs; ¹⁴for they are demonic spirits, performing signs, who go abroad to the kings of the whole world, to assemble them for battle on the great day of God the Almighty. ¹⁵("Lo, I am coming like a thief! Blessed is he who is awake, keeping his garments that he may not go naked and be seen exposed!") ¹⁶And they assembled them at the place which is called in Hebrew Armageddon.
>
> Revelation 16:13-16

America's time of warning stretches seven years from the end of The Jubilee Path to the 2024 eclipse. Then, five years (the number of grace) will pass before Apophis sizzles by Earth on April 13, 2029 (OMG, a Friday!), marking the start of the Tribulation. After those seven years of hellacious Tribulation, on April 13, 2036 (Easter Sunday), Apophis returns, portraying Christ's second coming. These three spans of years (the number of God's doing it) connect the 2017 end of The Jubilee Path to the 2036 end of The Shemitah Path by the number of years representing mankind's affairs fully summed up: 7+5+7=19.

2021 should begin what could be called a human race for integrity. America restraining of the mystery of lawlessness amongst the nations has been removed. Time to the Tribulation is short. And there are billions of people who need to follow Christ into the knowledge of truth, quickly, else their nearing deaths will be eternally doomful. We who know Christ need to be persuading as many as we can to follow Him onto the Argo for their delivery across this coming sea of despair. Although the entire world will sink into the abyss of deceit, be assured that, individual by individual, the Lord of Spirits is at the scene of this world's shipwreck, pulling honest, truth loving souls from the chaotic sea onto the calm deck of the Argo.

Everyone who knows the Lord must now courageously speak truth everywhere, following the example of America's Great Don, holding truth high in the public square, like the Mayor of Realville, confidently announcing it into the hearing of all. America's silent majority must vociferously broadcast truth from American sidewalks into all its marketplaces, showing their liberal neighbors that simple formula for more accurately estimating what is actually real. Voices of reason around the world, paralleling this host of signs in sun, moon, stars, history, and Biblical prophecy, need to broadcast an austere warning similar to that given Jerusalem before its AD70 devastation. Pay attention to the signs of reality! Become intellectually responsible! Ignore the symbols of liberal daydreams, lest reality becomes your nightmare. Raise countless voices up from godly homes, everywhere singing a pleasing song of crystal clear truth to the quickly returning Lord God of Truth.

Measurements

Quickly study and learn the truths of history and God's Word. The pieing, lying left is out to gaslight the world, now that America, the last bastion of resistance, has fallen down on the job. Speak truths fluently, respectfully, everywhere to everyone. Only individuals with good sense are now left to restrain the mystery of lawlessness; that restraint has now become a grassroots project amongst individuals. The world's street folk are now commissioned to stand down trouble by speaking up truth in every conversation, else the majority's silence will become its swan song.

> [16]Then those who feared the LORD spoke with one another; the LORD heeded and heard them, and a book of remembrance was written before him of those who feared the LORD and thought on his name.
>
> Malachi 3:16

If the lovers of truth and fellow man will now courageously stand strong in a censored and gaslighted world, picking up the restraint of lawlessness where American governance dropped it, helping their good-hearted but deceived neighbors learn how to discover truth, maybe the ultimate removal of lawlessness' restraint will need to happen by God's snatching His truth lovers Home, the Rapture.

The sense of this great "X" across America doesn't necessarily need to entail the crosshairs of some disaster shaking up from the New Madrid Fault Zone. For the crowd followers, it might well portray that Fault Zone busting over America's stiff necked, brass-forehead. But for the followers of Christ, it appears strikingly similar to the cross/Star-of-David debris field photographed by Hubble fourteen years earlier (the number of spiritual

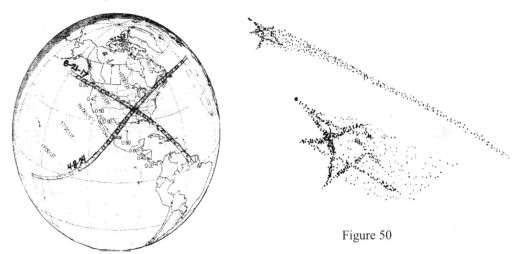

Figure 50

significance doubled). It could mean that America's restraint of lawlessness will not be completely finished until Christ's faithful truth tellers have been snatched from this world full of lying brats. Maybe the snatching away of His church (I Thessalonians 4:17) is portended by that great "X", should we who love truth decide to individually make our last stand. The crowd followers will stampede over the cliff into the abyss like a panicked herd,

Clear Signs of Trouble and Great Joy

not a one of them distinguishable from the rest. But each, individual follower of Jesus Christ will be personally distinguished by the highest of eternal honors...

> [4]they shall see his face, and his name shall be on their foreheads
> Revelation 22:a

That beats Andy's name on the sole of your boot any day!

After the Temple was measured for Ezekiel, the Jewish Temple was indeed rebuilt. Jerusalem was rebuilt following the measurement told Jeremiah, and its wall was expanded after the measurement told Zechariah. The rebuilding of the Jewish Temple today is in preparation, its measurement having been taken in Revelation. Biblically speaking, measurement means reality. We have been discussing very relevant, current history and Biblical prophecy measured in numbers of Biblical meaning, posted with signs in sun, moon, and stars. Reality is consistent.

So take note of patterns. Only hearts set on truth can see God's glory shining through the darkest chaos of gloom. Consult The Holy Bible. Its prophecies stitch empirically observable, astronomical signs into the fabric of history, giving us courage to stand firmly on the deck of the Argo like millions and millions and millions of Donald Trumps, each personally deliberating truths from God's Holy Words, then courageously speaking wisdom into the public ear. And do know that those truths of God's Word are being verified in the sky, even as you read this, by the many testimonies of the late, great 1948!

Footnotes

1. Johnson, Ken Th.D. Ancient Book of Jasher, Biblefacts Annotated Edition. 2008. Biblefacts Ministries, biblefacts.org. Pg. 30
2. https://www.timetoast.com/timelines/the-fall-of-the-roman-empire--3. Daniel Todd, founder and CEO. Copyright 2007-2016.
3. https://en.m.wikipedia.org/wiki/Byzantium
4. http://www.history.com/topics/ancient-history/byzantine-empire. Copyright 2016. A+E networks, Corp.
5. Other sources say the Empire seized control of Jerusalem in 1516. Both dates are technically correct. The Ottoman-Turks conquered and occupied Palestine swiftly in the last few months of 1516. But the Mamluk Caliphate's formal control over Palestine was not defeated until late January, 1517. When it collapsed, Ottoman control became undisputed.
6. https://en.wikipedia.org/wiki/Ottoman_Empire#Rise_.28c._1299.E2.80.931453.29. Wikimedia Foundation, Inc.
7. https://en.wikipedia.org/wiki/Orhan. Wikimedia Foundation, Inc.
8. https://en.wikipedia.org/wiki/Murad_I#Establishment_of_sultanate. Wikimedia Foundation, Inc.
9. Cahn, Jonathan. The Harbinger: The Ancient Mystery that Holds the Secret of America's Future. Frontline, Charisma Media/Charisma House Book Group, 600 Rinehart Rd, Lake Mary, Florida 32746. 2011. Pgs 153-77.
10. Brown, Simon. "Founding Fibs: The Religious Right Is Still Trying To Hijack George Washington's Legacy". May 9, 2014. https://www.au.org/blogs/wall-of-separation/founding-fibs-the-religious-right-is-still-trying-to-hijack-George-Washington's-Legacy.

Measurements

11. Holmes, David L. The Faiths of the Founding Fathers. 2006. Oxford University Press. Pg. 134.
12. Washington, George W. Washington's Inaugural Address of 1789. A Transcription. https://www.archives.gov/exhibits/american_originals/inaugtxt.html
13. The American Heritage Dictionary of the English Language. New College Edition. 1976. William Morris, Editor. Houghton Mifflin Company. At "deism". Pg. 348.
14. Biltz, Mark. When Will the Tribulation Begin? Where are we on the Biblical Calendar? (What Time is It?). Video, 1:05:11. 2021. El Shaddai Ministries.
15. https://en.wikipedia.org/wiki/Aliyah. Wikimedia Foundation, Inc.
16. Bullinger, E.W. Number in Scripture: Its Supernatural Design and Spiritual Significance. (Originally published by Eyrie & Spottiswoode (Bible Warehouse) Ltd., 1921). Republished 2014. Alacrity Press. www.alacritypress.com. Pg 185.
17. https://en.wikipedia.org/wiki/Amin_al-Husseini. Wikimedia Foundation, Inc.
18. Kristallnacht. http://www.history.com/topics/kristallnacht.
19. Kaiser, David. What Hitler and the Grand Mufti Really Said. http://time.com/4084301/hitler-grand-mufi-1941/. October 22, 2015.
20. Goldberg, Jonah. Liberal Fascism: The Secret History of the American Left from Mussolini to the Politics of Meaning. Doubleday. The Doubleday Broadway Publishing Group, a division of Random House, Inc., New York. www.doubleday.com. 2007. Pg. 9.
21. Ibid.
22. Beck, Glen. It IS About Islam: Exposing the Truth About ISIS, Al Qaeda, Iran, and the Caliphate. Threshold Editions/Mercury Radio Arts, Simon & Schuster, Inc. 1230 Avenue of the Americas, New York, New York 10020. 2015 by Mercury Radio Arts, Inc. Pgs 67-69.
23. Ibrahim, Raymond. Exposed: Obama Helped Decade-Old Plan to Create IS: The Resurrection of the Caliphate was Planned and Exposed to the World Nearly Ten Years Ago. November 6, 2014. http://www.frontpagemag.com/fpm/244702/exposed-obama-helped-decade-old-plan-create-raymond-ibrahim
David Horowitz Freedom Center.
24. Bullinger, EW. Pg 167.
25. https://www.newsinfo.inquirer.net/721613/mecca-crane-collapse-act-of-god-engineer. Agence France Presse. 12:35PM September 13, 2015.
26. Naeye, Robert. "A Stellar Explosion You Could See on Earth!" https://www.nasa.gov/mission_pages/swift/bursts/brightest_grb.html NASA's Goddard Space Flight Center. May 26, 2010.
27. Friedlander, Blaine P. Jr. "A Bright New Star Will Burst into the Sky in Five Years, Astronomers Predict". The Washington Post. https://washingtonpost.com/news/speaking-of-science/wp/2017/01/06/a-bright-new-star-will-burst-into-the-sky-in-five-years-astronomers-predict.
28. Parks, Jake. "Two Stars Won't Collide Into a Red Nova in 2022 After All". D-Brief; discovermagazine.com. 09-07-2018. http://blogs.discovermagazine.com/d-brief/2018/09/07/two-stars-won't-merge-explode-red-nova-2022/#.XJi7knbow.
29. Ibid.
30. Ibid.
31. Bullinger, E.W. Pg 189.
32. Biltz, Mark. Video at 0:29:00.
33. Crispe, Sara Esther. "The Secret of Elul". Chabad.org Magazine, a division of Chabad-Lubavitch Media Center. 1993-2017. http://www.chabad.org/theJewishWoman/article_cdo/aid/424-441/Jewish/The-Jewish-Heart.htm.
34. Google results show 1 Elul, 2017 to be August 23, 2017. The Great American Eclipse was August 21. A solar eclipse is caused by the moon passing precisely between the earth and the sun.

Clear Signs of Trouble and Great Joy

This happens only the day the waning moon becomes a waxing moon. The Bible prescribed each Jewish month to begin on the day of a new moon. Priests closely scrutinized the moon to determine the day of the new moon, for it was an empirically observed determination rather than a calculated one. A solar eclipse definitively identifies the new moon, which accurately identifies the first day of a month, in this case, Rosh Chodesh Elul, which then was definitely August 21, 2017.

35. Cancel Culture, Facebook, Twitter, crony-capitalism, potty-mouthed entertainers, deceptive news, etc. all aim at modifying the social paradigms of a population in order to conform them with attitudes of servitude towards authorities. The most visible manipulation of social paradigm is happening worldwide as the sight of masked faces everywhere inspires the ominous sensation of crises mixed into the relieving sensation of "obey the government directed solution". That mix is meant to adjust the old sensation of "you are smart enough to take care of yourself" into "you are a hater with no concern for anyone else". Liberals think they have the right to own every public paradigm. Therefore, they employ an organic messaging system to reformulate public paradigms, because fundamentally transforming America involves the fundamental transformation of your mind. Liberals are truly evil, selfish, thieving people, lusting to own your thoughts and usurp your will.

36. Goldberg, Jonah. Pg. 43.

37. Josephus, Flavius. The Wars of the Jews; or The History of the Destruction of Jerusalem, Book VI, Ch V, Secs 2-3. Complete Works. Translated by William Whiston, AM. Kregel Publications, Grand Rapids, Michigan 49501. 1960.

38. In 2010, Roberts based his decision to find the Affordable Care Act constitutional upon his pronouncement of its individual responsibility payment being a tax, rather than a penalty, just two days after his own Supreme Court ruled it to be a penalty, not a tax. The district court system, culminating at the Supreme Court, can not even hear a case on a tax until that particular tax has been levied. And at the time of Chief Injustice John Robert's lie, this "tax" had not been levied. He could not have even been ruling upon it if it were a tax. And that doubles his deceit. Ten years later, during President Trump's impeachment pie hearing before the Senate, Roberts, presiding over the hearing, refused to read Senator Rand Paul's question because it contained the name of the Democrat's Dixie Whistler. There is no legal barring of a whistleblower's identity from such a hearing. Robert's refusal was based upon his own personal ambition rather than Constitutional law, another lie! Then, when Texas brought before the Supreme Court a suit against four of Biden's vote-cheating states, Roberts lied about Texas not having legal standing in the theft of the US Presidency, by fraud, from every American citizen and all fifty states, Texas and its citizen's included. And that forms standing! Reality's own evidence reflects Roberts' face in the ocean of left-wing lies.

39. http://blog.chrisify.com/2017/08/the-seven-salems-of-eclipse-coincidence.html

40. Bullinger, E.W. Pg 192.

Chapter 8
The Late, Great 1948

And Joshua said to all the people, "Behold, this stone shall be a witness against us; for it has heard all the words of the LORD which he spoke to us; therefore it shall be a witness against you, lest you deal falsely with your God."

Joshua 24:27

The main focus of the Bible is Jesus Christ leading your following soul into eternal life, else your wandering soul casts off into in the chaotic abandon of outer darkness. All of The Bible's other themes build a bridgework of evidence and reasoning to carry that point across this world's chasm of deceit, ignorance, and unawareness dug by mankind's reckless communications and inattentiveness. If you aim your eyes and ears at finding the truth, you will find Jesus Christ, because, as the Truth, His aim is to find anyone desiring to know truthfully. We who have believed Christ Jesus know by experience the truth and faithfulness of God; we recognize His interconnections. We see His fingerprints all over the affairs of our lives. We don't believe Him because we see those fingerprints. We recognize the fingerprints because we believe the Bible.

Let's consider a few fingerprints empirically evidencing the deliberately designed purpose of the universe for displaying the truth of God's Bible, the reality of His work, and the glory of Jesus Christ's restoration of all things. Leaving His fingerprints all over everything is part of His search for you. Those fingerprints testify to the deeds He has worked into human affairs for effecting your eternal health and welfare.

Israel evidences God's faithfulness, not by their faithfulness, but by their faithlessness attracting His grace. God faithfully led Abraham's descendants into Egypt. He faithfully brought their nation out from being slaves to Egypt. He faithfully led them into their own land. He faithfully drove them out of their land when they rebelled. He faithfully returned a remnant of them to the land. Then He faithfully scattered them throughout the world for failing to receive their King, who faithfully rose Himself from the dead. Now He has faithfully returned Israel's remnant to their land for a second chance to acknowledge Him as their King. He will faithfully destroy evil's grip upon human affairs, and then He will faithfully rule in righteousness over all the earth from Jerusalem, as He promised He would rule from there. All of this was prophesied in the Bible. Most of it has been seen in history. The coming of The King of Kings and Lord of Lords we shall see soon.

Earth's sky full of science's mechanistically material universe is stacked with God's covenant stones, testimonials to His involvement in the fundamental events of Israel's history and of everyone's Doorway to salvation, Jesus Christ. Many signs made by specific

Clear Signs of Trouble and Great Joy

patterns of total lunar eclipses testify to the purposes God planned for Israel and Christ Jesus. In the last chapter we discussed a truly enormous pattern of meaningfully measured relationships between Biblical prophecy, history, and celestial events occurring over the last few centuries. But the signs we will now consider are testimonial markers which began occurring at Israel's national conception within Egypt thirty-six-hundred years ago, and continue occurring through our present day and into the future. Like evidentiary fingerprints, I AM WHO I AM has stamped these markers onto the skies above the formative experiences of Israel (His witness to the nations), the testimonial events of The Head of the Church, the resurrected Jesus Christ, God's witness of both Heaven and Hades, testifying to each of us personally.

History made a pareidolic display of a certain nation's formation in 1948. It is an important nation today, having captivated the world's attention, recently, by rattling its nuclear sabers and demanding the right to exist without threat or harassment. Three years after World War II, separate governments were formed out of the Japanese annexed Korean peninsula. The socialist Democratic People's Republic of Korea became a sovereign nation in 1948. The following year began the presentation of four eclipsed full moons in a row, each one on a major Jewish holyday. Additionally, we find a stitch of North Korea's history embroidered into the Shemitah Path. The Japanese Empire annexed the Korean peninsula in 1910, a Shemitah year. But North Korea wasn't given a book of oracles anciently prophesying the very history which made it a nation. Nor does North Korea have a prophetic chain linking the entire world's future to its land. There was no sky full of total lunar eclipses occurring on North Korea's most significant holydays. (Those eclipses were happening on Jewish holydays.) And the weave of North Korea's history into the Shemitah Path extends through only one stitch of time. No sequence of North Korea's historical events interrelates by meaningfully measured timespans. Pareidolia can't make giant imagery. It can only make small, vague, and irrelevant shadows, like images of Jesus on potato chips, buttered toast, and such. Pareidolia is short sighted and almost meaningless.

But the 1948 formation of Israel happened the year before testimonial eclipses began in the skies of Israel's own major holydays, rather than with eclipses on North Korea's holydays. And the Oracles given Israel long ago prophesied that the Israelites would again become a nation in their land for receiving their Messiah and King after wandering throughout the world. There is a simple difference between pareidolia and message. Pareidolia finds an image within a few otherwise chaotic details. Message is discovered in numerous, coherently organized, repetitive details which, by their own essence, clearly rise to the conceptual level of a language complete with a set of illustrative portraits (the ancient Mazzaroth). The major events developing the essential conditions the Bible predicts for the end of this evil age were accompanied by the highly organized and repetitive patterns of history and eclipses we have been analyzing. We will now analyze a set of three-hundred-eighty-four total lunar eclipses patterned into thirty-two simple counts of 19/48 and the Biblically relevant events over which those 19/48 counts occurred.

Before we begin, let's notice the involvement of some numbers carrying Biblical meaning into this set of 19/48 signs. Recall God's overall objective presented in the Bible is to restore His Creation to a state of righteousness, that is, complete and perfect order. It will be a new beginning for His Creation. New beginning is represented by the number eight. God's Creation is represented by the number four. Eight times four, a new beginning for Creation, is thirty-two, the number of 19/48 signs found between the time of The Flood and the end of Christ's millennial reign. There are twelve total lunar eclipses (the number of perfect order) involved in each set of three tetrads (the number of God's doing it) comprising every 19/48 sign, a total of three-hundred-eighty-four total lunar eclipses (8X4X12) blinking out God's testimony to His restoration work embedded within mankind's historical events.

The ancient nations quickly heard of Egypt's devastation by the God who led its Israelite slaves into a desert barely able to support a solitary wanderer. They knew hundreds of thousands of Israelites survived out there forty years. Thus Jericho saw signs of trouble when Israel crossed the Jordan on a dry riverbed,

> [Rahab] came up to them on the roof, [9]and said to the men, "I know that the LORD has given you the land, and that the fear of you has fallen upon us, and that all the inhabitants of the land melt away before you. [10]For we have heard how the LORD dried up the water of the Red Sea before you when you came out of Egypt...[11]And as soon as we heard it, our hearts melted, and there was no courage left in any man, because of you;
> Joshua 2:9-11

The other Canaanite nations foresaw the same trouble coming,

> [1]When all the kings of the Amorites that were beyond the Jordan to the west, and all the kings of the Canaanites that were by the sea, heard that the LORD had dried up the waters of the Jordan for the people of Israel until they had crossed over, their heart melted, and there was no longer any spirit in them, because of the people of Israel.
> Joshua 5:1

Israel's mere presence in those days testified of I AM WHO I AM's reality to these nations. Five hundred years later, Solomon recognized that the purpose God served in Israel was to show all the world I AM WHO I AM is the God of everything:

> [59]Let these words of mine, wherewith I have made supplication before the LORD, be near to the LORD our God day and night, and may he maintain the cause of his servant, and the cause of his people Israel, as each day requires; [60]that all the peoples of the earth may know that the LORD is God; there is no other.
> I Kings 8:59-60

Israel's testimony is not to their subjective selves. It is not about who they are. It is about Who made them what they are today. God's promise to make a great nation out of Abraham was unilateral. God would do it out of His own faithfulness. But, to each of Israel's generations throughout the following centuries, God's promise of peace in the land was bilateral. To the faithful generations He gave it; from the unfaithful ones He took it.

Clear Signs of Trouble and Great Joy

Surely a final generation would come for whom God would fulfill the promised glory of Israel's ultimate testimony, their perpetual, peaceful existence in the land God promised to Abraham. But even that generation will have to call on their God in the midst of troubles before the promise is fully bestowed upon them.

> ⁶But if you turn aside from following me, you or your children, and do not keep my commandments and my statutes which I have set before you, but go and serve other gods and worship them, ⁷then I will cut off Israel from the land which I have given them; and the house which I have consecrated for my name I will cast out of my sight; and Israel will become a proverb and a byword among all peoples. ⁸And this house will become a heap of ruins; everyone passing by it will be astonished, and will hiss; and they will say, "Why has the LORD done thus to this land and to this house?" ⁹Then they will say, "Because they forsook the LORD their God who brought their fathers out of the land of Egypt, and laid hold on other gods, and worshiped them and served them; therefore the LORD has brought all this evil upon them."
>
> I Kings 9:6-10

So, Israel's relationship with God has long been dicey. Humans are not typically faithful to God. Every generation of man has seemed to aim at the subjective, Progressive "nothing my forefathers knew is smarter than what I know" attitude, that "I've seen it all, I know it all, so either get out of my way or do what I say" myopia. The Israelites' smarmy "intelligence" kept assimilating Canaanite Baals and Ashtaroth into their faith. They were looking for the face of their God on Canaanite potato chips even while their God was waiting to make Himself seen through His faithful work alone. So God kicked them out of the land.

> ¹⁶ Ephraim is stricken,
> their root is dried up,
> they shall bear no fruit.
> Even though they bring forth,
> I will slay their beloved children.
> ¹⁷ My God will cast them off,
> because they have not hearkened to him;
> they shall be wanderers among the nations.
>
> Hosea 9:16-17
>
> ³ Direct thy steps to the perpetual ruins;
> the enemy has destroyed everything in the sanctuary!
> ⁴ Thy foes have roared in the midst of thy holy place;
> they set up their own signs for signs…
> ⁹ We do not see our signs;
> there is no longer any prophet,
> and there is none among us who knows how long.
> ¹⁰ How long, O God, is the foe to scoff?
> Is the enemy to revile thy name for ever?
> ¹¹ Why dost thou hold back thy hand,

The Late, Great 1948

why dost thou keep thy right hand in thy bosom?

Psalm 74:3-4, 9-11

The Northern Kingdom, Israel, was poetically called "Ephraim". Since the time of Jeroboam there was only a short dynasty of northern Kings who even approached doing right before the Lord. They were its only kings from the royal lineage of David. But not even did they direct Israel back into union with Judah, whose kings were all from David's lineage, except Athaliah, the feminist usurper. So Ephraim wandered for the next twenty-six-hundred-and-seventy years before I AM WHO I AM signaled for them.

> [7] Then Ephraim shall become like a mighty warrior,
> and their hearts shall be glad as with wine.
> Their children shall see it and rejoice,
> their hearts shall exult in the LORD.
> [8] "I will signal for them and gather them in,
> for I have redeemed them,
> and they shall be as many as of old.
> [9] Though I scattered them among the nations,
> yet in far countries they shall remember me,
> and with their children they shall live and return.
> [10] I will bring them home from the land of Egypt,
> and gather them from Assyria;
> and I will bring them to the land of Gilead and to Lebanon,
> till there is no room for them.
>
> Zechariah 10:7-10

Today, Israel testifies to more than God's restoration of a nation for their remnant. Their current testimony to God crosses a line of even greater irrefutability than did Judah's ancient return from Babylon. Biblical minimalism loves to accuse the Jewish priests of forging the Old Testament while they were in Babylon during most of the sixth century BC. Of course, these liberal minimalists have no actual *information* supporting this accusation, as liberals are rarely able to support their accusations with any evidence at all. Maybe they gazed into burnt toast to find their proposition. But, unlike the entirely unfounded nature of a scoffers' propositions, the history at which the Israelite people arrived in 1948 irrefutably testifies to The Bible's truth.

> [19] Behold, at that time I will deal
> with all your oppressors.
> And I will save the lame
> and gather the outcast,
> and I will change their shame into praise
> and renown in all the earth.
> [20] At that time I will bring you home,
> at the time when I gather you together;
> yea, I will make you renowned and praised

Clear Signs of Trouble and Great Joy

> among all the peoples of the earth,
> when I restore your fortunes
> before your eyes," says the LORD
>
> Zephaniah 3:19-20

No argument can be raised against the reality of Israel now being a nation in the very land those ancient prophecies said it would be a nation. Ephraim has been gathered from the nations before our scientifically modern eyes. Yet by nothing more than subjectivity, skeptics and scoffers still despise Israel's current existence, accusing it of being only coincidental. They unscientifically proclaim, by no authority other than their own consensus, that there is no God to have brought the Israelites home, "We've never seen evidence of God!" Their subjective minds don't even know what God's evidence might look like. Those Israelites who returned from around the world to become a nation again in 1948 are a very clear evidence of God. So, to avoid facing it, liberal minimalists accuse the Bible's writers of making up narratives and prophecies. Yet the Bible's last words were penned almost nineteen-hundred years before Israel was indeed reestablished for our very own scientific eyes to empirically observe it being what the ancient Bible said it would be where it would be. About that, subjectivity can only cry "Coincidence! Potato chip! Pareidolia!"

Science preaches the knowing of reality through empirical observation. Israel is now the empirically observable fulfillment of Biblical prophecy. Moreover, the eclipses on the full moons of Israel's major holydays directly following its reestablishment and upon its recapture of God's Holy City were also very empirically observable. Even the history developing those two events is measurable in Biblically meaningful numbers. How enormous must a potato chip get before people realize they are observing reality?

In the world of literature there is The Bible. Then there is everything else. The Holy Bible flowed off the pens of particular individuals as much as did any other book. It is as much a presentation of its writers' experiences, thoughts, and attitudes as is any other book. In fact, its distinction is not in its being produced by some forty or so individuals. Many other books are also anthologies. Its distinction isn't even in its claim of being Divinely inspired. Other books also claim Divine inspiration, the Bhagavad Gita for instance, and The Koran, The Book of Mormon, etc. The Bible's distinction is found in past and current history rationally validating its claim of Divine inspiration. No other book is accompanied by such validation. Even the stars testify to it (Genesis 1:14, Psalm 19:1-4, Luke 21:25). The Bible's distinction is in its words having been led forth by The Holy Spirit from the life experiences of dozens of individuals sprinkled throughout fifteen centuries to write an amazingly unified, coherent, and focused message for even our current generation. Its distinction is in its validation made of every soul who's dared to live in real accord with its prescriptions. From Genesis through Revelation, the events and continuous existence of Abraham's genetic lineage play a supportive role to Jesus Christ, the lead character in the story of your eternal life. That is the Bible's distinction. And the next signs we are about to discuss, as the distinct Book said there would be signs, shout the indisputable distinction of The Holy Bible just as Psalm 19:1-4 proclaims the stars do.

According to the Bible and other ancient, Jewish history books, and according to a

The Late, Great 1948

sense which can be made of archaeological data, long ago mankind encountered a communicating intelligence delivering a critical message about everyone's need to eternally survive death. The Mother Goddess, many archaeological artifacts related to her, the Mazzaroth having become the Zodiac, and the key aspects of the Mazzaroth having been spun into anciently perceived gods indicate how mankind once knew of a particular message given to them regarding some kind of eternal survival to which they should pay attention. That's a rather large proposition to support. But we've already discussed the nature of revelation in Chapter 2 and the reality of Biblical prophecy that only revelation can explain. The evidence discussed in Chapter 9 will undergird that previously supported proposition like bedrock.

Ancient liberals progressively distorted a divinely given message into numerous lines of idolatry, some still around today. Those liberal distortions have seduced today's scoffers and atheists into accusing the Bible of being only a rewritten facelift of what earlier generations completely made up. But this seduction is only a modern paradigm crafted in the minds of men who threw more *information* into their dumpsters than what *information* they actually pondered. It is now a paradigm falsified by the very signs about which you are reading. Reality's *information* shows the Bible's central theme, the saving work of Jesus Christ, was known, at least in essence, long before people were believing those anciently twisted liberal ideas.

God gave us His Bible through men of a higher attitude than that of liberal twisters. They more humbly searched for the truth, noting and honoring what the more righteous and faithful people of their ancestors had to say, struggling against their own personal biases and subjectivity, to reach a realistic understanding of their experiences. They didn't force every thought to bow down to their own pareidolic, personal imaginations, like we do today. They knew what they saw. They recalled what God revealed to their fathers. So they understood what they felt. They wrapped faithful minds around real experiences because they treasured *information* for its guiding evidences. They were a lineage of historians, not tale bearing spinners of twisted yarns. Their minds the Holy Spirit used to write His Bible. And they wrote other books, too.

Recall from the last chapter that the ancient book of Jasher assigned the date, 1948AM, to Abraham's birth. Whether or not those more objective historians recorded this particular date with impeccable accuracy is neither here nor there. That they wrote "1948" instead of any other number as the particular date of Abraham's birth is very here, very there, and rather prophetic. What are the chances an obscure, ancient, Hebrew history book, narrowly rescued from the AD70 destruction of Jerusalem, would date the birth of its national father using precisely the same number which also dates the rebirth of that very nation God promised to their national father? Potato chips are for parties. The Bible is for believing.

Add the enormous volume of Biblical prophecies about Israel's rebirth for receiving its Lord and King to that testimony, then observe those perspicuous eclipses on Jewish holidays blinking at Israel's recent reestablishment as well as its reception of The Holy City from which The King of Kings will reign. The chance of such an enormous amount of meaningful concurrence just occurring by Darwin-like happenstance narrows to a virtual impossibility, especially considering Genesis 1:14's statement about the sun, moon, and stars having been made for signs. For this nation is not just any nation, as North Korea is.

Clear Signs of Trouble and Great Joy

It is the nation which the most widely distributed Book in human history calls to the witness stand regarding I AM WHO I AM, His faithfulness, and mankind's faithlessness. Indeed, it was the nation of which The Book states,

> He has not dealt thus with any other nation,
>
> Psalm 147:20

not even North Korea. Israel existing as a nation today bears witness to the truth; I AM WHO I AM authored The Holy Bible.

So, we're now going to observe the number of Israel's testimony, 1948, appearing amongst patterns of total lunar eclipses set over historical events key to that unique nation's testimony, just like the stone Joshua set up testified to their covenant with God. Then we will observe those same patterns monumentalizing key events of Christ's church, the undeniable evidences of Jesus' resurrection, the setting of the Tribulation stage, Jesus' amazing return, and His Millennial reign. Although these patterns might appear pareidolic, remember, pareidolia is seeing patterns where there are no patterns. Seeing patterns which actually exist is called "empirical observation", eyes to see, as The Bible states it. Refusing to empirically observe actually existing patterns is more subjective than even pareidolia. It's called "foolishness". The correlation of these 19/48 patterns with historical events relevant to the design and purpose of Israel, Christ, His followers, and Christ's return effect a testimonial message in and of itself. The denial of their relevance will require immense explanation by some means other than "coincidence". I'll leave it for you to decide whether God has written this in the skies for a message corroborating His Holy Word, or whether my mind is just toying with an impossibly enormous bag of potato chips.

Several folks have pointed out that we recently arrived at a year 1,948 years after the continual burnt offering was cut off in AD70. Fair enough. 2018 came with much speculation. But it also passed the same as any other year. At least seemingly it passed the same. 2018 came three years after the last year of three, highly significant, clearly perspicuous holyday-tetrads which began in 1949, the only three holyday-tetrads forming a 19/48 pattern between them. It is indeed odd that 2018 came 1,948 years after Jerusalem and its Temple were destroyed. January 31, 2018 began the first of three total lunar eclipses occurring six months apart. These three sets of threes shout of God's working something enormously significant. Of course, I suppose skeptics could destroy that set of threes by demanding we add a fourth to them, the three 1,948 year epochs having occurred from the Day of Creation to the year before 1949. And for once, maybe they'd be right.

Recall that nineteen is the number meaning the complete summation of man's affairs. Forty-eight is four, the number of Creation, times twelve, the number of complete order. We see in 19/48 the complete summation of man's affairs for restoring complete order to God's Creation.

From the beginning of the first year of the 1949-50 holyday-tetrad to the end of the first year of the next one spans 19 years: January 1, 1949 through December 31, 1967. The beginning of the second year of the 1967-8 holyday-tetrad to the end of the 2014-5 one spans 48 years: January 1, 1968 through December 31, 2015. Sixty-seven years span the begin-

The Late, Great 1948

ning of 1949 to the end of 2015. The midpoint of the second holyday-tetrad divides those sixty-seven years into 19 and 48 year spans.

This sixty-seven year timespan began the year after Israel became a nation (1948), and the meeting point of its nineteen and forty-eight year periods fell at the end of the year in which Israel regained control of God's Holy City, Old Jerusalem. The time span ends with the final year of the holyday-tetrad having occurred during the fundamental transformation of the nation God rose to restrain lawlessness.

Israel existing as a nation with Jerusalem as its capital and the removal of the hindrance of lawlessness are necessary circumstances for the Tribulation event summing man's affairs before Christ's return to reign over all Creation in righteousness. Coincidentally, or not, the Old City of Jerusalem was returned to Israel's possession sixty-seven years after nineteen completed centuries of Christ's grace holding back the wrath of God, 19 and 67. Sixty-seven is also the nineteenth prime number. Knock, knock, knock, anyone? Another set of three "coincidences"?

Are these patterns only pareidolia? Are their precise relationships with Israel, God's Holy City, the release of lawlessness onto the world stage, and all the Biblical prophecies relevant to these circumstances only Darwinian, non-designed, purposeless happenstances? Or is this "colossal potato chip" testifying to something extra special about Israel and Jesus Christ for the entire world?

When Christ rules from Israel war will be no more (Micah 4:1-4). The human lifespan will increase tenfold (Isaiah 65:20). Weeds and trash trees will give way to cypress and myrtle forests (Isaiah 55:13). The world will again be beautiful and truly a joy to live in, as was The Garden of Eden. The global warming Eve started at the forbidden tree will be harnessed and tamed for mankind's benefit by The Giver. And every question of why such a loving God has not yet put an end to evil will have been appropriately answered. It is together a grandiose proposal made by the Bible. So, it will need exceptional verification.

Maybe God intended the meanings of nineteen and forty-eight to portray the significance of Israel possessing Jerusalem for Christ's righteous order restored to Creation in place of mankind's chaos having been summed up. The division of the 19 and 48 year periods at the end of the year, 1967, cryptically doubles the significance of Israel's capture of Jerusalem, God's Holy City, from where Christ will end mankind's evil. This sixty-seven year pattern is packed with Biblically relevant meaning. All thirty-two of these 19/48 patterns of sixty-seven years are immense stacks of rocks along the path God blazed for mankind's story. They are observable on the NASA lunar eclipse list through much patient work.

Let's do a very short, mental experiment. How coincidental would all this have been if The Book of Jasher had dated Abraham's birth at 2002AM, which was also 2002BC, while the nation of Israel was reestablished AD2002. That would yet be enormously coincidental. But the number 20 has no Biblical meaning for bearing significant correlation with the demise of evil like 19 has. Neither does 2 (02) bear any relevant significance like 48 does (the number of Creation raised by the number of complete order). Get the picture about 1948? Get the message about 1967? And if you think I have only made up spiritual meanings for numbers capable of making pareidolia seem like message, then consult E.W.

Clear Signs of Trouble and Great Joy

Bullinger's book on the Biblical meanings of numbers written nearly fifty years before these prominent events with their verifying patterns even began happening, and nearly a hundred years before their connections were noticed. If anyone could give those spiritual meanings to these numbers for fitting real patterns to the historical fulfillments of significant prophecies, it could only have been the Bible's wonderful I AM WHO I AM.

Passover is the commemoration of a dreadful day when I AM WHO I AM dealt the convincing blow to an ancient Pharaoh refusing to believe the meaning within a pattern of plagues. I believe Pharaoh's mistake can teach us enough to confidently make that statement, now that archeology has unearthed an estate sporting twelve tombs (one being a pyramid) in the ancient city of Avaris, abandoned at the very time Ipuwer lamented the bloodwater, the darkness, the destroyed crops, every Egyptian family burying a member, and the slaves leaving with all the Egyptian gold. Gee, what could Ipuwer have been talking about? Who built that estate? Death came in the night for all of Egypt's first born, those of both livestock and man. But while the death angel passed through Egypt's streets, it passed over every house having lamb's blood painted around the doors. The bloody "paint" was a sign which the Israelites hoped the death angel would note as surely as God hopes you will note these signs. For the Egyptians, the final summation of Moses' ten challenges to Pharaoh was bad. But for Israel, the final sum of those events were God's protective preservation in His arrangement of their earthly plights, Passover. Avaris was left with many bodies hastily thrown into Egyptian graves, as archaeologists have now discovered.

According to the Bible's predictions, the Tribulation will be an excruciatingly fateful spell. War, famine, disease, ravenous beasts, earthquakes, fire, enormous hailstones, even blood-water and asteroid and/or comet impacts will pare the human population to a meager remnant; God works for remnants, not for crowds, mobs, dog packs, or Forward storming hordes of Progressives. At the final moment of this horrific event, Israel's remnant faces the onslaught of the world's anti-Semitic, fascistically rabid, ethnically diverse mob. The cornered remnant returns in desperation to I AM WHO I AM, crying out in repentance from their unbelief and lack of awareness. Good for them! Bad for the mob. Jesus has been waiting for this call with Passover baited on His breath. He returns like a mother bear charging to the aid of her distressed cub. With a simple word of command the entire, malicious, Progressively diverse mob is eradicated. Game over! Nineteen sums badly for them. Ultimately, demise passes over Israel again. They will have been given a second chance to acknowledge their King. Then God will raise this little nation to the head of all the nations, because, indeed, they will acknowledge Him. And so Christ begins His righteous reign of peace in Jerusalem -perfect orderliness for the entire world. It shall be a great 48 after a very rough 19.

> Then every one that survives of all the nations that have come against Jerusalem shall go up year after year to worship the King, the LORD of hosts, and to keep the feast of booths [tabernacles.]
> Zechariah 14:16

No wonder the spirit of the world is full of hatred for Israel! The world rejects the

The Late, Great 1948

Biblical narrative of Israel's final place at the forefront of the nations. Yet, mankind's rebellion, having been measured into three nineteen-hundred-forty-eight year periods, testifies that Israel's future place at the head of the nations will be God's doing.[1] Even more astonishingly, this great number, 1948, has been flying like ensigns over all of Israel's history, signifying that their earthly deeds will culminate in Christ ruling the nations from Jerusalem. These sixty-seven year spans of tetrads stack even more coincidence upon an already impossibly high heap of consistent, cohesive, coherent convergence…signs, rationally speaking.

When Joshua led the Israelites across the dried Jordan riverbed, they heaped in the middle of it a testimonial pile of rocks taken from the west bank which, after the waters returned to their flowing, only God could see. On the west bank they heaped a testimonial pile of rocks taken from the dry riverbed which everyone could see. At Shechem, where Joshua covenanted with the people, he erected a great stone as a witness to the covenant made there. Many other events and agreements of Old Testament history were marked by testimonial signs in similar manners. Their message is always about the stones having witnessed those events, and so they testify of those events to their observers. The Bible commands they were not to be taken down.

As you are about to read, these sets of tetrads particularly patterned into the numbers 19 and 48 are stacked above historical events thematic of Israel's being and Christ's reign. They perform the same function as did those stones Joshua set up. They were hung in the sky for testimonial markers of things God and man witnessed together. Moreover, these arrangements of total lunar eclipses were placed in the sky not by man's hands, but by God's very own, creative, prescient, non-Deist, omnipotent hand. These 19/48 patterns are noticeable only to us of these last days, who've come to the unsealing of Daniel's prophecies. So, for the sake of discussion, let's refer to them as signs of witness, and more gloriously, as tetrads-of-testimony. But unlike the stones Joshua stacked, these tetrads-of-testimony can not be taken down and thrown away by the hands of scoffers. Their mark has been made. Their mark shall stand.

Clusters of tetrads occur rather cyclically. No tetrads will happen for two to three centuries. Then fifteen to twenty will occur over the next three to four centuries, with most of them clustered in the middle two centuries. For example, the three centuries from 1600 through 1800 experienced no tetrads. The two centuries from 2200 to 2300 will experience no tetrads. But the intervening three centuries, from 1900 through 2100, are sprinkled with seventeen tetrads, the Biblical number expressing fullness of spiritual significance. There could have been sixteen. Or there could have been eighteen. But there were seventeen. Within this spiritually significant cluster of tetrads we find our 1949-2015 testimony of tetrads monumentalizing both Israel's reestablishment and the setting of the Tribulation's stage.

From the time of Abraham to AD3000 (the approximate time for the final Judgment Day, if this book's proposition about Christ's return matches what God has planned) there are thirty-two signs of witness. Thirteen of them involve five tetrads conforming with the arrangement of those five tetrads occurring from 1949 through 2015, the set of tetrads amongst which this 19 and 48 year pattern was first noticed. The remaining nineteen signs-

Clear Signs of Trouble and Great Joy

of-witness also form the 19 and 48 year patterns within a sixty-seven year timespan, but they have some basic dissimilarities with the other thirteen.

Once again we find the ominous number nineteen occurring in our patterns. The summation of man's affairs is a very basic element of the restoration of all things. And we noticed that the number of tetrads-of-testimony having the same pattern as the 1949-2015 one are the ominous number, thirteen. This would seem strange, considering the 1949-2015 tetrad-of-testimony is composed by those three holyday-tetrads marking Israel's return to nationhood, Jerusalem's return to Israel, and the mystery of lawlessness to be released to its own demise. But then, of course, the end of evil happens through a very ominous time. Or could something more be shining under the surface appearance of those thirteen? Regarding that number, E.W. Bullinger says,

> The Saviour, though without sin, was "made sin," or a sin offering, for His people. He was "wounded for our transgressions," and bruised for their iniquities. He was, in fact, "NUMBERED WITH THE TRANS-GRESSORS" (Isaiah 53:12). [the emphasis is his]
>
> Therefore this number is not only the all-pervading factor of SIN, but also of sin's atonement. It is not only the number which brands the sinner as a rebel against God, but it is the number born by the sinner's Substitute.[2]

When we peel one more layer of significant difference off those thirteen, orderly tetrads-of-testimony, we find that only one displays the 19 and 48 pattern amongst three special, holyday-tetrads. Moreover, those three tetrads mark the extraordinarily significant events of Israel's return to possess God's Jerusalem. The other twelve might involve one or two holyday-tetrads, but never all three forming the 19 and 48 year expression. We now see the Glorious One and the twelve patterned after Him, Christ and His Apostles.

OK. That's a cool coincidence. But when I considered what significant event might have occurred during each sixty-seven year span of these three-hundred-eighty-four winking, blinking eclipses, I found what coincidence entirely fails to explain. God stacked these signs of witness like heaps of testimonial stones over the following historical events:

1661-1595BC and **1643-1577BC**: Joseph had become vizier of Egypt after interpreting Pharaoh's dream, when these signs of witness began. Biblical minimalists will deny even the possibility of a Joseph, but the lone eclipse of February 25, 1362BC pops that balloon full of Old Chronology's hot air. The Biblical narrative is now more than just a possibility. It is attested in the heavens by these signs of witness aligning with David Rohl's empirically evidenced rationale for a chronology of ancient history trued by reality itself. His New Chronology finds Joseph being the vizier of Egypt with Jacob's entire family in Egypt, where they would begin growing into the nation God rose up for testifying to the twenty-first century, "scientific" world that I AM WHO I AM exists and is not a Deist. According to the Book of Jasher, Joseph began his viziership to Pharaoh 2229AM, which is 1666BC.[3] Then Joseph dies only a few years before these two closely interwoven signs end. The

The Late, Great 1948

seven fat years predicted by him began 1665BC, the lean years began 1658BC. These two intertwining tetrad's of testimony to Joseph's history in Egypt began the year following the birth of his twins, Manasseh and Ephraim, whose families became numbered amongst the tribes of Israel.

1585-1519: The Book of Jasher says

> ...it came to pass after the death of Joseph, all the Egyptians began in those days to rule over the children of Israel...[4]
>
> 4 And it came to pass in those days, in the hundred and second year of Israel's going down into Egypt, that Pharaoh king of Egypt died, and Melol his son reigned in his stead, and all the mighty men of Egypt and all that generation which knew Joseph and his brethren died in those days.
> 5 And another generation rose up in their stead, which had not known the sons of Jacob and all the good which they had done to them, and all their might in Egypt.
> 6 Therefore all Egypt began from that day forth to embitter the lives of the sons of Jacob, and to afflict them with all manner of hard labor, because they had not known their ancestors who had delivered them in the days of the famine.[5]

Jasher dates the rise of this new generation of Pharaohs and of Egypt's liberal children at AM2340, approximately forty-five years after Joseph's death. That year converts to 1555BC by use of the relationship we discovered between Anno Mundi and BC at Abraham's birth being 1948AM, the same as 1947BC.

1538-1472BC: This tetrad-of-testimony stood above the birth of the great prophet God raised up to lead Israel out of Egypt to becomeHis testimonial nation. Moses was eighty years old when God sent him to free Israel. The Exodus occurred in 1447BC, therefore Moses was born 1527BC.

These four signs-of-witness winking above Joseph's viziership and death, the rise of Israel's enslavement, and the birth of their deliverer, Moses, all intertwine. Signs-of-witness intertwine when another sixty-seven year sign begins before the previous one has ended. Three is the number of God. Since God is the ultimate state of being, there was nothing in addition to God before He began creating, metaphorically speaking (God exists outside the realm of time.) First there was God (3), then He made Creation (3+1=4). Accordingly, four is the number of Creation, this physical universe and everything functioning in accord with its physical laws. He would act graciously towards His creation's eventual turn away from Him, which concept is the next additional aspect, five, the number of grace. (We have five fingers on each hand for working graciously and five toes on each foot for walking graciously.)

Today, mankind is graciously being warned of the nearing end of God's patience for

Clear Signs of Trouble and Great Joy

His knowledge being thrown into our dumpsters while we mix our deceit into His Creation. The beginning chord of this warning was struck in today's heavens by the four holyday-tetrads of 1493-4, 1949-50, 1967-8, and 2014-5, each entailing four blood moons. The thirty-two signs-of-witness began with an intertwining of the first four. Together, these first four signs-of-witness and the four holyday-tetrads illustrate the intertwining of The Holy Bible's narratives and prophecies with historically known events. We can include Regulus, the king star, in this "coincidence", its being a system of four gravitationally connected stars which Jupiter triple-conjuncts concurrently with the four major events of Christ's work to separate evil out of His Creation: 1) at His birth, 2) at Jerusalem's destruction, 3) at Jerusalem's repossession by Israel, and 4) at Jesus' return to Jerusalem.

In Chapter 7 we saw the many times which God begins the new epochs of His story with sets of three, as if to indicate His doing of what followed. This set of four interwoven signs begin with three signs-of-witness intertwining Joseph's viziership to Pharaoh. The sale of Joseph into Egyptian slavery not only began this epoch of Israel's nationhood, but it also began the chain of seven empires, the heads of the dragon/beast (Egypt, Assyria, Babylon, Persia, Macedonia, Rome, and the Ottoman/Turk Empire) linking through history from Joseph's viziership to the end of the Ottoman-Turk Empire. This chain of heads, noticed by even the ancient Sumerians, reaches down through history to the August 14, 1914 Rosh Chodesh Elul total solar eclipse warning given to that seventh metaphorical dragonhead, calling it to give the Jews their homeland. Three years later, the Empire was forced to free the Promised Land in AD1917, the year in which the Jubilee and Shemitah Paths crossed, two years before the seventh dragonhead met its demise with the end of World War I in 1919, the number of the complete summation of man's affairs (19) repeated for emphasis.

1914 through 1919 was an enormously important span of time in God's developing story of man's affairs. It wasn't the demise of the dragon's seventh-head which was the summation of mankind's affairs. That head would rise again as the eighth-head. The demise of the eighth-head, the Antichrist's empire, during the Tribulation will be the complete summation of mankind's affairs. The final demise of the resurrected seventh-head is spookily foreseen in the number of the year in which that seventh-head was mortally wounded, 1919, prior to rising again for its final summation. And also note, the year of the seventh-head's initial demise followed the year of its Rosh Chodesh Elul warning by the number of grace, five, 1914 to 1919.

God is big enough to write meaning into the measures of history's years however much we demand to "scientifically" blow it off as pareidolia, foolishly throwing clear signs into the dumpsters of intellectual irresponsibility.

Whether it is only coincidental, or whether God led forth His story of mankind in such a manner that history's physical patterns would actually confirm His spiritual reality, the 1917 crossing of the Jubilee and Shemitah Paths in the year the seventh-head of the dragon/beast relinquished control of the Holy Land occurred thirty-two years before The unique sign-of-witness composed entirely of holyday-tetrads began winking and blinking its 19 and 48 testimony the year after the first three 1,948 year epochs of human history ended at Israel's reestablishment. Thirty-two tetrads-of-testimony stand above Israel's history from Joseph's viziership to the end of Christ's Millennial reign in Jerusalem, the city coura-

The Late, Great 1948

geously acknowledged to be Israel's capitol by President Trump, one-hundred years after the Ottoman-Turk Empire released control of it in 1917. These thirty-two tetrads-of-testimony bear witness to God working a new beginning for His Creation, 8X4, eight, the number of new beginning, times Creation's number, four.

You can't make this stuff up. Human hands can not make such meaningful measures of history any more than human hands could have made up The Shroud of Turin.

The empire of the Antichrist, the eighth head that will be of the seven, is stirring in this very time of our days. The Muslim Brotherhood was formed shortly after the dragon's seventh-head, the Ottoman-Turk Empire fell, and the Caliphate ended. The Brotherhood was established to revive the demised Caliphate. Turkey's current Prime Minister Erdogan, a member of the Muslim Brotherhood, maintains meaningful connections with the vanquished Ottoman-Turk Empire.

> Turkish President Recep Tayyip Erdogan has warned the world Jerusalem "belongs" to Turkey, harking back to the Ottoman Empire's control over the city for hundreds of years up until it was ejected in 1917...back in June [2020] Turkey's top religious figure vowed to continue Islam's global struggle for dominance "until Jerusalem is completely free" and called on the Muslim world to rally against Israel.[6]

Isis' dream of world domination is not over. It will merely pass to a different Muslim horde, if need be, maybe to Erdogan's horde. And radical Islam yet surfs the waves of Islamic refugees surging into Europe and washing up on shores throughout the world. We can not be sure how these threads of history will weave the future's Biblically prophesied Tribulation. But we can be sure they will. Currently, three tetrads-of-testimony intertwine while God has been directing the preparation of the world stage for Christ's return to defeat evil for a thousand years of peaceful righteousness. God stacked these testimonial markers in the sky; they are real; use the process described in Appendix 1 to discover them for yourself. Now let's continue observing how they have stacked up to our time.

1433-1367BC: The people of Israel had been wandering in the desert for forty years when God led them into a conquest for Canaan. *Information* from the Bible dates the Exodus at approximately 1447BC, as we noted in the last chapter. Therefore, Joshua led the congregation of witnesses across the Jordan approximately 1407BC. This began the fulfillment of God's promise to Abraham, which was completed forever in AD1948, the number displayed by each of these testimonial patterns. To the Israelite congregation, Moses assured God would do it. And throughout the next thirty-four-hundred years God has left His fingerprints all over these people and this land of Canaan. David Rohl describes the discovery and partial preservation of the testimonial covenant stone Joshua set up at Shechem (Joshua 24:26) as witness to the people's covenant with God while this very pattern of testimonial eclipses was winking and blinking its witness of God's covenant in the skies overhead. The testimonial stone was discovered in the early twentieth century and restored to its base. But

Clear Signs of Trouble and Great Joy

it was soon yanked down and cast into an excavation dump for its connection to Biblical narrative. But after the war, a different archaeological team recovered the lower half of the stone (it had broken, and the top was lost) and set it back onto the socket Joshua prepared for it.[7] Its final return to its proper place occurred under the glorious tetrad-of-testimony begun 1949 and ended 2015.

1028-962BC: Four centuries after the conquest of Canaan began, the congregation of witnesses desired a king, as Moses prophesied they would (Deuteronomy 17:14-20). After passing the congregation through the maniacal hands of the liberal Saul, God gave them into the more righteous hands of David. David is the first king of Israel's never ending dynasty. The last king of David's dynasty will be the eternally reigning, wholly righteous Jesus Christ, coming soon to God's City in Israel. David began his reign approximately 1010BC and reigned to 970BC. He unified the people into one nationalism soon to support the coming greatness of Israel.

981-915BC: Solomon reigned after his father's death. He expanded the kingdom to the maximum extent of its borders, as they are described at Genesis 15:18-20 and I Kings 4:20-21. The nation also reached the zenith of its glory and wealth under Solomon. He built the First Temple to I AM WHO I AM. But in spite of his famed wisdom, he lived foolishly. He overtaxed and enslaved the people. Consequently, after his death, his son, Rehoboam, led Israel into tragedy through the foolishness he learned from Solomon, his dad. Seven years before this tetrad-of-testimony ended, Rehoboam caused a rebellion by increasing taxes even above what his dad had been over-exacting. He refused to heed the advice of the more historically seasoned "old white guys" of his father's administration. They counseled him to become the people's servant, rather than to make the people his servants (I Kings 12:6-7), like our old, dead white-guys counsel our Progressives. Rehoboam also lusted more for the counsel of his liberal friends, who told him to increase the taxation and forced labors of the people (I Kings 12:8-11). By over-taxation, the nation was split into the Northern Kingdom, thenceforth referred to as Israel, and the Southern Kingdom of Judah. God revealed His cause of this division (I Kings 12:21-24), eventually trapping the unworthy liberals within their own, subjective self-interest.

923-857BC: This tetrad-of-testimony began blinking and winking over the early years of the Divided Kingdom. Although the division was the working of God, He worked it through Rehoboam's greed as a consequence of Solomon's introducing worship of Ashteroth, Chemosh, and Milcom for the satisfaction of his many wives. God warned him not to enable idolatry by multiplying wives to himself. God would rather have blessed a united kingdom with great prosperity for its faithful worship of I AM WHO I AM only, having a king in love with one queen. God isn't the totally unconditional stoop today's Laodicean church makes Him out to be. He demands faithful response and an actual desire to reflect His character. Therefore, He began sending prophets to call the kings and people to make Israel great again by repenting

The Late, Great 1948

of their liberal idolatries and returning to the faithful worship of only I AM WHO I AM. Elijah, the first of these prophets, began his ministry approximately 870BC. Under this tetrad-of-testimony Elijah called all Israel together at Mt. Carmel. With all of Baal's prophets on Mt. Carmel, he presented to Baal the challenge of striking up a fire to consume the sacrifice his prophets would offer. He presented the same challenge to I AM WHO I AM, except he drenched his sacrifice with water. Baal's sacrifice never caught fire. But fire descended from heaven, and in an instant, licked up the sacrifice to I AM, and all that water with it. The prophets of Baal were put to death that day. This sign-of-witness was blinking overhead.

905-839BC: Elijah didn't die. Like Enoch, he was taken into heaven. Under this sign-of-witness, Elijah chose Elisha for his successor, and was then taken onto the sweet chariot swinging low. Elisha received double Elijah's spirit. He prophesied with tears in his eyes that Hazael would become king of Syria. When Hazael asked why the tears, Elisha said he knew Hazael would badly mistreat the Israelites. It was this Hazael who defeated Joram and Ahaziah of Israel and Judah, the victory inscribed upon the Tel Dan Stele bearing the name of King David. Its discovery became the straw breaking the back of the twentieth-century minimalists' accusation that King David and the Israelite empire never existed.[8] This Tel Dan Stele is like a tiny "manufacturer's tag" sewn onto God's beloved House of David, a bitter tag upon which Biblical minimalists to this day choke. It is like God is discussing truth with His deniers, having thrown down, in the far past, markers of the very points He now makes through our archaeological discoveries today. And we will see in the next chapter the most astounding marker ever buried in dirt.

These first nine tetrads-of-testimony were hung over the most berated and denied Biblical events of Israelite history. Today, this history is attested beyond denial by archaeological and astronomical evidences discovered during the last two centuries, with the most compelling evidences having come to light in the last few decades. It is as though God, Himself, through the exposition of evidence, is steadily elevating His warning for mankind to end their stubborn, subjective accusations against His Holy Word, and to prepare for His truthful righteousness coming soon in Jesus Christ. He has hung signs in the heavens to be testimonial heaps of stones above the purposes He designed for Israel and their coming King. From their people, through the line of David, comes your only opportunity to live a blissfully glorious eternity, if you so choose to follow the truth Jesus Christ told.

847-781BC: Joash, Ahaziah's son and the eventual successor to the throne, was only a baby when Ahaziah died. When Ahaziah's mother, Athaliah, saw he had died, she slaughtered all of the royal family, except for Joash, whom his aunt hid. Athaliah then usurped the throne of David. This event is prophetic of what is soon to come in our own days. For the Antichrist will rise up and usurp Christ's throne in Jerusalem. As Athaliah put to death all but a remnant of the royal family, the Antichrist will execute all but a remnant of God's saints. But in the seventh year of her reign,

Clear Signs of Trouble and Great Joy

Jehoiada the priest brought Joash out of hiding, slew Athaliah the usurper, and made Joash king, just like Christ, our high priest, will return in the seventh year of the Tribulation, cast the Antichrist into the lake of fire, and take up His rightful reign.

Before we proceed, let's note that, from the beginning of the Davidic kingdom through its rebellious idolatry and division to the year 781BC, five tetrads-of-testimony intertwined in the heavens. Moses clearly warned the children of Israel what would happen if they worshipped false gods. David's kingdom hadn't lasted eighty years before "the wisest of men" was constructing altars to the false gods of his royal harem. Moreover, the kingdom split, further incensing God. (Indeed, He caused the split because of their idolatry.) And if an usurper sitting upon Judah's throne angered God, the Northern Kingdom marching to a whole succession of usurpers brushed elbows with the limit of His patience. They even spurned the temple God established in Jerusalem as the place to worship and sacrifice, building a temple and altar in Samaria, a truly liberal action. The only reason this entire promised land full of tradition-rejecting liberals was not turned immediately into a smoldering clinker was by the patient grace of God. These intertwining tetrads-of-testimony winking over God's nation of bumbling people are five in number, the Biblical number representing grace. For over four-hundred years God restrained His anger with Israel's idolatry before He drove them from the land. He graciously returned Judah to their homeland after they had spent seventy years in Babylon. (And note the ten signs-of-witness hung over these historic events of God's nation of witnesses, the number of perfect completeness.) Then…

377-311BC: Alexander the Great conquers and unifies the Middle East, Anatolia, and Greece. His success resulted in Greek becoming the international language of the day. The significance of this development is not just that there became an international language, but that the one which became international was the clearest and most precise language of all that day's languages, the language best able to deliver the subtle details of the soon to be written New Testament. The angel prophesied to Daniel about Alexander. Minimalists, scoffers, and atheists tried to date the Book of Daniel's writing to the late second or early first century BC, after Alexander had become history, because of the clarity of its prophecies.

> [3]Then a mighty king shall arise, who shall rule with great dominion and do according to his will. [4]And when he has arisen, his kingdom shall be broken and divided toward the four winds of heaven, but not to his posterity, nor according to the dominion with which he ruled; for his kingdom shall be plucked up and go to others besides these.
>
> Daniel 11:3-4

Above Alexander's God-given success flew this tetrad-of-testimony. Eventually, The New Testament would spread throughout the Middle East and the whole of the Mediterranean world written in the accurately expressive Greek language having been made the international language by Alexander the Great's success. But first…

The Late, Great 1948

290-224BC: Philadelphus II (285-247BC), a Ptolemy king ruling Egypt, commissioned the translation of the Hebrew Scriptures into Greek.[9] The work was begun around the middle of the third century BC and is said to have been completed in the second century BC. Although this project was part of the hellenization of the Middle East, it marked the beginning of The Holy Bible's international history, even though its international beginning was made by the translation of the Samarian text of the Old Testament into Greek, rather than the more faithful Masoretic text. The Samarian text had been maintained and transmitted by the rebellious, idolatrous Northern Kingdom. Its differences from the Masoretic Text are interesting, but not overwhelming.

Seemingly right on cue with Christ's birth, who became the Savior of not only Israel, but internationally, of the entire world as well, the testimony of the signs soon to occur in the years of our Lord (AD) wink above the works and evidences of Jesus Christ and His Church.

AD162-228 and **AD180-246:** These two signs-of-witness intertwine over the bloody sport the Romans made of the early Christians. The martyred Christians testify to Christ's reality, but not so much by their torture for believing in Christ. History shows how surprisingly willing people are to die for falsehoods they sincerely believe, even as willingly as they are to die for actual truths. Consequently, their witness serves less as evidential testimony to non-Christians and more as an encouraging testimony to Christians about suffering for the One who suffered for us. The evidential matter regarding these horrific deaths speaks to the genuineness of their commitments to Him. Revelation's letter to the angel of the church at Smyrna exclaims

> [8]And to the angel of the church in Smyrna write: "The words of the first and the last, who died and came to life.
> [9]"I know your tribulation and your poverty (but you are rich...) [10]Do not fear what you are about to suffer. Behold, the devil is about to throw some of you into prison, that you may be tested, and for ten days you will have tribulation. Be faithful unto death, and I will give you the crown of life. [11]He who has an ear, let him hear what the Spirit says to the churches. He who conquers shall not be hurt by the second death."
>
> Revelation 2:8-11

Their patience and perseverance has been an encouragement to Christ's followers ever after. They are a testimony to the world about the gospel's ability to grip sincere souls facing unimaginable horrors more effectively than deceit drags insincere souls into complete delusion. About these witnesses and those who followed in their bloody steps through the ages is written the renowned Foxe's Book of Martyrs.

AD267-333: Whether Constantine the Great really became a Christian or merely

Clear Signs of Trouble and Great Joy

gave his men over to a Christianized superstition is yet debatable. But more importantly, he and Licinius, his rival/colleague until AD324, decriminalized Christian worship in AD313 by their co-authored Edict of Milan. The Christian blood drenching the Coliseum floor evaporated from Roman entertainment, leaving that old stage stained in the same bloody hue soon to adorn the Antichrist's coming tirade.

AD390-456: Christianity is made the official state religion of the Roman Empire.

> Christ's martyrs and confessors of the first three centuries, in their expectation of the impending end of the world and their desire for the speedy return of their Lord, did not contemplate the great and sudden change about to befall the Roman Empire. The renowned, early church theologian, Tertullian, considered the Christian profession to be irreconcilable with the office of a Roman emperor. Nevertheless, church clergy, laity, and the commoners of the Roman world quickly acclimated to a new order of things. The church recognized in it a reproduction of the theocratic constitution of God's Israelite people under the ancient covenant.[10]

This should seem to have been a "yippee" moment for Christianity. Instead, it brought the worst out of Christian people. Driving the bus by what you believe might seem only normal once your hands have taken hold of the steering wheel. Nobody would sit in the driver's seat to drive according to what someone else believes. Pagan hands had been steering the Roman bus down some very bloody allies, at least according to Christian assessment. Now Christians had the wheel. It was their turn to steer the bus. And they felt placed in the driver's seat by God Himself:

> If we have died with [Christ], we shall also live with him;
> [12] if we endure, we shall also reign with him;
>
> II Timothy 2:11b-12a

> ...for thou wast slain and by thy blood didst ransom men for God
> from every tribe and tongue and people and nation,
> [10] and hast made them a kingdom and priests to our God,
> and they shall reign on earth.
>
> Revelation 5:9b-10

However misguided their interpretations of such Scriptures may have been, the nature of their response was understandable. Try on the shoes of these Christian emperors succeeding Constantine. The life, death, and resurrection of Jesus Christ constituted the basic meaning of their lives. His resurrection was only a little less ancient history to them than the scientific revolution instituting the meaning and way of our lives is ancient history to us. They believed they had in hand the text of God authenticating their ideas. Would you not institute what you believed to be com-

The Late, Great 1948

munication from God at least as wholeheartedly as our modern governments institute the icy mechanisms of the "scientifically" described materialistic universe? We condemn those Christian emperors for using the rule of law to squelch paganism and elevate Christianity, but we applaud our Supreme Court for elevating the assumptions of evolution and squelching the revelations of Creationism. We abhor those Christian emperors for laying the penalty of death on everyone refusing to acknowledge the Christ who lived, died, and rose again in their very recent past. Yet we glorify our Supreme Court for raising the penalty of death upon today's mortal sin of being an unwanted, prenatal neighbor. Gee!? Maybe we're wearing their shoes.

So let's not be casting stones at Theodosius and the emperors following him for the way they drove the Roman bus for Christ. We're driving our bus just as fascistically, and even more hellaciously, for Satan, believe it or not. They applied their faith to driving against paganism just as intently as we apply paganism to drive America's bus against Christianity.

But this was not the first time in history men with knowledge of and worship for I AM WHO I AM reigned over an earthly government. The first time it occurred, there were also tetrads-of-testimony hanging in the sky. We discussed that time above, 1028-962BC and 981-915BC, King David, Solomon, and the zenith of the Israelite empire. Both empires, first the Israelite and then the Christianized Roman Empires, ended in disaster. Neither the Word of God nor the doctrines it prescribes for both Judaism and Christianity spoil good situations; people do that.

The Christianization of the Roman throne seems to have been a fulfillment of Daniel's prophecy about the great statue that was struck on the foot by a stone carved out of a mountain, but not by human hands (Daniel 2:45). At least it was a fulfillment of one of that Scripture's Sethite-like dual prophecies. During the fifth century AD, the western half of the Roman Empire slowly crumbled into dust. Eventually, Constantinople was left to carry on the eastern half of the empire for another nine-hundred years until it was conquered by the metaphorical seventh-head of Revelation's dragon/beast, the Ottoman-Turk Empire. We do not usually recognize Byzantium as the continuation of the Roman Empire, but it prided itself in the past glory of Rome; its people dressed like Romans; they legislated like Romans; they played like Romans, and they were Christians like the Romans eventually came to be. Only they did it with enough Oriental accent that we forget Byzantium was the Roman Empire surviving in the east under Christian rule.

Therefore, it is little wonder that the next three tetrads-of-testimony hang over tremendously meaningful, seemingly inconsequential occurrences of Byzantine history.

AD795-861 and **AD824-890:** The Shroud of Turin inspired image of Jesus Christ returned to Byzantine coinage after nearly a century of radical iconoclasm had ravaged Christian artwork. This might seem to be a truly trivial event. But to the current scoffers and those who refuse to believe Jesus, these simple little coins testify about the historical existence of the greatest Christian relic half-a-millennium before the date proclaimed by normal science to be that of its "forgery". Until AD1204, a full

Clear Signs of Trouble and Great Joy

bodied image of Christ on a linen cloth was known to exist in Constantinople. The historical descriptions of it very well describe The Shroud of Turin. The details of the face struck onto Byzantium's gold coins match the details of the faint face appearing on The Shroud. Such correlations are used by scientists and archaeologists, alike, to formulate knowledge out of theories and theories out of observations, unless, of course, those correlations have anything to do with substantiating The Holy Word of God. Then they are treated not so much as correlations as they are treated like mere coincidences, anomalies, or worse: dumpster fodder, never to be seen or heard again.

But the Shroud inspired image of Christ restored to those coins is not the entire import of this pair of sign's intertwining. Justinian II first issued the coins in AD692, almost a century-and-a-half after The Shroud was rediscovered in a sealed vault over the main gate of Edessa. Maybe these intertwining tetrads-of-testimony would have stood over that event if the little coins, themselves, were the main point. But less than thirty years after the first coin was struck, Byzantine Christianity stumbled into an iconoclastic stupor. Ignorance does not serve holiness. Knowledge does. And knowledge is delivered through imagery, pictures, and other depictions of concepts, as well as, through linguistic communication. Stupidity thinks a picture is an idol. But the Lord God knows better. Unfaithfulness is idolatry; pictures are communication. Communication is important to the knowledge of God's people.

> [6] My people are destroyed for lack of knowledge;
> because you have rejected knowledge,
> I reject you from being a priest to me.
>
> Hosea 4:6

It wasn't that God showed disapproval for iconoclasm by hanging this sign-of-witness over the return of Christ's image to the coins rather than hanging it over the iconoclastic period. It is that He emphasized the importance of transmitting ideas to the community of believers through Christian images and expressions.

> [16]...those who feared the LORD spoke with one another; the LORD heeded and heard them...
>
> Malachi 3:16

AD900-966: That greatest relic, only four-hundred years before the date "scientists" accused of being that of its forgery, was carried from Edessa to Constantinople under this sign-of-witness. Constantinople had long desired to possess it. So it arrived in Constantinople under great fanfare and acclaimed miracles. This wasn't a casual incident happening within the shadowy cloak of forgetfulness. Some passing antiquities dealer didn't just forget and leave it there, like many atheists would rather accuse E.T.'s mom of doing. It left enduring, detailed, evidential marks on the annals of history, there at Constantinople, to which normal scientists would rather pay no

The Late, Great 1948

mind. So, like a worm on a hook, the C14 date caught lots of suckers, and demonstrated how people are destroyed by their lack of knowledge and failure to fully communicate.

AD1305-1371: The Shroud of Turin emerges into our commonly perceived history (or should we say, misperceived history) under this testimony-of-tetrads. The great Ian Wilson and others have now corrected the misperception about The Shroud having become first seen at the little Lirey church in the 1340s. A wealth of evidence verifies its existence long before then. But extreme jealousy over The Shroud set into motion six centuries of subjective scoffing and unfounded accusations against its authenticity. The subjectivity of more recent considerations imagined many forgeries and copies to explain away the sparse, historical footprints of The Shroud's movements here, there, and yon through time. But like we acknowledge the proof to be in the pudding, the truth is in those small details. Ian Wilson, maybe the world's foremost scholar on The Shroud, has collected and presented all of history's pesky little details of its movements over the centuries. They stack like bricks into a solid foundation supporting its true antiquity. Moreover, The Shroud's authenticity has now been scientifically verified beyond any possible doubt. (Except for, maybe, E.T.'s mom.) So the fourteenth century exhibitions at the Lirey church launched The Shroud of Turin into our twenty-first century, "scientific" world. Consequently, it has testified both of Christ's resurrected reality and of normal science's biased failure to even think in accord with evidence upon any matter involving the Bible.

AD1457-1523: in 1492, Ferdinand and Isabella issued the Alhambra Decree meaning to expel all practicing Jews from Spain. Maybe that doesn't sound like a big deal to us, today, but it was a big enough deal for I AM WHO I AM to hang four total lunar eclipses in the Passover and Tabernacle skies of the following two years. And if that doesn't pique our interest, then the repeal of the same decree in the year following Israel's capture of Jerusalem, just months after four more total lunar eclipses had winked above four more Passovers and Tabernacles, should be somewhat more thought provoking. These two sets of holyday-tetrads, plus the 1949-50 one, are the three which defined this entire system of signs as being confirmational indicators of God beginning His process to end evil. Their composition is a bit coincidental. But one of those rather coincidental sets of eclipses occurring the year after Israel became a Biblically prophesied nation, while another occurring the year after America's discovery and Alhambra Decree's issuance, and a third beginning the year Israel captured Old Jerusalem and ending in the year of the Alhambra Decree's official repeal, is no less than purposeful design. They are not at all just coincidences. Sorry, Charlie Darwin! Your mechanistic-like universe is neither purposeless nor non-designed.

And surprise, surprise, surprise! Above the first of that signaling set of holyday-tetrads flies one of I AM WHO I AM's signs of witness! How far can coincidence possibly stack on top of already impossible levels of coincidence and

Clear Signs of Trouble and Great Joy

not demand being acknowledged as designed communication?

Columbus discovering America (from the perspective of the European world of his time) is no back-burner issue. Even though The New World was the Native American's Old World, the European's Old World was deeply changed by Columbus discovering the New World. It doesn't matter how much we try to whitewash and convolute history. Columbus' discovery of America was as paradigm shifting for fifteenth century Europe as it was for the unfortunate Native Americans. So the Manifest Destiny of Andrew Jackson's era grew into the Progressive deceit of Barack Obama's scandals, spreading the same sense of loss the Indians' experienced to everyone who fundamentally understood and loved America's Exceptionalism.

Since, and even before, America's founding, the faithful efforts of godly people continuously struggled against the arrogant deceit of unrestrained, governmental tyranny. Columbus' discovery of America developed into God's glove for slobber-knocking Hitler's Final Solution only seventy years before Barack Obama gave Iran an inside track to the development of nuclear weapons, a pallet of US dollars for spending along the way, and an ideological smoke screen for covering up its vows to scrape Israel into the sea and destroy America. Today, the American political establishment is restoring anti-Semitic tyranny.

AD1515-1581: the Ottoman-Turk Empire took control of the Holy Land in 1517, beginning Jerusalem's Jubilee Path. It had defeated the Roman Empire, which had defeated the Macedonian Empire, which had defeated the Medo-Persian Empire, which had defeated the Babylonian Empire, which had defeated the Assyrian Empire, which had defeated the Egyptian Empire, each of whom controlled the Promised Land at least some time during its stint at being a world empire. Many eschatologists, including myself, believe these chain linked empires were the mountains and kings represented by the seven heads of the beast (Revelation 17:9-10). The Ottoman-Turk Empire achieved the first criteria for being the seventh of these empires in 1453, when it broke down the walls of Constantinople, turned its churches into mosques, and renamed it Istanbul. The Empire achieved the second criteria of its distinction two years after this particular tetrad of God's testimony began flying over the world's skies. History and testimony, alike, make it clear that this empire was the dragon's seventh-head. The Jubilee Path crossing the Shemitah Path exactly eight Jubilee's later, in 1917, the year Jerusalem was set free while the head of this dragon was being mortally wounded, is immensely significant. By the light of the First Aliyah, anyone paying attention could have read, clear as a bell, the 1 Elul, 1914 shadowy line drawn across The Empire's dirt. The Empire's loss of Jerusalem three years after its 1 Elul eclipse of 1914, wrote, "Mene, mene, tekl, parsin," across The Empire's course of history.

AD1949-2015, 1967-2033, and 1985-2051: ALL HAIL JESUS CHRIST! KING OF KINGS!! LORD OF LORDS!!! The very year after those three 1,948 year epochs of human history end, the one and only 19 and 48 sign of witness comprised of three

The Late, Great 1948

holyday-tetrads began. Those holyday tetrads-of-testimony intertwine with the next sign-of-witness beginning in the year Israel wins control of God's Holy City, 1967. Then those two intertwine with a third sign-of-witness beginning in 1985. Under the first (and second) sign-of-witness, Israel takes control of God's Holy City, Old Jerusalem. Under the second (and third) sign-of-witness, Apophis rendezvous with Earth, portending the beginning of the Tribulation. Under the third sign-of-witness, alone, Christ returns to reign from Jerusalem, according to what the unique tetrad has to say in harmony with Apophis. The third tetrad-of-testimony, the one under which Christ returns, begins thirty-six years after the first of this triplex of intertwining signs began in 1949. Thirty-six is not only 9X4, the summation of man's affairs (9) within God's Creation (4), but it is also 3X12, God's working (3) of complete order (12). And that is precisely what Christ's return does: it sums up man's chaotic affairs in Creation for Christ's reign of complete order.

How prophetic could celestial motions be? The following history and prophecy occurs under these three intertwining tetrads-of-testimony.

In 1967 Israel won control of Jerusalem, God's Holy City from where Christ will reign, completing God's preparation of Israel for His endgame plan. Then al-Qaeda set into motion the first step of its plan to control the world, flying two airliners into the World Trade Center and another into the Pentagon on September 11, 2001. God replied with His own warning on September 11, 2015, lowering the boom of a giant crane onto Islam's premier mosque at Mecca, exactly fourteen years, two hours, and thirty-seven minutes (rounds up to three hours) beyond the moment al-Qaeda's first airliner struck The World Trade Center. Then on May 14, 2017, the great President Trump announced America's recognition of Jerusalem as the official capitol of Israel, ending the five-hundred year Jubilee Path under the sign-of-witness to hang over Christ's return, the Path having begun under the 1457-1523 sign-of-witness. The KIC 9832227 not-a-new-star in the constellation of the ever present Truth will show America that, yes, it can repent from its deceitful, liberal harlotry of arrogant lies. On Friday, April 13, 2029, Apophis will blast past earth only twenty-five-thousand miles away, portending the beginning of Revelation's Tribulation. Then, during the Tribulation, the 2032-3 unique tetrad portrays the Tribulation's nature like a verifying signpost marking the ultimate pinnacle of liberal tyranny, the Antichrist demanding worship as God. And quite presciently, the unique 2032-3 tetrad will end the tetrads-of-testimony begun at Jerusalem's return to Israel's possession in 1967. Then comes Easter Sunday, 2036, the day after Passover, when Apophis' second rendezvous approximates Christ's return to dwell with mortal man as Lord and King. By then, if my proposition about this massive system of signs being THE warning of Christ's return is correct, evil will be a smoking pile of wreckage. Before the end of the third tetrad-of-testimony, The Age of Aquarius, The Waterbearer, Christ providing prosperity, peace, and righteousness to all mankind, will have begun, the topic of Chapter 9: Holy, Holy, Holy.

Three overlapping tetrads-of-testimony (the number of God's doing it) span Israel's reunion with God's Holy City and the signs-in-the-sun-moon-and-stars-

Clear Signs of Trouble and Great Joy

presaged return of Jesus Christ, The King of Kings and Lord of Lords, ruler of The Everlasting World Empire. His return to reign from Jerusalem will be the most blessed paradigm shift in mankind's mortal history. It will be the day evil is destroyed for righteousness to begin. This 19 and 48 pattern being found amongst the 1949-50, 1967-8, and 2014-5 holyday-tetrads confirms their prophetic nature. It highlights the enormity of change developing at that juncture of mankind's history. That the tetrads-of-testimony begins the year after Israel's return to nationhood in 1948 adds immense significance to those three 1,948 year epochs ended at 1948. These are not coincidental patterns. These are communicative patterns occurring at the most clearly obvious fulfillment of Biblical prophecy in our own times: Israel. In any other field of inquiry, this extent of empirically evidenced order observed amongst chaos would be unquestionably grasped as the certain indication of an irrefutable reality. Even one percent of this amount of order found in radio signals amongst the universe's background noise would establish SETI's proposition of extraterrestrial intelligence as certain truth in the eyes of the entire world. This enormous system of signs, if ignored by science, will verify that science is less a field of inquiry and more a field of iniquity.

With the eternal condition of billions of people to shortly become either heaven or hell, depending upon whether each thinks it is Christ or their own self who is to say what's true, what on Earth makes us think a God big enough to work the Tribulation process in accord with this much perspicuously ordered pattern would be neither capable nor desirous enough to warn everyone willing to listen? Maybe deceit on Earth makes us think that. The enormity of these patterns amongst historical events, Biblical prophecies, and total eclipses of both moon and sun can invoke only insanity as a basis for rejecting the clear message embedded in the coherence of their intertwining aspects. Eyes to see do see. Ears to hear listen. And still, SETI is paid more respect than the Bible. You can't make this stuff up.

AD2072-2138: Whatever event worthy of testimony occurs during this sixty-seven year timespan is yet to be seen. The Bible is prophetic of only one, specific temporal event after Christ begins His reign on earth, although it describes a few details of how the mess made by the Tribulation will be cleaned up and about what life will be like in a righteously ordered world. But that one specifically prophesied event, the loosing of Satan to lead a brief rebellion against Christ, follows the thousand years of peace. It can not be the object of this tetrad-of-testimony, or any of the others. Only two interesting points can be noted about this sign-of-witness. It begins thirty-six years after Apophis' second rendezvous with Earth. Thirty-six is 3X12, the number of God's working perfect order. Then, in Chapter 9, we will see an anciently revealed, prophetic timeline end under this sign-of-witness.

AD2477-2543, 2506-72, 2524-90, 2564-2630, 2600-01, and 2611-77: This set of six tetrads-of-testimony precisely spanning four Jubilees sheds a ray of light upon the future times over which they hang. Particularly in Zechariah, but also sprinkled throughout most of the other prophetic books of the Old Testament, there are references to Ephraim and Judah's return to nationhood for enjoying Christ's reign from Jerusalem. Israel will be elevated to the head of the nations during Christ's millennial reign, and mankind will enjoy the peace for which God's common folk have been longing. Moreover, According to The Book of Revelation, the dead in Christ will be raised to reign with Christ during these thousand years. It is mankind's time to shine in faithfulness before the Lord as a testimony to the universe about Christ's restorative success. Man's number is six. Not five. Not seven, eight, or four, but six. How prescient! We will return to these six intertwining tetrads-of-testimony again in Chapter 9, for they also lay on an anciently made, prophetic timeline.

History offers a plethora of events to see under each of these thirty-two signs. A lot happens around the world in sixty-seven years. We could have chosen any one of those myriads of other events to discuss rather than the ones noted above. Yet it is more than significant that each of the particular events we did find, each one testimonial to the activities developing God's reign over man's affairs, actually did occur under one of these signs. Isn't it more than strange that all of those signs-of-witness occur in correlation, first, with once disputed events of Biblical narrative now verified by empirical evidence, and then with events testimonial to the Biblical truth about the resurrected Jesus Christ, His Church, and His Biblically revealed plan to reign on earth from Jerusalem? Whether or not this set of events represent subjective pareidolia, they are all events involved in God's salvation plan. And each is cohesive with the fundamental purposes of Israel and Christ.

The Bible says all souls exist for eternity, at least it does when read without distorting its literary devices beyond the simpler sense they make themselves. Alas, eternal existence does not necessarily mean bliss. Nor must it mean Hell. It just means no end. Eternity would not be a happy condition to experience forever with a bunch of Hitlers, Jack-the-Rippers, Lizzie Bordens, Karl Marxes, and Progressive accusers working mayhem into everyone's affairs. How and when I AM WHO I AM's perfect "place of abode" came to be mixed with such seeds of self-centeredness is beyond the scope of this book. But it did. God does not tabernacle with mayhem, whether caused by selfishness or anything else. Therefore, everyone desiring to abide with Him can not live for their pleasure alone.

So there has to be a sorting between the desirers of good and the despisers of righteousness. There must be categories. In this case, inescapable, indelibly divided, eternal categories -one for the seeds and fruits of chaos, and the other for the seeds and fruits of order, since those two natures do not mix. Yet, like thoroughly shaken oil and vinegar, they form a froth of mankind's current confusion. Speculation about how God uses this physical universe to separate evil from righteousness is for another book another day. But for this book, today, each of an individual's choices evidence God's discretion for sorting that individual into his or her proper, eternal category. At the commencement of Judgment Day,

> ...from his presence earth and sky fled away, and no place was found for them.
>
> Revelation 20:11b

Clear Signs of Trouble and Great Joy

Signs of God Witnessing Biblical History

Between 2000BC and AD3000, thirty-two times twelve total lunar eclipses occur in a chronological pattern spanning 67 years and displaying 1948, the year of Israel's Biblically prophesied reestablishment. Amazingly, every one

```
|----------------------------------------- 67 years ------------------------------------------|
oooo                       oooo                                                            oooo
|------------ 19 years -------------|---------------------------- 48 years ------------------|
```

of these thirty-two, sixty-seven year spans of 19/48 patterned eclipses occur over one of the following Biblically significant events:

1661-1595BC	Joseph becomes vizier to the Pharaoh of Egypt
1643-1577BC	and serves as vizier until his death.
1585-1519BC	Then a hostile Pharaoh rises up and enslaves the Hebrews.
1538-1472BC	Moses, whom God used to free the Hebrews, was born.
1433-1367BC	Joshua leads the Hebrews into Canaan where they become the nation of Israel.
1028-962BC	King David reigns over Israel,
981-915BC	and then King Solomon reigns over Israel. Then the greed of Solomon's successor
923-857BC	divides the kingdom. The prophet Elijah is sent to raise God's warning against
905-839BC	their division and idolatry. Elijah is taken to heaven, and Elisha replaces him.
847-781BC	Athalia, usurper of Judah's throne, is thrown down before these nations end by 586BC
377-311BC	Greek becomes the international language in which the New Testament will be written to the world.
290-224BC	The Old Testament is translated into Greek.
AD162-228	Christianity was illegal in the Roman Empire, and for 200 years Christians
AD180-246	were slaughtered in the Roman coliseums for entertainment.
AD267-333	Constantine decriminalizes Christianity and the slaughter of Christians ends.
AD390-456	Christianity becomes the official religion of the Roman Empire, and the slaughter of pagans begins.
AD795-861	120 years of "Christian" iconoclasm in The Byzantine Empire ends, and the Shroud of Turin
AD824-890	inspired image of Jesus Christ returns to Byzantine coinage.
AD900-966	The Shroud of Turin is ceremoniously carried from its ancient home of Edessa to Constantinople.
AD1305-1371	The Shroud of Turin enters the annals of acknowledged history at a little church in Lirey, France.

----------Earth enters the first of two epochs prophesied on the most mysterious pillar of its most ancient temple----------

AD1457-1523	Four total lunar eclipses on Jewish holydays hail the discovery of America.
AD1515-1581	The Ottoman-Turk Empire captures Palestine in 1517, 400 years before relinquishing it in 1917.
AD1949-2015	The newly reestablished nation of Israel is admitted to the United Nations,
AD1967-2033	and then, in 1967, wins control of Old Jerusalem, from where Jesus Christ will reign.
AD1985-2051	Jesus Christ returns in 2036 or thereabout.
AD2072-2138	An accurately prophetic path marked onto archeology's most mysterious pillar ends.

----------Earth exit's the first epoch (2138) and enters the second epoch (2477) prophesied on that ancient pillar----------

AD2477-2543	AD2536 ends the first half of Jesus' Millennial reign in Jerusalem according to the extensive
AD2506-2572	system of signs in the sun, moon, and stars (Luke 21:25). Archeology's most mysterious pillar
AD2524-2590	of Earth's most ancient monument displays a system of symbols portraying two timelines of
AD2564-2630	Earth's final history, each timeline entailing six of the final twelve of these thirty-two incredible
AD2582-2647	signs-of-witness. These signs-of-witness irrefutably tie that monument to the Biblical narratives
AD2611-2677	about the Great Flood, Enoch, Noah, and God's Holy Spirit speaking through ancient prophets.

"The heavens are telling the glory of God; and the firmament proclaims His handiwork." (Psalm 19:1)

Figure 51

The Late, Great 1948

  The purpose for the universe will have been worked by Judgment Day, that great Day of Final Sorting. On that day God sorts all who are vinegar by the spiritual sum of their temporal lives into no place found for them. It's not that He's a mean-weenie. It's that He's a giver. They longed to deny His Holy Bible and His claim upon reality; they strove to socially distance from the Almighty and His chosen ones; they masked their mouths against speaking His truths; they thought of wickedness as good and chaos as desirable. So, for all eternity they get what they sought: a place without His Word, His presence, His people, His truth, or His order (a.k.a. righteousness), a place where they will suffer the horrors of eternal chaos as a fitting reward for their arrogance. But the Author of The Word and Designer of Order brings to reign with Him everyone who cries out from sincere desires to be joyful oil instead of that bitter vinegar of their birth. They laid down their right to accuse; they picked up their duty to follow; they drew close to God; they spoke His truths and thought on His righteousness. Therefore, by His grace Jesus turns their vinegar into oil as easily as He turned water into wine. Note again the five tetrads in the 1949-2015 holyday sign-of-witness, the number of grace, 5. It could have been four, or six tetrads. But God made it 5.

  Of these thirty-two tetrads-of-testimony, thirteen involve a particular pattern made of five tetrads, no more, no less. But the 19 and 48 pattern of only one of those thirteen is formed by three, holyday-tetrads, 1949-50, 1967-8, and 2014-5. Seeing the purpose of our universe involve God's separating the temporal froth of its badly shaken, eternal souls into pure oil and bitter vinegar, seeing these tetrads-of-testimony relate to events raising up the key players God will use in that sorting process (Israel, The Bible, The Church, and Jesus Christ), and seeing that His grace is the key process sorting desirers of righteousness into eternal joy, then notice how the frothy thirteen are made of a Glorious One, composed of three holyday-tetrads, and a like-patterned, yet not so gloriously composed twelve. In it we see Jesus Christ and His twelve apostles, as well as, I AM WHO I AM and the twelve tribes of Israel. This isn't like seeing Jesus' face on a potato chip. There are billions of potato chips amongst which we might find His face a few times. There is only one sky-full of celestial lights in which we do find this immense system of patterns precisely relating mankind's history to the message of The Holy Bible. In the number of these tetrads-of-testimony we see doom and gloom (13) turned to righteousness and order (12) by The Glorious One who laid down His position of being Almighty God to humbly participate in the froth of our fleshly struggles. This One and those twelve are all composed of the same ordered pattern of five tetrads (the number of Christ's grace) making the 19 and 48 message, man's affairs completely summed up (19) for restoring complete order to His physical creation (12X4).

  Either Darwin's non-designed, purposeless universe is a colossally enormous, unfathomably intricate coincidence beyond even that of finding *Moby Dick* written in a pattern toasted onto a potato chip, or else I AM WHO I AM intimately designed the universe for the purpose of declaring the glory of Jesus Christ, God's Holy Word given to mankind. Either these patterns display the salvation He brings to anyone willing to believe the Bible for what it says, or they torment meaningless life forms by an inexplicable show of nonexistent glory in a purposeless universe truly cursed with the most cruel indifference towards all its life forms. These heavenly bodies more than just seem to be writing on our wall. They clearly

Clear Signs of Trouble and Great Joy

warn of coming trouble and tell of great joy in a language developed from their own extraordinary depth of convergence with The Holy Bible and history. From marking events to framing historic processes to enumerating meaning, these patterns consistently correlate with Biblical narratives, themes, and prophecies regarding eternal joy sorted apart from unending despair. Only by utter disregard for empirical observation and objective reason can anyone fail to see their purposeful design shining brightly out of Darwin's "purposeless, non-designed" universe.

> That goose egg guess Chucky laid,
> Of a non-designed, purposeless place,
> Cracked on heavenly signs God made
> And went drooling down ape-man's face.
> Not one of The King's prancing steeds,
> Nor any of His gallant men,
> Shall ever put Chucky's Humpty-Dumpty
> Back together again.

God's signs written across Earth's ceiling testify to a reality purposefully designed for the righteous reign of The Glorious One. Like Joshua raised up a pillar of testimony to the commitment made by his fellow Israelites to be God's people, I AM WHO I AM raised up thirty-two 19/48 patterns of eclipses as testimonies to the commitment He made to be the righteous God of all who desire to reflect Him and trust Him to carry their physical souls into new life (4X8).

Now let's follow *information* to the four faces of Revelation's beasts proclaiming, "Holy, Holy, Holy," soon to be that blessed ditty sung from Earth's skies above this communicative God of new life reigning in Jerusalem.

Footnotes

1. AD1948 passed, and man is waxing more evil than ever. Therefore, why do I propose that mankind's evil is measured into those three nineteen-hundred-forty-eight year epochs, seeing that evil continues after they have passed? The year, 1948, was a milestone in mankind's affairs. The reestablishment of Israel was the first circumstance necessary for Christ's reign on earth from Israel in the land God promised to Abraham, their ancestral father. That isn't the completed end of evil; it is the beginning of its end.
2. Bullinger, E.W. Number in Scripture: Its Supernatural Design and Spiritual Significance. (Originally published by Eyre & Spottiswoode (Bible Warehouse Ltd., 1921)). 2014. Alacrity Press. Pg. 162.
3. Johnson, Ken Ph.D. Ancient Book of Jasher, A New Annotated Edition. 2008. Biblefacts Ministries. Biblefacts.org. Pg. 107.
4. Ibid. Pg. 131.
5. Ibid. Pg. 136.
6. Kent, Simon. Ottoman Empire Redux: Turkey's Erdogan Tells the World 'Jerusalem is Outs'. October 4, 2020. https://brietbart.com/middle-east/2020/10/04/turkeys-erdogan-tells-the-world-

jerusalem-is-our-city.
7. Rohl, David M. Exodus: Myth or History? 2015. Thinking Man Media, 6900 West Lake Street, St. Louis Park, MN 55426. Pg. 324.
8. Biblical Archaeology Society Staff. The Tel Dan Inscription: The First Historical Evidence of King David from the Bible: Tel Dan inscription references the "House of David". https://www.biblicalarchaeology.org/daily/biblical-artifacts/artifacts-and-the-bible/the-tel-dan-inscription-the-first-historical-evidence-of-the-king-david-bible-story/, 11/08/2016. This story was included in a Biblical Archeology Review top ten list. The full story was originally published in 2011.
9. http://realhistoryww.com/world_history/ancient/Misc./The_Bible2.htm
10. Schaff, Philip. History of the Christian Church. Vol. III Nicene and Post Nicene Christianity; From Constantine the Great to Gregory the Great A.D. 311-600. 1910. 5th edition revised. Wm. B. Eerdmans Publishing Company. Grand Rapids, Michigan. Pg. 91

Clear Signs of Trouble and Great Joy

Dayenu.

Chapter 9
Holy, Holy, Holy

*It is good to give thanks to the LORD,
to sing praises to thy name, O Most High;
to declare thy steadfast love in the morning,
and thy faithfulness by night...
at the works of thy hands I sing for joy.
For lo, thy enemies, O LORD,
for lo, thy enemies shall perish;
all evildoers shall be scattered.*
Psalm 92:1-4,9

And round the throne, on each side of the throne, are four living creatures, full of eyes in front and behind: the first living creature like a lion, the second living creature like an ox, the third living creature with the face of a man, and the fourth living creature like a flying eagle. And the four living creatures, each of them with six wings, are full of eyes all round and within, and day and night they never cease to sing,
*"Holy, holy, holy, is the Lord God Almighty,
who was and is and is to come!"*
Revelation 4:6-8

    Multiple billions of creatures go chaotically about the earth denying the reality of I AM WHO I AM through closed eyes, accusing Him of all the evil they themselves are working, praising themselves and each other for every good the Lord has done, and giving glory, thanks, and honor to every sort of created thing, but never to Him. Four living creatures full of open eyes, in front and behind, surround the throne of God in Heaven, ascribing holiness to I AM WHO I AM, singing His praise, and giving Him glory, thanks, and honor. The difference between heaven and earth is enormous, for the time being.

    Ezekiel also saw these creatures (the ones full of open eyes) in a vision while he was living at Chaldea's Chebar River. They were under God's throne, maybe supportively, as it descended to Earth amidst fire and smoke. They seem to have had a particular relationship with our physical world, moving beside wheels while separating mankind's flawed Earth below from God's holy throne above, as if He moves on the praise, glory, honor, and thanksgiving of the living remnant of a dying creation.

    [19]*And when the living creatures went, the wheels went beside them; and when the living creatures rose from the earth, the wheels rose.* [20]*Wherever*

Clear Signs of Trouble and Great Joy

> the spirit would go, they went, and the wheels rose along with them...
> Ezekiel 1:19-20a

In John's revelation, the creatures surrounded both the throne of God in heaven and the twenty-four elders seated around God's throne (Revelation 5:6). They were humans seated between the creatures and God. John referred to them as "elders", not necessarily because they were old, but rather because they were experienced. They once lived amongst foolishness and had to successfully dodge its paradigms to maintain their righteousness. They learned to avoid emotional enticements, pursuing wisdom instead. And if by nothing else, then by faith alone they understood the Lord's paradigms to be actual reality. So the seats upon which these elders sit are engage their experiences of traversing evil's desert on a horse with the Lord's name. The elders represent everyone who abides in Christ, while living through this place of anti-God bias, to be raised up to share in God's throne with Christ. Many eschatologists interpret the elders as being Christ's twelve apostles and Israel's twelve patriarchs, specifically. But the Bible states that those who live in Christ will reign with Christ. These elders more coherently portray that reign. The four praising creatures surrounding both God's throne and the elders highlight God's reign shared with the remnant of man. In John's vision, the creatures did not move in unison with wheels in wheels, establishing connection between God's throne above them and the earth below. The connection between God and Earth was now enthroned, the elders, expressing the utter extent to which God bestowed His holiness upon the living remnant of the peoples having come to Him through Christ.

The Hebrew word Ezekiel used to denote "living creatures" means *alive, life, live, living, beast, creature, rawness of flesh, maintenance, appetite, years of age, lifetime, springing and flowing of water, company, troop, multitude, congregation,* and of course, it means these four *living creatures* he saw.[1] It obviously is a very illustrative term packed with ideas ranging from the aspects of temporal life on earth to the nature of these *living beings* of heaven above.

Ever since I first learned of these creatures during my adolescent years, I've had an abiding curiosity about their faces. The Bible seems to describe them purposefully, but not ceremoniously. It is almost as if the reader is expected to be familiar with both the faces and why those particular faces were their faces. Therefore, whatever the Bible means by those faces, however deep or shallow their meaning might be, should be found in the concepts of the people who first used them as symbols. But my search through the literature available to me found more speculation about the origins of and reasons for this set of faces than useful information.

For example, these faces came to represent the four Gospels of The New Testament by the third century AD. Some authors say Matthew's gospel is represented by the face of a man for its focus on Jesus' human nature and life. Mark is the lion by its focus on Christ's royalty. Luke is the ox by its focus on Christ's service and sacrifice. And John's gospel is an eagle, for his stories explain the spiritual aspects of Jesus' teaching. Others speculate the man's face is used because man dominates the animals, the lion's for its being the king of beasts, the ox's because it is the strongest domesticated animal, and the eagle's because it is the highest of the birds. Or again, the gospel of Matthew opens with the genealogy of Jesus, therefore a man; Mark's gospel opens with a voice crying in the wilderness, therefore a lion; Luke opens with Zacharia's sacrifice, therefore an ox; and John's gospel, which addresses the divine nature of Christ more than the others, is the eagle, thought by the ancients to have been the only animal able to stare into the sun, the symbol of the divine.

Those ideas are cute, if nothing else. But they don't offer much information about the ancient origin of these faces and what they meant to their originators. If the authors of the Bible used the faces to portray the cherubim in some interrelationship between the throne of God and the being of mankind, then what they proposed did not come only from emotive speculation. It most likely came from some ancient understanding, possibly one revealed to a very early prophet of God. Otherwise this set of symbols would not have lasted the weathering of time. Their significance is implied by the enormous number of other symbols shared between the Bible and early Mesopotamian cultures. That sharing is why scoffers accuse, without evidence, the Bible of borrowing Mesopotamian ideas. However, the proposition of these shared symbols coming to the Bible from sources earlier than even the Mesopotamians quite well describes the Biblical narratives of Noah, Enoch, and even Adam.

But these faces were applied to the gospels in the third century AD. By then, most of the ancient religious symbols had reached pallid states of normality, having had much of their source ideas obscured by millennia of thick cultural patina. Whatever the second century Christians meant by their application of the faces to the Gospels would not be as fruitful to my search as what Ezekiel may have understood about them five-hundred years earlier. And how could we ever determine what Noah, Enoch, and Adam may have perceived of them, unless we might be willing to think coherently about early Mesopotamian history in the light of the Bible, since it has been shown by essence of the celestial bodies, themselves, to be God's Holy Word to mankind.

Certainly the Israelites must have had some cultural recognition of these faces in Ezekiel's time, since communications usually employ recognizable elements. If God meant Ezekiel's vision to be intelligible, then surely something about this set of faces was culturally familiar enough to effect communication. Jonah and the great fish are good examples of revelation communicated through cultural imagery. Nineveh surprisingly repented at the preaching of a Hebrew. Hebrews were a people greatly despised by the Ninevites. One of Nineveh's deities was a fish/man combo, also honored by the Canaanite communities. Surely the stories of a man spat onto the beach from the mouth of a great fish would have spread through the local towns and down the trade routes like wildfire, preceding Jonah to Nineveh. It would have attracted enormous attention. The beeline he made for Nineveh evidently followed that spreading report, because the Ninevites immediately repented at the preaching of this man "sent" from the mouth of their very own fish-god. Imagery delivers immediately profound meaning. Successful communication is built upon well known imagery and symbolism. Ezekiel most likely understood some, then extant, cultural awareness which the four faces in his vision represented, otherwise God may not have used them.

I sensed a possible significance regarding some connection between these living

Clear Signs of Trouble and Great Joy

creatures and God's purpose for creation as expressed in the Book of Ezekiel, the gospels, and Revelation. But I more expected some essential meaning about men crying out to a hearing and capable God, or about His reply. It seems that the Bible's presentations of these cherubim with four faces are always in contexts of God and man's interrelationships. For Ezekiel, the context was the throne of God approaching man's Earth. For John, it was the remnant of man elevated from Earth to share in the throne of God. I never speculated that these faces might have some relationship with the constellations. Most of my curiosity about them had accrued long before I was aware of the Mazzaroth gospel, when I was still misperceiving the Zodiac to be nothing more than a set of forbidden, demonic symbols.

But these four faces cling to the gospel meaning of the Mazzaroth like cockleburs on a cowboy. As we saw in Chapter 4, the constellations lend their imagery to the communication of a way through death into eternal life. The most fundamental gospel themes are inherent in the four constellations represented by these faces, and especially in the eagle's face replacing the scorpion. Moreover, there are twelve main constellations in the Mazzaroth, so it naturally has four "corner" constellations like a clock has quarter hours at 3:00, 6:00, 9:00, and 12:00. The corner constellations became an ancient expression of the earth's four directions as well as its four seasons. The ancients painted them into their frescos, honored them with sacred symbols, and even elevated them to portrayals of their divines.

By their order amongst the other eight constellations, the four faces of Revelation and Ezekiel's beasts are easily identifiable in the Mazzaroth as corner constellations. Three identify with main constellations, but, for reasons which will become amazingly apparent later in this chapter, the eagle identifies with Aquila, and indirectly with Cygnus (deacon constellations). The face of the man identifies with Aquarius. The bull identifies with Taurus. The lion is Leo. With the exception of the eagle, they are all equally spaced in the Mazzaroth -every third constellation. Scorpio is the fourth equally spaced, main constellation. But, for some reason, it has been replaced by the eagle.

Yet, two other sets of four equally spaced constellations are also available for being corners: Sagittarius-Pisces-Gemini-Virgo and Capricorn-Aries-Cancer-Libra. What is so special about Taurus-Leo-[Eagle]-Aquarius, besides the eagle seemingly being a misfit?

While Noah and his family were building a giant ship to float out a global flood, the celestial corners marking the seasons were Taurus at the Spring equinox, Leo at the Summer solstice, Scorpio at the Fall equinox, and Aquarius at the Winter solstice, if we follow the precession of Earth's axis back to the Biblically implied date of the Flood (now knowing that the motions of the celestial bodies conclusively evidence the Bible to be God's holy communication to mankind). Maybe this arrangement of constellations became forever immortalized in mankind's mind by the horror of the Flood. Maybe these were the only cardinal corners of the Mazzaroth mankind knew at that point in time. As we have seen the amazing design of the heavenly bodies mark the historical events fulfilling Biblical prophecies relevant to the death, resurrection, and return of Jesus Christ for restoring all things, then maybe the symbols of these four constellations also delivered more concepts to the ancients than just seasonal weather changes. And maybe, for some related reason, the eagle represented the zodiacal corner which was actually the place of the scorpion's face. In fact, this substitution is enormously evidential of my proposition that the Mesopotamian symbolism and myths were twisted from man's earlier, actual contact with God, that contact having been presented in the Bible without any twist. Since man carried these four symbols with them throughout history, maybe they had become popularized by a more ancient meaning of much greater import than merely seasonal weather change. (Who needs the stars to tell the weather's changing when it's starting to get cold and snowy?)

The Biblical perspective of history acknowledges a high degree of intelligence and sophistication in mankind from the day of its creation, rather than the tree hanging, chin drooling, grunting ape-man of Darwin's fantasy. Biblically speaking, an intelligent mankind experienced the man-bull-lion-[eagle] as the constellations of the equinox/solstice sunrises and sunsets of their first seventeen-hundred years. It would have been their first memory of the lights in the sky, the signs setting the seasons and the appointed times, and the imagery of the eternal hope their Creator spoke to Adam, Eve, and against the serpent (Genesis 1:14). First memories tend to be lasting memories. Especially when they are punctuated by such a hellacious disaster as a global flood. Maybe the artwork in the sky spoke a concept larger than merely the passing of equinoxes and solstices indicating when to drop a seed in the dirt and wet on it.

Correlations with this celestial artwork strongly suggest that the ancient Mesopotamians perceived some semblance of a spiritual being coming from the seed of a woman, not so much as a propagation of the race, but more as the overthrower of life's degrading circumstances and processes, an overthrower of chaos, the breaker of the serpent's head. Over time, the generational memory of the overwhelmingly catastrophic flood would easily brew up a sense of chaos chasing order through the night, a fear of waking up one morning to find their whole way of life destroyed. The tower of Babel was their collective response to such fears. The tower, itself, was not God's problem with that response. Their collectivism was His problem. Collectives do not allow diverse ideas. Diverse ideas present more possibilities for discovering effective solutions than does the repetitive thinking of a few arrogant power pimps. God desired a world able to discover His clues, not a blindfolded and gagged world led by the consensus of a few elitists. His first utterance regarding this Babel problem was that they were all one people with one language, therefore, they could do anything (Genesis 11:6). Amongst the "anything" they could do was to develop the power of a bully-hood, that great evil of totalitarian, top down governance, which they did. God didn't make man to reflect elitists, government figures, pop-stars, movie stars, foolish athletes, or smiley faced, news fakers. He did not commission man to follow other men. Each human is born for the purpose of reflecting I AM WHO I AM, personally, and following Christ on her or his own two feet. So God destroyed Babel's collective by confusing their language, causing them to scatter around the world and seek solutions to their troubles more independently. Even within the great diversity that followed the collapse of their tower, the memory of the catastrophe which inspired the tower trickled down through history in the form of numerous flood stories scattered around the world.

> The survivors of the Flood in all the legends range from a single individual to a small group. All seem to have been forewarned of the impending disaster...It is uncertain how many separate catastrophe traditions exist, although

Clear Signs of Trouble and Great Joy

the total certainly amounts to many hundreds. They are broadly divisible into those which describe a terrible conflagration, and those which record an all-embracing flood, usually identified with Noah's Deluge.[2]

As the expanding population spread across the earth, each faction would have carried with them the former symbols they perceived to be important. So, we find in Western Europe early symbols thematically similar to those we find of very ancient Mesopotamia. We find essential similarities in various symbols sprinkled around the world down through time, like a few scraps of telltale debris cast forth from a mighty, slow motion explosion.

As we noticed in Chapter 4, the bison and the woman's interrelationship portrayed in the "Venus and the Sorcerer" fresco of the Chauvet Cave paintings suggest a special regard for "the woman" amongst their ancient religious symbols. The large cat and the bison of that portrayal suggest some conceptual association between Leo, Taurus, and the woman. Following that association, Andrew Collins notes how the Milky Way's Great Rift overlays this Venus' legs, such that the constellation, Cygnus, overlays her womb about where the Bison's head does the same.[3] Collin's suggested connection of Cygnus to this early depiction becomes quite interesting; as Cygnus is also found portrayed in the very early Middle East. The humanist perspective suggests the Chauvet Cave's Venus was regarded for her fertility alone. But the Cygnus connection suggests something more.

The Chauvet Cave artwork was found with no caption ascribing it to some fertility concept, and with no museum brochure describing that fertility concept. Neither is there any massive convergence of known, ancient expressions necessarily bearing some "baby factory" concept about this "Venus". However, the rest of the *information,* ignored by frantic clingers to Darwin's fantasy, suggests a much more coherent interplay of rudimentary understanding than simply some idea of a Great Momma Making Many Babies. The alignment of the figures in the Great Hall of the Lascaux Cave with constellations of the Zodiac, as Jegues-Wolkieweiz pointed out, and the zodiacal elements of this Venus and the Sorcerer of Chauvet Cave suggest a very ancient understanding of constellations being the portrayal of directions for escaping death's doom and chaos' scourge, a set of concepts badly distorted long before the Mesopotamians began writing history, a set of concepts orbiting Earth in Cygnus' area of the sky.

Gobekli Tepe, Nevali Cori, Catal Huyuk, Alaca Huyuk, Jericho, and even Jerusalem are just a few cities to which orthodox scholarship assigns histories far older than 4000BC, the approximate Biblical date of Creation. Orthodox scholarship bases its claims of geological epochs on *information,* so it says. But truly, the orthodox paradigm is formed from assumptions which must confine much *information* to eternally sealed dumpsters (not a very scientific approach). Those assumptions leave far more *information* unexplained than the scanty *information* they beg to support their guesses. One teacher of evolutionary fantasy boldly confessed to the trickery of dating methods and their assumptions.

> After explaining why rubidium-strontium isochrons were one of the best dating methods, the professor assigned us the problem of dating a rock using two sets of data. Neither my classmates nor I could get the dates for

that one rock to agree. After making us sweat for a while, the professor explained that he wanted to show us the method doesn't always work...At the end of my geophysics unit on radiometric dating, the professor was going over the long list of assumptions required to convert any measurement of radioisotope amounts into some estimate of age. Midway through the list of unwarranted assumptions and inconsistent results, the professor paused to joke that if a Bible-believing Christian ever became aware of these problems, he would make havoc out of the radiometric dating system! Then he admonished us to "keep the faith."[4]

Of course, keeping that faith involves keeping silent about dumpsters full of *information* its assumptions must lock down and seal up, else Bishop Bell's sauropods might spill the beans on when living dinosaurs were still being seen by man.

A 4000BC (or thereabouts) Creation theory is based not only on the same scanty amount of *information* that skeptical secularism engages, but more importantly, it explains the rest of the *information* which evolution theory must by all means ignore, hide in its dumpsters, and deny, lest their guessing be destroyed. Of course, Carbon-14 and other dating methods are not considered in the Biblical Creation paradigm, because they are not *information*; they are pareidolia carefully crafted to support imagination about billions of years. Proponents of evolutionary processes not only recognize the highly speculative nature of their "knowledge", they admit the insufficiency of its evidence. Simply stated, the secular paradigm is not merely underdetermined by the evidence it admits, it is succinctly falsified by the evidence it must by all means hide. The theory of a God-less, evolved universe fails to explain the massive evidence which completely determines the Biblical paradigm of a young Earth. That's why normal science must bring its own mop bucket and dumpster to every cosmological research project. Likewise, key "facts and data" of the Carbon-14 dating process, also projected into the past by mere speculation, would have been drastically affected by the nature of a global Flood and its lingering affects (for example, "global warming"). Alas, the simply observable reality of endless billions of square miles of multiple geologic layers laid down in planar form of relatively homogenous constitution cry out for the only possible explanation -that of an unimaginably immense depth of water covering the globe through one, very short moment of geological time, washing up all Carbon-14 considerations of everything before it, and for a few centuries, of anything following it.

It is a little amazing how modern historians and archaeologists go searching through ancient histories looking for verification of their theories, yet they drop those same ancient histories like hot potatoes whenever they attest to a global Flood, or to the great length of life before and shortly after that Flood. As amusing as was all the scoffing at King David's existence in the face of the Tel Dan Stele, even more amusing is the scoffing at the Great Flood in the face of some six-hundred histories, legends, and myths of a great flood from sources all around the world, sources other than the Bible. That's not at all consistent. Consistency, if you recall from Chapter 2, is a necessary criteria of scientific statements. And the ever expanding accumulation of new geological observations articulate the Bible's Flood narrative in reality's own terms, while, at the same time, they invalidate the uniformitarian theory of ageless eons dreamed up during the naïve, bumbling infancy of geology. Ever more

Clear Signs of Trouble and Great Joy

observations are demanding science to play by its own rules of Tentativeness towards its evolution paradigm. But in order to preserve that cute, little ape-man born of geological naiveté, it won't. The duplicitous treatment secular historians give to evidence demonstrates the abiding insincerity of their search for evidence to support what they want to believe about the past, an insincerity extending from worse than just infantile naiveté. Their mental behavior smacks of a global flood of bias…outright deception, spoken truthfully.

When the reality of a Global Flood enters the picture through acknowledging all *information* as a whole, then the explanation of reality's picture requires an Almighty God having the ability to bring about that Flood, and having communicated His plan to someone. This is the narrative of the Bible and many of the worldwide flood stories. And when *information's* implications are thusly acknowledged, the Venus-and-the-Sorcerer connection with big cats, the bison, and Cygnus begins to hint at some core set of ancient concepts involving man, lion, bull, and eagle.

The Bible's thesis offers the perfect explanation for why the man, the bull, the lion, and the eagle have since time immemorial represented the four corners of the heavens and the earth. But to see this explanation, one must open his eyes to the Bible's proposition about itself being the true message of the Holy God to a myopic mankind living in deliberate denial of Him. The Bible says the sun, moon, and stars were made for signs. That proposition is soundly supported by the massive system of patterns the celestial bodies make, rationally and scientifically evidencing the purposeful design of the universe for displaying Biblical messages. These four symbols are amongst those signs.

> [4] He has caused his wonderful works to be remembered.
> Psalm 111:4

In the first chapter of Genesis, we find Adam interacting with God on an intellectual level. God commands the man to till the Garden, to reproduce abundantly, and to differentiate carefully between everything they could eat and the one thing to definitely not eat. He led the animals before Adam and accepted the names Adam gave them. We don't know how long or how deeply involved were Adam and Eve's interactions with God prior to biting into what God said not to eat. So the proposition that God may have told Adam at least some of the names He gave to the stars is not unreasonable.

> [4] He determines the number of the stars,
> he gives to all of them their names.
> Psalm 147:4

As we've discussed before, to the ancients, names had meaning. It is indeed possible that, after their fall from obedience, Adam knew more about mankind's salvation than just some seed of some woman eventually crushing some serpent's head (Genesis 3:15). Further information from other early Hebrew histories imply that, by the time of the Flood, the Sethites (the lineage of families from Adam through Noah) had a detailed knowledge about the motions of the constellations and celestial bodies, an understanding of the meaning they were created to convey by at least the names God gave the stars, and an ominous perception of a coming, worldwide doom.

According to the ancient Hebrew history, the Book of Jasher, Adam's great grandson, Kenan[5] was exceedingly wise and knowledgeable. He became a great teacher and leader of all the people. By his wisdom and knowledge he discerned that God would eventually destroy mankind in a flood. So he wrote his knowledge on stone tablets to keep it safe.[6] Enoch was born in the seventh generation from Adam. The Bible states that he walked with God so closely that he didn't die. He was taken up to Heaven alive (Genesis 5:21-24), the same way Elijah was. All of the ancient Hebrew histories refer to the Sethites' effort to acknowledge God and live in accord with what their parents, Kenan and Enoch, for instance, taught. Recalling that the Hebrew term for righteousness involves a rational, investigative nature, that intellectual responsibility aimed at knowing things as they really are, it is not surprising to find a carefully, yet not perfectly, preserved historical and cultural knowledge having been passed down to Noah by Sethite teachers. Noah, in turn, taught his children, who successively passed their knowledge down to Abraham (although maybe only one was left to teach him, Melchizedek). There has always been a remnant faithfully preserving the past at the same time there has been a horde liberally twisting it to their own tastes. Eventually, Moses may have tapped into Sethite preserved history, amongst other sources, when developing his narrative. Maybe he found seeds and roots of old Sethite stories on the dusty, ignored, basement shelves of Egyptian and Midian libraries. For within the rational nature of righteousness is that penchant to know things in accord with what can be observed, reality proclaiming itself to be true by its own *information*. Therefore, every effort to live righteously would value knowing history accurately as a moral obligation. So a righteously thinking type of people, finding it imperative to preserve *information* accurately, might even find themselves entrusted with the Oracles of God. The discovery of some deep respect for recording history and knowledge by an ancient lineage of people is no more surprising than is the coherence of ancient details written into those Hebrew history books.

The Book of Jubilees introduces us to Enoch's role as a great teacher. The Bible and other ancient Hebrew histories indicate that Enoch lived his entire life during the lifetime of Kenan, Adam's great-grandson. Even Adam was alive during most all of Enoch's life. It wouldn't be rational to think they didn't know and interact with each other on a familial basis. Of course, science profoundly scoffs at this purview of history. But recall, science must apply mop and dumpster to *information* before its guesses will survive the scourges of evidence. Hebrew history humbly scoffs at science's great *information* filled dumpsters.

Kenan became suspicious that God would eventually drown mankind in a flood, as evil was escalating beyond what God could tolerate. Adam was also predicting an end of their world. But Enoch would have the best knowledge of approaching doom. He was given a vision of "...what was and what will be..."[7] Nothing was not revealed to him, his book proclaims. Indeed, the memory of him spread throughout the Middle east, Anatolia, and Egypt as being the first scribe, the inventor of writing, and the messenger of God. The Greeks portrayed him as Hermes, and the Romans as Mercury. Is it then so surprising that the ancients would memorialize such things about some real person of Enoch's connection with the widespread story of the disastrous Flood? Josephus adds to the coherent testimonies

Clear Signs of Trouble and Great Joy

of Enoch-like prescience amongst early man,

> All these [Sethites] proved to be of good dispositions. They also inhabited the same country without dissensions, and in a happy condition, without any misfortunes falling upon them, till they died. They also were the inventors of that peculiar sort of wisdom which is concerned with the heavenly bodies, and their order. And that their inventions might not be lost before they were sufficiently known, upon Adam's prediction that the world was to be destroyed at one time by the force of fire, and at another time by the violence and quantity of water, they made two pillars; the one of brick, the other of stone: they inscribed their discoveries on them both, that in case the pillar of brick should be destroyed by the flood, the pillar of stone might remain, and exhibit those discoveries to mankind; and also inform them that there was another pillar of brick erected by them.[8]

It would be nice to know where the land of Siriad was, where Josephus said the pillar remained to his day. Some folks think these pillars are the great pyramids of Egypt. But that doesn't seem coherent with the available *information*. Anyhow, where they might be isn't relevant to the point we're constructing, which is the wisdom and understanding taught by the prominent figures of the Sethites and their penchant for engraving warnings onto pillars.

Regardless of the Sethites' struggle to maintain, teach, and live by their wisdom and understanding of God, not all of their contemporaries did likewise. The penchant of the horde, in those days, was to fall away and make things up as much as the Sethites' penchant was to maintain time tested beliefs and seek more knowledge. So, before the Flood there was falling away, preaching, revival, and falling away again. Each effort for revival produced only a fading success. By the time of Noah, little wisdom and understanding of God was left, even amongst his Sethite relatives, who after all, were human more than Sethite. Consequently, their world was washed away by such a devastating cataclysm that we hardly have a clue about their culture beyond the sketches made by these ancient Hebrew histories and a few trinkets occasionally found in coal deposits.

That the Sethites held a particular and strong interest in the sun, moon, and stars, even after the Flood, is a reasonable conclusion to draw from the Bible's statement that the celestial bodies were created by the Sethites' God for the purpose of making signs and indicating seasons. Thousands of ancient observatories scattered around the globe, built for noting particular motions of the celestial bodies, support that conclusion. And a Sethite tradition of carving reliefs onto pillars and frescos is a reasonable proposition to draw from Josephus' account of Kenan's tablets and the two "pillars of doom" he says Sethites made to preserve their knowledge. So it really isn't a stretch of the imagination to think these pre-Flood Sethites developed and disseminated symbols as visual aids for incorporating a particular understanding of the heavenly bodies into their teaching.

> And [Kenan] grew up and he was forty years old, and he became wise and had knowledge and skill in all wisdom, and he reigned over all the sons of

> men, and he led the sons of men to wisdom and knowledge...and he turned some of the sons of men to the service of God.[9]
>
> And the soul of Enoch was wrapped up in the instruction of the Lord, in knowledge, and in understanding...And Enoch rose up according to the word of the Lord...and he went to the sons of men and taught them the ways of the Lord, and at that time assembled the sons of men and acquainted them with the instruction of the Lord...And he taught them everlasting life.[10]
>
> And Methuselah acted uprightly in the sight of God, as his father Enoch had taught him, and he likewise during the whole of his life taught the sons of men wisdom, knowledge and the fear of God, and he did not turn from the good way either to the right or to the left.[11]
>
> And [Enoch] was the first among men that are born on earth who learnt writing and knowledge and wisdom and who wrote down the signs of heaven according to the order of their months in a book, that men might know the seasons of the years according to the order of their separate months... And what was and what will be he saw in a vision of his sleep...he saw and understood everything, and wrote his testimony, and placed the testimony on earth for all the children of men and for their generations.[12]

Enoch's book was written to our current generation. That might be a little audacious to say. It might be a little surprising to hear. However, we are either blessed by the most enormous set of coincidental, highly repetitive astronomical alignments correlating with key events regarding some silliness about salvation set forth in a goofy old religious book, or else those astronomical alignments were arranged by The Almighty God to be signs posted over events verifying the truth of His Holy Bible, revealing the reality of salvation, and warning that the time to abandon liberal guessing and follow the truth is very short. If the latter is true, then the Biblical understanding of Enoch is correct, and the Mesopotamian/Greek/Roman understanding is garbled.

The Book of Enoch went lost to the Western World for nearly two-thousand years. During that time Western Christendom spread its tepid version of reality around the world, subjectively believing Enoch's book went lost to all mankind. But in 1773, the Scottish explorer, James Bruce, discovered the Book of Enoch in Ethiopia and brought three copies of it to Europe and Britain.[13] Its opening paragraph was found to read:

> The words of the blessing of Enoch, with which he blessed the elect and righteous, who will be living in the day of tribulation, when all the wicked and godless people are to be removed. And he began his story saying: Enoch, a righteous man, whose eyes were opened by God, and who saw the vision of the Holy One in heaven, which the angels showed me. And I heard everything from them, and I saw and understood, but it was not for this generation, but for a remote one which is to come.[14]

Clear Signs of Trouble and Great Joy

Maybe it is just more coincidence that the generations finding and studying Enoch's book now were the generations to whom its warning was written. Shortly, we will draw some concepts from the Book of Enoch which will quite astonishingly suggest a melding of Noah's post-Flood family, that proposition about the Mazzaroth being the gospel told by the constellations, Gobekli Tepe's Pillar 43, and Hebrews 1:1...

[1]In many and various ways God spoke of old to our fathers by the prophets,

remembering that Enoch is Biblically introduced as being a prophet, and Noah spent many years in his acquaintance.

Where the pre-Flood Sethites may have done their teaching and living can't be determined for certain. From ancient times until today the peoples of Eastern Turkey have regarded the region between the Euphrates and Tigris headwaters to be the place of The Garden of Eden and pre-Flood Sethite activities. For that to be the case, these two rivers would have necessarily been returned to their approximate courses after the horrendously, catastrophic Flood, which creationists say laid down miles of sedimentary deposits over the entire earth. But, the location of the pre-Flood Sethites is not that important to the thesis of this book, although it is a greatly interesting topic. Sethite concepts of life and death and the metaphors contained in their celestial symbols are this book's importance. Since Noah was a Sethite brought up in the company of the greatest Sethite, Enoch, he would naturally carry those concepts and symbols across the Flood to the place of his landing, another popular topic.

The same ancient peoples of Eastern Turkey, and most of the other Middle Eastern cultures, from the dawn of written history to this day, have attributed the landing of Noah's great ship and the history of his first few generations to have occurred in the general region around the headwaters of the Tigris and Euphrates in southeastern Turkey. Whether or not it was near to the lost Garden or the region of the pre-Flood Sethites, four rivers flowing out of that giant swell in the earth might have been a compelling attraction to an old man and his family who remembered the homeland washed away in the deluge.

Gobekli Tepe is nearby.

Most historians view human development as proceeding from lemurs through Darwin's chin drooling ape-man to Gobekli Tepe, where scientism drops to its knees in solemn worship of a few grunting dullards who finally discovered how to mix seeds, dirt, and water for growing their food after hundreds of millennia of picking their nuts and berries while chasing rabbits. Would such a simple perception as dropping a seed into a wet hole really have driven the tremendous ambition necessary for recent, rabbit-chasing chin-droolers to immediately discover the large scale engineering and stonework required for the construction of the monumental glory that is Gobekli Tepe? Would any lineage of creature so slow of mind as to spend hundreds of millennia just picking his nuts for survival simply wake up one morning smart enough to plant a seed in a hole and wet on it, let alone to create the mystery carved onto Gobekli Tepe's Pillar 43? Probably not. That poor critter would most likely have forgotten where he planted the seed before he could discover it sprouting.

We've discovered the Bible to be tested and found true in the historical fulfillment of its prophecies (the life, death, and resurrection of Jesus Christ, the destruction of Jerusalem and dispersal of the Jews around the world, the reestablishment of Israel as a nation in possession of God's Holy Jerusalem, etc.). We have discovered the complex motions between the sun, moon, and earth forming a truly massive system of patterns correlating with those fulfillments of Biblical prophecy. We have even discovered recent history relevant to Israel and Jerusalem to be measured out by Biblically meaningful numbers while correlating with essential Biblical prophecies and extraordinary astronomical motions. Moreover, we've noted this phenomena to smack of putting evil to its end for the glorious restoration of all things lost in The Garden. This systemic intertwining of history, prophecy, and indelibly set signs in the heavens is precisely what we should expect if the Bible is the true, objective Word of the loving, communicating God, whom The Bible says created the celestial bodies for signs, and whom it says named the stars. Just as we hypothesized in Chapter 3, if that God were real, then this system of presciently warning *information* would exist.

Else what man or men could possibly have written such a monumental Book in complete accord with those very real signs and events yet to occur so many centuries into their future? And not just a few unrelated signs and events, either, but numerous ones whose extensive interrelationships thematically cohere with that Book's principal narrative? Chin drooling nut pickers? Someone from their hapless descendents? Could the son of ape-man... really??

But wouldn't the safe passage of a little family through a horrendously catastrophic Flood not drive its first few generations of children to build a monument the size and glory of Gobekli Tepe, and then humble its later generations to writing into a prescient Book the inspirations given to them by God Himself? They had the skills to build a massive ship and the faith to be stirred by that God. They showed thankfulness to that inspiring God by building an altar and sacrificing some of their few, rescued animals (Genesis 9:30). Gobekli Tepe sits in Noah's post-Flood pathway to our history like a well fit paving stone. Some Christians have been viewing Gobekli Tepe in this light. And I think they're right, since *information* also indicates the same conclusions.

Other people see Gobekli Tepe in the light of a horrific disaster. Graham Hancock proposes that Gobekli Tepe monumentalizes a great, world wide catastrophe and presents a warning of its possible reoccurrence. If this were the case, the imagery monumentalizing the survival of a globally catastrophic event and the warning of worse to come most likely would leave a lasting impression upon the symbols and motifs of its survivors. Motifs would be carried away from the monument through time and twisted into a variety of religious concepts by liberal attitudes taking advantage of memory's generational fade.

A few of the earliest religious symbols we know around the Mediterranean reflect this possibility. Gobekli Tepe, Novali Cori, and related settlements prominently represented the bull, the big cat, and the vulture/eagle, amongst others, for instance the fox, and of course, the snake. Frescoes of women giving birth in the fashion of Catal Huyuk's "leopard lady" are also common. But those seem more involved with birth than authority. It was later that this lady took her seat between lions, an ostensible display of authority having likely originated from the idea of mankind's Savior-Seed bearer twisted into an idol for worshipping. Is twisting not mankind's perpetual course of liberal thought? Has almost everything

Clear Signs of Trouble and Great Joy

other than his actual Creator not been the objects of his worship? Symbolism apparently twisted from Gobekli Tepe's images appears around the Middle East, Europe, Australia, and Central America not so subtly, and almost everywhere else more subtly. Even a glimpse or two of it can be discerned amongst North America's indigenous peoples.

That might be the reason the Bible prominently represents idolatry as a woman. Eve chose the forbidden fruit for Adam. The overfed Mother Goddess led astray half of Noah's progeny. Note the harlot God instructed Hosea to marry for her symbolic value, the woman of Zechariah's vision carried in a jar from Jerusalem back to her Babylonian place (Zechariah 5:5-11), and Revelation's harlot in scarlet riding the seven-headed beast (Revelation 17:3-4). The prophets consistently referred to Israel's idolatry as harlotry.

It wasn't necessarily that Gobekli Tepe represented the woman for worship. The sighting stone of Gobekli Tepe's Enclosure D immortalized the woman as the bearer of The Seed for saving mankind, while at the same time using a pair of snakes to represent her legs in the birthing position. The symbolism carved onto the focal point of Enclosure D appropriately portrays the concept of a faulty woman of death bearing a divine seed that saves. Then, by other descendents also spreading from Gobekli Tepe over the centuries, this seed bearer was twisted into the earth-enlightening, reigning woman we now find represented in ancient artifacts strewn around Europe, the Middle East, Australia, and elsewhere. We even have a very representative example of how this twisting process not only can, but does work. The humble, God-fearing Mary of St. Paul's time is now regarded as the immaculately conceived Mother of God receiving worship from the Catholic Church. To the first century church she was merely the honored and beloved mother of Jesus.

The Mother Goddess' early recognition didn't mean the male figure was absent from religious imagery, but only subordinated to the woman by a portion of Noah's descendents. Those stylized, central pillars of every enclosure at Gobekli Tepe seem to represent men. They interestingly resemble a possible "priest and king" motif that became an ancient, prehistoric theme. The female presence at Gobekli Tepe does not dominate men, as it came to do elsewhere. It is merely there on Enclosure D's sighting stone, in her human faultiness, giving birth to mankind's most significant personality.

Ea later appears in early Mesopotamian culture as the God of the fresh waters. Today's historians, acculturated with feminist biases, accuse "a male god", often this one specifically, of supplanting the Great Mother Goddess, who was, in the Mesopotamian times, represented by a strikingly svelte Inanna. But allowing *information* to speak for itself, the female images at Gobekli Tepe did not sit amongst lions in authoritarian pomp as did the later Mother Goddess, Inanna, Ishtar, and their intellectual offspring. Nor did they stand on the backs of lions. Maybe it was the female images who horned into Ea's space. And maybe her power was in the form of twisted truths, the enlightenment of the snake, which accompanied this "great" mother around the world. The man amongst the bull, lion, and eagle bears a significant testimony to that probability. The woman with the bull and lion of "Venus and the Sorcerer" hints at her replacement of that man.

When we focus on Enclosure D's sighting stone, it astoundingly shows what Gobekli Tepe's builders were expecting this woman to bear. And what we see represented is not an homage paid to just any childbirth. The survivors of the catastrophe knew from their Sethite ancestors that their salvation from evil would be through the seed of a woman. It would be easy for their descending generations to confuse the seed of the woman for meaning propagation in general, rather than a woman eventually bearing The Crusher of The Serpent's Head, in specific.

But The Crusher wasn't forgotten. He was their power. The admixture of power and child bearing evidently got confusing. Certainly confusion is a liberal goal. And bear in mind, it is we of the current, liberal generation of mankind who most project the sole aspects of fertility upon those ancient images. Nothing from the past necessarily demands that single aspect of only propagation from the female images. Our liberal scientists haven't simply forgotten The Crusher of The Serpent's Head; they purposefully deny Him. The ancients' symbols didn't come to us with museum brochures attached. We wrote those liberally.

So there should be little surprise at finding elements of God's Word amongst early Mesopotamian symbols. God spoke to Adam, inspired Enoch, and directed Noah to safeguard His creatures. Many Sethite generations gave their attention to the God Adam and Eve experienced, and to Enoch's teaching. Noah was himself a preacher of righteousness who crossed over the horrific Flood, carrying this Sethite knowledge of God to our side of that Flood's paradigm shift.

Today's historians accuse the Bible's authors of copying Mesopotamian themes and narratives when an alternative explanation not only exists, but is even irrefutably evidenced. They make their mistake by prohibiting the God of the Bible from explaining any phenomenon simply because they do not want their theories to allow for His existence. But evidence not only allows for His existence, evidence calls for it. Moreover, evidence not only falsifies their theories of His non-existence, it seamlessly melds with that immense number of deeply coherent signs we've been discovering in science's beloved mother Nature. There can be no honest denial of the evidential conclusion. Together, evidence and those extensive signs demand a more realistic explanation of the Bible's existence than the imaginative hoax science accuses the Bible of being. Normal science launches its challenger of God's reality high into the freezing chill of evidence to their contrary. The coming reality of Jesus' feet on the Mount of Olives will soon hurl science's challenger back to Earth in a smoking trail, too.

The Bible, the Book of Jasher, and even Josephus represent the Sethites as teachers and preachers of their knowledge of God through the generations from Adam to Noah. Noah built the altar and worshipped God immediately after leaving the ark. He didn't leave his Sethite knowledge and traditions behind, like liberals have done. The Sumerian, Babylonian, Assyrian, and other Middle Eastern liberals distorted Noah's post-Flood teaching into their own imagery. It's what liberals do. Like today's liberals distort news for their own gain, the Mesopotamian liberals distorted the teachings of Enoch and Noah. In fact, the Sethite line running from Noah through Abraham were Mesopotamians distorting memories, too. Abraham's own father was a merchant of idols. The liberally twisted up knowledge of God is why I AM WHO I AM called Abraham out of that culture of deceit to preserve God's story from anciently faked news. Why should we be so surprised to find Biblical elements where Sethite memories have been twisted up by liberal imagination?!

The early, Sumerian seven-headed dragon with one head slain became well described by the Bible, not because the Bible copied, but because Noah taught and the Holy Spirit

Clear Signs of Trouble and Great Joy

moved upon whomever It would for delivering God's Word to humanity (Hebrews 1:1). Gobekli Tepe is very busied with serpents. Eve was busied with a serpent. God embedded the first prophecy about salvation in a curse upon the serpent. So that multi-headed dragon's reality lived down through the ages in various stories to our own time, from where we can see it depicted in history, astronomical phenomena, and The Word of God. The first head of the seven-headed dragon's rise through actual history began at Joseph's entrance into Egypt. The Rosh Chodesh Elul, 1914 eclipse fairly warned its last head. World War I stripped the Promised Land from its grip before striking the wound depicted in those ancient Mesopotamian portrayals. The Bible didn't copy. It accurately presents God's knowledge of the serpent's reality which the ancient Mesopotamians personalized with the liberal flavor of their own imaginations. The Bible and the Mesopotamians both acquired their *information* from the same primitive source: Noah, who learned it from Enoch, who received it from God.

Yes, little Sadie, thank God for *information*.

This dragon is pictured in the stars under Leo's claws. The Book of Revelation proclaims that it will rise again as an eighth empire, having come from the seven, to meet its final demise in that terrible winepress under Christ's feet. Reality shows us, today, an Islamic ambition to revive the Caliphate lost to World War I; Recep Tayyip Erdogan dreams of reviving that Caliphate within a neo-Turkish empire. As uncanny as this seems, it is far more spiritual. Everything is spiritual behind the veil that is this physical universe. The Bible is spiritual. History is spiritual. The future is spiritual. You and I are spiritual, whether or not we desire to admit that truth. Else there would not be that massive convergence and correlation of *information* discussed in the preceding five chapters. Nor, I presume, would there be a Gobekli Tepe if the physical realm was not a veil knit for hiding a spirit realm behind mostly mechanistic material. Yet the activities of the spirit realm have left patterns and finger-prints all over the physical universe and human history to shine forth truth in spite of liberally twisted and spun-up deceit. This system of signs is an intersection of celestial patterns with historical events for stitching together the infinitely better sense made by God's Holy Word.

Like the snake and the dragon symbols are spread around the globe, the swastika is another ancient symbol extending from the early generations of people near to these eight survivors of a global flood.

> The 'swastika' (from ancient Sanskrit svastika, meaning 'well-being') is in reality the ancient motif of the four rivers of Eden flowing from the heart of paradise.[15]

The Biblical framework avails this symbol's ancient meaning. The general area around the Mesopotamian headwaters was pivotal to mankind gaining a new foothold on survival after the Great Flood. They carried away from there pre-Flood, Sethite memories of Eden in a nostalgic, "good luck" sense of a lost paradise out of which flowed four rivers.

The most ancient depiction of a swastika comes from just north of Gobekli Tepe and around the northeastern corner of the Black Sea in southeast Ukraine.[16] Hinduism has used the symbol prolifically from ancient times, when it also began using the single lock of hair hung from the back of an otherwise bald head. Not only was the earliest depiction of a swastika found just north of Gobekli Tepe, but the single-lock hairdo was earliest depicted just somewhat south of Gobekli Tepe. A bust was found at ancient Novali Cori. Its face was entirely broken away. The head was entirely bald except for that very Hindu-like lock of hair. And interestingly, the base of the lock, where it attached to the head, was given by its artist the distinct shape of the heads on those Gobekli Tepe snakes. The swastika preceded written history into Northern Europe and westward all the way through the Iberian peninsula. It even made its prehistoric way into the Americas. The bald-head/lock-of-hair 'do, the snake, and the woman went south to India. The snake and the woman spread into Australia. As all the world was repopulated by the children of eight individuals, we find the swastika and various other ancient symbols of this between-the-rivers area sprinkled around the globe from very early times.

Early Australian art bears numerous, close similarities to Gobekli Tepe imagery, including the identical portrayal of the "seated communicators", as some interpret Gobekli Tepe's two half-circles facing each other with a bar-like line separating them. And we will shortly visit an ominous Central American image inversely paralleling Mesopotamia's sages clothed in fish skins bearing apparent buckets of blessing. Space forbids the presentation of all mankind's early symbols spreading out from this obvious cultural beginning to places flung far around the world. That's for another book another day. Simply note here that a number of consistencies between Gobekli Tepe's imagery and imagery sprinkled around the world do exist. Are they evidence that the early origins of man interconnect at this ancient, Mesopotamian monument as the Holy Bible says it does? Or, are we to believe another astronomically colossal coincidence occurred soon after the sons of Darwin's ape-man stopped picking their nuts and berries to wet on seeds they buried in the dirt? The Bible's Genesis record stitches all of these oddities into a coherent and meaningful history glorifying God even unto the stars having been purposely arranged for speaking His good news, while Chucky's darling little ape-man can't even peel the first banana of his story.

Recognizing that the Bible-like Mesopotamian tales are remnant memories of very ancient realities passed down by careless generations of liberal twisters is more intellectually responsible than accusing the Bible of copying Mesopotamian themes dreamed up by ape-man has-beens who, for millennia, couldn't figure out how to bury a seed in a hole and wet on it. If it weren't for this overwhelming system of correlations amongst very specifically patterned eclipses, known Israelite history, and The Holy Bible's various discussions about signs in the sun, moon, and stars, then the proposition that Bible authors copied Mesopotamian themes would be more plausible. But the system of correlating patterns in the sky is too immense to be mere coincidence. The historical events they stood over are too prophetically interrelated to be non-designed and purposeless. Its rationality rises to the level of intentional communication, rock-piles along life's path. Consistently occurring patterns will always mean communication, at least that idea drives SETI's endless examination of the universe's background radio waves. SETI scientists would wet their britches in jubilation if they found even a thousandth of this highly organized pattern within the radio noise through which they rummage; even they know that the existence of a communicative pattern always means the

Clear Signs of Trouble and Great Joy

existence of a communicator. This immense system of interrelated, prophetically oriented correlations verifies the proposal of a Divine Communicator behind history's "anomalous" fulfillment of The Holy Bible's prophecies. It corrects the misconception about the Bible borrowing Mesopotamian themes.

Could the four faces of the cherubim have surfed those waves of misconceptions to us from the crest of Gobekli Tepe?

History, as corrected by this post-Flood paradigm, coherently supports the Biblical *information* about God having been the One who named the stars (Psalm 147:4). Maybe Frances Rolleston was on the right trail even if she did pull a few linguistic rabbits from her hat. Man has had a very long preoccupation with the heavens from his early history. Ancient observatories are scattered across every continent. And similarities about specific constellations and their general meanings are strewn throughout the world's various, far flung cultures like shards of an anciently shattered pot. Recognizing these correlations while considering the artistic expressions at Gobekli Tepe, Catal Huyuk, Novali Cori, and their likes will finally help us to understand why the eagle replaces Scorpio amongst the faces of those four, corner constellations lent to the living creatures of Ezekiel and Revelation. It will also shine forth the glory of God's divine inspiration of The Holy Bible.

I promise, you will be amazed.

Vultures, eagles, and hawks are raptors, so named from the Latin, *rapio: to seize, carry off.*[17] Artwork at Catal Huyuk portrays the practice of excarnation, the pre-burial removal of flesh from the bones of the dead. The nature of the vulture is employed for this dirty work (or yummy work, depending on whose perspective is engaged). In a fresco at Catal Huyuk, vultures are seen gathering to feed on a headless body exposed atop a tower made to lay the dead out for the vultures. The vultures fly the devoured flesh into the sky. Contrary to the earlier conclusions of James Mellaart, the practice of excarnation was not performed at Catal Huyuk as much as it was merely portrayed. Further excavation there has revealed that the large majority of interments were made flesh intact. Evidently, excarnation was more the representation of a belief than it was a common practice.[18]

And why were their dead always depicted as headless figures? Ancient Anatolians perceived a connection between the head and the soul. This connection was often represented by a disc in their artwork. In the Catal Huyuk fresco of excarnation, the dead man's head, portrayed as a disc, is set atop an adjacent tower where another pair of vultures have perched. But they're not dining. They spread their wings over it in caring protection.

A motif from the Aratta area of a southern Ukrainian people thought to predate the Catal Huyuk culture adds sense to the famous vulture carved on Pillar 43 of Gobekli Tepe. A ladder reaches through two, slightly downward curved lines, evidently representing horizons. Below the lower line, at the foot of a ladder, at the base of the fresco, is a reptile, the crea-

Figure 52

ture of the underworld. Between the two lines are three people. Two of them are to the left of the ladder walking towards the west, the culturally understood direction of the dead, the direction of the setting sun. The other person is to the right of the ladder, East, their direction of life, the direction of the rising sun. Above the upper line (horizon) and resting at the ladder's top is a very large disc in which there is that common, ancient, spiral motif. But here it loops only once. Attached to the four "corners" of the disc are what seem to be four flying birds viewed from above, each with its beak touching the disc's circumference. And on the west and the east side of the disc stand two more birds, also facing the disc, as if all these birds are caretaking the disc, or shuttling it off into the sky, or entering it, as if it were an upper realm of some kind.[19]

    This ancient Ukrainian motif offers enough mental food for long contemplation. It offers interesting correlations with the Catal Huyuk depiction of the tower bearing the disc/head where the vulture/eagles have taken it into their care. Catal Huyuk and the Aratta area are several hundred miles apart. Gobekli Tepe is situated between them.

    Gobekli Tepe's famous Pillar 43, the Vulture Stone, is also arranged into three vertically stacked registers. But its top register depicts only one waterfowl accompanying one of three horizontally arranged buckets, handbags, or houses, depending on who's describing them. (I think they look more like the cab sections of 1948 Ford pickups.) Between them and the top of the pillar are short, wavy lines, and below them, two more rows of those same wavy lines extend from over the head of the east-facing[20] main vulture/eagle to the back of an east-facing, long legged waterfowl. An "I" shaped figure stands before its beak as if paying it respectful attention. In the curve of the waterfowl's lap is a snake half curled around a horizontal, "I" shape. On the west edge of the pillar is the prominent, spread winged vulture/eagle holding a disc on its east pointing wing. To the east of it and below the long legged, waterfowl is what some describe as being a vulture/ eagle chick, also facing east. All of these birds face east, not as the stone is actually oriented, but as if it were a map, maybe a map of life. But the snake in the curve of the waterfowl's lap faces downward. The snake's tail proudly mimics the waterfowl's head and beak, falsely portraying an east facing bird. Slightly below and in front of that mockery lies the horizontal "I", as if having fallen for the snake's deceitful imitation. These four symbols, the waterfowl, the downward facing snake with its waterfowl mimicking tail, the horizontal "I", and the upright "I" occupy the upper east side of the Vulture Stone, seemingly in

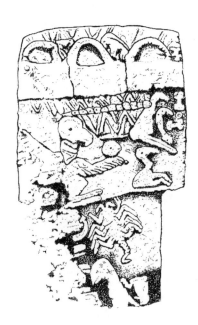

Figure 53: Gobekli Tepe's Pillar 43

Clear Signs of Trouble and Great Joy

isolation.

If we allow the Biblical narrative of Noah's Flood to intermingle with similar symbolism found in ancient Northern Mesopotamian artwork, and even in obviously related symbolism scattered throughout the world's various myths, legends, and histories, then these four symbols seem to form an astounding key to the entire set of Pillar 43 imagery. But we need more description before we can see the keys operate.

On the west side of the bottom register is a snake curling outward from Enclosure D's wall and curving upward onto the pillar. Below it is the face and feet of a fox also leaping out from the wall, as if both enter the pillar's scene from outside Enclosure D. To the immediate east of the snake and the fox's face is a scorpion facing the feet of that prominent vulture/eagle in the above register. Below the scorpion, just beyond the tip of its stinger, is another vulture/eagle giving a ride to a headless man on its back. Poor chap.

Mystery attracts attention. The figures on this pillar have attracted many explanatory stories. Some are plain silly. But others better correspond to available *information*. Recalling that Gobekli Tepe's "nephews and nieces" of Catal Huyuk portrayed the dead as chaps who've lost their heads into the care of vulture/eagles, the most common understanding of the Vulture Stone is that it deals with the problem of death: the headless chap below died, and his body is being carried into the sky on the back of a vulture, while the confident vulture/eagle above guides his soul on its protective wing. But are we prepared to understand just how deeply available *information* carries Pillar 43's symbols into the very same meaning of death as presented in The Holy Bible? And could that depth of meaning be a clue as to why this monument was buried? And to why the eagle supplants Scorpio in the faces of the cherubim?

If we ignore anti-Bible biased paradigms, which, after all, are truly falsified by every shred of the very real *information* filling up normal science's dumpsters, the meaning of Gobekli Tepe opens up like a blooming flower. And that is not a temerarious proposition, since Gobekli Tepe is located in a historically, very Sethite area. The proposition helps explain the mystery of why Gobekli Tepe was covered over. Even Graham Hancock hypothesized that it might be a buried, monumental communication to later generations regarding a catastrophic event which he believes the Gobekli Tepe builders experienced. But he does not attribute the disaster to the Biblical Flood.

The pillar itself might beg to differ with Mr. Hancock.

We bury time capsules containing clues of our current culture and events whether or not we place in them any script beyond a simple greeting. We sent Voyager 1 into deep space with clues of our late twentieth century lives, along with some representative script. The Great Flood would have been a more significant event to its survivors than the Voyager leaving the solar system was to us. And its surviving Sethites, knowing very well why the Flood happened, being the lineage of the God-followers, would certainly have had a warning to share with forthcoming generations. The imagery in Pillar 43's top register is conceptually consistent with the main elements of the Biblical Flood account. No general information of the Biblical account is missing from the imagery in that register. Although its imagery does not flesh out the Biblical sketch of the Flood, it can accommodate every Biblical detail of the Flood. No elements of the register are foreign to the Biblical account. It forms a coherent portrayal of The Great Flood perched atop a related message in the symbols of the next two registers. Pillar 43 bears all the aspects of an artistic time capsule when we humbly allow the Bible enough dignity to speak its two cents.

Consensus dates Gobekli Tepe to 9600BC. Consensus dates the Bible to after the middle of the second millennium BC, at the earliest. Therefore, consensus will bear no thought of Pillar 43's symbols representing Biblical themes. However, reality was not created by consensus. Thomas Kuhn demonstrated how consensus has perpetually gone astray from reflecting reality in science's paradigms. Dumpsters full of rejected *information* testify to consensus' myopic version of reality. Bear in mind that nothing of the past was either scripted or prescribed by the consensus of today's "scholars".

It is entirely possible that the Bible accurately expresses many ideas long held by people more ancient than the Mesopotamians, Sumerians, and Arattans.[21] And some of those ideas are expressed on this pillar.

Close interrelationships of the pillar's symbols and Biblical themes exist. They are objectively observable. And moreover, all of the pillar's symbols sensibly involve Biblical themes, even the fox. So, we will follow them as the nature of true science means for humble minds to do, rather than make up popular ideas while ignoring controverting evidence, as normal science teaches its students to do. Let's allow evidence itself to objectively attach whatever interrelated Biblical meanings might actually correlate with the symbols of the first register. Then evidence from the second register will resoundingly confirm our proposition, not that it will really need any confirmation at that point.

Each of the three handbags, or buckets, or maybe houses (or 1948 Fords), are accompanied by an animal figure, each perched on the cowling of one of the old pickup cabs. From right to left, the first has been described as a frog, or some reptile, like maybe a turtle. The second is somewhat more obviously a mammal with long horns arching over its back, probably an ibex. The third is obviously a long legged, long necked waterfowl, similar to the one facing the upright "I". Hypothetically consider that, in the context of the first register, Noah's ark is metaphorically represented by these three containers, as if together they represent the saving vessel carrying a cargo of birds, animals, and creeping things. Noah constructed the ark with three levels, as God instructed him to…

> [14]Make yourself an ark of gopher wood; make rooms in the ark…make it with lower, second, and third decks…[19]And of every living thing of all flesh, you shall bring two of every sort into the ark, to keep them alive with you… [20]Of the birds according to their kinds, and of the animals according to their kinds, of every creeping thing of the ground according to its kind…
>
> Genesis 6:14a,16b,19a,20a

We can see the division of the ark into three decks conceptualized across the top register, each deck represented as a room, one with a bird, another with the animal, and the last with a creeping thing. The association of basic concepts is, so far, quite complete, nothing missing, nothing extra.

Clear Signs of Trouble and Great Joy

Above our metaphorical ark are wavy lines like what the ancient Mesopotamians used for representing water. Below it are more wavy lines, as if our ark is seen floating upon a great expanse of water. More correlation. (We're doing what science is supposed to do, search for correlations, coherence, and order amongst seemingly random *information*.)

A MONUMENT TO NOAH'S ARK

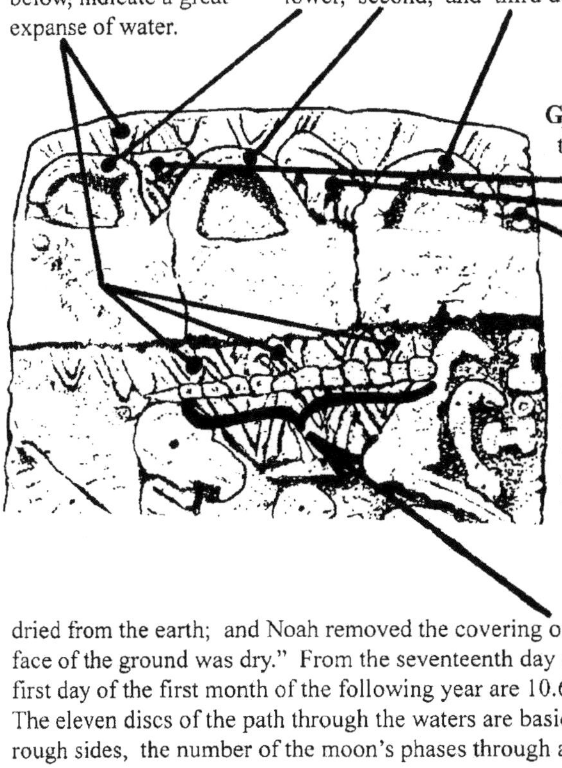

Throughout ancient Mesopotamian history wavy lines were used to depict water. These lines, being above and below, indicate a great expanse of water.

Genesis 6:13-16 - And God said to Noah, "...Make yourself an ark...make rooms in the ark...make it with lower, second, and third decks..."

Genesis 6:19-20 - "And of every living thing of all flesh, you shall bring... birds... animals... every creeping thing of the ground."

Genesis 7:11 - In the six hundredth year of Noah's life, in the second month, on the seventeenth day of the month...all the fountains of the great deep burst forth, and the windows of the heavens were opened.

Genesis 8:13 - In the six hundred and first year, in the first month, the first day of the month, the waters were dried from the earth; and Noah removed the covering of the ark, and looked, and behold, the face of the ground was dry." From the seventeenth day of the second month of one year to the first day of the first month of the following year are 10.66 lunar months, eleven, rounded up. The eleven discs of the path through the waters are basically squared off, each having four, rough sides, the number of the moon's phases through a lunar month.

Figure 54

From the forehead of the main vulture/eagle to the back of the waterfowl's neck are eleven squared off discs. The waters of The Great Flood burst forth on the seventeenth day of the second month of the six-hundredth year of Noah's life (Genesis 7:11). The floodwaters were completely receded from the earth by the first day of the first month of Noah's six-hundred-first year (Genesis 8:13). Either by a little math work, or by a simple count, we discover that the waters of the Great Flood, represented by those wavy lines, overwhelmed the earth and floated Noah's boat for eleven lunar months, the number of squared off discs lined across the wavy lines. Eleven squared off discs not only supply more correlation with the Biblical Flood narrative, but that same correlation answers the question of why these discs are squared off. The moon goes through four phases during each lunar month. These indeed are moons, each with four sides representing the moon's four phases. They proceed from the planning head of the Main Vulture/Eagle to The Waterfowl's guiding protection at the disaster's end. Why a waterfowl and not another vulture/eagle? Cygnus is a waterfowl, a swan, or often you see it portrayed as a goose. It is also known as The Northern Cross, upon which Christ died. God plotted the end of all life by the Flood, and planned the sacrifice of Christ, His Son (Cygnus overlays the planning head of The Great Eagle), for the redemption of the remnant aboard the ark. There, speaking truth and spiritual security to the attentive at the end of the Flood, is a staple of ancient man's diet, a waterfowl, as Christ is The Truth, a staple of saved man's diet, The Bread of Life. Even the plotting and planning of the Flood and the guiding security given those who listened and thus were carried through the eleven months of floodwaters appear on Pillar 43. Pillar 43's symbols correlate with The Holy Bible's narrative, basic concept for basic concept, nothing missing, nothing added!

The pillar even depicts the fall of those who reject warnings, giving their attention to the deceptions of the liberal deceiver's beak-faking, theory-making tail rather than to the instructions of the Traditional Truth Teller passed down through the generations. These symbols speak the basic narrative of God's Holy Word about Noah's Flood and the importance of attending God's *information* as the relevant part of theory formation. It is a warning etched in stone and buried for us, not only of another physical disaster, but even more, a warning against listening to liberal deceit shouting down critically important *information*, sweeping God's *information* out of our public squares, censoring it from our children's education, and throwing it into "science's" dumpsters.

Don't be of the fallen "I" listening to the mocking snake tail! Stand in sound attention paid to all of the *information* given us by I AM WHO I AM. Even little Sadie is wise enough to thank Him for *information*.

Later in Mesopotamia, a popular form of ritual blessing was portrayed as involving handbags quite similar to these on Pillar 43. Andrew Collins and Graham Hancock both note that similar handbags can be identified in artwork around the world. (Bing "handbags in ancient art".) Biblically speaking, to find the symbols of Gobekli Tepe and its related cultures repeated in other areas of the world is no surprise. Noah and his immediate family were at one time the only humans alive. Their culture was the only human culture. All people since descended from them. God scattered the growing population throughout the earth (Genesis 10:9) only a couple centuries after the Flood. Relevant and conceptually simple perceptions from that earliest community at Gobekli Tepe would have been carried along many of their trails, especially from this obviously significant monument.

It is easy to understand the concept of "blessing" so popularly ascribed to these handbags. The complete correlation between Pillar 43 and the Biblical Flood account all but irrefutably proves that Noah's immediate family carved those handbags onto Pillar 43. As representations of the ark, they communicate God's salvation of a remnant from His inescap-

Clear Signs of Trouble and Great Joy

able end of mayhem. God's preservation of a remnant would be perceived by that remnant as a "blessing" from the realm of spirits. Put yourself in their situation; it is easy to imagine the perception; it is human nature. And those three handbags atop the main pillar of what, for at least a few generations, would have been known as THE central monument of THE surviving world would deeply etch that simple, "survival by the blessing of God" impression about those "handbags" into everyone's minds, which then was carried throughout the world.

The human mind tends to "get the picture" when it encounters cohesively interrelating *information* being coherent with previously known concepts. (Only liberals pie what's obvious.) This is quite the same way the mind perceives written text when, knowing the alphabet, it encounters all of these squiggly little marks neatly ordered into sequential rows extending down a page. You understood that sentence not by pareidolia, but simply by the consistent arrangement of the carefully shaped squiggles being coherent with what you know about the alphabet, spelling, word meaning, and grammar. By the same kind of coherence, these handbags imply "Ark", because the waves imply "waters" through which the line of squared off discs imply "eleven moons", while the three figures imply "birds, animals, and creeping things". The Waterfowl facing the upright "I" implies the "upright survivors" listening at the beak of the caring Waterfowl. But the toppled "I" listened to the fake beak of the downward facing snake's tail, and, therefore, perished for their allocation of attention to the uselessness of faked information.

Do I read too much into the consistent coherence between every one of this stone's symbols and the basic elements of the Biblical narrative? Science is made from observing the consistency of coherent interrelationships. I don't have enough faith in mere coincidence to reject the possibility that the Biblical Flood is depicted in the top register of Pillar 43. We've previously discussed the enormous amount of *information* converging in the Mazzaroth, Biblical prophecy, meaningfully measured history, and signs in the motions of celestial bodies. Man could never make that up. Its coherence with Biblical prophecies and themes confirms God's purposeful stacking of the stars, deliberate inspiring of The Word, and faithful influencing of history's path. And now we've found more convergence upon the top of the world's most ancient monument. Pillar 43 looks conspicuously like a sanctified, time capsule buried for future eyes to see what the survivors of a divinely caused cataclysm knew about the God who produced that destruction, spared them, and is now warning us.

Two explanations of this pillar's vultures, headless man, scorpion, conspicuous disc, and vulture chick of the lower registers have been offered by two great thinkers. Andrew Collins and Graham Hancock serendipitously unlocked the door to the most astonishing explanation of an archaeological discovery ever offered. The explanation found through the doorway they opened was discovered the very same way the Rosetta Stone "explained" Egyptian hieroglyphs.

Andrew Collin's associate, Ron Hale, just as serendipitously supplied the catalytic ideas which threw open the unlocked door concealing the Pillar 43 mystery. I merely stepped through the door and turned on the Mazzaroth's light. In its illumination, Mr. Hancock's analysis adds a truly astounding dimension to Collins and Hale's observations of the pillar's symbols. Together, their analyses raise the Vulture Stone's assumed "dealing with death theme" to explanatory heights far more enigmatic than what the academicians of orthodox science and secular history will be willing to rationally process. They will scramble to their dumpsters bearing all of this *information*, refusing to admit that it is *information*. But for the rest of us who don't pie life's tidbits, the explanation rises to crystal clarity from the rock solid support of undeniable evidence.

Yet, each of Collins' and Hancock's explanations, taken by itself, stops one informational step short of assembling the truly astounding coherence the entire Gobekli Tepe monument makes when viewed in the illumination of Enoch's Mazzaroth. Many different theories have been proposed about Pillar 43 representing constellations of one portion of the sky or another. Andrew Collins perceived the main vulture of Pillar 43 as representing the constellation Cygnus, embedding the Milky Way (in which Cygnus flies) into the pillar's sense.

The Milky Way is another concept reaching out from ancient history's darkness. To many cultures it represents a death path for departed souls, fitting the pillar's perceived theme of death. Then Ron Hale scaled a star map of Gobekli Tepe's 9600BC sky to the size of the system of symbols carved on the pillar. He overlaid the map's celestial north pole onto the disc caringly supported by the main vulture's wing. The overlay is based on Mr. Collins' assumption that Siberian shamanic teachings relate the entrance into the afterlife to the Celestial North Pole. He suggests the belief was passed down from the Gobekli Tepe people. The aforementioned Aratta people lived on the geographical path from Gobekli Tepe to Siberia. The Catal Huyuk people lived on the geographical path to the west through Anatolia. The Word of God reveals that mankind spread around the world from the Middle East, which would explain why several Gobekli Tepe symbols are found around the world. Others have not failed to notice the basic similarities of many symbols and myths spread around the world, and so, have attempted to explain the phenomenon. Carl Jung said they were all the expression of humanity's shared consciousness. More orthodox explanations try to discover how population movements could have influenced different peoples as their paths crossed, which better explains the great diversity of details in the myths, but must beg astronomical coincidence to explain the remarkable similarity of their basic fundamentals. Once again, only the Bible offers the most logical explanation. People flowed out of the Middle East to fill up an unpopulated world, taking with them the early basics of a few spiritual truths, then liberally distorted them into myths.

Mr. Hale aligned Cygnus inside the head of the main vulture of the middle register with the Celestial North Pole placed over the disc on the vulture's wing. Scorpio then overlaid the left side of the scorpion in the lower register, and Libra overlaid its right side.[22] The feet of Ophiuchus, the serpent bearer, overlaid the scorpion's head region. Serpens, the snake Ophiuchus restrains, stretches out to seize the Northern Crown at the vulture chick's back. Below the pillar's scorpion symbol, Centaurus aligns over that unfortunate, headless rider of the lower vulture. In the middle register, Hercules overlays the rider's head/Celestial-North-Pole perched upon the Main Vulture's wing, and Bootes appears directly on top of the vulture chick.

This sight was too much for an old skeptic like me to contain. I immediately recognized every one of those constellations by its meaning in The Mazzaroth's depiction of the gospel as discovered and popularized one-hundred-seventy-five years before Gobekli Tepe

Clear Signs of Trouble and Great Joy

was known to anyone. But I was not willing to believe it so immediately, like I also wouldn't believe Mark Biltz' four-blood-moon proposal without verification. I set my astronomy software to November 14, 9600BC, and scaled my own star map of Gobekli Tepe's sky to the same size as a photocopy I made of the Vulture Stone. Mr. Hale is most precisely correct! Try it. You'll like it. Those constellations remarkably overlay Pillar 43's symbols! The alignments are consistent, complete, and coherent. It was so "Rosetta Stone" obvious! It is so like I AM WHO I AM's penchant to communicate over the ages through symbolism.

Mr. Collins and Mr. Hale further noted a few constellations that lay where there were no symbols on Pillar 43, such as most of Ophiucus, Serpens, and the Northern Crown mentioned above. They note Aquila, Sagitta, and Delphinus within the Main Vulture's right wing, and Pegusus at that wing's tip. They note Lyra at the tip of the Main Vulture's beak. They note that the Waterfowl, the snake, the upright "I", and the fallen "I" lie between Ursa Major and Ursa Minor. Underneath the headless vulture rider in the lower register, they note the Southern Cross. Of Bootes, the two Ursas, Centaurus, and the Southern Cross, Mr. Collins and Mr. Hale specifically indicate almost a complete lack of understanding as to why these particular constellations correspond so well with those symbols. But the participation of these constellations in the gospel theme of the Mazzaroth makes those correlations crystal clear. Pillar 43 and the Mazzaroth are the very epitome of how consistency and coherence between two sets of corresponding symbols can testify to a complex message given from a single source to different peoples at different times. This is an obvious verification of the Holy Spirit's function between God and man as portrayed in The Holy Bible.

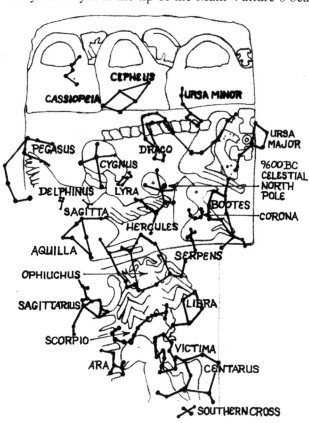

Figure 55: The constellations on Pillar 43

Many Christians since Frances Rolleston's work on the Mazzaroth believe God taught the gospel to the Sethites by the mnemonic usage of star names and constellation pictures. For two centuries they believed that Noah carried the prophetic knowledge of the gospel in the stars through the Great Flood, preserving that information for later generations,

Holy, Holy, Holy

Pillar 43's symbols as understood by the cultural similarities of Catal Huyuk, Noveli Cori, Jericho, and other early sites:

The head is the place of the soul. A bodiless head depicts the soul's existence after death.

Spread-winged vultures/eagles attending a bodiless head depicted the soul's shelter and escort to the entryway of the place for the dead.

The Celestial North Pole is the entrance to the place of the dead.

But a chick depicts new life.

The scorpion depicts the existence and process of death.

Vultures/eagles gathering around headless bodies depict the physical state of death and decomposition.

A headless body depicts a dead person. On Pillar 43, it is a man.

The meaning of the constellations in the far north, Winter sky as deciphered by Frances Rolleston a century-and-a-half before Gobekli Tepe was discovered:

Cygnus, the ever presence of Christ, man's Savior and provision, the intentions of The Great Eagle.

The Kneeler, Hercules, is Christ having entered and returned from the place of the dead, and He holds the keys to death's locked gate (Rev. 1:18).

Bootes is the resurrected Christ quickly returning to reap men's souls.

Aquila, the sacrificed Christ, shelter in the two wings of The Great Eagle.

Ophiuchus, Christ restraining evil from seizing the Crown.

Libra, Christ's righteousness sufficient to pay sin's wages.

Victima (Lupus), Christ receiving death, the wages of our sin.

Centaurus is Christ's righteousness driving Him to that sacrifice.

The Southern Cross, Christ's death-of-a-sinner makes believers the new life-of-His-righteousness.

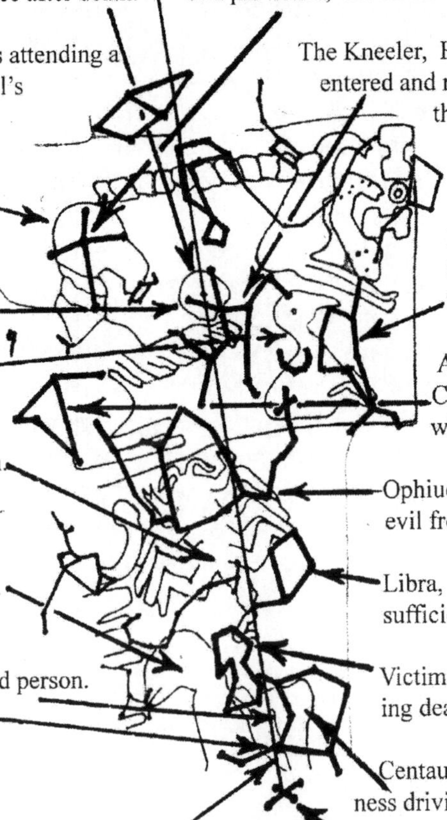

Post Script:
Notice from the Sagittarius and Scorpio Timelines, the Polar Meridian at the Winter Solstice rises from the position of the Southern Cross upwards, bisecting the disc/head on the wing of The Great Eagle, plowing through Draco to stand before Cepheus, the King, like we are raised by the cross and carried by Christ's intentions past sin's demise to stand in eternal life before our new Father.

Figure 56: Meaning of Pillar 43 symbols compared to Mazzaroth meanings

Clear Signs of Trouble and Great Joy

definitely a Sethite thing to do. In Chapter 4, we speculated the possibility that the gospel as told by The Mazzaroth was shown to Enoch who taught it to the world of his time. But we also noted the lack of any written or archaeological evidence from deep antiquity supporting such a concept. But now empirical observation is requiring us to backpedal from that presumed lack of evidence. With remarkable precision, the Mazzaroth, foretelling the death and resurrection of Jesus Christ, overlays Pillar 43's symbols depicting death and preservation thereafter.

The headless vulture rider is known from other ancient artwork of related cultures to depict the course of death. Centaurus is known from the Mazzaroth to depict Christ's death. The Southern Cross below it further defines the manner of His death. Collins noted the connection between the headless man and the disc on the Main Vulture's wing above him. The line of that sense travels through the scorpion, representing the familiar sting of death concept so well known from the Bible. Overlaying the left set of legs on the pillar's scorpion is Scorpio, which represents the same sting of death in the Mazzaroth. Over the Scorpion's right legs lays Libra, measuring the price to redeem mankind from the sting of death as paid by the headless chap perched on the Southern Cross and riding the vulture's back.

Collins speculated that the disc resting on the Main Vulture's wing was the soul/head of the poor chap below being carried to the Celestial North Pole. The ancient Siberians believed eagles carried the souls of the dead to the Celestial North Pole, where they were to enter the afterlife. The later Greeks characterized it as the Gates of Hades. Collins cites as evidence the right angle formed by the boundary between the middle and lower register intersecting a line which can be drawn from the poor chap's neck through the scorpion to the center of the disc perched on the Main Vulture's wing. Onto that disc is aligned The Kneeler, the Mazzaroth's depiction of Christ's departed soul having descended into Hades, and then having returned with power over death, the keys of Death and Hades. So it is that The Kneeler bears the guardian of Hades' gates, Cerberus, in his left hand, having, himself, entered Hades and returned.

> ...I died, and behold I am alive for evermore, and I have the keys of Death and Hades.
>
> Revelation 1:18b

Thus Christ crushed the head of death, the seed of the serpent, preventing evil from seizing God's crown, the eternal reign of righteous life. Ophiuchus, the serpent bearer, with a foot crushing Scorpio's head, overlays the head region of Pillar 43's scorpion, while holding back Serpens, the snake, who is stretching out to seize that Northern Crown from in front of Bootes, the constellation overlaying the vulture chick. In the Mazzaroth, Bootes represents Christ's quick return.

For this restraint of evil and defeat of death a new song of praise and glory is sung to the Lamb of God, Christ crucified. Lyra represents that song of praise for mankind's purchased salvation; it is aligned at the tip of the Main Vulture's beak. Aquila, Sagitta, and Delphinus, more constellations portraying Christ's self-sacrifice and resurrection, overlay the Main Vulture's body. Cygnus, representing the continuous presence of Christ, overlays the Main Vulture's head, as noted above. And Pegasus, the winged, white horse bearing all of the blessings and bounty of victory, flies just off the Main Vulture's right wingtip.

Aboard the ark floating out a global flood were those birds, animals, and creeping things. I wondered about the bird's odd pose. Maybe the pose is meant to situate it within the outline of the ark's three "bucket" handles. And the great horns of the ibex really are not necessary to distinguish the portrayal of "animal". The reptile and the snake are the only creatures facing downward, sharing the obvious concept of death. These oddities lose their oddness to the three constellations overlaying the three blessing buckets representing the ark. Aboard the ark was the Sethlite line of people in Shem through whom God would bring mankind's Savior, Jesus Christ. Cygnus is not only a waterfowl, it is also known as The Northern Cross, the death of Christ paying for mankind's sin, sacrificing for its survival. Could that bird's odd position maybe allude to a waterfowl chick just hatched. Cassiopeia, the Seed bearing woman not only overlays its portion of the ark, but also situates itself directly under it. And the ibex' horns are unrivaled amongst the animals (moose and elk sport antlers, not horns). Recalling that horns portray divine power amongst the ancients, what better animal to perch upon the blessing bucket overlaid by Cepheus, the constellation representing the King of kings. Overlaying the bucket carrying the downward facing reptile is Ursa Minor, the constellation representing God's little flock, His chosen people through whom mankind's Savior was born (the Waterfowl chick of the first blessing bucket). Skeptics and scoffers rant about the ark's inability to deliver a load of life across a global flood. But on Pillar 43 the eye-witnesses of that Flood etched their testimony to God's providential hand preserving the ark through enormous risk to achieve a particular purpose.

> [18]For Christ also died for sins once for all, the righteous for the unrighteous, that he might bring us to God, being put to death in the flesh but made alive in the spirit; [19]in which he went and preached to the spirits in prison, [20]who formerly did not obey, when God's patience waited in the days of Noah, during the building of the ark, in which a few, that is, eight persons, were saved through water. [21]Baptism, which corresponds to this, now saves you, not as a removal of dirt from the body but as an appeal to God for a clear conscience, through the resurrection of Jesus Christ, [22]who has gone into heaven and is at the right hand of God, with angels, authorities, and powers subject to him.
>
> I Peter 3:18-22

The correlations between Frances Rolleston's gospel in the stars (the Mazzaroth) and this Pillar 43 are simply beyond just stunning. They are obviously deliberate. Confer Figures 55 and 56; note where the body of Ursa Minor intersects the path of eleven moons; what significance might that bear?

Just as the Rosetta Stone deciphered Egyptian hieroglyphs, the Mazzaroth deciphers Pillar 43. And behold! Pillar 43 portrays the good news of Jesus Christ! Even the Seed bearer is present in Gobekli Tepe's sermon.

Etched onto Enclosure D's sighting stone is an unabashed portrayal of a woman's

Clear Signs of Trouble and Great Joy

legs spread in birthing position, the sighting hole being her vulva. The woman is bearing her Seed into Enclosure D. The Seed she bore is from the mind of The Great Eagle. It is Cygnus in the head of the Main Vulture. The planned death and resurrection of the Seed would eventually defeat the sting of death and crush the head of the serpent's power over humanity. Cygnus symbolizes God's promised Seed of the woman. The growing population of the remnant who survived the Flood carried the Pillar 43 representation of this prophecy down through the centuries. But they twisted and bent it into the shape of their own imaginations until, eventually, modern liberals accuse the Bible of copying Mesopotamia's misrepresentations of Gobekli Tepe's Biblical truths.

The sighting stone
Figure 57

    The points at each of the etching's ankle, knee, vulva, other knee, and other ankle are roughly in the configuration of Cassiopeia's stars, the Queen depicted in the Mazzaroth who is positioned in The Milky Way at the place from which Cygnus flies. Cassiopeia clearly represents the woman bearing the Promised Seed, Cygnus, flying the length of the Milky Way into the purpose at the Southern Gate for which He was born.

    But the liberal members of Noah's descendants ignored Cygnus and twisted Cassiopeia into the portrayal of The Queen of Heaven receiving the serpent's enlightenment for mankind as The Mother Goddess, Inanna, Ishtar, Isis, Asherah, etc. She even appeared in the primitive artwork of Australia's aborigines with other elements of Gobekli Tepe and the Bible combined into the rainbow serpent. They form a subset of symbols taken from this Gobekli-Tepe/Mazzaroth/Enoch/Noah coherence: the rainbow (sign of God's covenant, Genesis 9:13), the snake, and the Seed bearer.[23] This most ancient Australian depiction of a mother goddess not only sports a birthing-position/Cassiopeia correspondence to Enclosure D's sighting stone, but she is thatched with the ancient representation of snake skin. (Note how the body and legs on the sighting stone are apparently made of two snakes.) Is it just coincidental that we find the "astronaut" of the Central American Olmec sculpture wearing a serpent dangling off the back of his head and holding forth one of those ancient handbags of Mesopotamian depiction, also found on Pillar 43?

    We see pieces of this Gobekli Tepe puzzle scattered around the world. Even the North American Indians' path of souls to the place of the dead is the Milky Way, central to the set of constellations shining the light of the gospel upon Pillar 43's symbols.

    So it is not at all astonishing that Bernice's Hair replaced The Desire of the Nations in the Zodiac. It is not astounding that The Mother Goddess is yet with us today in modern liberals' beloved Mother Earth and in their high-strung feminism. She was with the band of men who arrived at Eridu to build the tower of Babel for making a world empire. And, today, she still rides the back of the scarlet, seven-headed beast, drunk on the blood of the saints, so effectively was the Seed bearer twisted into the Queen of the Heavens.

    But the woman's legs carved on the sighting stone in the birthing position (apparently upside down) is not the only significance to note. This stone was situated in Enclosure D's northern wall. Through its sighting hole, Deneb, the tail star of Cygnus, could be seen

Holy, Holy, Holy

Some examples of Gobekli Tepe symbols found around the world.

The three symbols representing the ark on Pillar 43, and the handbags of the gods held by Mesopotamian and Olmec priests are curiously similar. These "handbags" are found sprinkled around the world, as is the swastika.

The earliest swastika depiction was found North of Gobekli Tepe, just beyond the Black Sea. Throughout history, it has been widely used in India as an allusion to joy, and blessing. Some have suggested that the swastika is an ancient allusion to the four rivers flowing from the Garden of Eden.

This symbol was photographed on the chest of an Australian shaman in the 1930's. An identical symbol was found carved on Gobekli Tepe's pillars six decades later. Some similarities between Gobekli Tepe and ancient Australian art are striking, especially the similarities between their mother goddesses.

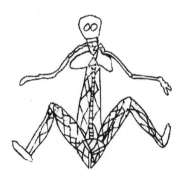 

The ancient Australian mother goddess and the Gobekli Tepe mother goddess are identically poised and almost identically caricatured. The close identification of the Australian's with the rainbow snake, the snake head on Gobekli Tepe's, and the apparent representation of her legs by snakes on Gobekli Tepe's sighting stone allude to an association between Eve, the mother of all living (Gen 3:20) and the serpent in the Garden of Eden.

Figure 58

Clear Signs of Trouble and Great Joy

Victory Story Told in the Milky Way

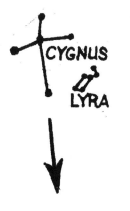

CASSIOPEIA

In the northern Milky Way, Cassiopeia gives birth to Cygnus, The Seed God promised would crush the serpent's head.

CYGNUS
LYRA

Cygnus flies down the Milky Way towards the scene where He will deal the fatal blow to evil and the sting of death. Lyra, the praise lifted up to God by everyone He will save from eternal death, accompanies Him.

SAGITTARIUS
SCORPIO
ARA

Now represented by Sagittarius the archer, at the southern end of the Milky Way, The Seed stands poised with His arrow drawn towards Scorpio, the sting of death. This scene occurs on the solar ecliptic where the sun passes between Scorpio and Sagittarius once a year. As precession brings the sun through this scene earlier each year, eventually its passage will begin occurring at the Winter Solstice. Every year, on the Winter Solstice, the sun will move further towards Scorpio, leaving Sagittarius like a shot arrow. With that arrow shot, Ara, the flaming altar of God's wrath, is poured onto everyone denying Christ, The Bible, and the truth.
At the Winter Solstice today, the sun is poised to represent the launching of the arrow.

Figure 59

Holy, Holy, Holy

setting in 9600BC.[24] When I set my astronomy software to Gobekli Tepe's 9600BC horizon facing north, I was astounded at the sight which blessed this monument's builders. Through it, Cygnus is seen flying in a circle around the Celestial North Pole. Daily it brushed the northern horizon, as if to represent Christ's first coming, then rose back into the sky representing His return to heaven, "His own place" of Hosea's prophecy. What makes this scene particularly beautiful is that Cygnus flies around the Celestial North Pole head first, in proper flying form. Moreover, Cygnus is a swan, which is a waterfowl. The two waterfowl on Pillar 43 not only correlate with the expanse of water represented there, they also correlate with Cygnus' depiction of Christ in the Mazzaroth. In the days of Gobekli Tepe's builders, Cygnus portrayed Christ's death and resurrection ever with us in its daily motions, its saving wings spread over all mankind, ever circling in the Northern sky.

> [7] How precious is thy steadfast love, O God!
> The children of men take refuge in the shadow of thy wings.
> Psalm 36:9

The ancients would have understood it, being offspring of Enoch, the *information* source for all this imagery, the prophet to whom God showed everything. Someone would have remembered some aspect of a concept so enormously important. And so The Lord God's Holy Spirit had ancient mental material preserved in a few conservatively traditional minds for shaping into The Holy Bible. The Bible didn't copy the Mesopotamians. The Mesopotamians twisted up its previously inspired truths.

Apart from Cygnus' motion around the Celestial North Pole, its position in the sky portrays a flight down the Milky Way towards the Southern Gate between Sagittarius and Scorpio. The Southern Gate is the scene of death's eternal destruction in preparation for the eternal life of Christ's believers. Just up the Milky Way, from where Cygnus proceeded, is Cassiopeia, the Seed bearer. Her legs are yet in that "W" birthing position from where Cygnus, the promised Seed of the woman has flown down the Milky Way towards its purpose of defeating death and evil. When Cygnus overlays the head of Pillar 43's Main Vulture, it keys the Mazzaroth's constellations depicting the death and resurrection of Jesus Christ onto the Pillar 43's symbols. Those carved symbols depict the death of a poor chap and the guidance of his soul to the place of the dead in the Celestial North Pole, around which Cygnus flies daily. Peering through Enclosure D's sighting stone reveals Cygnus circling the Celestial North Pole. For there at the Celestial North Pole, as evidentially believed in those days, The Kneeler would someday own death's gatekeeper. Today, those with eyes to see and ears to hear know The Kneeler as Christ having returned from death bearing the keys to its gate. "North" is a very significant, intangible symbol of Pillar 43 and Gobekli Tepe.

Bear in mind that, from Gobekli Tepe, the lineage of the struggling faithful spread out over time. For several decades at least, the growing population would have assembled at this, the only monument of their time. Could it have become a mount of assembly for the seminal roots of all the nations when there were only several hundred or a few thousand souls? A traditionally minded people would have tried to maintain their memories of Enoch,

Clear Signs of Trouble and Great Joy

his prophecies, the Flood, and the God they served, conserving their memories of the condensed gospel God presented to Adam and Eve. To these conservative Sethites, Gobekli Tepe must have been their monument to Enoch's Lord of Spirits, the ruler of righteousness.

The Venus and Sorcerer of Chavaut Cave strongly hints at Gobekli Tepe's imagery carried westward. Note the similarity of the legs and the area of Venus' vulva with those on Gobekli Tepe's sighting stone. But still, orthodox historians will call on Bulova bones so they can insist that Chavaut Cave preceded Gobekli Tepe by millennia.

Figure 60

From Gobekli Tepe, also, proceeded through time the followers of Ham's liberal child, Canaan. Canaan discovered the stone onto which was carved the pre-Flood knowledge of the snake's "enlightenment" given to the woman. Although Pillar 43 had no symbol carved for Cassiopeia of the Mazzaroth, her constellation lays just below the waterfowl chick sitting on the leftmost bucket of the top register representing the ark safely carrying life across the Great Flood. Surely it must have seemed to the liberals that the promised Seed would symbolically come through her enlightenment, instead of coming through her literal childbearing. Consequently, those who would eventually become Australia's aborigines composed the rainbow snake and a Mother Goddess out of the Biblical rainbow's correspondence with Gobekli Tepe symbols. Others moving further through South America to Central America composed the Olmec "astronaut" holding forth his bucket of blessing. Both are snake oriented images. The Mother Goddess spread throughout the Mediterranean world, too, with snakes twisting around her arms. The snake stole the throne of God in the hearts of men through the twisting and leftward listing of Noah's post-Flood, liberal children as they spread from Gobekli Tepe.

Was this Gobekli Tepe the Bible's "mount of assembly in the far north"? Did Noah there monumentalize the gospel shown to Enoch, making it a prophecy of Mount Zion set up in the far north? And did Noah's liberal children there enthrone the snake to be mankind's guiding light slithering onto God's seat atop the mount of assembly in the far north, establishing the harlotry of idolatry's persistent embattlement of God at this "new beginning" of mankind? The basic symbols used in both god-concepts, the eagles and birds of the sky gods and the snakes and women of the earth gods, were at that monument on this mount far north of Israel.

Jerusalem is not far north of Israel, Judea, or even of the Negev. Most of Israel is actually north of Jerusalem. Even so, one time in the Bible we find Mount Zion referred to as

> Mount Zion in the far north.
>
> Psalm 48:2

"Mount Zion" is a metaphor of Jerusalem's being the place of The King of Kings' death, resurrection, and reign. Gobekli Tepe sits far north of Jerusalem on a mount where

Holy, Holy, Holy

Kukulkan, Quetzalcoatl, and Gobekli Tepe

Did South and Meso-America's serpent worship originate from Gobekli Tepe? If the Biblical narrative is history, then all humanity originated from the family that rode out the Flood in a boat. Their culture would have affected all cultures, at least in a few subtle elements.

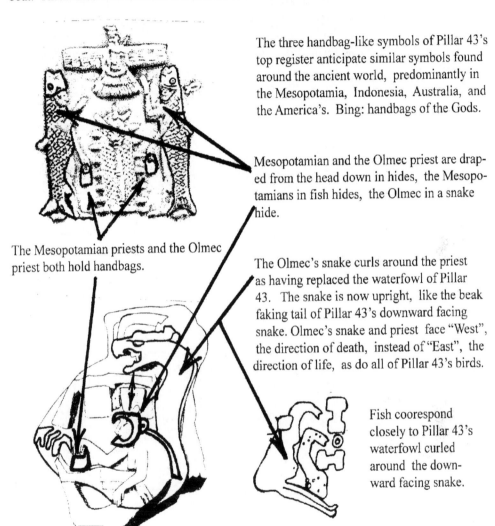

The three handbag-like symbols of Pillar 43's top register anticipate similar symbols found around the ancient world, predominantly in the Mesopotamia, Indonesia, Australia, and the America's. Bing: handbags of the Gods.

Mesopotamian and the Olmec priest are draped from the head down in hides, the Mesopotamians in fish hides, the Olmec in a snake hide.

The Mesopotamian priests and the Olmec priest both hold handbags.

The Olmec's snake curls around the priest as having replaced the waterfowl of Pillar 43. The snake is now upright, like the beak faking tail of Pillar 43's downward facing snake. Olmec's snake and priest face "West", the direction of death, instead of "East", the direction of life, as do all of Pillar 43's birds.

Fish coorespond closely to Pillar 43's waterfowl curled around the downward facing snake.

The early disparity between the sky-gods and earth-gods is apparent in these differences.

Figure 61

419

Clear Signs of Trouble and Great Joy

there evidently was an assembling together for observing the prophesies of that King's birth, death, resurrection, and reign carved onto its pillars in symbolic imagery. But also coherent with Biblical themes is the spiritual attack upon Gobekli Tepe as seen in the usurpation of its symbols by Noah's liberal grandbrats. They depicted the snake giving enlightenment to mankind through the woman, Eve's mistake (according to the faithful), the "wisdom" taught to Cain (according to liberal brats). We've seen how her set of concepts descended this mount of assembly to spread throughout the world in all the Rainbow snakes, Kukulkans, Quetzalcoatls, seven-headed dragons, and the many snakes accompanying Ashterahs, Ishtars, Isises, Inannas, and Mother Goddesses, right on down to the "heavenly" queens of today's liberal feminists striking out against every Biblical tradition like vipers. Of the liberal brats' god, Isaiah states,

> [13] You said in your heart,
> "I will ascend to heaven;
> above the stars of God
> I will set my throne on high;
> I will sit on the mount of assembly
> in the far north."
>
> Isaiah 14:13

These are the only two references in the Bible to some mount in the far north. The perfect alignment of Mazzaroth constellations onto Pillar 43 and the anciently bifurcated trails of two god-concepts, battling down from prehistory to our own day, correlate with the Bible's narrative of Noah, his faith, his preaching and elevation of God amongst his immediate descendents. Ham's mocking his father, Noah, for being drunk further evidences the seed of deep division. Images reminiscent of the three "handbags" of Pillar 43's top register were sprinkled around the world by the scattering of mankind from Babel, where the tower was under construction. Throughout history snakes and eagles became prominent motifs almost everywhere. Eagle wings were attached to horses, bulls, lions, and people communicating the spiritual nature of a sky god, while the serpent slithered away from Gobekli Tepe, Mother Goddess in tow, twisting out new concepts of Earth gods and animism. At the earliest time after the Flood, Gobekli Tepe stood on its mount as a monumental beacon of God's plan regarding man, beaming forth a story of deception, salvation, and restoration by using the imagery of snakes, eagles, and gospel ladened constellations. Gobekli Tepe's circumstances easily avail the concept of some mount of assembly in the far north made for expressing all the hope and glory of Mount Zion.

Let's return to the Revelation 12 portent discussed in Chapter 7. Revelation 12:14 states that the childbearing woman whom the dragon attacked was given "…the two wings of the great eagle…" by which she escaped the dragon. The Greek text of this verse applies the definite article to both the "wings" and the "great eagle", while precisely noting two wings. Koine Greek was a very precise language. As we noted in the last chapter, God led Alexander the Great to spread the Greek language around the Mediterranean world as the vessel for adequately communicating the glorious subtleties of God's ongoing work. (God stacked a testimonial monument of lunar eclipses over Alexander's success.) The definite article used by Koine Greek expresses specificity regarding the concept to which it is attached. It means specific wings instead of just any wings. It means a specific eagle instead of just any eagle. And note further that it is not just THE eagle specified, but it is the GREAT eagle. Could the Holy Spirit have meant all of this precision about those two wings and that great eagle of Pillar 43 on the mount of assembly in the far north when inspiring John to use those expressions? Just saying...

Christ is the King coming (Bootes overlaying the vulture chick) to destroy all evil from the place of His beloved faithful, and then to set up righteousness to reign forever (the Northern Crown following immediately behind the vulture chick). He is The Waterfowl speaking into the "ears to hear" of the upright "I". Christ, with undefeatable strength, keeps His chosen ones, the redeemed of the world (Ursa Major to the right) and the redeemed of Israel (Ursa Minor to the left) safe from deception and spiritual destruction (the half curled snake mimicking, with its tail, a bird beak speaking to the fallen "I", those who refuse to hear the call of The Waterfowl). From The Great Eagle the Great Flood proceeded (the beginning of the eleven "squares" at the Main Vulture's head), and by Him the Flood receded after eleven months (the end of those eleven "squares" at the back of The Waterfowl's neck). By Him the riders of the ark were protected and carried safely through the Flood (the three figures on the three handbags, the waterfowl, the animal, and the creeping thing).

Lyra, The Southern Cross, and The Northern Crown align onto Pillar 43 without any carved symbols to receive them, and yet they clearly participate in the pillar's message by the context of their positions amongst other carved symbols. Other constellations likewise align by context onto the pillar's blank spaces, presenting messages by their coherence with meanings indicated by constellations overlaying the carved symbols around them. Every constellation of the Mazzaroth within Pillar 43's section of sky either lends meaning to or receives meaning from the concepts made of the constellations which do overlay a symbol. The consistent coherence between the Mazzaroth and Pillar 43 evidences a knowledge held by Gobekli Tepe's builders about all of the Mazzaroth's constellations. The complete correlation between the Mazzaroth and Pillar 43 makes that pillar's symbols as understandable as a chapter of a very familiar Book. It's a Rosetta Stone sort of thing.

The first of those constellations without symbols which we must consider is Victima, the unfortunate dog taking the sharp end of Centaurus' lance in the heart. It lays appropriately between the stinger of Pillar 43's scorpion and the headless vulture/rider overlaid by Centaurus. Victima slain by the Centaur's lance represents Christ, the headless vulture rider, taking upon Himself the sting of death. The fox jumping onto the pillar out of the enclosure wall is appropriately nearby. Some scholars have thought this fox and the snake above it were partially covered up by a later addition to the enclosure wall. But could their entrance onto the pillar out from the wall symbolize Christ and Satan's interjecting themselves from the spirit realm into the physical realm, making our temporal fracas between good and evil, life and death, salvation and destruction? And note that the two central pillars of the enclosure wear fox hide loincloths. Did something sad happen to a fox?

[21]And the LORD God made for Adam and for his wife garments of skins,

Clear Signs of Trouble and Great Joy

and clothed them.

Genesis 3:21

I don't mean to imply that these two pillars must necessarily represent Adam and Eve, although below we will discuss an observation which suggests they might. It is quite possible that the Sethites at least conceived of Adam and Eve's first garments as being fox-hides. God Himself sacrificed whatever beasts grew those hides (Genesis 3:21). And if the beasts sacrificed were foxes, then Victima's being a dog relates the sense of that sacrifice covering their nakedness to Christ's sacrifice on the cross covering their sins. The two pillars could represent priests, or possibly priest and king, which is the whole office of Christ, while the fox hides could be an allusion to Adam and Eve being clothed by God's doing. That possibility is certainly coherent with the conceptual interplay between the fox-hide/salvation/Victima metaphor, Pillar 43, and the Mazzaroth. And the possibility increases our understanding of why Christ giving Himself up in death is portrayed amongst the constellations as the dog at Centaurus' spear point.

The Sethites who carved Pillar 43 may have first perceived of Victima as a fox, the first animal to lose its hide in order to relieve some of humanity's distress. A generational drift following Noah may have carried Victima away from Gobekli Tepe's fox into our modern day as a dog. This would explain why the fox is such a prominent image at Gobekli Tepe. And note that the fox is the third of only three creatures entering the Vulture Stone's surface from under the cover of Enclosure D's obscuring wall, (the Main Vulture and the snake being the other two.) The fox, in fact, leaps into the scene, for above it, a serpent had slithered in. First, the deceiver of Eve entered the perfectly created, material world. Then the fox jumps into the snake distorted mess as Victima, the sacrifice to cloth mankind in mercy for its wrongful trek down a serpentine path. Victima's location at the scorpion's stinger of death, driven there by the tip of Centaurus' lance, verifies this set of symbols as representing Christ's death on the cross to rescue from death those who stand upright before Christ's face to receive direction from His Holy Word. The "I" standing receptively before The Water-fowl's beak is the key to seeing Christ's death and resurrection on Pillar 43 through the symbols of the Mazzaroth sky as viewed in the far north.

The next constellation without a symbol to note is Ara. It will carry a concept from a different message keyed onto the pillar by the sneaky-snake deceived "I". Ara is not for the riders of Christ's back. It is for the spitters-into-Christ's-face.

It is most sad that the same process which has brought mankind such well fed, clothed, comfort and relief from physical distresses has also taught mankind to spit all of the purposeless, non-designed Hell of Darwin's fraud into the face of God's truthful Bible. It is sadder yet that the God-given process of crying out to heaven in the form of religious observance became a wide variety of excuses for warring and murdering instead of encouraging and loving, just a different way of spitting. God's Bible expresses the purpose for which this physical reality was designed. It is the place for testing every individual born for his or her correct category at death, heaven or hell. Each test determines the final place for a soul: eternal life or eternal death. Those who believe and follow The Waterfowl ride to their sanctuary upon the very back they caused to be scourged. But those who spit in God's face are not born to safety upon that back which they also caused to be scourged. They are left to perish in The Lamb's flaming wrath, Ara overturned upon a world full of self serving spitters.

The Holy One, by submitting Himself to a gruesome death for offering salvation to everyone, earned the authority to overturn God's wrath upon every spitter-into-God's-face for their spurning His excruciating sacrifice. As the "I" before The Waterfowl's beak is all of the saved who humble themselves so I AM WHO I AM can speak truth to them, Lyra is the expression of praise, glory, and honor sung from the beak of The Main Vulture escorting humbled souls into eternal life. But of the fallen ones stooping to hear from the fake beak of the snake's tail, Ara, the flaming altar of wrath to overturn upon a wicked world is the expression spat from the beak of the vulture carrying the tortured and scourged Man of Salvation, crucified on the Southern Cross. By His death for no sin of His own, Christ earned the right to throw the flaming wrath of Ara onto everyone spurning His sacrifice.

Why an altar and not a vile, such as Crater overturning onto Hydra's back? Altars are instruments of sacrifice. They contain fire. The fire will burn up either the sacrifice or the one failing to sacrifice. By the time of the end, the fire will have been completely affective. Those spitting into Christ's face refuse to sacrifice themselves to receive every Word of God's lips. They are consumed by flames of judgment for their unrepentant denial. Those who believe Christ voluntarily climb onto their death ride during their temporal lives, dying to themselves and rising to Christ before their souls depart their bodies. They are ridden out of the place upon which Ara will pour and delivered to the place where Christ has gone before. When Christ returns with all those who chose to die with Him, He will pour wrath upon the arrogant fallen who chose to live for everything other than Him, thus dieing alone. It's as simple as if you don't board the Argo you drown in a flood of flaming trouble.

Read the signs; observe the prophesied history; or just use Teddy's Theory and take my advice, get humble. Stay humble. Take Jesus at His Word. And, for Pete's sake, don't spit!

By what rationale can all this meaning be seen in the Vulture Stone? Pareidolia? No. Pareidolia is seeing Nebraska man in a pig's tooth; pareidolia is seeing millions of years in the fossils of the same beasts American Indians depicted in their petroglyphs. It is seeing the very orangutan-like Lucy having human feet merely because a trail of modern homo-sapiens-sapiens' footprints were discovered in the region where that probable, orangutan fossil was found. Pareidolia is seeing epochs of time in layers of dirt merely because not all fossil types are found in every dirt layer. It is seeing eons of time in the same fossilization process which a recent steer skull showed to occur quite quickly. It is seeing Timex watches in the current, chemical composition of rocks and the present CO_2 ratios of Bulova bones. Read Thomas Kuhn's explanation of how normal science stays lost for the sake of its own pareidolia…excuse me…paradigms.

Those paradigms of science are the pareidolia, not this consistent, cohesive, coherent system of actual patterns found on Pillar 43 being thoroughly congruent with the patterns of The Mazzaroth displaying the same gospel written into The Holy Word of God. Pareidolia is seeing patterns in chaos where there really are no patterns. But these symbols purposefully carved onto Pillar 43 exist in very organized patterns precisely correlating with those organ-

Clear Signs of Trouble and Great Joy

ized constellations of The Mazzaroth telling the orchestrated plan of salvation for every chaotic soul who will deal with evidence humbly enough to believe what God communicates wherever He communicates it. The same principle of rationality was employed to decipher Egyptian hieroglyphs by the Greek text also written onto the Rosetta Stone is here used to decipher Pillar 43 by The Mazzaroth ladened with the gospel of The Holy Word of God. Why should that same "Rosetta Stone principle" now be denied simply because we find it deciphering the reality of I AM WHO I AM, the Great Flood, Jesus Christ, The King of Kings and Lord of Lords, and His Holy, Holy, Holy Word?! Only fools deny for the right hand what they accept for the left hand. This wonder is not pareidolia. It is message.

But as we noted above, two attempts to explain the Pillar 43 phenomena by the Zodiac fell one step short of full clarity. We have gone a step beyond Andrew Collins and Ron Hale's correlation of Cygnus with The Vulture Stone and found the gospel of Jesus Christ imprinted there. Now let's consider Graham Hancock's attempt to explain the Vulture Stone by some correlation with Sagittarius. What might lay one step beyond his proposal?

Collins and Hale criticize Paul Burley and Graham Hancock for their suggestion that Sagittarius overlays the main vulture of Pillar 43. They complain that Sagittarius does not follow the wide spread of Cygnus concepts around the world, indicating that it is Cygnus which unlocks the Pillar 43 mystery by its relationship to the ancient path of souls leading into the afterlife (The Milky Way). Sagittarius seems not to be a player in that path. Nor is it related to the Celestial North Pole like Cygnus is. Moreover, the Cygnus and Sagittarius regions of the star map are quite different. The scale necessary for each to overlay the carved symbols on the pillar is different. Collins and Hale say coherence conclusively shows the pillar to be concerned with the Cygnus system of constellations. So how could Burley and Hancock's Sagittarius, being so different, possibly correspond to Pillar 43 as well?

I employ a problem solving technique psychologists Allen Harrison and Robert Bramson, Ph.D. describe as synthesis.[25] To a synthesist, conflict is humility's opportunity to find a solution rather than arrogance' opportunity to create an advantage. The idea of synthesis scares most Christians because they perceive it to be the snake's way of corrupting truth by compromise. But I am not referring to synthesizing concepts by altering *information* to create fits out of misfits. That is compromise, at best, which most often is somewhat deceitful. Compromise is always the cheap solution extended from opponents only willing to partially sacrifice. Compromise cuts corners. Synthesis looks for true *information* wherever it can be found; it never leaves any behind nor throws any into a dumpster. Self-sacrifice is necessary to accept implications made by discovered *information*. A synthesist never enters a conflict wondering which *information* set of which opponent will need to be discarded in the end. All *information* is the smoke of reality no matter who has interjected it into the argument, or why. Every piece of *information* has an important point to make, no matter how small, no matter how obscure, no matter who has been its proponent. To a synthesist, solutions are found within *information,* not in within consensus. To the synthesist, *information* must be followed to wherever it leads.

Therefore, we synthesists engage in conflicts to discover *information,* period. Opponents, nemeses, and even outright enemies bear *information,* too. Nobody is totally false. So debate and deliberation are always more useful than swords and war. Synthesists don't care

Holy, Holy, Holy

who brings the *information* to the table; we only care that all available *information* is on the table, placed into its proper spaces within the puzzle reality assembles itself through the natural fit made by its own pieces. The truth always comes together in the eyesight of genuine truth seekers.

The synthesist doesn't care about whose paradigm suggested which *information*. The synthesist is looking for the overall paradigm formed by all *information* brought together. We willingly abandon any paradigm which requires hiding or disposing of real *information*, for reality receives all its *information* like a momma bear receives all her cubs. The synthesist thinks outside of man's paradigms to allow reality to form its own paradigm. That's Rhinnie's Principle. And quite simply, it is the only way a jigsaw puzzle is properly assembled. It is a prejudice towards reality, a bias towards truth; neither my way, nor your way, nor any other way matters except reality's way. For reality presents itself by its own *information*, therefore, sufficient *information* will always form itself into reality's paradigm. It is a KIC 9832227 sort of thing.

Then, since all *information* is reality's *information*, why reject your opponent's *information* just because he's a bit foesome? Rejecting his facts and data is understandable, since facts and data are just some man's impressions of *information*. *Information* presents reality, not my facts and data, nor my foe's facts and data. But everyone possesses at least some useful *information*. Therefore, using also the *information* your opponents bring to the debate avoids paradigm traps like those Thomas Kuhn revealed to be deluding scientific theories, and it breaks the ever worse trap of becoming your own unrealistic fool, a spitter-into-God's-face, an impeacher of reality.

Obviously, the synthesist becomes everyone's enemy. It's a lonely job. But here at Pillar 43, Collins and Hale's *information* amazingly synthesizes with Burley's and Hancock's when the Bible's penchant for presenting multiple meanings through single symbols (that ancient Sethite tradition) is allowed to play into the mix. I imagine Collins, Hale, Burley, Hancock, most all Christians, all spitters, and about every archaeologist, geologist, and paleontologist is mentally pelting me with rotten tomatoes and other sorts of organic "quality" veggies.

Yet Biblical imagery masterfully intertwines reality's *information*, stirring up metaphor, creating symbolism, and delivering multiple meanings and dual prophecies. Expressions contained in numerous prophecies regarding Israel's return from exile in the sixth century BC also speak to circumstances of the Israelites' return from being scattered around the world to become a nation again in 1948. It is only by understanding the basic plan of God in terms of its human history that the fulfillments of dual prophecies can be properly discerned. Usually there is some telltale concept in either the text itself, or in the context of the prophecy, or in the historical circumstances keying one aspect or the other of a dual prophecy to its proper, fulfilling event.

Maybe Pillar 43 provides its own contextual key between Collins' Cygnus message and Hancock's Sagittarius message. If so, and if the Sagittarius message is found to be consistent and cohesive with The Word of God, associating a Biblically coherent meaning with Pillar 43, then we will have found even more validation of Gobekli Tepe's being the ancient Sethite monument bearing the deeply prophetic knowledge of Enoch taught by Noah

Clear Signs of Trouble and Great Joy

to the ancestors of all today's mankind on the mount of assembly in the far north.

Let's propose that The Waterfowl and the half curled snake in its lap are the keys to the dual prophecy. The Waterfowl faces to the viewer's right, as does the chick, the main vulture, and that tiny waterfowl upon the leftmost bucket/handbag ('48 Ford) in Pillar 43's upper register. As read like a map, East is to the right, the direction of the sunrise, the direction of life. But the snake faces downward. It is the only downward facing symbol carved onto this pillar except for the reptile on the rightmost bucket/handbag (Ford). Those form the clue. The scene of prophesied disaster is etched onto Pillar 43's lowest register.

But even more suggestive, the snake's tail mimics The Waterfowl's head, and it mimic's a pointing to the right, portraying a double deception. The snake truly faces downward, not to the right. And it is truly a snake, not a waterfowl. Deception is one of The Bible's main themes about what has come to be man's natural condition. God is true though every man is false. The Deceiver is the mocking, imitating, lying Antichrist which the Apostle John, by inspiration of the Holy Spirit, said is in the world already (I John 4:3) convincing fools that they are true and God is false.

The Waterfowl's "I" stands upright to hear truth from that bird's beak. Salvaging attentive listeners from the destruction of deception is The Main Vulture's objective for the head/soul upon its outstretched wing. This is why it is The Great Eagle. It also is what God did for the rider's of Noah's Ark portrayed at the top of the pillar. He does it by the self imposed sacrifice of the vulture rider in the bottom register. To Noah, that sacrifice was prophetic. To us, it is historic.

But the downward oriented snake curls around a toppled "I" presenting a foreign concept to all of this Waterfowl/Great Eagle/truth/salvation message. Its "I" has fallen onto its side, toppled by fake news from a fake beak formed by the butt end of a deceiver. Since the truth spoken to the upright "I" keys the Cygnus set of constellations onto the pillar to tell The Gospel, maybe the stinky lies the snake hisses to the fallen "I" key Sagittarius' set of constellations onto the same pillar for telling the destruction of The Gospel's detractors. This set of constellations represents Christ destroying death, that is, Sagittarius drawing aim on Scorpio's heart with His arrow. And with that shot, Ara will overturn onto the world full of evildoers, the unrepentant deniers, liars, and spitters-into-Christ's-face. The headless chap, then, represents all the deniers of Christ who will ride death alone forever.

Eerily, when Sagittarius and Scorpio are scaled to overlay the Main Vulture and the vulture chick, constellations of a very different message correlate with the pillar symbols to portray not only the destruction of evil, but also, just as Mr. Hancock proposed, the very era in which evil's destruction will occur. Spookily, that era of destruction portrayed upon Pillar 43 is ours, just as Mr. Hancock deciphered it to be.

The topic of the pillar has now switched from the righteous Sacrifice producing life for the upright listeners to the authority of the Sacrificed One to end the fallen followers of the beak-faking snake's tail. The poor, headless chap in the lower register now represents all those who deny the truth of reality: Jesus Christ. They walk away from a merciful connection with I AM WHO I AM, their only source of life. Instead of accepting the Sacrifice's sufficiently weighed payment for their sins, they must climb aboard that vulture's back and eternally pay for their own evil. They will spend their own paychecks on eternal flames,

Holy, Holy, Holy

having sacrificed their eternal joy for a short life of self-service. They are dead people walking. The only entity above them is now Ara pouring out God's wrath.

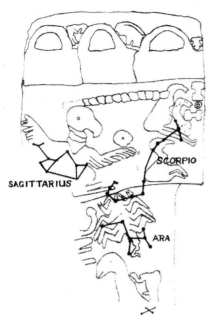

Figure 62

The stinger of the pillar's scorpion symbol is now overlaid by the constellation, Ara, in the position of pouring its wrathful flames onto that lower vulture's living-dead riders. Sagittarius, as we know by The Mazzaroth, is Christ destroying all foulness by the authority of His own self-sacrifice. For those who refuse to spit upon Him, Sagittarius' arrow struck down Aquila to pay for their sins. But, as Aquila falls, Sagittarius has strung another arrow and aimed it at all the world's foul refusal to simply believe Jesus and abide in His Word. This time He aims the arrow straight at Scorpio's heart, Antares, now overlaying the vulture chick. No resurrection for those dead! The sting of death and the deceit of the snake go hand in hand. The disc on The Great Eagle's outstretched wing now represents the sun warning us to look up, for our redemption draws near. Or our destruction draws near; take your choice; God gives to all what they choose.

So, how can we just leap to thinking the disc is the sun warning us of the end? Where does *information* say that? Actually, Graham Hancock didn't just leap there. And I didn't just follow him. The following is *information's* bridge.

Above, we noticed that Cygnus flies down the path of the Milky Way from its birth at Cassiopeia. Cygnus is The Seed of promise born of woman for the purpose to which it flies. Its purpose intersects the earth's history at two different points for accomplishing two different steps of its goal: 1) to plant and 2) to harvest. Both are prophesied on Pillar 43. The first we've discussed: Jesus birth, death, and resurrection planting the opportunity for the eternal life of anyone who accepts the truth of Jesus Christ. The second lies beyond Cygnus, on down the Milky Way into The Southern Gate, the area of sky between Sagittarius and Scorpio, the scene of Christ shooting death in the heart. This conceptual arrangement brings into Enclosure D the view of Cygnus flying from Cassiopeia, through the sighting stone, between the twin central pillar worshippers, South through The Southern Gate to overturn Ara onto a world full of truth deniers. In great Sethite-like fashion, the entire of Enclosure D is richly laid out for the presentation of multiple themes. And in Sethite fashion, Pillar 43 is laid out to present both points of Christ's physical intersection with Earth's flow of history.

We've now accumulated enough *information* to fully appreciate the wonder of how the Waterfowl and the mocking snake act as two keys indicating which piece of sky Pillar

Clear Signs of Trouble and Great Joy

43's symbols represent for prophesying each of Christ's first and second intersections with Earth's history.

The constellation, Cygnus, is a swan, a waterfowl. Andrew Collins points out that a line extending from the neck of the poor chap below upward to the disc on the Main Vulture's wing forms an accurate approximation of Cygnus by intersecting the boundary between the lower and middle registers. The disc on the Main Vulture's wing would be in Deneb's position, the tail star of Cygnus. Recall that Deneb was seen through the sighting stone dipping below the horizon and rising again daily, a continuous depiction of Christ's death and resurrection, the very theme of The Kneeler overlaying the disc on the Main Vulture's wing. When Cygnus, the waterfowl, is placed in the head of The Great Eagle the death/resurrection story of The Mazzaroth aligns onto Pillar 43. They are the constellations Ron Hale noted.

Now, erase all of that from your mind in order to see how the other key aligns onto Pillar 43 the horrifying prophecy of Christ's return to defeat evil. We need to extend the line Andrew Collins drew from the neck of the headless chap through the center of the disc. It is actually the polar meridian at the Winter Solstice. But his line needs to be shifted to the headless chap's right armpit for the more accurate position of that meridian. When I saw the need for that adjustment, I thought the coherence that Collin's line passing through the neck was lost, until I noticed from overlaying Pillar 43 with Winter Solstice star charts that the meridian actually rises from the center of the Southern Cross below the headless chap and extends through the pit of his right arm, the arm God offers for our help, security, and salvation. It passes on through the disc to the northern horizon of Gobekli Tepe's 9600BC sky-chart set to 6AM Gobekli Tepe time, December, 21, the Winter Solstice. At the southern end of this meridian, while Sagittarius overlays the Main Vulture and Scorpio overlays the vulture chick, Ara is pouring its flames onto the poor headless chap. This Sagittarius/Scorpio/Ara set of constellations imply Christ's defeat of evil.

But how does it imply that Christ's return to defeat evil occurs in our time? The most obvious clue comes from the disc on the Main Vulture's wing. Graham Hancock proposed it to be representing the sun moving through the Southern Gate, depicting the era when the sun's position during Winter Solstice is moving out of Sagittarius into Scorpio. And that, he said (and is right), is today! The imagery on the pillar indeed forms a caricature of the sun's passage through the Southern Gate targeting our time, as that time. But the sun does not pass through the central region of Sagittarius (as depicted by the disc on the wing) when Sagittarius is overlaid onto the whole of the main vulture. Even if it did, the sun, as depicted there at the Winter Solstice, would have occurred in the early nineteenth century. That's not now. No global catastrophe happened then.

When Scorpio's star, Antares, is overlaid onto the vulture chick's eye with Sagittarius inside the main vulture's body, the polar meridian rises from the headless chap's armpit through Ara, touching the tip of the scorpion's stinger, rising through Eta Scorpii in Scorpio's tail, and continues on through the solar disc. The polar meridian bisects the disc on the main vulture's wing. The disc now represents a much more discernable position for the sun. Moreover, the center of the disc is appropriately where the sun would be located relative to Scorpio on a sky chart of some future Winter Solstice day. But the sun will be standing in

Holy, Holy, Holy

this position on a Winter Solstice nearly six centuries from our time. Mr. Hancock's theorized cataclysm needs to be a little closer than six-hundred years into the future to be sensational to us now. And we thought maybe this set of constellations overlaying the pillar would certainly portray the same end of evil about which our extensive system of signs forewarn. Have we found this pillar to be spraying its indications all around our time like a bad marksman?

Or maybe we've discovered the world's first lenticular image. We will return to the sun standing at the center of the disc shortly, but first, let's look at the pillar from a different angle. Of course, we can't hold Pillar 43 in our hand and turn it back and forth to see its picture move. But there seems to be two different perspectives involving this Sagittarius/Scorpio/Ara set of constellations highlighting two different epochs quite the same way two different sets of constellations fit the pillar's images. We don't want to be like liberals, reading our own imaginations into Pillar 43's imagery. Therefore, if these two different focuses were actually intended by the pillar's carvers, the pillar's own symbols will be found keying two different fits.

Notice the dot carved at the tip of the main vulture's wing. If we allow the overall shape of the main vulture to suggest that dot to be Gamma Sagittarii, a.k.a Alnasl, the tip of Sagittarius' arrow, and the eye of the vulture to be Lambda Sagittarii, the top of the more familiar Sagittarius teapot, then we find Sagittarius accurately depicted by the whole of the main vulture (see Figure 63 on the following page).

Let's refer to Sagittarius' more familiar teapot image. The main vulture's odd, clownish feet are now explained by their correlation with Epsilon Sagittarii, where the teapot spout meets the teapot's base. Up the side of the teapot, where the spout meets the top, Delta Sagittarii overlays the leading edge of the vulture's wing. Phi Sagittarii, where the teapot top meets the top of the handle, marks the vulture's other shoulder. Sigma Sagittarii, the upper corner of the handle, rests on the trailing edge of the other wing. Pi and Ksi Sagittarii, two stars in the back arm of the archer define the tip of the vulture's other wing. The spread-winged vulture accurately defines the main body of Sagittarius. And it depicts Sagittarius in its precise, celestial position relative to the polar meridian once every day.

The intent of that overlay is verified by the polar meridian during the Winter Solstice extending from the headless chap's armpit through the scorpion's stinger and the solar disc on the vulture's wing, running through the very same points on the pillar symbols as it did with Scorpio overlaying the snake, chick, scorpion, and vulture feet. That meridian forms a pivotal reference for our lenticular imagery between Sagittarius and Scorpio. And since the meridian is now seen extending through the spout of the teapot instead of through Eta Scorpii, we know it is representing two different times.

Amazingly, when Sagittarius is set to precisely overlay the main vulture in accord with those two marks made on the pillar, over the course of seven-hundred Winter Solstices, the sun will pass from the beginning of the line of moons above the vulture's head to that line's end at the back of the waterfowl's neck. Pillar 43 symbolizes that small piece of sky to scale, and the line of moons symbolizes a short strip of the solar ecliptic. When the polar meridian at Winter Solstice is then set to coincide with the line running from the headless chap's armpit through the disc on the vulture's wing, with Sagittarius still precisely over-

Clear Signs of Trouble and Great Joy

THE SAGITTARIUS TIMELINE

1453 The burgeoning Ossman beylik defeats Byzantium to become the seventh dragonhead. A sign-of-witness begins in 1457. America's discovery and the Jews' expulsion from Spain in 1492 are marked by a holyday-tetrad. Another Sign-of-witness begins in 1515. Then the Jubilee Path starts as the Ottoman-Turk Empire seizes the Promised Land in 1517.

1775 God raises up America to restrain the mystery of lawlessness "science" would cause by replacing empirical evidence with assumptions to fabricate an anti-God evolution myth which would foster the draconian Nazi, communist, and Progressive governances of today. Enoch's book was rediscovered for the Western World in 1773, and in 1781, Frances Rolleston was born. She rediscovered the gospel in the stars, which became the "Rosetta Stone" for interpreting Enoch's prophecy carved onto this pillar.

1877 The Shemitah Path starts after five years signalled grace the world will need to endure this path to its end.

1948 God displays His faithfulness in keeping His promise to Abraham of Israel in the Promised Land.

2138 The Sagittarius Timeline ends with the last sign-of-witness to occure before the final six intertwine through the Scorpio Timeline.

The downward facing snake is Draco. The disc in front of the faked beak of its tail is the sun's 1948 position when both are slipped back onto the timeline with Draco in its Mazzaroth placement of the gospel overlay.

The Sagittarius Timeline approximates a seven-hundred year epoch, the number of spiritual significance.

The Jubilee Path.

The Shemitah Path.   **Jesus returns!**

This many thematic coincidences are simply impossible for nature to randomly produce or for man to creatively fashion. So much correlation between the past, the present, and the future so anciently and accurately carved onto Pillar 43 could only have occurred by inspiration from The Holy Spirit of God. And that is precisely how The Holy Bible proposes it happened:

In many and various ways God spoke of old to our fathers by the prophets.

...no prophecy ever came by the impulse of man, but men moved by the Holy Spirit spoke from God.
(Hebrews 1:1 and II Peter 1:21)

The Polar Meridian
(Notice how it rises from the Southern Cross.)

Figure 63

430

Holy, Holy, Holy

laying the main vulture, the sky-chart's sun aligned onto that strip of the solar ecliptic makes it a prophetic calendar. As the precession of the earth's axis advances through each year, the sun at each year's Winter Solstice will advance along that strip of the solar ecliptic.

It now becomes easy to locate the times implied by the pillar and notice the prophetic nature of its calendar.

This section of the solar ecliptic obviously begins at the back of The Great Eagle's head. There, the lower edge of the marks representing the Flood waters suggest the beginning position of the sun's movement through this epoch of Winter Solstices. The sun stood at that beginning of the pathway during the Winter Solstices shortly before AD1453. Recall from Chapter 7 that this is the time when the sultanate set up by Murad arose to become the seventh dragonhead, the Ottoman-Turk Empire, upon defeating the last vestige of the Roman Empire. The rise of the seventh dragonhead began the endgame of this chess match between God's righteousness and the chaos of the Accuser/Deceiver/Destroyer.

And I suppose it should be thought as merely coincidental that the ancient Sumerians had depicted the dragon with its last head slain. But I doubt it was merely coincidental.

So, as the years pass and the Winter Solstice sun moves along this section of the solar ecliptic marked onto Pillar 43, we find prophesied, not one particular year or point in time, but a seven-hundred year epoch spanning from the rise of the final dragonhead to the beginning of the reign of The King of Kings after He will have defeated the eighth head of the dragon. Along this trackway the Winter Solstice sun shines from the rise of the seventh dragonhead through the Jubilee Path begun at 1517, to the rattling of Israel's dry bones at the first Aliyah of the Shemitah Path begun at 1882, and on into, and through, the Tribulation. The discovery of America, America's rise to nationhood, the rediscovery of the Book of Enoch, and the rediscovery of the Gospel in the Stars are there. The rise of secular nihilism is there, with its twentieth century governments of murder and mayhem recently restrained by America and its allies. The end of Biblical history's three 1,948 year epochs is there at the reestablishment of Israel as a nation with Jerusalem being its capitol amongst all of those signaling, lunar eclipses on Jewish holydays. The discovery and decipherment of Gobekli Tepe is there, accompanied by the verification of The Shroud of Turin's authenticity. The discovery of this immense system of signs praising the glory of God's design and purpose for the universe is there. The beginning of the Tribulation is there. And its end. The sun's movement from standing in Sagittarius at the Winter Solstice to standing in Scorpio is there, that dawning of the age of Aquarius. And then this seven-hundred year section of solar ecliptic stretching through the Southern Gate ends at the back of the waterfowl's neck, approximately one century into Christ's reign of righteousness.

Let's be sure to notice that this path of history marked onto Pillar 43 began at a tetrad-of-testimony, 1457-1523, and it will end at another, 2072-2138 (there are nineteen decades from 1948 to 2138, the number of the complete end of man's affairs). After 1371 there will be only twelve more tetrads-of-testimonies to AD3000. Six of those twelve happen after 1453 and before 2139, practically the entire time portrayed by the sun's crossing this path. That tidbit of *information* will meet some interesting correspondence later in this chapter.

I think it is at least somewhat significant to note the general time of the Winter

Clear Signs of Trouble and Great Joy

Solstice when the polar meridian bisected the sun on this path. Possibly, it is a moment of history deliberately prophesied by the pillar. The sun stood bisected by the polar meridian on the Winter Solstice of 1775, two years after James Bruce rediscovered Enoch's book for the Western World, the year before the Declaration of Independence was made by the nation God established for hindering Draco's mystery of lawlessness, and six years before the birth of Frances Rolleston, who rediscovered the gospel expressed in the stars, the "Rosetta Stone" astonishingly translating Pillar 43's symbols.

> From what I find, it appears that Frances Rolleston was the first person, in more modern times, to resurrect the study of the stars. Indeed it also appears that since the time of Ptolemy's *Amalagist*, God's people had lost interest in the stars. Thus, the heavens seemed to be silent from the early AD's to the 1800's…At which time, a humble, French woman brought the truth of the stars to light once again.[26]

1948 has a special indication on the path. It isn't marked there, but it is marked on the pillar. The Southern Cross and Draco, located on the pillar as shown in Figures 55 (pg. 410) and 63 (pg. 430), are constellations of the Cygnus/gospel sky-chart overlay. Compare the downward facing snake to Draco. That downward facing snake is the forward part of Draco; Draco's formational stars are even carved onto Pillar 43's snake. The disc set in front of that snake's beak-faking tail is a very ancient Mesopotamian representation of the solar disc. When that disc is moved with the snake into Draco's position upon the path of moons, it becomes the sun standing in that position at the Winter Solstice of 1948. It is the earliest prophecy of Israel's reestablishment for receiving The Great Eagle, Jesus Christ, when He returns. God's Holy Spirit made the reading of this pillar to be an evidential matter of decipherment by using the scientific process of following ALL evidence to where it leads (which scientists do not do, witness their dumpsters filled with *information* they don't desire to follow).

Israel Raised Above All Nations

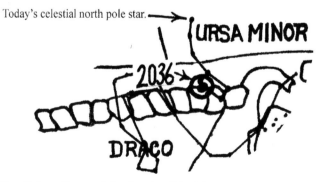

Recall from Chapter 7 that Israel will be lifted above all nations at Jesus' return, as it is portrayed by Apophis' April 13, 2036 flyby. Recall from Chapter 4 that Ursa Minor represents the little flock, Israel, in the Mazzaroth, as Frances Rolleston deciphered it. Compare Figures 55 and 63; notice that the 2036 Winter Solstice sun falls upon the Sagittarius Timeline precisely where Ursa Minor first intersects it.

Figure 64

As we might expect, the year Apophis' portrays to be that of Jesus' return is indicated on Pillar 43 where Ursa Minor, the Mazzaroth's little flock, Israel, first intersects the path of moons. Recall from Chapter 7 that Israel will be lifted above all the nations when Jesus returns. On the 2036 Winter Solstice, the sun will stand where Ursa Minor, that constellation portraying Israel, first contacts the path of moons.

Without words, the constellations make a powerful speech!

Also, the 1775 sun stands on the timeline right where Draco first intersects it (see Figure 63, pg. 428). This is the time when geology, in its infancy, began to imagine its no-God guess for beguiling the world into making up its own utopia. Man's struggle to effect utopia put the mystery of lawlessness on steroids, empowering the very nightmarish governments at the top of our current world's Babel of towering assumptions. God rose America to restrain those lawless nations, which it did until it finally joined them on November 3, 2020. There, where Draco crosses the timeline, the sun stands at America's founding (see Figure 63, pg. 430). We can't make this stuff up. This pillar demonstrates itself to have been designed by the very great prophet, Enoch. That demonstration empirically evidences Enoch to have been a very real, living human being.

Francis Rolleston's rediscovery of the "Rosetta Stone" interpreting the symbols of an ancient message carved for a distant generation onto a pillar of a monument buried conspicuously like a time capsule is in no way trivial. This "Rosetta Stone", which she rediscovered, was crafted by the same prophet of the people who carved and buried Pillar 43 at Gobekli Tepe. The rediscovery of the book written by Enoch, who was shown the *information* depicted by the "Rosetta Stone" which has now been found correlating in every detail with the conspicuously buried pillar, is in no way irrelevant. The restraining effect of a God-fearing nation of independence is imperative amidst a foolhardy world goose stepping through lawlessness towards its fascistic, Tribulation style Waterloo. These three historical relationships to this monumental time capsule occurred around the year the sun was bisected by the polar meridian during the Winter Solstice of 1775, maybe not precisely, Mr. White, but very coherently. It is the voice of the heavens of Psalm 19 proclaiming the glory of God's work without using words.

Within His glorious work is The Holy Spirit, throughout time, moving upon the intellectually responsible hearts and minds of respectful people humble enough to communicate the clues He gave for discovering a most imperative message. Only I AM WHO I AM, The Great Eagle, can make this stuff up. In many and various ways His Spirit spoke to our fathers of old. And in many ways our fathers symbolized or wrote His communication for our understanding. A collection of their writings has been, through the centuries, called The Holy Bible, The Living Word of God, that book of concepts upon which today's Progressive liberals spit. Against such spitting I AM WHO I AM, today, is graciously warning, for tomorrow, He will be stomping it in the great winepress of His wrath.

So, let's swing on this same polar meridian back to Scorpio's overlay of the vulture chick above the scorpion and before the downward facing snake. When we place the star chart of Scorpio onto Pillar 43, at the Winter Solstices, the sun passes through the pillar's disc instead of along the path of the Flood's eleven moons. The disc then represents a calendar, the same way we saw the path of moons do it. If the disc were placed here to represent some future year in particular, might that year be the one in which the sun stands at the center of the disc on some future Winter Solstice, bisected by the polar meridian? All of the

Clear Signs of Trouble and Great Joy

The Scorpio Timeline

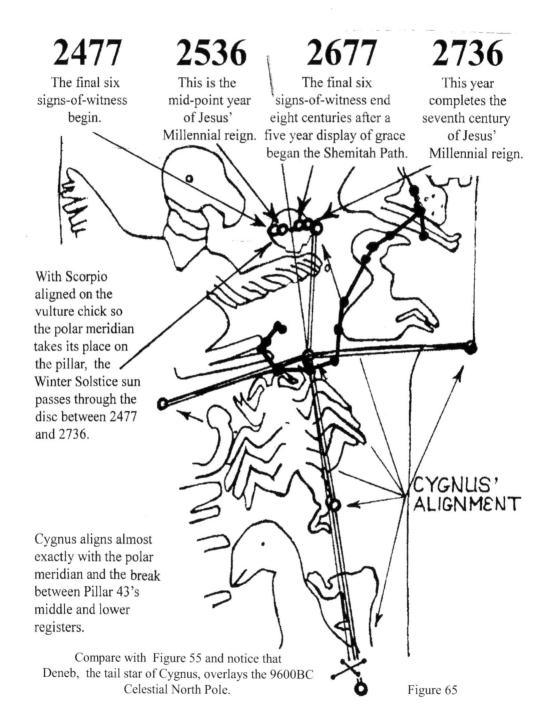

2477 The final six signs-of-witness begin.

2536 This is the mid-point year of Jesus' Millennial reign.

2677 The final six signs-of-witness end eight centuries after a five year display of grace began the Shemitah Path.

2736 This year completes the seventh century of Jesus' Millennial reign.

With Scorpio aligned on the vulture chick so the polar meridian takes its place on the pillar, the Winter Solstice sun passes through the disc between 2477 and 2736.

Cygnus aligns almost exactly with the polar meridian and the break between Pillar 43's middle and lower registers.

Compare with Figure 55 and notice that Deneb, the tail star of Cygnus, overlays the 9600BC Celestial North Pole.

Figure 65

imagery seems to spiral into this singular point of time. It is the position of the sun during the Winter Solstice of AD2600.

Recall the thirty-two tetrads-of-testimony from the previous chapter. Recall their curious correlations with historical events relevant to Israel, Christ's Church, The Shroud of Turin, Christ's return, and whatever might occur during Christ's Millennial reign. Recall how those thirty-two signs-of-witness end with six tetrads-of-testimony interlinking through a two-hundred year span of time which entails the AD2536 midpoint of Christ's Millennial reign. They begin AD2477 and end AD2677. Those six tetrads-of-testimony surrounding the middle of Christ's reign of peace occur while the sun moves through Pillar 43's disc over the course of a little more than two-hundred Winter Solstices.

Every tetrad-of-testimony that will happen during the Millennial reign of Christ is a meaningful element of these two prophetic calendars marked onto Pillar 43. The next one to occur marks the end of the Sagittarius Timeline, 2072-2138. Then the last six interlink through the Scorpio Timeline at the midpoint of Christ's Millennial reign. The two timelines on Pillar 43 are amazingly interconnected by each entailing six of the final twelve tetrads-of-testimony to occur before the great Day of Judgment. Twelve is not only the number of complete order (perfect righteousness), as we noted in Chapter 8, but it also is the number of Jesus' apostles who carried the good news of coming righteousness around the world. Moreover, the tetrad-of-testimony immediately preceding these final twelve occurred while the little Lirey church displayed the Shroud of Turin for all the world to witness, circa 1340, testifying to the empirically observable evidence of the living Jesus Christ, His Shroud bearing the photograph of His resurrection. That one and these twelve make thirteen, the total number of Jesus and His apostles, the number of tetrads-of-testimony formed in the same pattern as that glorious one of 1949-2015.

So here we have multiple, 19/48 patterned lunar eclipses testifying to the timelines presented upon an ancient pillar to mark the end of mankind's rebellion and the middle of Christ's millennial reign of peace and righteousness. We find man's earliest temple, the stars in the heavens, and The Holy Bible all speaking of the same redemption by the same One who died and was raised to redeem all who will simply believe.

Lest it all be a coincidence. And if it is just a coincidence, what good would it do to speak of its enormity? For the enormity of this coincidence simply piles on top of the already enormous coincidence of earth's fantastically precarious position in the universe for the support of life as we know it -the Goldilocks Zone. Secular scientists desperately try to discover life outside the Goldilock's Zone, because that zone so powerfully argues for creation by God. Indeed, they've found some bacteria and small life forms thriving in oven-hot waters, freezing cold areas, and even in highly acidic zones. But no buffaloes roam there. Birds dare not fly those zones. Only a fool inept as man would think that a few exotic bugs represent the entire ecosystem. The overwhelming rest of us critters are more delicate than peaches and eggs, needing Goldilock's place for survival.

Earth's solar system is precariously placed within a galaxy precariously positioned in a universe perfectly balanced to even exist. Men pie and lie and dissemble *information* into saying whatever they want to hear, like "God is dead!" But, the birds, animals, and creeping things Noah brought across the Flood, evidenced by this ancient pillar, and all the

Clear Signs of Trouble and Great Joy

fishes that swam in the flood below his ark certainly required Goldilock's "just right" zone. That reality can not be pied by the observation of a couple bugs in boiling water and a bitty fish or two taking comfort in battery acid. Physical existence is either made of an insane quantity of coincidences spoofing up some "Holy God, the Creator" idea and penning some crusty old Bible (Buddha has no overabundant coincidences, nor does Allah, Krishna, Mythra, Thor, Zeus, or any other of man's kooky imaginations), else physical existence has been ordered by The Holy God of The Holy Bible to speak a warning for us to know I AM WHO I AM by His evidences rather than by our own machinations.

We can now discuss another measurement which could not sensibly be presented in Chapter 7 for the lack of an *informational* foundation laid in Chapter 8. Recall the patterns of five, the number of grace. Five tetrads-of-testimony interwove over the history of Israel's divided kingdoms before they were expelled from the promised land. Five heads of the dragon rose and fell before the rise of the Roman Empire, under which the Christ, Our Lord of Salvation was born. Five years intervened from the time Orhan's eldest son died to the time Orhan also died, leaving the beylik to Murad, who set it up for becoming the seventh dragonhead as verified by Daniel 12:11. Jerusalem's Jubilee Path crossed the fifth shemitah of the Shemitah Path two years before the seventh dragonhead received its mortal wound, which was five years after its Rosh Chodesh Elul, 1914 warning to repent. Five years after America's Rosh Chodesh Elul, 2017 warning, during the shemitah year, 2022, the KIC 9832227 not-a-new-star will shine its "yes you can find the truth" light out of the constellation of The Ever Present Truth upon America's response to its 1 Elul warning. Five years after the 2024 total solar eclipse completes the giant "X" across America comes Apophis portending the Tribulation. Five days after the exact midpoint of Apophis' two consecutive passes by Earth winks the total lunar eclipse of the Antichrist usurping Christ's throne, the rise of the eighth dragonhead to persecute Israel again. Eight shemitahs into the Shemitah Path began Hitler's Holocaust persecution of the Jews; eight shemitahs after the end of that persecution began the five shemitah fast-track to the Tribulation.

Five years before the Shemitah Path began with the gathering of the Jews into the Promised Land was AD1877.

So what!?

1877BC happened to be the year God promised to make Abraham into a great nation. AD1877 was the fifth year before God's promised process to end evil began, necessitating the reestablishment of the nation of Israel, the fulfillment of that 1877BC promise, and the grace Israel will need to endure the Tribulation, the end of the process. And there, where the 1877 sun stood upon Pillar 43's Sagittarius Timeline, Draco's tail overlays the prophetic path, whispering themes of evolution, disdain for The Holy Bible, accusation of I AM WHO I AM, and glorification of man's rule over other men, the people-conquering nihilism which led to the most horrific governmental slaughters of human life in history, and will once again hoodwink mankind into a worse slaughter during the Tribulation. Although man's hands carved Pillar 43, man did not make up its prophecies. Man machinated evolution to murder the God who created Pillar 43's prophecies.

[18]For the wrath of God is revealed from heaven against all ungodliness

and wickedness of men who by their wickedness suppress the truth. [19]For what can be known about God is plain to them, because God has shown it to them. [20]Ever since the creation of the world his invisible nature, namely, his eternal power and deity, has been clearly perceived in the things that have been made. So they are without excuse; [21]for although they knew God they did not honor him as God or give thanks to him, but they became futile in their thinking and their senseless minds were darkened. [22]Claiming to be wise, they became fools, [23]and exchanged the glory of the immortal God for images resembling mortal man or birds or animals or reptiles.

[24]Therefore God gave them up in the lusts of their hearts to impurity, to the dishonoring of their bodies among themselves, [25]because they exchanged the truth about God for a lie and worshiped and served the creature rather than the Creator, who is blessed for ever! Amen.

Romans 1:18-25

Then, precisely eight centuries (new beginning) after AD1877, at AD2677 the last of the tetrads-of-testimony will end. While the sun will be standing inside the Scorpio Timeline's disc of Pillar 43 on AD2677. The Millennial reign of The Great Eagle will be outshining the sting of death at the tip of Scorpio's tail. Death will barely exist amongst the mortal remnant of man on Earth at this time when Aquila's face outshines Scorpio's stinger.

A great process will be concluding by AD2677, and another will be at its beginning. Death will have been defeated. Mostly. But not entirely. Although mortals will yet be living and dying, death will be rare; life will be long; suffering will be faint; joy will be richly abundant. The Accuser/Deceiver/Destroyer, Satan, is to be loosed from his imprisonment in the bottomless pit to come forth and deceive the nations after a thousand years of great joy has transpired (Revelation 20:7-8). But it is hard to imagine that people will be so susceptible to the despair of dirty deeds after a thousand years of almost entirely righteous joy. Might some process begin at or after AD2677 which will lead the nations finally embracing Satan at his release, similar to the process nihilism was beginning to work at 1877 which, eventually, would lead once Christian nations to trashing the Bible and murdering Jews?

The human propensity to accept deceit and despise righteousness needed two-and-a-half centuries for developing scientific myths into cultural beliefs diametrically opposed to the truth of the Bible. That development, of course, has led human propensity even further from the truth about right and wrong; ("Who's to say what's right, man, who's to say what's true?" c. 1960s.) Eventually moral depravity became the bumper crop of scientific myths like Olathe sweet corn becomes a bumper crop of good fertilization (well spread BS). Science's shameful fertilization of moral depravity has led mankind onto a stage prepped for the Tribulation scene.

That same human propensity might arise in the latter centuries of Christ's Millennial reign to, in a similar manner, prepare a cultural bandwagon of deceit for Satan's release from confinement. Cultural attitudes shift slowly. The greatest of changes are the most subtle, unobservable processes that sneak into effect attitudes which would have been resoundingly rejected if they had been immediately proposed. Witness the ways Progressives slith-

Clear Signs of Trouble and Great Joy

ered into controlling the cultural degradation of a once very traditional America. People are like frogs, never noticing the ever changing temperatures of their cultural/political waters. AD2677 might be the beginning of a three-hundred-fifty year attitudinal decline transforming the joy of the Millennial culture into a more cynical doubt useful to the Accuser/Deceiver/Destroyer for leading one final rebellion against Christ when he is released.

The speculation of a social process setting up Revelation's prophesied rebellion at the end of Christ's Millennial reign is a pretty wide chasm to cross upon the little evidential bridgework we have. Over the centuries following Christ's return, the Winter Solstice sun will have passed Scorpio's stinger, having brightly outshone death for the time it was there. But also, much of the world population will have been growing towards the end of its new longevity of life even as more of its population will be entering that wonderful stage of young adulthood adorned with all the liberal splendor of rebellious encroachment upon traditional ideas. Soon death will be increasing and kids, grandkids, and great-grandkids will be weeping; little hearts of weak beating wondering if ole Jesus in Jerusalem is all He was made out to be when beloved Nanna and Pappa have breathed their last. Young minds ponder new thoughts after they've perceived old wisdom to have grown boring. Novice hearts explore unusual sentiments when traditional meaning has told its stories. This is human nature.

The oldest man recorded in the Bible lived only nine-hundred-sixty-seven years (Genesis 5:27). By the end of Christ's Millennial reign, death will have returned from its holiday, rising rapidly through the ranks of beloved grandparents, flooding the world with a strange level of grief within a Covid-19-like, thankless hysteria. It might be just that scene into which the Accuser/Deceiver/Destroyer is then released with the impeachment of Christ baited on his every breathe.

We can't know these things for sure. But we've seen a very good demonstration of deceit working horror into a thankless population full of irrational fears, and of how quickly foolish fear is willing to impeach good sense.

For the lovers of righteousness, blessing always follows trouble. It's what's for thanksgiving! Even so, this rebellion will be major trouble for the righteous, vexing them deeply, though neither snuffing them out nor crushing them. Lo! Their trouble will be squashed immediately (Revelation 20:9-10) by one, intense flame of righteous indignation breathed in one word from Jesus Christ.

For skeptics, no bridge is complete enough to cross because skepticism is about blocking bridges, not crossing them. If skeptics actually crossed bridges, they wouldn't be skeptics. Instead, they would be smelling the flowers of springtime; they could enjoy the safety of Passover. But that involves change, a.k.a. repentance, which skepticism guards against. Avoiding repentance is why they are skeptics.

Graham Hancock reasoned correctly about Pillar 43's warning. He even reasoned correctly about the indicated era of the Tribulation catastrophe. But Pillar 43 had more reasoning to present of its own, reasoning of the great joy to follow the catastrophe, the great joy that is the entire point of the catastrophe.

Pillar 43 is a warning to our generation, buried by the direct descendents of the great prophet, Enoch, who wrote his book of warning to the generation of the end, the very generation which unearthed his great-grandson's Gobekli Tepe and marveled at Pillar 43. Enoch's warning was not about running for the tall grass. It was not a warning to dig bomb shelters, stock up on beef jerky, pinto beans, and Twinkies. Nor was it a warning to lock, load and hunker down for battling every stranger over the possession life's moldy pittances. It is a warning to run to Jesus Christ and stand in submission to the truth of all things as He teaches them, abiding in the attitudes He lived, sacrificing for the welfare of those having less than you have. It is a warning to convince your neighbors, friends, and loved ones to do the same. Thus, I write to you. For when the wheels come off this life's normality, blood, flesh, and bone will find no safety. Only a physical remnant will survive God's wrath, but their souls and spirits will have been squared with the truth. So the coming Tribulation should not be a disparaging prospect, seeing that everything of this life WILL die eventually. For by any measurement, the greatest cause of death is conception itself; every conception necessitates another death, bar none. Therefore, running to The Great Eagle for a ride to safety upon His wings should be a natural thing to do. Pillar 43 depicts that wisdom.

It is solid evidence that the Bible got the stories of the Flood and Noah's ancestry correct, and that The Book of Enoch is indeed Enoch's work, and that he was shown all things by the messenger of the Lord of Spirits, including the constellations and their meanings, and that the Book of Jasher is also history, and that the Mesopotamians twisted history and the knowledge of I AM WHO I AM almost beyond recognition.

Even Pillar 43's number bears witness to its testimony. Thirty-six is the Biblical number of God's doing times the number of complete order (3X12). On God's side of Earth's affairs, forty-three is the number of God working complete order (3X12) into a great, spiritual significance (36+7). Does that not describe Pillar 43?

But on the other side of Earth's affairs, it is the number of testing (40) increased by the number of God's doing the testing (40+3). As the Shemitah Path turned onto the fast track leading into the Tribulation, September 11, 2001 tested America in the first year of President number 43 (Jonathan Cahn discusses that test in *The Harbinger*). At the writing of this book, America is under its Rosh Chodesh Elul testing, having begun in the 40th year after this book's author pulled the Mauser away from his head and committed to a path of following evidence to wherever it might lead, a forty-year test of whether he would follow evidence faithfully. Then, three years after America's Rosh Chodesh Elul warning began, this book was complete and ready for publishing (except for the inclusion of the results of Joe Biden's greatest, voter fraud organization in the history of American politics). God signs His work by making discernable measurements within and all around it. Pillar 43 is now one of God's great tests for whether or not anybody cares enough to see the prophesies He carefully measured onto it: Pillar 43 was discovered in 1993, forty-thee years before the destruction of evil that it prophecies will happen in 2036.

Got that, skeptics?

Secular historians should reset their Timex rocks by The Bible's narrative, else forever be known as nothing more than pieing, brass-foreheaded skeptics. After all, my fellow scientific truth seekers, isn't science's central purpose to follow ALL EVIDENCE to WHEREVER it leads, even across undesired bridges of repentance leading to Jesus Christ?

Do we need more validation before crossing that bridge with our Timex rocks and

Clear Signs of Trouble and Great Joy

Bulova bones reset? Note that, on Gobekli Tepe's 9600BC Winter Solstice sky chart of Sagittarius/Scorpio/Ara, the polar meridian runs beyond the disc through Draco to Polaris, which is ever so slightly "West" of that polar meridian. This close proximity is Polaris' relationship to the polar meridian all day every day, today, because Polaris is now the North Pole Star, continually in that close proximity to the meridian it attained only once a day when Sagittarius, Scorpio, and Ara illustrated God's outpouring of wrath onto the workers of evil and chaos, as those constellations took their positions around the polar meridian overlaying the pillar's main vulture, vulture chick, and scorpion. As the stars rotated around the polar axis throughout Gobekli Tepe's day, Polaris swung quite far from the meridian, and then back to it daily. It assumed today's polar star position only at those moments of each day when Sagittarius, Scorpio, and Ara took their positions upon the pillar's symbols to prophecy God's wrath to be poured onto evil. Moreover, Polaris is "the last cub (or lamb)" of the Little Bear (or the Little Flock), the Mazzaroth constellation representing God's faithfulness to His Israelite people amongst whom He will reign, now that He has returned them to being a nation.

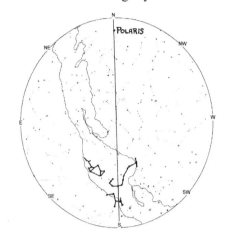

Figure 66: Polaris 9600BC

Reset your Timex rocks, dudes! Please, skeptics, take it from a fellow skeptic, now a follower of *information* to wherever it leads, cross the bridgework of evidence to Christ Jesus. He is the only doorway through which to escape the coming eternal despair.

We have finally accumulated enough *information* to discover the answer to my lifelong search for why the Eagle takes the scorpion's place amongst that set of corner constellations represented by the faces of the man, the bull, the lion, and the eagle. In light of the gospel overlay of Pillar 43, the Great Vulture/Eagle portrays Christ's care for the departed souls of those who turn to Him seeking righteousness, truth, and eternal life. The disc on its wing represents the death of the headless chap below and the souls of all those dying in His nature. It is positioned at the Celestial North Pole, through which the ancients believed the deceased entered the place of the dead. The Kneeler, who descended into death and returned with the keys of death's gate, overlays the disc, portraying Christ's victory over the sting of death to give the souls of His followers eternal life. In the early centuries of Christ's Millennial reign of righteousness, a time when death will be rare, the sun will be standing above Scorpio's tail at the Winter Solstices, washing Scorpio's stinger out of mankind's picture by its brilliantly shining righteousness. Six signs-of-witness intertwine through exactly four Jubilees, from AD2477 to AD2677, while the Scorpion's stinger is washed out in the brilliance of Aquila's glory, portraying Christ's victory over death as He reigns during a time of minimal death through those two hundred Winter Solstices marked by Pillar 43's disc resting on the wing of The Great Eagle. As that Victor, the face of The Great Eagle takes the scorpion's place in the representation of the four corners of the Mazzaroth, the four faces upon the cherubim, The Man, The Bull, The Lion, and The Great Eagle.

This message of salvation and purging has been known by the most faithful Sethites since the days of Enoch before the Flood.

> God opened the eyes of the righteous Enoch, so that the angels could show him a vision of the Holy One in the heavens...I, Enoch, alone saw the vision, the end of all things, and no man will see these things as I have seen them...Then the archangel Michael took me by my right hand, and lifted me up and led me forth into all the secrets of mercy and justice. And he showed me all the secrets of the ends of the heaven, and all the chambers of all the stars, and all the luminaries, whence they proceed before the face of the holy ones...Now, my son Methuselah, call all of your brothers and all of your mother's children to me; for the Word calls me and the Spirit is poured out upon me, that I may tell you everything that will happen to you in the future.
>
> 1 Enoch 1:2a; 19:3; 71:3-4; 91:1

Consequently, we find Noah's ark and the Flood depicted at the top of Pillar 43, obviously a Sethite monument to the three most significant events in mankind's history, The Great Flood, the first coming of Christ Jesus as the sacrificial Lamb, and His second coming as the Lion of Judah, the reigning King of kings and Lord of lords. Fifteen times The Holy Word of God alludes to Christ through various metaphors of the eagle. The Lord of Spirit's working both our salvation from and His destruction of evil is portrayed by the constellations overlaying The Great Eagle on Pillar 43. So also the disc on The Great Eagle's spread wings has adorned human history from ancient times forward as the winged sun, still used worldwide as a symbol, today. And note again the Apostle John's careful description of the wings which carried the Revelation 12 woman to safety as "THE two wings of THE GREAT EAGLE". (Revelation 12:14; emphasis is mine.) Koine Greek was a precision language; its definite article points to definitiveness. THE two wings of WHICH great eagle? Ezekiel 17 mentions two great eagles "with great wings", one symbolizing God, and the other, idolatry. Could one of those have been the one John meant, probably the one symbolizing God? At least in The Holy Spirit's mind, could John's reference have been alluding to Gobekli Tepe's great eagle with both wings spread? And could His inspiration of Ezekiel's allusion to great eagles with great wings have originated from His inspiration of Noah to carve onto Pillar 43 a representation of God flying the soul off to safety on the two wings of that great eagle? Pillar 43 is great evidence of God's Holy Spirit like a bear track is great evidence of a bear.

Seamless correlation between the Mazzaroth, Pillar 43, and Biblical themes extracts from these ancient symbols the death and resurrection of a strongman who will break the bonds of death and guide souls unto eternal life. That theme spread throughout the world. It was twisted into many different forms perceived today as the dying/rising god theme.

Today's deceitful buriers of *information* highlight those anciently diversified twists, Osiris, Tammuz, Baal, Attis, Dionysus, etc., to accuse the Bible's authors of having copied myths and legends to make up a Jesus. But here, on Pillar 43, the visual testimony of

Clear Signs of Trouble and Great Joy

orderly and ancient symbols aligns with the voiceless speech of The Mazzaroth in the heavens (Psalm 19:3) to testify that the gospel of the Holy Word of God was known (if not well understood) in Noah's day, long before there came any shifty, whispering minds spinning myths and legends barely reminiscent of Enoch's real history and salvation narrative the later Bible would present correctly, yes, by inspiration of The Holy Spirit.

And as Jesus' Daddy is the very best at piling on, to that testimony Pillar 43 adds a warning of God's planned action against evil and all its doers and twisters just in time for wiser people to repent of their doings and to get out of the way of The Deceivers' wrecking train.

The ancient Hebrew Book of Jasher told us there was such a warning. Even Josephus said two pillars of warning yet stood in his day. Could it be that Gobekli Tepe left such a major and early impression on the souls of the ancients that stories about pillars of warning went twisting through history from Pillar 43? Was Gobekli Tepe so elevated in honor by Noah's immediate generations that its faint memory reached the Bible as the mount of assembly, Mount Zion in the far north?

Embedded in pre-Sumerian history was a "worldwide" culture of several people who rode out The Great Flood in a ship blueprinted by God. Accounts, stories, and legends of that Flood spread throughout the world after the infamous Tower of Babel incident. Many characteristics of their ancient astronomy spread with them into diverse cultural elements sprinkled around the world, each becoming in its own way a distortion of the truth told by Enoch's Mazzaroth. But a few held more carefully to the ancient representations considered by the Sethites to be sacred history. It was they who discerned the saving work of Christ in the Mazzaroth's four corners of their day, Aquarius, Taurus, Leo, and Scorpio (to be defeated by The Great Eagle, Aquila).

Regarding the face of the man, Ea is actually an Akkadian title. The Proto-Sumerians knew him as Enki. He was wisdom, and the Mesopotamians believed he created the heavens and the earth. For this his epithet became "Ea, the genius who built everything."[27] As we noted earlier, he was the god of the underground waters which welled up to water the earth, as Genesis describes the antediluvian conditions (Genesis 2:6). He was anciently represented by all of the Mesopotamian cultures with two streams of water cascading from his shoulders in which fish were swimming upward.

> [21] But there the LORD in majesty will be for us
> a place of broad rivers and streams,
> where no galley with oars can go,
> nor stately ship can pass.
>
> Isaiah 33:21

Of course, Ea was the Mesopotamian distortion of the true Water Bearer -Christ Jesus- who died so that we, His fish, could abide in the living waters of new life. Aquarius, The Water Bearer, pours a stream of water into His fish's mouth.

At the time Gobekli Tepe was in use, according to Biblical chronology, the sun stood in Aquarius at the Winter Solstice, the season representing the fullness of death's process before Spring begins the renewal of life. Many of the ancient gods and goddess's had a personal beast. Ea's beast was the goat-fish.[28] We're more familiar with the goat-fish as Capricorn, the constellation immediately preceding Aquarius in the Mazzaroth. Capricorn is the constellation portraying Christ as the resurrected, life-giving sacrifice.

From the time of Creation until shortly after the Flood, again, according to Biblically implied dates rather than theories concocted from Timex rocks and Bulova bones, Taurus charged out of the Spring Equinox. I don't think an aurochs could be about anything other than a pair of giant horns backed by intensely focused, brutal force. As we noted previously, the aurochs shared the rhino's mythical nature: see it, charge it; hear it, kill it! So the Sethites passed "the horn" down to the Mesopotamian cultures who passed it down to nearly all following cultures as the symbol of power, and in particular, divine power. But the bull was also a sin offering, as secular historians are sure to overemphasize about Zoroastrianism. The bull symbolized the force of Christ's righteousness focused through humility at the chaos of sin destroying man. It was the ultimate power and authority above all that exists laying that power aside to plow The Way of escape for those entrapped by sin.

> [5]Have this mind among yourselves, which is yours in Christ Jesus, [6]who, though he was in the form of God, did not count equality with God a thing to be grasped, [7]but emptied himself, taking the form of a servant, being born in the likeness of men. [8]And being found in human form he humbled himself and became obedient unto death, even death on a cross. [9]Therefore God has highly exalted him and bestowed on him the name which is above every name, [10]that at the name of Jesus every knee should bow, in heaven and on earth and under the earth, [11]and every tongue confess that Jesus Christ is Lord, to the glory of God the Father.
>
> Philippians 2:5-11

Leo, the lion, that big cat, represented kingship from the earliest of man's imagery. Regulus, the king star, is in Leo. The Mother Goddess, impersonator of God, eater of the wrong fruit, sat flanked by two lions, an obvious "Berenice's Hair" sort of thing to do. Later Mesopotamians depicted Inanna, the successfully dieted and more svelte Mother Goddess, as standing on the back of a lion (good for the lion that she lost those extra pounds). The lion is everywhere depicted in connection with kingship and governance. It has always been regarded as the king of beasts, even in the pages of the Bible. And many times we've visited the discussion about The King destroying Hydra in this constellation of Leo. Until shortly after the Flood, the Summer Solstice sun stood in Leo.

We searched for why the Eagle's face represented the Fall Equinox sun standing in Scorpio at the time of the Flood, and we discovered far more than we expected. Since the scorpion was no part of God's eternal plans, but the Eagle's victory over it was, the Eagle represented the Fall Equinox sun standing in Scorpio from Creation until the Flood, thanks to Enoch's righteous ear tuned towards The Lord of Spirits, like the "I" standing alertly at The Waterfowl's beak. We have thoroughly discussed God's process for defeating sin represented by the Eagle's face taking the place of the Scorpion's. The ancient Sethites knew the

Clear Signs of Trouble and Great Joy

process well enough to carve it upon the world's most amazing monumental pillar. The physical reality is that the sun actually stood in Scorpio, the sign of death, during the Autumn Equinox. But through the life giving work of God becoming a man, the power of man's sin would be broken so that man could mount up on the two wings of The Great Eagle into perfectly righteous life after death, rather than burrowing into the dirt of eternal chaos like the downward facing snake faking a bird beak. We have the promise of eternity while we are yet mortal. While we are yet sinners we are seen as righteous through His grace. When the sun stands over Scorpio's sting of death, Aquilla is portrayed, the Conqueror of death.

But these would not remain the seasonal corners for long, after the Flood. The precession of earth's rotational axis acts like a clock, slowly turning the band of constellations through the equinoxes and solstices over many, long years. Eventually, the constellations of those four ancient cultural symbols would no longer be the corners of heaven and earth. Then, centuries of murder, mayhem, and socio/political/religious misguidance diverted man's sight from why these faces were prominent symbols. As their constellations turned away from the four corners of the celestial sphere, the faces lost their glorious meaning in the dust of the past. Their message was forgotten. Only their magnificence was remembered.

And let's not think that the ancients didn't quickly notice that slow change of precession. Precession produces one degree of change every seventy-two years. That's enough difference to notice in a couple decades using a pair of well placed rocks. As the constellations march through the seasons, the point in the heavens around which all of the stars perpetually rotate moves in a circle, advancing one degree every seventy-two years, sometimes passing over, or very near, a star along the way, but most often not. This movement of the celestial North Pole outlines a simple story told by the stars it has passed throughout the ages.

But there is a major disparity between Biblical history's timeline, the position of the Celestial North Pole as depicted on Pillar 43 and verified by the 9600BC sky charts, and Thuban's becoming the North Pole star soon after the Flood. The order of these celestial motions does not march consistently through the Biblical concept of time.

The earth's axis may have been yanked[29] from pointing into The Kneeler, before the Flood, to pointing at Thuban in the tail of Draco, after the Flood, by the same, proposed astronomical event that may have caused the Flood, if indeed the chronology implied by The Holy Bible does mean an actual 4000BC creation date. This proposal of a displaced axis is not without evidence. It is an element of a few sound cataclysm theories. Of course, their theoretical chronologies do not fit Scriptural chronology, but that has been the condition of all scientific speculation ever since Draco's Hutton, Darwin, and company imagined today's "science" of origins by gazing too deeply into their Timex rocks and Bulova bones.

But regardless of speculation about when and how the celestial North Pole moved from The Kneeler to Thuban, it is not speculation that indeed it did. Over the next four millennia it has wandered to pointing in our days at Al-Ruccaba, a.k.a. Polaris, on the tip of Ursa Minor's tail, the little bear depicting God's faithfulness to His little flock, Israel. How apropos.

The story told by the movement of the celestial North Pole is a brief of the world's spiritual affairs. At the time the man, lion, bull, and eagle symbolized the constellations of the celestial corners, the stars revolved around The Kneeler. He represents mankind's Savior who holds the authority over Hades' gates in His hand with His foot on Draco's head. It's interesting that early man expressed Creation's corners by those four faces even after precession moved them away from the actual corner positions. The symbols retained enough cultural significance to be picked up by the Bible for representing the faces of those four cherubim surrounding God's throne. Of course, The Holy Spirit who directed the authors of the Bible knew full well what those faces meant and why. The Ultimate Author of the Bible designed the gospel message of Jesus Christ into them for proclaiming holiness around the throne of God being shared with the remnant of man.

Eventually, the stars came to circle Draco's tail at Thuban, the Celestial North Pole shortly after the Flood. God engaged the Flood to end the great depth of wickedness on Earth. But He did not restore Earth's pure nature of innocent perfection. He commissioned mankind to make and keep laws in the post-Flood world in order to restore his own modicum of decency. But man's actual proficiency in righteousness was yet only a hope. Ancient governments were as criminal as today's governments. Man was yet false, therefore he was not capable of governing correctly. Sethite wisdom would need to be taught to the following generations and the rule of law carried forward by a Savior for life to be good.

Ancient Hebrew histories and Josephus say that Canaan, Ham's son, found the wisdom of Cain (taught by the fallen, rebellious angels) carved on a stone for preservation through the Flood. He revived that wisdom of the snake, as it came to be called. Genesis is sure to note that Canaan's families spread abroad. His meddling again with the snake spread with them while the snake's tail, Thuban was the turning point of the stars. From the Mother Goddess of deep antiquity through the Minoans, the Myceneans, and on to the Greeks, the snake and the woman have been depicted in the relationship of illumination given to man as received by woman. The rainbow snake is as ancient as Australia's Mother Goddess. It most likely ventured to South America on the minds of early migrants to become Quetzacoatl and Kukulkan.

Slithering snakes carved throughout the Gobekli Tepe monument add more coherence to the theory that it was Noah's monument to the story of God's work towards mankind's salvation. And meanwhile, just as the tail of that snake on Pillar 43 mimicked The Lord of Spirits, when all these snakes were slithering out to confuse the recovering population of mankind, all of the stars of heaven were beginning to revolve around Thuban in the tail of Draco, mimicking the Waterfowl's beak, once again speaking deceit to the fallen "I".

Even Abraham's dad was an idolater, so extensive became the liberal twisting of the Sethite's traditional message just a few hundred years after that Great Flood. However, I AM WHO I AM is never without a few faithful followers. The Book of Jasher tells how Abraham was sent as a child to be raised by Melchizedek in Jerusalem until he was forty. Possibly that is a legend. But it is probably history. Abraham paid to Melchizedek a tithe after defeating the kings of the Sodom/Gomorrah coalition. His paying that tithe is better explained by the relationship Jasher describes between the two. Even while reality's light was flickering off in the rest of the world, God was yet harboring His flame of light in Abraham's heart for spreading it abroad through the descendents of this Man-of-Faith born

Clear Signs of Trouble and Great Joy

AM1948, 1948BC.

God brought Israel forth from the generations of Abraham in a succession of similarly astonishing events. Of course, today we are too "intelligent" to fall for such "astounding fairy tales". "We" have never seen anything in our days that indicates how special Jacob was, or that the Bible was God's Oracles given to him! And Israel being witness to God is just plain laughable to "geniuses" such as "we". Or so we have been trained to think by academicians lacking so much integrity that they discard actual, bona fide observations in order to proclaim they have never seen evidence.[30]

Yet, today, the existence of Israel as a nation testifies to us concerning the faithfulness of I AM WHO I AM (1877BC/AD1877/AD2677) and the reality of Biblical prophecy. The very timely unearthed Gobekli Tepe also testifies. Signs in the celestial skies testify. Israel now having God's Holy City as its capitol testifies. All of the stars of the sky now orbiting the tip of Ursa Minor's tail testify to what the Mazzaroth constellations testify about God's loving care for Israel.

As the Sethite line proceeded down through history using eyes and ears to see and hear, The Holy Spirit of God refined and detailed for them the nondescript message He first gave to Adam and Eve, adding to it the prophetic details represented in the mnemonics of the Mazzaroth He gave Enoch, defining it on the monument He gave to Noah, setting it into the promise He made to Abraham, teaching it by the Law He gave through Moses, completing it in the Oracles He gave His Chosen People, until He spread around the world through His church the mystery of the gospel and has today begun fulfilling His promises to set Israel at the head of the nations for Christ to reign from Jerusalem.

> [2] For behold, darkness shall cover the earth,
> and thick darkness the peoples;
> but the LORD will arise upon you,
> and his glory will be seen upon you.
> [3] And nations shall come to your light,
> and kings to the brightness of your rising.
> [4] Lift up your eyes round about, and see;
> they all gather together, they come to you;
> your sons shall come from far,
> and your daughters shall be carried in the arms.
> [5] Then you shall see and be radiant.
> your heart shall thrill and rejoice;
> because the abundance of the sea shall be turned to you,
> the wealth of the nations shall come to you.
> [14] The sons of those who oppressed you
> shall come bending low to you;
> and all who despised you
> shall bow down at your feet;
> they shall call you the City of the LORD,
> the Zion of the Holy One of Israel.
> [15] Whereas you have been forsaken and hated,
> with no one passing through,

Holy, Holy, Holy

I will make you majestic for ever,
a joy from age to age.

Isaiah 60:2-5, 14-15

[20]Thus says the LORD of hosts: "Peoples shall yet come, even the inhabitants of many cities; [21]the inhabitants of one city shall go to another, saying, 'Let us go at once to entreat the favor of the LORD, and to seek the LORD of hosts; I am going.' [22]Many peoples and strong nations shall come to seek the LORD of hosts in Jerusalem, and to entreat the favor of the LORD." [23]Thus says the LORD of hosts: "In those days ten men from the nations of every tongue shall take hold of the robe of a Jew, saying, 'Let us go with you, for we have heard that God is with you.'"

Zechariah 8:20-23

"God Forbid!" you say?

Even Earth's northern celestial axis now testifies of the people amongst whom Christ's throne will sit. If we do not have the eyes to read this writing in the sky, then the script written to us shall be what God wrote on the wall to Belshazzar, "Mene! Mene! Tekel parsin!"[31]

The Great Rift of the Milky Way runs between Sagittarius and Scorpio. It is a zone of interstellar dust hiding not only an enormous number of stars, but the very center of our galaxy, as well. Sagittarius and Scorpio are strategically placed on either side of it for their part in telling the restoration story. Not only does Sagittarius represent our Lord defeating evil, launching His arrow through the Southern Gate into the heart of Scorpio, the sting of death, but this battle is so placed that, from Earth's view, the center of the galaxy lies in the middle of its scene. It rather seems like maybe The Almighty's eternally final victory over evil is the purpose of the physical universe. The splendor of the sun lights up the scene of Christ's victory over evil once a year on its trek through the constellations. And for two-hundred years the sun will light up that scene at the Winter Solstices during Christ's reign of righteousness while six signs-of-witness lock arms in testimonial solidarity! It is the heart of an even more astounding scene very soon to begin appearing in our skies.

The Lord gave Israel a quick victory over its neighboring Arab enemies in 1967. By that victory Israel won control of Old Jerusalem, the city the Almighty God, from ancient times, chosen to be His very own. The enormity of the Six Day War can not be emphasized enough. It was marked by the occurrence of four total lunar eclipses in succession on the most meaningful Jewish holydays of 1967 and 1968. Moreover, the pogrom Ferdinand and Isabella decreed in 1492, also having been marked at its inception by the same kind of holyday-tetrad, was officially repealed immediately following the 1967-8 holyday-tetrad. If these occurrences were not unusual enough to be honored as a message about Israel developing into the place of the King of King's reign, then at this very occasion in the 1960s began the move of the Winter Solstice sunrise past Alnasl, the star representing the tip of Sagittarius' arrow, to where it would five-hundred years later be shining over Scorpio's

Clear Signs of Trouble and Great Joy

stinger in the midst of six interlocking signs-of-witness, each blinking out 1948 in coded form, the number of God's faithfulness to His promises given man (6) as a man (Jesus Christ). The attestation to that new beginning carved onto the most astounding ancient monument ever discovered is sufficiently unusual to ring an alarm in every alert mind.

The 1960s were the dawn of the age of Aquarius, not because the song was all that good, but because at the Spring Equinox, the sun's rising began slowly shifting from Pisces towards Aquarius as the Winter Solstice sunrise began its move into Scorpio. The holyday-tetrads, the Jubilee Path, the Shemitah Path, and the testimonies of the late, great 1948 stack into one monumental sign revealing the incredible nature of our days, at this dawn of the age of Aquarius, to be the pouring of living water onto His fish after Ara's flames have scorched evil with its doers off the face of Earth. And why else was Gobekli Tepe discovered during the very age that its Pillar 43 depicts to be the time of this truly profound paradigm shift? Is this why the Sethite Book of Enoch was rediscovered just two centuries ago and found to be stating in its very first paragraph…

> These are the words of the blessing of Enoch, to bless the elect and righteous, who will be living in the day of Tribulation, when all the wicked and godless will be destroyed. God opened the eyes of the righteous Enoch, so that the angels could show him a vision of the Holy One in the heavens. "From them I understood that the vision I saw was not for my generation, but for a far distant one."[32]

Although Enoch's book was brought out of Ethiopia in 1773, it laid in a British museum until 1893 before it was translated and disseminated, precisely one century before Gobekli Tepe was in the midst of being unearthed (1992-4). Was this because it was written to the generation of the Tribulation? Did Gobekli Tepe's builders feel it necessary to bury the Pillar 43 message for its preservation until our generations were born? If the answers to these questions are yes, then the efforts of Gobekli Tepe's builders worked as intended. Being Sethites full of great honor for Enoch, they, too, would have perceived Pillar 43's message to be for the generation of the Tribulation. I AM WHO I AM, being the God of the Sethites, The Creator of the Luminaries, The Stacker of the Stars, obviously directed the transmission of Enoch's book and the discovery of his children's pillar to our current times. I do not have enough faith in coincidence to believe that all this *information* does not evidence the reality of God's revelatory communication to a desperately foolish mankind.

Is it just coincidental that the Bible predicted Israel's return to the promised land? Is it just coincidental that Israel's return to nationhood occurred in the midst of all those total lunar eclipses on Jewish holydays? Is it just coincidental that the Bible says the celestial bodies were made for signs and that they proclaim God's glory and work? Is it just coincidental that the Zodiac perfectly represents the gospel in its imagery? Is it just coincidental that Pillar 43 pictorially presents the story of The Great Flood and aligns with the Mazzaroth to both tell the gospel and warn of the end of wickedness at this very dawn of the age of Aquarius in which we live today? Is it just coincidental that the Lascaux Caves,[33] the Maya Calendar, and Gobekli Tepe all bear the same forward gaze upon our very time of celestial change? Is it just coincidental that a truly enormous number of signs in the sky and measurements of Jewish history and fulfilled Biblical prophecies point to a Tribulation time that will end as the Winter Solstice sunrise is launching like an arrow towards Scorpio's heart as Christ returns to defeat evil and bless mankind's remnant? Is it just coincidental that the sun's passage from Sagittarius into Scorpio through these Winter Solstices was beginning when Israel won possession of God's Holy City to be His throne? Is it just coincidental that Enoch said he was shown all that was to befall mankind, which categorically includes the gospel and the Tribulation? Is it just coincidental that Noah was a Sethite teaching from his Sethite great-grandfather Enoch's prophecies? Is it just coincidental that Gobekli Tepe was built in the region where the Bible says Noah's ark landed? Is it just coincidental that the Bible's Mazzaroth keys the message of Pillar 43's symbols to God's plan of salvation like the Rosetta Stone keyed the translation of Egyptian hieroglyphs by ancient Greek? Is it just coincidental that both Enoch's book to the people of the Tribulation and Noah's Gobekli Tepe time capsule were discovered in our "scientific" age when these signs scream a warning of the coming end of all deniability of God's Holy Word (KIC 9832227)? Is it just coincidental that Pillar 43's Mazzaroth-deciphered message depicts evil's destruction at this time of the sun's passage into Scorpio? Is it just coincidental that Pillar 43 speaks dual-prophecy as masterfully as does the Holy Bible? Is it just coincidental that the Christmas sign of Jupiter's three conjunctions with Regulus began again in 1968 as Israel received God's Holy City where Christ will reign, and then it will again begin seventy-two years later, two years after Christ has returned to reign from Jerusalem?

No! It's not at all coincidental! Labeling evidence as "coincidence" weaves theories into paradigms for carpeting over reality. These are not coincidences. They are revelatory realities forming a bridgework of evidence for anyone's rational consideration to cross upon! Thank you, Holy God, for that, as even little Sadie is wise enough to say.

The asteroid, Apophis, possibly a remnant of the gravitationally massive object which may have pulled earth's axis into a wobble, causing a devastating Great Flood, will soon fly by the earth twice, seven years apart, before and after four total lunar eclipses occur on dates expressive of Tribulation meaning. All of this together is extremely evidential. It is far too extensively patterned to be just coincidental. But only eyes to see will see evidence. Eyes too blind see only coincidence. Soon blind eyes will clearly see chaos, and forever they will see nothing but chaos. Avoid that! See Christ! Accept Christ. Abide in Him. Board the Argo and be carried across the coming flood of trouble or perish under Ara's shower of flames.

After those devastating seven years have ended, and the glorious reign of Jesus Christ has begun, the sun will be moving into the house of Scorpio. For the next thousand years of righteous government on Earth the Spring Equinox sun will stand in Aquarius, the bearer of life giving water to the righteous remnants, The Face of The Man. The Summer Solstice sun will stand in Taurus, the sacrificial bull driving away evil, The Face of The Bull. The Autumn equinox sun will stand in Leo, the regal King, The Face of The Lion. And the Winter Solstice sun will stand in Scorpio, death defeated by The Great Eagle, the victor over death, The Face of The Eagle, Jesus Christ, to whom every knee will in the end bow and every tongue confess that He, in reality, is Lord of lords and King of kings.

Clear Signs of Trouble and Great Joy

Who can help but contain their praise and adoration at such an enormous reality as Christ Jesus' love for the humble, gathering them under His saving wings where they shall experience the wonders of this coming age of righteousness and all of the eternal bliss He has planned for their lives evermore? Throughout the thousand years of Jesus' reign from Jerusalem, like the living creatures in God's Celestial Throne Room, the man, the bull, the lion, and the eagle will be continuously circling Earth as the corner constellations singing, "HOLY! HOLY! HOLY IS THE LORD GOD ALMIGHTY!"

> [9] And the LORD will become king over all the earth; on that day the LORD will be one and his name one.
>
> Zechariah 14:9

> [1] The heavens are telling the glory of God;
> and the firmament proclaims his handiwork.
> [2] Day to day pours forth speech,
> and night to night declares knowledge.
> [3] There is no speech, nor are there words;
> their voice is not heard;
> [4] yet their voice goes out through all the earth,
> and their words to the end of the world.
> In them he has set a tent for the sun,
>
> Psalm 19:1-4

Footnotes

1. Strong, James. Strong's Hebrew and Greek Dictionaries. Quickverse, A Division of Findex.com, Inc. Omaha, Nebraska. At H2416.
2. Allan, D.S. and Delair, J.B. Cataclysm! Compelling Evidence of a Cosmic Catastrophe in 9500BC. 1997. Bear & Company. One Park Street, Rochester, Vermont 05767 Pg 150-1
3. Collins, Andrew. Gobekli Tepe: Genesis of the Gods. The Temple of the Watchers and the Discovery of Eden. 2014. Bear & Company, One Park Street, Rochester, Vermont 05767. Pg. 74.
4. Sharp, Doug & Bergman, Dr. Jerry, compilation editors. Persuaded by the Evidence: True Stories of Faith, Science, & the Power of a Creator. Parker, Dr. Gary E. "I Preached Evolution: A Biology Professor's Story." 2008. Master Books, PO Box 726, Green Forest, AR 72638. Pg. 260.
5. Jasher calls this one Cainan. The Bible calls him Kenan. There are far too few manuscripts of Jasher to determine whether the error is on the part of the author or a corruption by a later copyist. We shall refer to him as Kenan.
6. Johnson, Ken TH.D. The Ancient Book of Jasher. 2008. Biblefacts Ministries. Biblefacts.org. Jasher 2:11-14. Pg 9.
7. The Ancient Book of Jubilees. Translation by Charles, RH. 1917. 2013. Introduction and appendices by Dr. Ken Johnson, TH.D. Biblefacts Ministries. Biblefacts.org. Jubilees 4:19. Pg 25.
8. Josephus: Complete Works. Translated by Whiston, William, A.M. Kregel Publications, Grand Rapids, Michigan 49501. Twelfth printing, 1974. Pg 27. Antiquities of the Jews. Book I. Chapter 2, Section 3.
9. The Ancient Book of Jasher. 2:11,14b Pg 10.

10. Ibid. 3:5-6,26. Pg 11.
11. Ibid. 4:3. Pg 12.
12. Ancient Book of Jubilees. 4:17,19. Pg 25.
13. Lumpkin, Joseph B. Introduction to The First Book of Enoch. The Books of Enoch: The Angels, The Watchers and The Nephilim. 2011. Fifthe State Publishers, Post Office Box 116, Blountsville, AL 35031. Pg 17.
14. Johnson, Ken Th.D. The Ancient Book of Enoch. 2012. Biblefacts Ministries, biblefacts.org. Pg. 14 1:1-2
15. Rohl, David. From Eden to Exile: The 5,000 Year History of the People of the Bible. 2009. Greenleaf Press, Lebanon, Tennessee. Pg 28.
16. https://aratta.wordpress.com/the-history-of-the-swastika. Cradle of Civilisation: A Blog About the Birth of our Civilisation and development.
17. Ullman, B. L. and Henry, Norman E. Latin For Americans. Second Book. 1962. The Macmillan Company. Pg. 478.
18. Hodder, Ian. The Leopard's Tale: Revealing the Mysteries of Catalhoyuk. 2006. Thames & Hudson Ltd, London. Pg. 125.
19. Hooker, Heather Lee & Tim. The Ancient Aratta Civilization of Ukraine, Older than Sumeria. 2009. T&H Hooker & Megalithomania. (Video -see Youtube). 00:05:35.
20. I wish to avoid confusing whose left or right, the viewer's or the pillar's, therefore I refer to directions as if the pillar were a map. Pillar 43 is arranged nearly North to South from its right to its left. Although that arrangement might bear a very interesting expression, my use of "East" and "West" are in no way meant to be a reference to the pillar's physical arrangement.
21. Aratta was a legendary land of Sumerian lore filled with precious gems, lapis lazuli, silver, gold, and unimaginable wealth accessible only by a difficult journey through treacherous mountains. The Hindu Mahabharata also mentions an Aratta. And the Bible mentions a very lofty mountain, "With an anointed guardian cherub I placed you; you were on the holy mountain of God; in the midst of the stones of fire you walked," (Ezekiel 28:14) most likely not Aratta, but possibly an intersection of related ideas.
22. Ancient zodiacs of the Mesopotamian region and many of later times represented only eleven constellations, which perplexed me greatly until Mr. Collins and Mr. Hale opened this door for our minds to step through. In those zodiacs having only eleven constellations, Scorpio grasping Libra in its claws forms one constellation instead of two. We see that very concept depicted here on the Vulture Stone where both Scorpio and Libra overlay the scorpion symbol.
23. Virgo represents The Seed Bearer amongst the constellations in today's depictions of the Mazzaroth gospel. It represented Sala I to the Mesopotamians. Most likely Sala I came to the Mesopotamians through the Hurrians, an ancient people appearing from the darkness of prehistory in this same area anciently attributed to The Garden of Eden, Enoch, Noah, and the beginnings of prehistory after the Flood. One of the Mazzaroth's more confusing aspects is the occurrence of multiple portrayals of the same personage. The point of each depiction is less the identity of a personage and more the essence of that personage's effects. Usually the effect will involve the sense of bordering and nearby constellations. Thus, Christ is portrayed in Victima, Centaurus, Sagittarius, Aquila, Capricorn, Aquarius, Perseus, Auriga, Orion, Aries, Taurus, Cancer, and Leo as the sacrifice freeing us from sin, the sacrifice defeating death, the sacrifice become eternal life, the bearer of life's necessities, the breaker of the dragon's head, the comforter of the redeemed, the hunter of the prowling lion, a.k.a. Satan, the purity of the sacrifice, the sacrifice driving evil away, the unbreakable embracing of the redeemed, and the final defeat of evil's primal cause. Casiopea is regarded as the Bride of Christ in current analyses of the Mazzaroth. Maybe most thinking about this constellation fails to notice the

Clear Signs of Trouble and Great Joy

correspondence of its shape to the woman's birthing position. Maybe those who do notice that think it too risqué to engage in their discussions. But here at Gobekli Tepe we now witness the thought of ancient man, only a generation or two after the Flood, viewing Casiopea quite definitely as the birthing of the One who is to crush the serpent's head. Sala I, Virgo, is more representative of the blessing from the Seed rather than of birthing it.

24. Collins, Andrew. 2014. Pgs 85-6.
25. Harrison, Allen F. and Bramson, Robert M. Ph.D. Styles of Thinking: Strategies for Asking Questions, Making Decisions, and Solving Problems. 1982. Anchor Press/Doubleday, Garden City, New York. Pg 20.
26. https://godsamazingstarsecret.blogspot.com/2019/featured-teachers-of-mazzaroth-part-3.htm
27. Lafayette, Maxilillien de. Encyclopedia of Gods and Goddesses of Mesopotamia, Phoenicia, Ugarit, Canaan, Carthage, and the Ancient Middle East. Times Square Press. New York. Berlin. 2015. Pg 148
28. Black, Jeremy and Green, Anthony. Gods, Demons and Symbols of Ancient Mesopotamia: An Illustrated Dictionary. University of Texas Press. Austin, Texas. 1992. Fifth printing, 2003. Pg 75.
29. Allen, D.S. and Delair, J.B. Pgs 191-2.
30. Most people will contest my proposition that normal science provides these pre-4000BC dates. They will insist that it is actual science providing them. But referring to our discussions in Chapters 1 and 2, science is not a consensus. It is an investigation. Any investigation must consider all available *information* honestly. As we noted in Chapter 1, massive amounts of real *information* are rejected and hidden from public view in order to prop up evolution. Moreover, I have presented a massive amount of *information* in Chapters 4 through 9 which more than supports the theory that the Holy Bible is the Holy God's Word. Theories based on the Bible's information have a place for every piece of the *information* discarded by science, as well as all of the *information* science acknowledges. The paradigm emerging from the Bible, given its proper place in the puzzle, explains what science today finds unexplainable, which includes the young earth theory.
31. Daniel 5:24-27. Daniel interpreted the message to Belshazzar as saying, "Weighed! Weighed! Found wanting!" It is never good to be on any other page than the one I AM WHO I AM is on. And He gives discernable clues about that page for those with eyes to see and ears to hear.
32. Johnson, Ken Th.D. Ancient Book of Enoch. 2012. Pg 14. 1:1-2.
33. Recall from Chapter 4 the teamwork of Jegues-Wolkieweiz and Geneste. They discovered that the Summer Solstice sunset shines onto the bull of the first hall in the Lascaux Cave system. Evidently the bull was painted there to receive that light. Throughout the ages, this Summer Solstice sunset has shined forth from various houses of The Mazzaroth. But after 2040, for the remainder of earth's Biblically prophesied history, that evening light shining onto the bull of the Lascaux Cave will be shining forth from the House of the Bull, Taurus!

Dayenu.

Chapter 10
What Then Shall We Do?

The chief aim of all investigations of the external world should be to discover the rational order and harmony which has been imposed on it by God.

Johannes Kepler

The most delicate matter in the physical universe is the accuracy of reality's reflection made by the human mind. The most critical matter in the physical universe is to what the mind pays its attention. These signs made by highly ordered patterns of eclipses correlating with historical events significant to the very core of The Holy Bible's prophecies and salvation narrative DEMAND attention. Their expressions are undeniably designed. Their reality is irrefutable. Their correlation with and witness to The Holy Bible is empirical. They are reality's *information*. Inattentiveness to any of reality's *information* is the bull in every mind's china closet.

Consequently, never again will I hear that moldy old "The Bible isn't a scientific textbook" babble without giving it immediate, sharp, irrefutable challenge. "Science" derives from a Latin word meaning "knowledge". Science is the search for knowledge. Knowledge does not bar the concept of God from explaining phenomena. Liberalism does. The Bible speaks prolifically about "the knowledge of God", i.e., the science of God. It fills its pages with intellectual, attitudinal, and emotional descriptions of God's character and being, as well as evidentiary reports of His movements within, impressions upon, and inspirations of human affairs. It presents, describes, and evidences a spirit realm stretching infinitely beyond the limits of mankind's five, myopic senses, a realm which explains the spiritual insights we can perceive by our sixth sense (should we be honest enough to acknowledge the existence of that sense). The Bible instructs science about the most important concept for knowing the origin of the universe and life: following ALL evidence (no dumpsters allowed) to the truths ALL evidence constructs, no God-explanation barred. The Bible appropriately calls this mental condition "eyes to see and ears to hear" a.k.a., faith and good sense. It doesn't matter what normal science calls it.

Through faith and good sense the natural philosophers were discovering the glory of God found in the designs and patterns of their physical surroundings. They understood The Holy Bible to be a special item of their physical world -God's Word reduced to physical writing upon physical paper. They knew of its spiritual origin, accepted its physical container, and followed its enlightening lead into mental paradigms receptive of the spirit realm's existence for better understanding the physical realm. But "science" was that new paradigm crafted especially to whitewash all perspectives discovered by natural philosophers

Clear Signs of Trouble and Great Joy

and to kill off the God natural philosophers feared, else we would earn Bachelor of Natural Philosophy degrees, instead of every imaginable degree of B.S.

Paradigm is a framework of understanding within which observed *information* is processed into facts, data, and ultimately, knowledge, i.e., science. The Bible directs us to the evidences paradigm needs for science's accurate reflection of reality, just as any good textbook will do. Prophecies fulfilled by the later occurrence of their prophesied events, and the correlations of signs in the sun, moon, and stars with fulfilled, as well as, yet to be fulfilled prophesies, Biblical themes, and measurements meting out concepts through the use of Biblically meaningful numbers are just the beginning of the Bible's prescribed scientific observations for building accurate mental paradigms. The Bible's pages are full of prescriptions for creating order (righteousness) in your emotions, mind, and affairs. Orderly emotions, thoughts, and actions sharpen the sixth sense for making even more refined observations of informative patterns. More refined observations see the spirit realm's affects left upon this physical existence like Sherlock Holmes could see finger and foot prints. As a good textbook should, the Bible more than hints at the origin of our universe, life, and the spirit realm. Just as scientific textbooks do, it objectively presents those matters, evidencing the truth of its instruction, prescribing tests for the verification of that truth, and warning of lab explosions caused by ignoring its lab instructions.

> [18]For the wrath of God is revealed from heaven against all ungodliness and wickedness of men who by their wickedness suppress the truth. [19]For what can be known about God is plain to them, because God has shown it to them. [20]Ever since the creation of the world his invisible nature, namely, his eternal power and deity, has been clearly perceived in the things that have been made. So they are without excuse; [21]for although they knew God they did not honor him as God or give thanks to him, but they became futile in their thinking and their senseless minds were darkened.
>
> Romans 1:18-21

Truly, the nineteenth century "scientists" were intellectual thieves. The Bible was available to them as much as was the dirt in which they chose to root for "evidence" disproving the Bible's message. Yet they refused to consider The Bible as evidence. Its fulfilled prophecies were known to almost three millennia of witnesses preceding them. But these infantile "scientists" weren't interested in constructing their paradigm from all evidence followed to wherever it would lead. They were interested in sorting to use only those evidences supporting their precontrived, new fantasy about some God-less nature baloney, and then throwing the rest of the evidence, hopefully before seen by the public, into their dumpsters, because it contradicts those intellectual schemes they publicly palm off as "science of origins" and "nature of the universe".

Natural philosophers had, for a few centuries before the arrival of these intellectual thieves, been discovering that nature's details actually were glorifying God and evidencing His Word. Using the Bible like a scientific textbook, they were well on their way to proving the Apostle Paul correct in his letter to the Romans, "…His eternal power and deity has been clearly perceived in the things that have been made." That reality needs no scientific

What Then Shall We Do?

measurement for millions of people who've been, over the millennia, perceiving Him in nature. Alas! The new "democratic" form of politics planned to overthrow God's authority of king making in order to raise up man's authority to elect God's opponents as their dictators. The clean break from God they desired required hiding all empirically observable evidence of the universe's design, purpose, and meaning, since the universe so clearly displays God. Therefore, infantile science's eighteenth century "researchers" could not follow evidence to wherever it might lead. Evidence keeps leading to the Bible. Indeed, they were intellectually searching for a godhood of man, a new Tower of Utopia. The Bible calls it "Antichrst". And behold! Today, with our own eyes we are watching Antichrist's tower being assembled from herd-mentality and scientific BS (boardroom schemes).

Consequently, these new "scientists of origins" sorted evidence as much or more than they searched for it. They kept all evidence that was twistable into supporting their imagined paradigm of no god, no design, no purpose. All evidence falsifying their desired paradigm of God-lessness they hid (and still hide) from their students and the general public in handy-dandy dumpsters of inattentiveness. Witness how the sauropods of Bishop Bell's tomb deny the early theories of dinosaur extinction millions of years ago. Notice how the Anasazis in Utah, the Freemonts in Colorado, and Jayavarman VII joined the testimony of human experience with dinosaurs by also carving dinosaur depictions. Then notice "science's" dumpster treatment of those tangible, actually existing depictions of human observations of real dinosaurs. Explain Pillar 43's prescient, deep intersection with The Holy Bible's salvation narrative by anything other than true affects made by the spirit realm. Explain science and academia's denial of Pillar 43's prescience by anything other than shear, foolish inattentiveness, a.k.a. sorting real evidence to hide, because "out of sight" makes it "out of mind".

Holding to falsified paradigms while hiding falsifying evidence is intellectual theft, otherwise known as subjectivity, bias, and deceit. It is today's scientific way.

The world hasn't been told this truth (as if to say it isn't reasonable enough for intellectually responsible, honest thinkers to discover themselves). Instead, people have been taught to live with elephants in their intellectual rooms, paying them no mind, asking about them no questions, feeding them not a speculation, casting them not even a glance. From kindergarten to doctoral dissertation, children are publicly taught nineteenth century subjectivity about some god-less ape-man being their great granddaddy a few thousand times removed. Today's culture has been so steeped in scientific lies that even the overwhelming majority of Jesus' believers follow herd-mentality against the Bible's authority of being The Textbook of Critically Important, Spiritual Knowledge (science of spirit).

But there remains a secret knowledge amongst man. It isn't like that of a secret society. It isn't held secret by its knowers. It has been kept secret by the world's institutions of "learning". Even though it is the real science about the physical universe, those institutions don't want it known so they can teach their BS (bungled suppositions) instead. But those who think on the secret knowledge know it is real by every evidence that is a real evidence (many folks fake evidence), even though they believe it by faith. The general public's lack of knowledge about it no more falsifies it than the knowledge of quarks' six flavors are falsified by the public's general unawareness of that. If knowledge has to be widely

Clear Signs of Trouble and Great Joy

known in order for it to be real, then would jets not fly and rockets not rock since there are so few jet engine and rocket scientists? We've been safely flying on jumbo jets for several decades now, and most rocket launches have made it into space. So, I guess knowledge doesn't really need to be generally known (have consensus) in order to be real. Knowledge exists by neither consensus nor the popularity of its knowers, but only by the affirmation of reality's own *information,* the smoke of existence.

This knowledge by *information* isn't a secret because it has been held secret by its knowers. Through the span of twenty centuries, every generation had at least a few people, and most often many more, trying their best to spread it around the world, sharing it with anyone who would listen, prince or pauper. It has been kept secret because the overwhelming majority of people have paid it no attention, even while dishonest scientists deliberately hide it behind a wide variety of mental antics. Certainly, great numbers of the general public have heard of it, like they've heard of how jet and rocket engines work. But they presume the "secret" knowledge does not affect them. They've been trained by "scientists" to pay it no attention, as if keeping it out of their minds will eliminate it from the reality of their futures. Jesus' empirical feet soon standing upon the Mount of Olives will correct that misperception.

Jesus said the knowledge of His approaching return would be discernable by signs in the sun, moon, and stars while mankind's events grow more turbulent (roaring seas and distressed nations). Should that knowledge need validation by consensus of the general public any more than jet engineering or subatomic physics should need widespread public understanding to be real? We are quite happy to know jet engineering follows empirical evidence more than popular consent, especially when riding two-hundred tons of aluminum careening forward, near the speed of sound, thirty-two-thousand feet up. Even folks on the ground rejoice over the secret knowledge of jet engineering, seeing how disastrously a jet can dive through their living rooms. Jesus directed us towards making an empirical observation for knowing when His return draws near, rather than towards the feckless discourse of cold-nor-hot preachers marinating in the public consensus of a subjectively contrived "science". The Holy Spirit wrote into His Scientific Textbook the lab test for discovering the approach of Jesus' return.

Mankind's current knowledge base has available within it all of the Biblically prescribed elements of reality necessary for mapping where we are in time's procession through God's plans, regardless of how few people know some of its key elements. Surely Daniel's prophecy has been unsealed! NASA's website details every eclipse of sun and moon between 2000BC and AD3000, making available those signs in sun, moon, and stars which God's Scientific Textbook said could be found. Unadulterated history books detail numerous elements of past realities. So do archaeological artifacts. It is then possible, although not popular, nor consensually agreeable, to empirically observe history broken into three 1,948 year epochs (ceremonious epochs, maybe) followed immediately by a 19/48 year pattern formed within the chronological arrangement of twelve total lunar eclipses, each happening on a major holyday of the little nation reestablished in 1948, just as The Holy Bible prophesied it would be reestablished twenty-five-hundred years earlier. (Is that not more than a little curious?) It is now possible to empirically observe the occurrence of every known 19/48 year

What Then Shall We Do?

pattern of eclipses over developmental events of that reestablished nation, of the followers of its Messiah, of the evidence of the Messiah's resurrection, of His return to reign, and of His reign. It is possible to empirically observe the landmark events preparing God's Holy City for that Messiah's reign by noting a pattern of Jubilee years found on history's map, counted in Biblically meaningful numbers that correlate with the essential nature of their landmark events. We have empirically observed seven-year beats of time measure the gathering of the reestablished nation's people for receiving their Messiah. We've seen two total solar eclipses, each occurring on the Jewish day calling for repentance, strike paths of totality across two nations interrelated by history, Biblical theme, and archeological evidence. We have seen the gospel written in constellations interpret the most ancient temple known to man, translating the messages of its most mysterious pillar. And we've seen that pillar prophesy the end of man's affairs for us who are living at what very much seems to be the end of evil's age.

I assure you, none of this accords with popular consent. None of it is widely known. But like the greatest of jets, it gloriously flies. Like a rocket, it escapes Earth's gravitational deceit. Every one of its elements accords with empirical observations of real literature, physical things, and historical events. God inspired Moses to write in the Bible about the empirical nature of the sun, moon, and stars for knowing the spiritual essence of humanity's dirty deeds to be ended by God's restorative work.

Empirical evidence has guided solutions to some of mankind's most vexing problems. It enabled the construction of towering cities, sound bridges, and safe air travel. It developed industries to feed and cloth the world population, and to treat its illnesses. In many simple ways, empirical evidence laces our physical lives into one common reality.

So the postmodern suggestion of every person having "his own" reality crumbles at every busy, street intersection where everybody MUST precisely obey the highly common, certainly, inflexible meanings of red-lights and green-lights, else make or become road-kill. Each postmodern, personal reality is chewed to powder at breakfast, lunch, and dinner tables where every individual had better eat real food and drink real water, or else, really die (real slowly). Restrooms are more evidence of common realities nobody should attempt to escape by postmodern BS (boorish subjectivism). Death extensively evidences the fecklessness of postmodern BS (boastful speculations). Nobody is excused from reality, however liberally they distort it inside their own minds. Even bank records around the world evidence the commonality of all our ties to reality -each and every one of us has only as much as we each really do have, regardless of how much every one of us would rather have. Reality is what it is, regardless of how any mind desires to personally perceive, believe, or attempt to deceive it.

The animal world displays this same necessity of humility towards reality. The mouse that spurns the sight of a cat is metabolized by the cat. The bunny who disregards a shadow crossing the meadow soon melds with an eagle chick. And the lioness that always lolls under a shade tree will eventually feed that tree.

Just as reality is what it is, I AM is Who He Is.

Only man's mind bucks reality to snub every thought of I AM. Only arrogance concludes there is no God. Or at least it makes up its own version of God. Arrogance rejects all

Clear Signs of Trouble and Great Joy

possibilities of a larger, self defined, spiritual reality overarching this small speck of a physical universe, no matter how well its overarching nature is empirically attested, simply because arrogance wants to fly its own reality into the face of God's irrefutability. Since arrogant people can not see or taste spiritual reality as immediately as they can see red or green lights, or like they can taste breakfast/lunch/dinner, they think they can loll around in reality's shadows make believing whatever gives comfort to their souls, instead of hunting for God's expressions about why things really are what they are.

They despise The Holy Bible. They embattle its God because He proclaims the way things are regardless of the way they dream things to be. So man's arrogance created scientists to throw God's evidence into dumpsters, like over glorified garbage men, simply because God made reality to evidence The Bible instead of Darwin's goose-egg guess.

> 7 For what great nation is there that has a god so near to it…
> Deuteronomy 4:6

Israel, through all its prophecy-fulfilling history, is observably a great nation, again, today. And it will shortly be raised to the head of all the nations.

> 19 I call heaven and earth to witness against you this day,
> Deuteronomy 30:19

Immense patterns in the celestial bodies and their concurrent motions correlating with the themes, narratives, and prophecies of The Holy Bible are now empirically observable, while Darwin's Humpty Dumpty guess gets none. Even so, who will look?

> 4 Where were you when I laid the foundation of the earth?
> Tell me, if you have understanding.
> Job 38:4

None of us today saw the past happen. Nobody saw Earth's foundations laid. Nobody saw radioactivity decay or isotopes' transportation through rocks from the laying of those foundations until today. Yet we still make up whatever we want to believe about matters we can never examine firsthand.

> 18 Have you comprehended the expanse of the earth?
> Job 38:18

Only the few "secret" knowers acknowledge, consider, and deliberate in accord with ALL of reality's available evidence and *information,* no twisting, stretching, or obfuscating, and no trips to science's deceitful dumpsters. These knowers honestly consider reports written by others who did experience the past firsthand. The rest of mankind, without comprehending the expanse of the earth, proclaim that the experiencers-of-the-past didn't know what they were talking about, and then falsely contrive what the past's real witnesses never saw.

What Then Shall We Do?

> ^{33} Do you know the ordinances of the heavens?
> Can you establish their rule on the earth?
>
> Job 38:33

We've learned the physical laws of the heavens for creating comfort on Earth. But have we paid a lick of attention to the spiritual laws of the heavens for creating peace and security? Have we considered the messages reflected in the motions and positions of the heavenly bodies? Have we honestly tried their counsel to receive mankind's Only Solution? Or have we left our feet stuck in Progressive BS (bum solutions)? Where's SETI now that the real patterns have been found speaking of God and His own Scientific Textbook?

> ^{27} Is it at your command that the eagle mounts up
> and makes his nest on high?
>
> Job 39:26

Most of reality happens beyond our influence and command. Viral infections, firestorms, earthquakes, asteroid strikes, and political lawlessness leave us breathless over what we cannot control. Still man's consensus goes against taking shelter in the shade of The Great Eagle's two wings. Science and academicians commanded The Great Eagle to stay out of sight. But there He is, now verified by the rest of His Creation to yet be nesting on high!

> ^{2} Shall a faultfinder contend with the Almighty?
> He who argues with God, let him answer it.
>
> Job 40:2

Following God's prescriptions makes a godly civilization. But contention with God over every point has shaped modern Earth's savagization. 24-7, countless mouths flap in the wind against God. But not one speaks the answers to life's more vexing troubles. Shortly, they all shall be stopped, praise God!

> ^{9} Have you an arm like God,
> and can you thunder with a voice like his?
>
> Job 40:9

Extend it, if you have one; try thundering if you lust for failure. Or settle your minds on the peacefulness of better reason, and keep your arms to yourself.

> ^{18} "Come now, let us reason together,"
> says the LORD:
> "though your sins are like scarlet,
> they shall be as white as snow;
> though they are red like crimson,
> they shall become like wool."
>
> Isaiah 1:18

Clear Signs of Trouble and Great Joy

If there were one thing that could make everything right for anyone who desires goodness, it could only begin with gracious mercy, that thing everyone needs. Although everyone falls short of what's right, some have, by His mercy, been born again into the new culture of an eternally blissful and honest reality.

> [16] Then those who feared the LORD spoke with one another; the LORD heeded and heard them, and a book of remembrance was written before him of those who feared the LORD and thought on his name. [17] "They shall be mine, says the LORD of hosts, my special possession on the day when I act, and I will spare them as a man spares his son who serves him. [18] Then once more you shall distinguish between the righteous and the wicked, between one who serves God and one who does not serve him.
>
> Malachi 3:16-18

Earth has never been without a remnant fearing the Lord and thinking on His name. They, too, are His evidence. God's cloud of witnesses has hung in Earth's sky since the day Eve bit the apple. They've given their testimonies for our education. We can hold those testimonies and read them in The Holy Bible, the books of Jasher, Seder Olam, Jubilees, and an extensive list of other literature written throughout the ages down to our very own day, such as *Foxe's Book of Martyrs*, *Pilgrim's Progress*, *The New England Primer*, *Four Blood Moons*, *The Harbinger*, and even this book you're now reading. Testimony is carved all over the Gobekli Tepe monument in harmonious unison with the Bible written centuries later, and in correlation with signs and messages written in the constellations and the motions of the sun, moon, and Earth. I AM WHO I AM is a God of testimony. Testimony is a product of evidence, the lyrics of empirical observation's tune. And now science's most important tune, empirical observation, is heard playing, also, to the lyrics of The Holy Bible.

Unfortunately, the existence of a Biblical prophecy about the approaching Tribulation is solid evidence of mankind's propensity to employ dumpsters in his reasoning more than empirical observation. God gave man the freedom to use either imagination or observation for determining what to believe. Tribulation prophecy indicates that man will eventually drink his own imagination down to its last, horrific dreg.

Therefore, first, let's not imagine that this incredibly extensive system of signs prophesied in The Holy Bible (Genesis 1:14 and Luke 21:25) will convince most folks to repent of their arrogant contention against God and return to humbly serving The Lord of Spirits, Jesus Christ. Jesus said the way to destruction is broad and traveled by most, while the way to eternal life is narrow and far less traveled. He desires all to be saved (II Peter 2:10). But He knows most will perish (Matthew 7:13-14). The day of their end draws near. Massive numbers of souls are within a couple decades of physical death and eternal doom. In comparison, Covid-19 will seem to have been a birthday party. God has been plainly evidencing this coming end. He will not be judged as the fault of their horrific fate (Romans 3:4). The doomed will certainly accuse Him. What's new? He's been accused by "science" for two-and-a-half centuries. But the end will show the truth of His new beginning. For He has given the world an enormous system of crystal clear warning signs about the coming

What Then Shall We Do?

destruction of contentious people preceding His restoration of faithful people. The evidence God has been pouring before our twenty-first century eyes declares the guilt of every soul who rejects, to his bitter end, the truth Jesus Christ told in His Daddy's Holy Bible. Therefore, no excuse is available for any who perish. God produced sufficient, empirical evidence of His reality, the Bible's truth, and the approaching destruction of truth deniers.

Second, the truth is that every person's need for God's grace is immense. Every individual human is directly and personally liable for Jesus' death. When Jesus asked His Father to remove from Him the cup of torturous death-by-crucifixion if possible, He was evidencing the real involvement of your imperfection in His gruesome death. Had you lived without error from conception through forever, then you would have been mankind's perfect remnant saved by your own behavior. Jesus would not have needed to die to save a remnant, for you, yourself, would have been the remnant of mankind which God sought. Then Jesus could have just taken perfect, little you and flown off to heaven, leaving all of the earth-full of sin tarnished souls to perish forever, no cross, no grace, no hope, no change.

But that's not the way God's story played, because you sinned too. So do I. Everybody does. Each and every one of us is personally and individually liable for Christ's death. Christ had to die for the remnant nobody made of themselves. We're all guilty. Our blood is on His cross. From there it will either condemn or commend you in accord with your decision to accept or reject His blood on your shoulders.

Accepting His blood makes your sins white as snow, bright as the Spirit driven truth. He will throw all evidences of your every sin into normal science's dumpsters, and He will give to you a free pass down His narrow path to life. It is each person's own responsibility to follow Jesus, bearing his own cross, with Christ's blood on his own shoulders, else bear the full guilt of spilling that blood and perish without a cross, without hope, unchanged. Of this God warns. The evidences of His Holy Bible, history, and numerous signs made in sun, moon, and stars call out for your good sense.

Third. The world full of God deniers point fingers of shame at the church's bad behaviors. The Spanish Inquisition was not the only fascistic movement of the church. Today's church is filled with fascistic attitudes, behaviors, and judgments. They call it "denominationalism". Fearing the world's scorn, the church judges its own culture and traditions in accord with fascistic, herd mentalities rather than by every word of The Holy Bible. However much the church covers up its age-old BS (banal superstitions), group-think eventually comes shining forth its ruse. The God deniers' criticism of the church is more often proven right than wrong by the herd behavior of today's Laodicean "faith".

Today's church doesn't see Laodicea when looking in the mirror. Mankind's rebellion against parental knowledge and discernment drew a battle line through the church, as well. The church distanced itself from the "moldy traditions" of the "embarrassing past", replacing God's marriage institution with mankind's marry-anything foolishness, and dropping its inclination to reflect Biblically defined morality to pick up Carter's wide variety of little "justice" pills. Ours has become a church of slogans and shtick, looking the religious part and feeling good about itself without entirely believing or following God's Holy Word. It takes the form of religion, but denies the power of Biblical truth.

Therefore, it is not surprising to find much of today's church supporting a woman's

Clear Signs of Trouble and Great Joy

right to kill prenatal boys and girls while convincing the surviving remnants of her offspring that they are nothing more than glorified apes who, somehow, can do, be, or think whatever they want. Only at adulthood do these surviving remnants of gestation realize the truth of their limitations, resulting in a desperate detour of their minds into accusing traditionalism for those limitations and promoting Progressivism as the salve.

> And the LORD said, "Behold, they are one people, and they have all one language; and this is only the beginning of what they will do; and nothing that they propose to do will now be impossible for them."
> Genesis 11:6

The Holy Bible, The Mazzaroth, and Noah's Gobekli Tepe intersect at the deepest of tradition. We find them saying the same things today they said when first made. Even baseball and apple pie somewhat change over time. But nobody can tamper with God's communication to mankind. He was the same God at Creation that He will be on Judgment Day. Therefore, what He said at any time rings equally true throughout all time. He is the most traditional of all conservatives. History, these signs in the sun, moon, stars, and The Holy Bible knock on the Laodicean Church door, "God has not changed. Why have you?"

Since man has changed his culture, church needs to fit in? We were created in God's image. Why should the church reflect man's collectivism?

These signs call for the actual recognition of Jesus Christ as the head of the church, not men, not women, not church leaders, preachers, reformers, or any other category of tamperer. Christ is the Head of His church by His Holy Spirit leading each individual into the truth as presented by The Holy Bible, that age-old, scientific textbook.

Here the knocking? Open your door. Buy His salve and gold. Dine healthily.

Finally, these signs speak of Christ's return to those who wash their robes in The Lamb's blood flowing down their shoulders from His cross, who submit to God's Word and abide in its unadulterated message. Eventually Christ must return because The Holy Bible says He will return. Therefore, eventually some generation is going to experience His return. Is it so irrational to think we are the generation of that return simply because He said nobody, including Himself, would know the day or the hour? It is rational that the closer His day and hour approaches the more its year or decade will be discernable by the signs He said would precede it.

> [28]Now when these things begin to take place, look up and raise your heads, because your redemption is drawing near.
> Luke 21:28

It is written, "…redemption is drawing near." It is not written, "…might be…" nor, "…should be…" or, "…could be…" but, "…is…" And it is written, "…when…" not, "…if…"

I'm not sure if you've noticed, but people have been "fainting with fear and foreboding of what is coming on the world" (Luke 21:26) for quite some time now. WWI must

What Then Shall We Do?

have been a foreboding experience, which even bore its own flu pandemic. The sight of Hitler's death camps shocked the world. Then within months was seen the horrors of Hiroshima and Nagasaki. Eight Shemitahs later, al Qaeda delivered their world changing 9-11 attack. A deceitful media tarred, feathered, and hung President Bush in effigy for six more years before the housing market collapsed, ushering in the Marxist President who enabled a worldwide jailbreak of jihadis, a bag of "Hands up! Don't shoot!" riots, and a whole lot of lawlessness at the top of America's justice system. What comes after the Covid-19 panic and liberally burnt up cities to fry piggy, little minds in the next blanket of lies? Waves of Muslim refugees have been washing over Europe again and again. China builds islands for military assault. The nations are distressed in "perplexity at the roaring of the sea and the waves" (Luke 21:25). Iran chants, "Death to Israel; Death to America", while slowly closing its grasp on nuclear weapons.

 I'm sure the Mongolian raids of Medieval history were foreboding to the peoples in their path. So also the European raids must have been foreboding to the American Indians. But "and" is a little, big word. "There will be signs in sun…moon…stars, **and** upon the earth distress of nations." Neither the Mongolian raids nor Manifest Destiny occurred during such a set of perspicuously arranged astronomical phenomena as we've discussed in this book. But WWI, WWII, 9-11, Covid-19, and "pigs in a blanket of deceit frying their minds like bacon" are interwoven with an enormous number of signs in our sun, moon, and stars. "And" expresses an important part of any message wherein it is used.

 If, to you, the signs we've discussed do not rise high enough to be those of which Jesus spoke, then what exactly are you looking for? A dumpster? If you merely don't wish to think Jesus' return is now drawing near, then what signs in sun, moon, and stars would you expect to see around the time you're thinking His return will be drawing near? And why has this very real mass of signs so perspicuously insinuated His return during a time so matching the descriptions His Bible gave, meeting the measurements foretold to Daniel, fulfilling the major prophecies of Israel's restored nationhood? If there is to be a more perspicuous system of signs rising to the description of Genesis 1:14, Psalm 19:1-4, and Luke 21:25, they will need to assimilate out of nowhere, because there aren't any more signs observable in NASA's cataloguing of God's sun, moon, and stars.

 So also Christ's return in another era would make a mess out of Daniel's "time, two times, and half-a-time" prophecy, leaving the essence of its measurements chaotic and senseless. Whatever unit of measurement Daniel's prophecy might then require would be anticlimactically meaningless in comparison with the otherwise highly meaningful number of perfect order raised by the power of God's doing it (10^3) which actually does correspond to this enormous system of signs. We've seen these signs correlate with the Bible to evidence good sense about history at the same time. We can only employ dumpsters for what eyes to see can see should we desire to plant our faith without regard to the glory told in our heavens, going out day by day and night by night, pouring forth speech and declaring knowledge without words.

 Jesus was born amongst many verifying evidences. That is the way God works. He gave prophecy regarding the time of Christ's birth. Approaching that prophesied time, He hung a star in the heavens to signal His birth. Wise men saw it. Angels came pronouncing

Clear Signs of Trouble and Great Joy

His birth to simple folks. Prophecies regarding the conditions of His return are fulfilling today. Highly perspicuous signs have been discovered in today's sky. Stories are reverberating throughout the Middle East about angels leading people to Christ. Godly men have been pouring forth books of expectation about Christ coming soon. Now we can look into the sky and know why.

God desires to forewarn His loyal servants because they need to forewarn the less attentive; it's the way community works. The loyal must wake up the sleeping and challenge the rebellious. When the time for harvest has come the water must be turned off the field, the harvest equipment must be lubed and tuned up, and the servants must enter the fields with harvest in mind, rather than planting. And now, our time is appearing to be made of that harvest essence.

But there is yet a bigger reason for His servants to be informed. Occasionally, the physical world more than just brushes elbows with the spirit realm. It did when Eve bit the apple. It did when Enoch walked with The Lord of Spirits. It did when God rained upon Noah's neighborhood, and when Abraham was obedient, and when God led the Israelites out of Egypt into Canaan. It did when God kicked Israel out of the Promised Land, and when Mary bore Christ into the world, and when He destroyed Jerusalem for its non-repentance. All of those events entailed paradigm shifts effected by an infringement of the spirit realm upon the physics of our physical realm.

The resultant paradigm shifts were as much shifts of mankind's spirituality as they were physical shifts, and sometimes more so. The recorded experiences of those shifts always involved spans of time where the separation between the spirit realm and the physical realm diminished, such that the physical veil hiding the spirit realm became more translucent, sometimes almost transparent, as spiritual activity heightened in the material universe. Enoch's neighbors searched for weeks to find him after he was taken to heaven. Noah didn't round up the animals for his ark; God led animals He selected to Noah, and He sheltered them under the shadow of His wings spread over the ark on that great expanse of water. Abraham and Moses both conversed with God many times. Prophets buzzed around Israel and Judah with warnings directly from God. Angels brought tidings of Christ's birth. Images were seen in Jerusalem's AD70 skies, and impossibilities happened around it's doomed Temple.

As Christ's return approaches today, the spirit realm is again beginning to poke through our physical veil. But this time Earth is soiled filthy with lies swallowed whole heartedly by mental pigs fully wallowed in scientific mythology, rooting for whatever does not evidence I AM WHO I AM, swallowing whatever makes them feel powerful. The spirit of Antichrist will continue to infringe upon mankind until that climatic intersecting of the physical Mount of Olives by Jesus' celestial feet.

His servants must greet that spiritual intersection with the spiritual work of prayer and understanding, rather than bewilderment and denial brewed from scientific paradigms. Prayer is critically important for God to work His effects upon the world. It isn't like God is empowered by prayer. He is the Almighty. He has all power already. He doesn't need your prayer for His power. It is like, we who love Him must invite His effects into our realm, else, He is Hitler entering without knocking or Biden taking what the voters didn't give. He

What Then Shall We Do?

is not a forcer. He is not a fascist. He is a giver. He works in response to request. Those who request Him get Him. Those who invite His will get His will.

It is not our 401K's, stock accounts, and business inventories that we should ask Him to safeguard against the upcoming Tribulation event. It is our minds, our hearts, and our paradigms, so that we can maturely know what to request and do, when, and why. The importance of prayer regarding today's material/spiritual intersection rises with the passing of every day. There needs to be prayer for a knocking upon neighbor and nemesis' doors alike. Prayer will be ultimately important during the Tribulation. The Lord God will have many minds sealed with the Holy Spirit for praying and doing maturely. We might be astonished, but we must not be alarmed. We must prepare to be of the right mind.

Do I know beyond all doubt that Jesus Christ returns in or around 2036, instead of maybe 2072, or say, 2677, or even 10677? No. Jesus' statement that we would not know the day or the hour of His coming could have metaphorically meant that we couldn't know the year or even the epoch of His return. If that is the case, His instruction for us to look up and know that our time of redemption draws near could only be metaphorical as well, for knowing His return draws near would breach that metaphor of not knowing at all. But Jesus said we could know His return draws near by observing the signs of it drawing near. Regardless of the obvious inconsistency of such speculation about not knowing, I do reserve my right to become the two-hundred-forty-third error. If I become that error, I will humbly apologize. But I will never regret speaking forth this amazing system of patterns correlating with God's Holy, Holy, Holy Bible.

After acknowledging the observable train wreck skepticism makes out of God's messages, belief should be able to confidently proceed from what eyes to see do see. And eyes to see do see signs. Many, many interrelated, concurrent signs. So I believe these empirically observable evidences for their coherence with God's Word. I am not superstitious enough to ascribe them to mere coincidence. I know how science's dumpsters disaffect rational belief. Consequently, I remain inclined to let evidence (of which God's Word is the soundest) lead me to wherever it goes. I believe with certainty the greater possibility that we are literally knowing our redemption is drawing near by the indications of celestial signs current events, and Biblical prophecies. I know by my own empirical observations that patterns of celestial alignments surpassing all possibilities of mere coincidence extend from the irrefutable reality of this material universe being intersected by God's spirit realm. (You can search and see, too.) I know by study and analysis that these signs amazingly correlate with historically fulfilled Biblical prophecies, with events relative to The Holy Bible's salvation narrative, and with current events foreboding an eerie distress of bad nations on the rise.

I might be wrong. But I might be right. Yet, I am not who these signs are about. Jesus is. His right will be our excruciating end should we judge Him wrongly. There will be the haven of joy, peace, and righteousness in The Holy Spirit for everyone who judges Him in accord with all *information*.

Thank you, God, for *information*!
Amen, little Sadie, amen.

Clear Signs of Trouble and Great Joy

Dayenu!

Appendix I
Analyzing Tetrads

Nothing in this physical existence can be seen without at least a little effort, even if that effort amounts to no more than the muscle tension required to move your eyeballs. However, the efforts needed to "see" electricity and how it might apply to the performance of useful tasks took a lot more effort than just glancing around. "Seeing" subatomic particles and their properties required building underground particle accelerators miles in diameter. Therefore, we should not think the great effort required to "see" these signs might rationally disqualify them from either being signs nor from being observable any more than the tremendous effort it took to "see" the periodic table should somehow disqualify chemistry from being science. So let's get started on getting you to seeing some signs.

Log the Tetrads

Step one is to develop a tidy list of tetrads. NASA's lunar eclipse catalog becomes handy at this point (indispensable, actually). You can search the www.nasa.gov homepage for the right buttons to open the catalog, or you can do it the easy way: Bing "nasa lunar eclipse page" and click on the link to https://eclipse.gsfc.nasa.gov/lunar.html. On the first page of that site, look for the link reading "Five Millennium Catalog of Lunar Eclipses: -1999 to +3000" and click it. Now, scroll down about three pages to the catalogue's table of contents.

You will find the table of contents, a.k.a. index, sectioned into fifty pages for your convenience. Each page displays a century of eclipses. It will pay you a small dividend of time to note the far right column of this table of century pages. It displays how many tetrads occurred during each century. Twenty centuries had no tetrads, so you won't have to scan almost half of NASA's five-thousand years of eclipses. I didn't notice this column when I made my list. Oops! There went a couple weeks of lost time!

Now, click on the first century-page listed as having tetrads in it.

Each century page begins with a section of statistical data about that century's eclipses. Notice that, amongst this data, NASA conveniently lists the years of the century's tetrads. I didn't read this information, either, when I prepared my list. Yes, I just dove in and scanned the entire catalogue from beginning to end; I confess that naiveté. So the month I spent should have taken only a week or two. Log those tetrad years onto your list. Leave ample space to note the month and day of both eclipses for each year. Those dates will be important later. Begin your list like this:

1991BC   1990BC

Clear Signs of Trouble and Great Joy

We have seen more than sufficient connection between celestial signs, Biblical prophecy, and actual history to go ahead and ignore the secular "BCE" obfuscation of reality. Notice that the BCE label still centers around the same event as does "BC": Christ's birth. The universe exhibits far too much order to have been the result of happenstance. "Science" exhibits far too much bias to have been the result of objectivity. The universe's order is observed correlating with The Holy Bible's Jesus Christ rather than Darwin's little ape-man. Therefore, let's log the years in correlation with what the universe says about itself: BC and AD.

Now you must add the month and day of each eclipse to your list. Locate in the eclipse list each year which you found in the preamble. The seventh column of the list indicates the eclipse type: partial, penumbral, total, etc. Seeing four eclipses labeled "T" listed one after the other will confirm that you've located that tetrad. Log the months and days for each year into your tetrad list. The first line of your list should now look like this:

1991BC 12-22; 1990BC 06-17, 12-12; 1989BC 06-06;

Whoa up, now. Did we find a mistake in NASA's information right off the bat? Actually, I think we found a mistake in presuming every tetrad entails only two years. Every tetrad indeed occurs in a two-year span of time, but three times in the five-thousand year list, a tetrad entails at least a few days of three different years. Although these "three-year" tetrads are not significant to the analysis we will be performing, they are revealing of what we are researching. The very first tetrad we find in NASA's lunar eclipse list is a three-year tetrad. By the time you are finished preparing your list of tetrads you will have found three of these three-year tetrads, not two, or four, but three. Again, we've discovered a count of three occurring three times. By now in our exploration of God's signs in the sun, moon, and stars, we can just consider this triple display of threes to be par for the course. So I chose to make my three-year tetrads stand out by center justifying them.

Be careful to not miss the AD900-901 tetrad if you are working off NASA's enumerations of tetrads. Notice the index indicates that the AD0801-0900 century interval contains eight tetrads, but the century page only indicates seven. This discrepancy occurs because one of the eight tetrads entails two eclipses in the 0801-0900 century and two in the 0901-1000 century. So you will find the first two eclipses at the end of the 0801-0900 page, and the final two at the beginning of the 0901-1000 page. You will find this same difficulty duplicated for the AD2600-2601 tetrad. Be sure not to miss either of these two tetrads, one will be significant to your analysis.

After you've begun to log the tetrads, your list should begin looking like this:

1991BC 12-22; 1990BC 06-17 12-12; 1989BC 06-06

1661BC 04-12, 10-06; 1660BC 04-01, 09-24;
1643BC 04-22, 10-16; 1642BC 04-12, 10-05;
1625BC 05-04, 10-28; 1624BC 04-22, 10-16;

Analyzing Tetrads

Note that the second tetrad is very relevant to your research, recalling from Chapter 8 that it begins the four interlinking signs-of-witness regarding Joseph in Egypt, Egypt's mistreatment of the Hebrews, entailing the rise of the first head of the dragon having seven heads (empires) interlinked throughout subsequent history to 1917. Also notice that it is a holyday-tetrad. And the tetrad following it may or may not be a holyday-tetrad, depending upon whether you could get across the bridge to smell the flowers on March 24, 1643BC. It's just like Jesus' Daddy to next stack a holyday-tetrad onto a three-year tetrad for marking Joseph in Egypt at the conception of the seven-headed dragon.

You will need to list one-hundred-forty-one tetrads, so use more than one sheet of paper unless you can write and read small print. Keep your list neat. If you format your list into columns of tetrads, leave enough room between the columns for making notations later. You might want to keep a clean original of your list and make copies of it for making notes. I listed my tetrads on a word processor. That keeps them neat and gives me as many fresh copies as I need for whatever notations are necessary. And be careful! It is amazing how quickly the eye can rebel after only a short time of tedious use. Double check your dates carefully.

This will take some time, so be patient, and enjoy the task.

Analysis for Holyday-Tetrads

This analysis doesn't get any easier, or less tedious. Fortunately, the Jewish Passover is established on the evening of a full moon. If God really was a mean-weenie, He would have established His holyday feasts on new moons so He could watch His people stumble around in the dark while celebrating.

Because Passover and Tabernacles are full moon events, we don't need to be concerned with the exact date of a Passover holyday, whether it is one date by Jerusalem time, but another by NY time. We only need to be concerned with which full moon was that year's Passover full moon. We discussed the difficulties of determining the Passover dates in Chapter 6, so let's not repeat it here. We found that any full moon occurring after March 27 and before April 19 will be a Passover full moon. Then it should go without saying that any total lunar eclipse occurring between those dates will be a Passover eclipse. But you will need to test the eclipses on your list occurring between March 18 and March 28 and between April 18 and April 27. I used the Passover calculator at http://www.covert.org/paschaldate.html. (And as I pointed out in Chapter 6, it is going to present March 19, AD843 as being a Passover full moon. Maybe Passover isn't as tightly tied to the Spring equinox as it is perceived.) Test the first full moon of each year of the tetrad. Every tetrad you find with both the first and third eclipses being Passover full moons will be a holyday-tetrad since the celebration of Tabernacles is on the sixth full moon after Passover. Mark the holyday-tetrads. I highlighted mine. Again, as we noted in Chapter 6, a few tetrads before the fourth century AD begin with an eclipse that can not be determined to be of a Passover full moon or not. Those with dates of March 26 or 27 and April 19 or 20, I asterisked as stronger possibilities, but still, I'm not aware of any way to know for certain.

Clear Signs of Trouble and Great Joy

Analysis for Tetrads-of-Testimony

This analysis gets even more tedious. So it also gets even more difficult to explain. You will need to start with a fresh copy of your tetrad list having the holyday-tetrads highlighted. Recall that a testimony-of-tetrads is determined by a particular pattern involving three tetrads. For ease of discussion, let's label those three tetrads A, B, and C in chronological order. The testimony-of-tetrads pattern spans sixty-seven years from January 1 of the first year of tetrad A to December 31 of the last year of tetrad C. Tetrad B will begin eighteen years after tetrad A began, and tetrad C will begin forty-seven years after tetrad B began. Any three tetrads found to have occurred in that pattern is a testimony of tetrads regardless of how many other tetrads occurred during the sixty-seven years spanning from the beginning of A to the end of C.

Beginning with the first tetrad on your list, you're going to test each tetrad for whether or not it begins a tetrad-of-testimony pattern. Mark that tetrad "A" so returning to it will be quick and sure. Add eighteen to the date of A's first year (for the BC tetrads, remember that you need to subtract instead of adding). If another tetrad begins on the resulting date, then mark it "B" and add forty-seven to the date of B's first year (or subtract if you're working in the BC era). If the resulting year is the beginning year of another tetrad, then mark it "C"; you've discovered the three tetrads forming a testimony-of-tetrads.

When you have discovered a testimony-of-tetrads pattern, double check your work. The difference between the first year of tetrad A and the first year of tetrad C needs to be sixty-five years by simple subtraction. If it isn't, then you made an error. Once you have verified the accuracy of your work, mark the testimony-of-tetrads on your list in some manner that makes the three tetrads an obvious set. In selecting a mark (such as braces or brackets) keep in mind that you are going to find a few instances of interlinking tetrads-of-testimony, as many as six, so choose a mark that will accommodate complexity.

Log each testimony-of-tetrads you find onto an index card. Title the tetrads by their year pairs, e.g. 1493-4, 1949-50, 1967-8, etc. Format your tetrad on the card in a manner which accentuates the three patterning tetrads A, B, and C. Include all intervening tetrads in some subordinated fashion. Highlight any tetrad on the card that is a holyday-tetrad. For example, you should produce one index card that looks like this:

```
                                                            AD1457-1523
            AD1457-8
              1475-6
                 1493-4 (highlighted)
                 1504-5
                 1515-6
              1522-3
```

These index cards will come in handy for categorizing the tetrads-of-testimony, so do not log any more than one onto each card. The cards will also be useful for accumulating interesting and significant historical events which occurred during the particular sixty-seven year span of each. For convenience, note the date range on the upper right hand corner as

Analyzing Tetrads

illustrated above. Be sure to designate AD and BC on every card. Time is one of The Lord God's belongings; He designs occurrences within it as He pleases; He has communicated His designs in accord with His good pleasure; and consequently, we find time masterfully measured and meaningfully bound to Israel and Jesus Christ. So don't be shy about BC/AD.

Clear Signs of Trouble and Great Joy

Appendix 2
Having Fun with Astronomy Software

You don't need expensive software to make these observations. Even rather close approximations will do. I downloaded my software free from https://sourceforge.net/projects/skychart/. I will describe the processes using the Cartes du Ciel software found there. It has its glitches, but it works sufficiently for observing these orderly patterns communicating the glory of God to attentive watchmen. We will do all of our observations with the right slide-bar of the sky chart at the top of its position.

The Star of The One Born King of the Jews
    In Chapter 3 we discussed the interesting motions of Jupiter around the quadruple star system, Regulus, in the Constellation Leo. Now we're going to see how to use the Cartes du Ciel Skychart software to observe Jupiter's motions yourself. Although this process appears very complicated, it is actually quite simple when you understand why you must do so many things to simply watch Jupiter go through a series of conjunctions with Regulus.
    Start the Skychart program.
    Set the slider-button of the vertical slidebar on the right of the window to the top position. This will orient the displayed stars onto a disc that represents the dome of the sky.
    Since the motions of Jupiter occur along the solar ecliptic, you don't need to worry about setting your point of observation to anywhere on the earth in particular. The solar ecliptic can be observed from any location. But, to feel like you were right there with the wise men, and to learn how to set your point of observation to a particular location, let's put ourselves in ancient Persia (Iran). Susa wasn't far from Chaldea, and it was yet an important city when Jesus was born. We don't know that this place was the home of the wise men, but it's close enough to serve our purpose. Wikipedia shows the coordinates of the historical sites it discusses. Go to "Susa" on Wikipedia, and find its coordinates in the upper right hand corner of the window: 32d 11m 26s N; 48d 15m 28s E. In your Skychart program, click the "Setup" button on the top row of buttons, then click on the second line of the dropdown box that reads "Observatory". Enter Susa's coordinates in the latitude and longitude boxes, then click "Apply". Check your work. Look on the world map in the lower part of the window for a tiny red circle indicating your point of observation to be immediately north of the Persian Gulf. That's Susa.
    Now that our "observatory" is close to the wise men's place, let's set the program's clock to their time zone. We don't really need to do this, since we're not concerned with the hours and minutes of the day. We are going to do it to pay respect to details. Paying respect to details makes one healthy, wealthy, and wise. But it can also make you late for what more

Clear Signs of Trouble and Great Joy

importantly needs done, so don't get washed away in a flood of details. Click the "setup" button again. Then click on the "Date / time" line. About half way down the "Date / time" dropdown box is the "more options" button. Click it. The "Time zone" box will now display. Click the "down arrow" to the right of that box. Use the slidebar to locate and select "Africa/Mogadishu" from the dropdown list. (Mogadishu is in the same time zone as Susa. Don't click "Apply" yet. Finally, let's set the date. The date setting is towards the top of the "Date / time" dropdown box, which should still be open. Enter "3" in the "Y" box, "8" in the "M" box, and "1" in the "D" box, then click the "BC" bullet to the right. The date of observation is now set to August 1, 3BC, Mogadishu time. Click "Apply" in the lower right hand corner of the "Date / time" dropdown box.

    You are now looking into the sky over Susa at about the same time the wise men would have been looking there. I think it is reasonable to conclude that the wise men knew Jupiter's motions around Regulus very well. For many centuries Jupiter and Regulus were regarded as the king planet and the king star by a people who watched the night sky more than we watch reality TV and fake-news. Most likely they knew what we are looking for to be only forty days from occurring. Let's "face north" by rotating the sky until the "N" indicating "North" is at the bottom of the star disc. Simply move the horizontal slider at the bottom of the window to the middle of the slidebar.

    We need to get Leo at the center of the disc, so we are going to step forward through time until Leo comes to that position. The rightmost box of the middle row of buttons on the control panel at the top of the window sets the time interval for walking the chart forward and backward through time. Use the down arrow to find and select "Hour" for your time unit, and make sure "1" is in the box just to the left of it. That will set your time interval to one hour. The second right-arrow to the left of the interval boxes steps the chart forward through time, one interval per click, and the left-arrow steps it backward an interval. Click either the forward or back arrow until Leo moves into the center of the chart. Now, let's enlarge your view. Click once on the leftmost magnifying glass in the middle row of control panel buttons. You are now ready to watch the universe's reality show that will actually be relevant to your life.

    Leo should still be in the center of your screen. Regulus is represented by the blue dot in the left half of Leo. Jupiter is the yellow dot to the left of Regulus. It is labeled. Now, set the time interval to "Day" and click the automatic advance arrow. Jupiter will begin advancing towards Regulus. But the whole display of stars will also be shifting to the left of the field of view with each daily advance. Jupiter will go through a conjunction with Regulus on September 13, 3BC. By November 2, Leo will be close to exiting your field of view to the left. Click the automatic advance button again to stop.

    I explored my software for a process that would hold Jupiter at the center of the screen and advance the stellar background so I wouldn't need to go through all the helter-skelter I'm about to describe. I'm sure it has such a process, but I couldn't find it. I had at hand the important detail of getting this book finished. So I moved on. That means, unfortunately, you will have to read on.

    You must now reposition Leo to the far right of the field of view so you can see these conjunctions. Select "Hour" for the time interval. Twelve clicks of the back-arrow

Having Fun with Astronomy Software

repositions Leo satisfactorily. Reset the time interval to "Days" and click the automatic advance button again. You will now see Jupiter enter retrograde motion in early December and begin approaching Leo again. Jupiter will pass by Regulus in mid-February, 2BC. In early April, Jupiter's retrograde motion will end, and in early May Jupiter will pass Regulus for the third time.

Reset your date to August 6, 3BC; center Leo in your display; set your time interval to one day. Now you can observe the conjunction of Jupiter and Venus, the queen planet, on August 12, 3BC, and then the conjunction of Venus and Regulus five days (clicks) after that. Venus' orbital path is much smaller than Jupiter's. Since Earth orbits between Venus and Jupiter, Venus will be found in conjunction with Jupiter at least twice a year. It probably wasn't Jupiter's conjunction with Venus that alerted the wise men. Most likely, they knew this conjunction happened regularly. It was probably that this conjunction occurred in Leo five days before Venus was in conjunction with Regulus and thirty-one days before Jupiter was in the first of its three conjunctions with Regulus.

Now reset your date to August 1, 15BC. Orient Leo to just beyond the right boundary of the sky dome (just below the west horizon), and advance through time in seven day increments. You will observe Jupiter's retrograde motion around Regulus again. However, if you set the date to August 1, 27BC, you will note that, although Jupiter was in conjunction with Regulus November 11, this was the day it was also beginning retrograde motion; it had not quite reached Regulus. Jupiter finally passed Regulus once in 26BC. Examine the months following August, 39BC. You will find only one conjunction between Jupiter and Regulus. Do the same for 51, 63, and 74BC. Only one each. But look at 86-85BC. Three conjunctions. Look at AD1884-5. Two years after the First Aliyah, Jupiter and Regulus were in conjunction three times. Look at 1967-8 and 2038-9. Again, three conjunc-tions.

That these triple conjunctions didn't happen in the exact years of Jesus' birth, the destruction of Jerusalem (cutting off the continual burnt offering), the First Aliyah's beginning of the Shemitah Path, Israel's capture of Old Jerusalem, and Jesus' return is not as much the point as that each of those particular, significant developments towards Christ's restoration of all things occurred close enough to a triple conjunction to be highlighted by it.

Viewing Gobekli Tepe in the Light of Its Rosetta Stone

The excursion through Jupiter's motions around Regulus should have given you plenty of practice to now appreciate seeing with your own eyes the *information* Gobekli Tepe's Pillar 43 has to present. You will need a few supplies from which to construct Gobekli Tepe's Rosetta Stone.

    1) Access to an ordinary MFC printer, and a good supply of paper.
    2) A clear, frontal photograph of Pillar 43.
    3) A black and a red Sharpie.
    4) A few clear, plastic report covers.
    5) A sufficiently good astronomy program, such as the Cartes du Ciel Skychart we used above.

Clear Signs of Trouble and Great Joy

Andrew Collins and Rodney Hale's enormously important paper: "Gobekli Tepe's Pillar 43: An Astronomical Interpretation" provides a very good overlay of the constellations onto Pillar 43. I developed my own overlay to confirm their work. If you also perceive a need to confirm the overlay, then Bing "Pillar 43" and select the frontal view picture with the best contrast. You may have to scale up or down the print of your chosen picture until it is a useful size. I scaled my picture to fill a standard letter size paper. Here's how. Divide the height of the picture you want to work with by the height of the picture you printed off the internet. My picture printed at 3 11/16 high (11/16 is .6875 by simple division). I want a ten inch high picture. 10/3.6875 is 2.72. That is the same as saying my desired picture will be 272% of the picture printed. So I set my MFC to enlarge the copy by 272%. I suggest working with as large a picture as possible to get a better view. Set your MFC to the percentage of enlargement you will need. Note the paper guides marked along the edges of your MFC glass. Orient your print-out in accord with those, and give it a try. You most likely will need to experiment and adjust, so be patient until you get a well centered, appropriately scaled image to use.

Now, set your astronomy program to 9600BC. Go to "Gobekli Tepe" on Wikipedia. Note the coordinates presented in the upper right corner of the "Gobekli Tepe" page. Enter those as the coordinates of your astronomy program, and select the same time zone you used above ("Africa/Mogadishu" on Skychart). Set your display to sky-dome. Advance your sky-dome through the day until you find all of the constellations from Cygnus to the Southern Cross, and from Pegasus to Ursa Major and Minor in the same display. You might need to search through the seasons to find your optimal view. Think about the process logically. If too many constellations are beneath the northern horizon, then the date of your sky-dome is too close to summertime. Try something closer to winter. Skychart gives the fullest view at early December, 9600BC, about 7:00AM. I moved my point of observation north eighteen degrees to get Ursa Major/Minor into view without eliminating Pegasus and the Southern Cross from view. I arrived in the far north, beyond the Black Sea, December 3, 9600BC, 7:16AM. Now, isn't it quite interesting that, in order to get the full view of the constellations speaking Christ's gospel onto Pillar 43 on the mount of Gobekli Tepe's "temple", we must travel to the far north (Psalm 48:2 and Isaiah 14:13)? Also noteworthy, an archaeological trail from Gobekli Tepe leads through this location in the far north, where the Aratta petroglyph was discovered. They could go to see the entire set of constellations covered by Pillar 43 from there.

Print your sky dome. It will help tremendously to print it with the constellations labeled and displayed with their lines. Mark all of your settings in the margin of your sky dome, so that if you want to get back to that particular display, you will not need to lose time doing it by trial and error (one of my favorite processes).

You must now scale your sky dome to your Pillar 43 picture. Measure the distance from the main vulture's eye to just underneath the headless vulture rider of the bottom register. This is 6 1/2 inches on my picture. Now, measure from Deneb, the bright star in Cygnus' tail (which will overlay the main vulture's eye), to the center of the Southern Cross. That's 4 9/16 inches on my sky chart (it most likely will be different on yours). Resizing the

Having Fun with Astronomy Software

chart to the picture by those points will place Deneb on the main vulture's eye with the headless chap sitting atop the Southern Cross. Since we are scaling the chart to the picture, divide the picture's distance by the chart's distance (6.5/4.5625 = 1.4247 for my project). I will need to increase the size of my sky chart 143 percent. I find it easier to print another copy of my sky chart and cut it out to more easily maneuver it on the MFC glass. Check your work. Do you have the full sky chart displayed?

Now, let's make an overlay. Place your resized sky chart in a clear report cover and shoot a couple staples through its outside edges to stabilize the system. Your overlay will be the most visible if you use a black Sharpie to mark the stars and a red Sharpie to trace the constellation lines onto the clear cover. I like bold, so I used a new Fine Sharpie. Use it neatly. Don't bother tracing the modern additions to the Zodiac, Telescopium, Pavo, Norma, etc. They don't belong to the Mazzaroth, the gospel as spoken by the stars. If you are using Cartes du Ciel Skychart, search carefully for Ara. It is there. The optimal setting I found for making my overlay had Ara overwritten by the label for Norma. Work your way around the chart so you don't need to rest your hand on any previous tracing and smear it. If you do need to rest your hand over a Sharpie mark, rest it on an index card, or something similar, to protect your work. Be sure to label your constellations. And be sure to include Draco. Draco marks a most interesting place on Pillar 43.

Pull the staples from your overlay. Remove your copy of the sky chart from the report folder and cut the crease of the folder so that it will fall in two pieces, one being your overlay. Align Deneb of Cygnus onto the main vulture's eye and the Southern Cross under the poor headless chap below. If you did your work carefully, you are now looking at Gobekli Tepe's Pillar 43 presenting the gospel of Jesus Christ.

When I have checked my work and found it acceptable, I like to protect my overlays with a layer of clear packing tape.

Constructing The Scorpio Timeline

Now, let's see Pillar 43 prophesy the return of Jesus Christ to banish evil from the world's throne.

Make another copy of the Pillar 43 picture that you used for the Cygnus overlay. A dot seems to be carved at the center of the disc on the main vulture's wing. Draw a line from that dot through the pit of the headless chap's arm extended towards the vulture. Your line should just touch the tip of the scorpion's stinger. It will soon become apparent why this line most likely represents the intended alignment of the polar meridian onto the pillar's symbols.

Let's now overlay Sagittarius onto the main vulture and Scorpio onto the vulture chick.

Set your skychart to Gobekli Tepe's latitude and longitude (37°13'23*N and 38°55'21*E according to Wikipedia) at December 21, 9600BC, 5:56AM (Africa/Mogadishu time zone). This is Winter Solstice morning at Gobekli Tepe. It happens to be the moment that Scorpio's tail star, Eta Scorpii, is crossing the polar meridian at the same time Zeta Draconis and Polaris in the northern area of the chart are both slightly West of it. The snake faking the waterfowl's beak is the pillar's representation of the front part of the constellation, Draco, from its nose to Zeta Draconis. The identification with Draco of the snake

Clear Signs of Trouble and Great Joy

symbol in front of the fallen "I" is certain. Not only is it the precise shape of Draco through its first curve, but all of the stars forming Draco are carved into the snake symbol at their proper positions. The fit is not only undeniable, but the inclusion of marks for Draco's stars makes it incontestably obvious that the carvers intended this symbol to represent that constellation. The polar meridian's position that we've noted is now rather confirmed by its cutting through Draco on the path of moons at the very point where Draco's curve represents the bird beak of the snake symbol. In Chapter 9 we proposed the beak-faking snake to be the key to viewing Christ's return on Pillar 43. It might be prophetic that Polaris hangs very near the polar meridian at this 9600BC Winter Solstice moment with the same slight inaccuracy that it revolves around the polar meridian today as our North Pole star. It seems deeply coherent with the immense system of signs we've studied from Chapter 3 through 9.

Locate the pit carved into the nose of Pillar 43's snake on the side of its head nearest the waterfowl's legs. If you want to see something cool and further confirmational, align Draco from your Cygnus/gospel-in-the-Mazzaroth overlay onto the beak-faking snake. Notice how well Draco's main stars align with the pits carved onto the snake all the way up its body. The snake on Pillar 43 is the front half of the constellation Draco. Its size is the same scale as the Cygnus/gospel-in-the-Mazzaroth overlay. But its carved symbol has been slipped from its proper position on the path-of-moons/solar-ecliptic to this position where its head is now at Scorpio's head in the Sagittarius/Scorpio/Ara overlay, kind of like the snake slithered from his rightful place to fake-assure Eve that he was truth and The Creator was deception. And Eve fell for it just like that fallen "I" under the bird beak faked by Draco's tail.

In the Cartes du Ciel software, the constellations expand as they near the horizon, and they shrink as they approach the zenith. Therefore, achieving the exact fit of the constellations will require much trial and error, and exactness may not be possible. Sagittarius was obviously meant to fall onto the main vulture, and Scorpio onto the vulture chick. It is rather apparent that the vulture chick's eye represents Antares, the red star of Scorpio's heart. So until we know better, we will use it for a reference point. Since the polar meridian is the focal point of both overlays, use a small square (like a mini carpenter's square) to draw a line perpendicular from the meridian to the center of the vulture chick's eye. Measure that line to the nearest 64th inch. Print your sky chart. Draw a line perpendicular from your chart's meridian through Antares in Scorpio. Measure that line also to the nearest 64th inch. Most likely, you will need a good magnifying glass and some patience to make those measurements, because you will need to carefully estimate the halfway point between your ruler's $1/32^{nd}$ inch marks. Divide the length of your pillar's line by the length of your skychart's line. My project yielded 2.957, meaning that I need to set my MFC to enlarge by 296%. After your skychart has been correctly enlarged, make a clear plastic overlay and try it out. Align Antares onto the vulture chick's eye and the meridian of your overlay onto the meridian you drew on your picture. The alignment should be quite close, although it probably won't be exact. If so, align the skychart's polar meridian exactly over the line you drew on your picture of the pillar. At this point, the constellations do not need to precisely overlay the symbols, (and they most likely won't because of the software effect described above.)

Having Fun with Astronomy Software

Note that Scorpio stretches from the waterfowl/snake through the vulture chick onto the head of the scorpion below. Sagittarius lays in the body of the main vulture. Below them, Ara pours its flames onto the headless chap. It is a very clear message about Christ vanquishing the sting of death and cleansing Earth of evil's servants. These constellations expand, shrink, distort somewhat, and slightly change positions on the symbols they overlay depending on whether we take our skychart from lower or higher in the sky. Their exact positions are not important; their obvious correlations with the symbols they overlay are. But the polar meridian is the focal point.

We will soon see why it must align with the center of the disc, the scorpion's stinger and the armpit. This section of the constellations overlays Sagittarius onto the main vulture as Christ launching His victorious arrow at Scorpio's heart, the sting of death. Scorpio interconnects the downward facing snake, the vulture chick, and the scorpion possibly to represent the process of deceit leading mankind's children into eternal death. Below, Ara, the flaming alter of wrath spat from the lower vulture's beak at the deniers/rejecters of Jesus' sacrifice, now overturns upon all who must ride the death vulture's back themselves for having spurned Christ's sacrificial ride made for anyone who chooses to accept its effect into their lives. With Draco keying these constellations onto Pillar 43, we find this pillar speaking of those who fall for the fake beak of the snake tail instead of standing strong before the honest, caring beak of the Waterfowl, born of Cassiopeia, as depicted in gospel overlay at the leftmost '48 Ford of Pillar 43's upper register.

Set your skychart's date to December 21, AD2600. If you want your chart to remain consistent with the 9600BC Gobekli Tepe time, then change your chart's longitudinal setting until the polar meridian precisely aligns onto Eta Scorpii. Otherwise, change your skychart's time of day until the alignment of the meridian and Eta Scorpii is achieved. The difference will be minor. Changing the time is easier and represents viewing this part of the sky from Gobekli Tepe at a different time of day. Changing the longitude represents moving to a location on Earth where the polar meridian aligns with Eta Scorpii at the same time of day we found it there in 9600BC. Prepare an overlay scaled to align the polar meridian with your line and Antares with the vulture chick's eye. Note that the meridian roughly bisects the sun on Winter Solstice AD2600 as it also bisects Eta Scorpii. Note that the sun is at the center of the disc on the main vulture's wing when Antares is approximately aligned with the vulture chick's eye on Winter Solstice, AD2600. Could AD2600 be intentionally indicated for some reason? It is near the end of those six, intertwining tetrads-of-testimony spanning the midpoint of Jesus' Millennial reign. Might it be indicative of that?

Constructing the Sagittarius Timeline

Now let's watch the sun move across the solar ecliptic marked onto the pillar by those eleven moons lined up through the Flood waters. Use the same copy of your Pillar 43 picture if you like. If you make another copy, draw the same polar meridian line onto it as instructed above. Notice the dot carved onto the pillar at the tip of the main vulture's wing. That dot represents Gamma Sagittarii, the tip of the archer's arrow, the spout of Sagittarius when it is represented as a teapot. The main vulture's eye is the star at the top of the teapot, Lambda Sagittarii. Draw a line through the wingtip dot to the main vulture's eye. The cros-

Clear Signs of Trouble and Great Joy

sing of that line with the polar meridian is a good point of reference for achieving a consistent alignment of Sagittarius with the polar meridian for each date we explore. When the polar meridian crosses this line at about 25% of the distance from the Gamma Sagittarii dot to the main vulture's eye representing Lambda Sagittarii, Sagittarius will accurately overlay the Vulture and the polar meridian will align with your line drawn on the pillar.

But achieving the proper alignment with the meridian by reference to the crossing of these lines each time we explore a different year is excruciatingly tedious and time consuming. It is much easier and quicker to note that when the polar meridian crosses Sagittarius at about 25% the distance towards Lambda Sagittarii, the meridian conveniently passes as slightly to the West of Eta Sagittarii, below the teapot spout, as it passes slightly to the East of HD165634, a yellow-giant star just above the spout. You will probably need to enlarge your software's sky chart to locate HD165634. You can either change the time of your skychart or use the longitude setting to step away from Gobekli Tepe's location until the polar meridian aligns between those two stars as described.

Set the date of your star chart to December 21, AD1400. Adjust either the time or the longitude until the meridian extends through the correct point of the spout of Sagittarius' teapot. Sagittarius will be so low in the sky that it is rather distorted, so set your latitude fifteen to twenty degrees south to place Sagittarius midway between the horizon and zenith. Enlarge your printed skychart enough to check the meridian's alignment for that 25% position. Scale the sky chart to your picture and make your overlay. Note that the sun stands at the beginning of the path of moons through the Flood waters. If you are using the free Cartes du Ciel software, note that between 1581 and 1582 the sun jumps through almost two-hundred years of precession. I presume that is a software glitch. Accordingly, I had to determine the sun's position at AD1400 mathematically. Mark and date the AD1400 location of the sun on your picture. Shortly after AD1400 the seventh-head of the dragon/beast rose up as the Ottoman-Turk Empire.

Let's try some other dates. Set your skychart to 1775 and prepare an overlay. Note that 1774-5 are the only years the sun is near perfectly bisected by the polar meridian at Winter Solstice. Recall that James Bruce had just returned from Ethiopia with three copies of the Book of Enoch which had been absent from Europe for almost fifteen-hundred years. The prophet Enoch is the only rational explanation for how Pillar 43 correlates with all of the skycharts we've been studying. His book explains that he was shown everything about the constellations and the plan of salvation including the return of the Messiah to end evil and reign in righteousness. Also note that this bisecting of the sun indicates the development of America's independence as the nation that would restrain the mystery of lawlessness until it is time for lawlessness to be unleashed. And note that six years after 1775, Frances Rolleston, the lady who discovered Pillar 43's Rosetta Stone, the gospel-in-the-Mazzaroth, was born. Her discovery and development of the Mazzaroth's meaning occurred a full century before Pillar 43 was uncovered. Only The Lord of Spirits can make this stuff up. Mark and date the sun's position at 1775.

Prepare an overlay for 1877. Mark and date the sun's position. Now align your Cygnus overlay onto this picture and notice that the 1877 sun is positioned at Draco's first curve, the fake beak of the snake tail. 1877 precedes the beginning of the Shemitah Path

Having Fun with Astronomy Software

leading into the Tribulation by five years, the number of grace needed for the world to walk that path with Israel. God's promise was given Abraham in 1877BC, and note that AD1877 is eight centuries (the number of new beginning) before AD2677, the year those six tetrads-of-testimony end during Christ's Millennial reign while the sun of the Scorpio Timeline stands within Pillar 43's disc at the Winter Solstice of that year.

Pillar 43 is beginning to look like a pictorial representation of the basic chronology of God's plan for the restoration of all things. The time represented by the Winter Solstice sun standing at Draco's tail on the path of moons is when evolution, "empirical science", and all their subsequent deceptions began rising in popularity to form the modern world's paradigm of rejecting the authority of God's Holy Bible. As the Shemitah Path was beginning, the nihilistic results of evolution were tying a Marxist blindfold over the world's eyes.

Do some other dates and mark the sun's positions. Do one at 1948. Notice that the downward facing snake and the disc in front of its tail-faking-a-bird beak are actually Draco with the disc representing the position on the timeline of the Winter Solstice Sun at Israel's reestablishment, 1948. Do another at 2036. Note that the sun stands in that year where Ursa Minor intersects the path of moons. Ursa Minor represents Israel in the Mazzaroth. The Holy Spirit told His prophet, Zechariah, that Israel would be raised above all nations when their Messiah returned for them. The enormous system of signs we have studied places Christ's return at 2036. And behold! Pillar 43 relates that year with the constellation representing Israel. Be sure to finally mark the position of the sun at 2138, one century after Jupiter's Christmas sign made two years after Apophis' returns to indicate Jesus' second coming. Note that this short path of solar precession through the Winter Solstice sky ends then, after seven centuries of prophesied time. Men couldn't have made that up without help from God's Holy Spirit.

Summary

These correlations could not be more coherent with the Word of God, Frances Rolleston's rendering of the constellations, and Pillar 43. This level of correlation has NEVER BEFORE been accepted as mere coincidence in any conceptual venue. This much correlation has always been considered to display the existence of a reality, even if the reality, itself, is not known. Just because The Holy Bible is central to all of this correlation is no reason to foolishly ascribe these immense correlations to mere coincidence. The Almighty Lord of Spirits created the universe purposefully designed to display the gospel of Jesus Christ. We would expect exactly that emphasis to be made by a Creator who is The Loving God of all things, communicative with His creation, and planning mayhem for what destroys it. Enoch was shown both the universe in its constellations and God's plans. He was the obvious designer of Pillar 43's message.

The only remaining mystery of Pillar 43 will be that of "highly educated" scholars denying the existence of what you've just explored. It is the world's greatest mystery of all, the mystery of how so much education can make such foolishness in otherwise intelligent men and women. Following *information* to whatever conclusion it makes is knowledge, a.k.a. science. Throwing any *information* in the trash is the mark of stupidity. Ask Rhinnie. Adjusting and even abandoning theories to accord with *information* is humility. Abandoning

Clear Signs of Trouble and Great Joy

information to ride theories into the dirt is arrogance. It is superstition. Modern science is so obstinately arrogant that it not only refuses to follow the Bible into the truth, it also refuses to follow its own Thomas Kuhn into humility. Sincere humility discovers the truth about reality. And the irrefutable truth is that Pillar 43 required far more skills than even a highly adept Neolithic people could craft. It required a real knowledge of the distant future, a knowledge of the wrath of the Lamb to be poured onto the scoffers of the world, and a knowledge of the fundamental concepts of the Gospel of Jesus Christ. It's knowledge (science) could only have come through God's Holy Spirit. It is a fingerprint on the universe's "mechanistic" material.

     Sorry, Charlie. We've now observed the universe's design for the purpose of displaying God's glory.

Appendix 3
Information You Should Peruse

You shall love the Lord your God with all your heart, and with all your soul, and with all your mind.
Matthew 22:37

(All referenced videos are found on YouTube -if they've not been censored by now.)

The Mazzaroth

Rolleston, Frances. Mazzaroth: Or, the Constellations (By F. Rolleston). [Followed By] Mizraim; Or, Astronomy of Egypt - Primary Source Edition. Nabu Public Domain Reprints.

Seiss, Joseph A. The Gospel in the Stars. (Originally published by Kregel Publishing in 1882.) 2005. Cosimo Classics. Cosimo, P.O. Box 416, Old Chelsea Station, New York, NY, 10113.

Sherstad, Dana and Barbich, B.J. Testimony of the Heavens: God's Redemptive Plan Preserved in the Stars, A Bible Study looking at God's Revelation to mankind as verified by the Written Word. 2011. Dana Sherstad and B.J. Barbich.

Banks, William D. The Heavens Declare: Jesus Christ Prophesied in the Stars. 1985, 2013. Impact Christian Books, Inc. 332 Leffinwell Ave., Suite 101, Kirkwood, MO 63122.

Bullinger, E.W. The Witness of the Stars. Reprint of the 1893 edition. Kregel Publications, a division of Kregel, Inc., P.O. Box 2607, Grand Rapids, MI 49501.

Enoch, Noah, and The Flood

Enoch. 00:55:02. 2013. Dr. Ken Johnson. PITN Prophecy Summit, Colorado Springs, Colorado. 2013.

Clear Signs of Trouble and Great Joy

After the Flood. 00:51:52. 2013. Dr. Ken Johnson. PITN Prophecy Summit, Colorado Springs, Colorado. 2013.

Noah's Testament: A Study in the Dead Sea Scrolls. 00:34:34. 2018. Dr. Ken Johnson. Lecture at Calvary Chapel, Lenexa, Kansas.

A Deluge of Evidence: Noah's Flood and the Historical Roots of Secularism. 0:50:11. 2015. Northwest Creation Network.
    0:25:30 polystrate fossil
    0:29:00 overfit valleys, valleys too big for their rivers to have carved
    0:35:00 some seemingly old things are not so old at all

The Creation Event Speaker John Mackay - Noah's Flood - The Evidence in Australia and Worldwide. 1:40:56. 2014. Creation Research.
    0:17:30 Ovid's account of the flood, Josephus' account of the flood, and the Armenian patron saint: Noah.
    0:34:00 correlations spread around the world
    1:16:00 from deformed tree fossils to colossal deposit of demolished trees
    1:36:00 how evolutionists respond to overwhelming evidence falsifying their theory.

What was the Post-Flood Period? - Dr. Kurt Wise (Conf Lecture). 1:14:09. Answers in Genesis.

Misconceptions about Noah's Ark and the Flood with Tim Chaffey. 0:09:58. Answers in Genesis.
    0:51:30 mutations cause disease and disorder, not benefit
    0:55:00 Darwin didn't know about molecular machines too complicated to have evolved.
    1:00:00 the frauds of Ernst Haeckel

A Closer Look at Noah's Ark. 1:08:39. Tom Hoyle. Apologetics Symposium. Northwest Creation Network.

'Handbags of the Gods' Ancient Civilizations and The Great Flood. The Lost History Channel. 04:31. Contemplates an origin of all civilizations from a common root as evidenced by widely disbursed, basically similar mythology regarding the handbag.

Allan, D.S. and Delair, J.B. Cataclysm! Compelling Evidence of a Cosmic Catastrophe in 9500BC. 1997. Bear & Company, One Park Street, Rochester, Vermont 05767.

Cooper, Bill, B.A. Hons. After the Flood: The Early Post-Flood History of Europe Traced Back to Noah. 1995. New Wine Press, PO Box 17, Chichester, West Sussex PO20 6YB, England.

Information You Should Peruse

Prophecy

The Missing Years: Prophecy and Chronology. 00:24:11. 2018. Dr. Ken Johnson. Calvary Chapel Johnson County. Lenexa, Kansas. Discussion of the Rabbinical attempt to obfuscate the Messianic prophecy of Daniel 9:24-27.

The Accuracy of Bible Prophecy. 01:08:08. 2014. Babylon or truth. Discussion of the astronomical odds against historical fulfillments of Biblical prophecies, and the historical events that were indeed the fulfillments of those prophecies.

Timed Prophecies 1948 and 1967. 1:02:17. 2013. Prophecy in the News.
    0:02:00 The problem Daniel 9:24-27 caused the rabbinical Jews and the curse they invoked upon anyone using Daniel 9 for researching the Jewish Messiah.
    0:11:30 The Daniel 9:24-25 prophecy calculated.
    0:20:15 The 1948 prophecies
    0:32:00 1967 prophecy.

The End Times by the Church Fathers. 0:53:37. 2016. Rocky Mountain International Prophecies Conference.

Biltz, Mark. Blood Moons: Decoding the Imminent Heavenly Signs. 2014. WND Books, Washington D.C.

Cahn, Jonathan. The Harbinger: The Ancient Mystery that Holds the Secret of America's Future. 2011. Frontline, Charisma Media/Charisma House Book Group, 600 Rinehart Road, Lake Mary, Florida 32746

Ancient Hebrew Literature

Biblically Endorsed Extra-Biblical Books. 00:14:03. Dr. Ken Johnson. Brief discussion of the other books The Holy Bible references as additional reading.

Testing The Book of Jasher. 00:10:15. 2018. 119 Ministries. A critical analysis against the authenticity of the recently acclaimed Book of Jasher.

The Book of Jasher. 00:11:31. 2016. Messiah Matters by ToraResource. Another critical

Clear Signs of Trouble and Great Joy

analysis against the authenticity of the recently acclaimed Book of Jasher.

Cepher Moments - Where Are the Lost Books of the Bible? 00:15:10. 2017. Cepher Publishing Group. Brief discussion of extra-Biblical literature as regarded in the final centuries BC and first centuries AD and the political affects upon today's purview of that literature.

Evidences of Biblical Truth

After the Flood. 0:51:53. 2013. Prophecy in the News.
    0:10:25 The canopy is mentioned in the ancient Hebrew texts.
    0:28:30 Black Granite Naos, an inscribed stone text mentioning the drowning of Pharaoh's army in a whirlpool.
    0:29:30 The Leiden Papyri describes a destruction of Egypt.

Egyptian Chronology and the Bible. 0:59:11. 2015. Northwest Creation Network.
    0:12:00 the Enlightenment shift
    0:17:50 six layers of the secular smoke screen
    0:22:30 Carbon-14 is a showcase, not a dating method (David Downs)
    0:25:00 Champollion's mistake remains the cornerstone of Egyptian chronology
    0:51:00 the discrepancies resulting in erroneous chronology

Nuclear Physics and the Young Age of the Earth - Dr. Jim Mason. 1:20:00. 2012. Creation Ministries International and Bible Discoveries TV.
    0:17:00 false dates of known rocks and assumptions involved in the dating process.

The Shroud of Turin as the Burial Cloth of Jesus? Dr. John Johnson. 0:58:02. 2016. Apologetics Forum of Snohomish County. Aspects of the Shroud that could not have been forged. The Shroud is filled with minute evidences, e.g., pollen, the cloth weave, scourge marks, missing thumbs, etc. that the skeptics and critics do not mention because they can not explain them.

A Remarkable Relic of the Resurrection: The Shroud of Turin and New Scientific Evidence. 0:47:18. November 12, 2017. Fr. Robert Spitzer. EWTN.

New Forensic Evidence Validates The Shroud of Turin and the Resurrection of the Person in It. 1:13:23.
    0:09:00 the extraordinary nature of the Shroud.
    0:29:30 the story of the bad carbon date

Information You Should Peruse

The Real Jesus: Paul Maier Presents New Evidence From History and Archeology at Iowa State. 1:28:19. May 11, 2013. The Veritas Forum
    1:27:20 Phlegon of Tralles writes of the darkness and earthquake recorded by Matthew at Jesus' crucifixion.

Sharp, Doug and Bergman, Dr. Jerry. Persuaded by the Evidence: True Stories of Faith, Science, and the Power of a Creator. 2008. Master Books, a division of New Leaf Publishing Company. www.masterbooks.com.

Bennett, Janice. Sacred Blood, Sacred Image: The Sudarium of Oviedo: New Evidence for the Authenticity of the Shroud of Turin. 2001. Libri de Hispania, Publications about Spain, PO Box 270262, Littleton, Colorado 80127-0005.

Guscin, Mark. The Image of Edessa. 2009. Koninklijke Brill NV, Leiden, The Netherlands.

Guscin, Mark. The Oviedo Cloth. 1998. The Lutterworth Press, PO Box 60, Cambridge CBI 2NT.

Wilson, Ian. The Shroud: The 2000-Year-Old Mystery Solved. 2010. Transworld Publishers, 61-63 Usbridge Road, London WS SSA.

Rohl, David. Exodus: Myth or History? 2015. Thinking Man Media, 6900 West Lake Street, St. Louis Park, MN 55426.

Dumpster Dining

The Age of Things: Does It Matter? - Dr. Kurt Wise (Conf Lecture) 1:02:11. November 20, 2017. Dr. Kurt Wise. Is Genesis History?

Out of Place Ancient Discoveries That Show History is Wrong. 1:11:21. 2009. Biblical Prophecy TV.

Forbidden History: Dinosaurs and the Bible. 1:04:21. 2009. A Restoring Genesis Film.

Michael Cremo - Extreme Human Antiquity Full DVD (Druta Karma.) 3:46:00. 2012. Michael Cremo is the authority on the archeology of archeology. His book, Forbidden Archeology: The Hidden History of the Human Race, is an eight-hundred page documentation of scientific misrepresentations and sometimes outright fraudulent portrayals of human artifacts discovered in archaeological environments where no humans could have been according to

Clear Signs of Trouble and Great Joy

the currently accepted paradigm of human origins. This is a long, rather monotonous video, so here's a few highlights to whet your appetite:

    0:46:00 Discovery of anatomically modern human jaw in a four-hundred thousand year old layer near Abbeville, France instigates a scrutinized excavation that uncovered dozens more modern human bones along with more tools.

    1:10:00 A shell bearing a carving of a simplistic human face was discovered near Walton-on-Naze, England in a two-million year old layer.

    1:33:00 The incredible case of disappearing information.

    1:35:00 At Miramar, Argentina an arrowhead found embedded in a three million year old toxodon femur instigated an excavation scrutinized by leading South American scientists and skeptics.

    2:09:00 Portugal's head government geologist discovered hundreds of human artifacts in twenty-million year old rock layers. The once prominently displayed artifacts are now locked away, out of sight, in The Museum of Geology in Lisbon. Very "unbiased science" indeed!

    2:13:00 The incredible case of the missing nineteen-million-nine-hundred-eighty-thousand years.

    2:23:00 Hundreds of human artifacts discovered in a thirty-million year old layer in Belgium were displayed in the Royal Museum of Natural Science in Brussels until their discoverer died. Subsequently, much of his documentation was "scientifically" destroyed, and the artifacts were "scientifically" hidden From public view. And we all thought "science" of origins was objective!

    2:40:00 The incredibly "scientific" process of knowledge filtering. The science filter in operation at the Museum of Anthropology at the University of California at Berkeley. The outrage of "objective scientists" over NBC's broadcast of Cremo's study into hidden archeology. In the end, the "objective scientists" invoked the FCC to fine NBC millions of dollars for the revealing broadcast. Fortunately, their bias filter was turned down by the FCC.

    2:50:00 "Objective science's" emotional attachment to Darwin's theory.

    3:05:00 If geologists, paleontologists, and archaeologists have indeed found human artifacts in layers of all geologic ages, then there should be non-scientific reports (newspapers, etc.) of artifacts found by common people by happenstance.

Forbidden Archeology Documentary 2018 - Mankind is Millions of Years Old. 00:39:37. 2018. DTTV Documentaries. Brief discussions of artifacts discovered within geological situations that would be impossible according to modern theories of evolution.

Malachite Man. 1:01:57. Dr. Don Patton. Several fossilized human skeletons found in a Cretaceous layer of a Utah copper mine in the 1930s.

Scientists Baffled - New Discoveries - Darwinian Evolution. 0:32:21. 2016. CBC Media

Information You Should Peruse

Group. Soft tissues and Carbon 14 in dinosaur fossils destroy millions of years of world "history".

What is Catastrophic Plate Techtonics? - Dr. Kurt Wise (Conf Lecture) 00:48:40. Answers in Genesis.

Comparing the Two Paradigms: Old Earth & Young Earth - Dr. Danny Faulkner (Conf Lecture) 0:57:24. 2017. Dr. Danny Faulkner. Is Genesis History.

Creation Astronomy: The Heavens Declare the Glory of God. 1:14:32. Spike Psarris. Apologetics Forum of Snohomish County. Northwest Creation Network.

Physics Disproves Atheistic Cosmologies. 1:25:48. Spike Psarris. Northwest Creation Network.

Cremo, Michael A. and Thompson, Richard L. Forbidden Archeology: The Hidden History of the Human Race. 2003. Torchlight Publishing, Inc., PO Box 52, Badger, CA 93603.

Evolution Pareidolia

Focus on Origins: Icons of Evolution. 1:50:39. 2012. Access Research Network.
    0:31:15 Darwin's strongest evidence was Haeckel's embryo fraud.
    0:56:35 Naturalism, the belief that every event must have a natural, material cause is philosophy pretending to be science, yet it is presented as science. It is indeed only a guess.

The Shortcomings of Radiometric Dating. 00:11:06. The Compass Magazine. Answers in Genesis.

Once Upon a Time: Understanding the Mythology Behind Radiometric Dating. 00:55:01. 2015. Dr. Tas Walker. Northwest Creation Network.

Science Confirms a Young Earth - The Radioactive Methods are Flawed. 1:00:58. Dr. Andrew Snelling. Answers in Genesis.

Fossils: History in Contention. 01:21:33. Chris Ashcroft, MS, Med, MTMS. Northwest Creation Network. Cedar Park Christian Schools.

Bergman, Jerry. Fossil Forensics: Separating Fact from Fantasy in Paleontology.

Clear Signs of Trouble and Great Joy

2017. BP Books, Bartlett Publishing. Bartlett publishing.com.

Ackerman, Paul D. It's a Young World After All: Exciting Evidences for Recent Creation. 1986. Baker Book House, Grand Rapids, MI 49506.

Woodmorappe, John, M.A., B.A. The Mythology of Modern Dating Methods. 1999. The Institute for Creation Research, PO Box 2667, El Cajon, CA 92021.

Miscellaneous

Thomas Sowell: Common Sense in a Senseless World. 0:56:57. Written and produced by Tom Jennings. Free to Choose Network.

Truth, Lies, and Science Education. 1:24:23 Paul Taylor MSeD. Northwest Creation Network.

Secrets of the Stoneage - Catalhoyuk 00:16:33. 2014. Cappa Docia. Some excellent portrayals of artwork.

Heiser, Michael S. **The Unseen Realm: Recovering the Supernatural Worldview of the Bible.** 2015. Lexham Press, 1313 Commercial St., Bellingham, WA 98225 LexhamPress.com.

Goldberg, Jonah. **Liberal Fascism: The Secret History of the American Left from Mussolini to the Politics of Meaning.** 2007. Doubleday. www.doubleday.com.

Savage, Michael. **Liberalism Is a Mental Disorder: Savage Solutions.** 2005. Nelson Current, a division of a wholly-owned subsidiary (Nelson Communications, Inc) of Thomas Nelson, Inc.

Beck, Glenn. **It Is About Islam: Exposing the Truth about ISIS, al-Qaeda, Iran, and the Caliphate.** 2015. Threshold Editions/Mercury Radio Arts, An imprint of Simon & Schuster, Inc., 1230 Avenue of the Americas, New York, NY 10020.

Cambray, Joseph. **Synchronicity: Nature & Psyche in an Interconnected Universe.** 2009. Texas A&M University Press, College Station.

Bibliography

Allan, D.S. and Delair, J.B. Cataclysm! Compelling Evidence of a Cosmic Catastrophe in 9500BC. 1997. Bear & Company. One Park Street, Rochester, Vermont 05767.

Allen, Richard Hinkley. Star Names: Their Lore and Meaning (formerly titled: Star-Names and Their Meanings). 1963. Dover Publications, Inc., 180 Varick Street, New York 14, New York.

Aratus. Phaenomena. http://www.theoi.com/Text/AratusPhaenomena.html. Theoi Project. 2000-2017. Aaron J. Atsma. New Zealand.

Armstrong, Karen. A History of God: The 4,000-Year Quest of Judaism, Christianity and Islam. 1993. A Ballantine Book. The Random House Publishing Group, Random House, Inc., New York.

Banks, William D. The Heavens Declare: Jesus Christ Prophesied in the Stars. 1985. Impact Christian Books, Inc., 332 Leffingwell Ave., Suite 101. Kirkwood, MO 63122.

Beck, Glen. It IS About Islam: Exposing the Truth About ISIS, Al Qaeda, Iran, and the Caliphate. 2015. Threshold Editions/Mercury Radio Arts, Simon & Schuster, Inc. 1230 Avenue of the Americas, New York, New York 10020. by Mercury Radio Arts, Inc.

Bennett, Janice. Sacred Blood, Sacred Image: The Sudarium of Oviedo: New Evidence for the Authenticity of The Shroud of Turin. 2001. Libir de Hispania, Publications About Spain, PO Box 270262, Littleton, Colorado 80127.

Bergman, Jerry, PhD. Fossil Forensics: Separating Fact from Fantasy in Paleontology. 2017. BP Books, Bartlett Publishing.

Bertrand, Jim. EC2016 -Jim Bertrand- The Very Latest Research of the Shroud of Turin. (video)

Biblical Archaeology Society Staff. The Tel Dan Inscription: The First Historical Evidence of King David from the Bible: Tel Dan inscription references the "House of David". https://www.biblicalarchaeology.org/daily/biblical-artifacts/artifacts-and-the-bible/the-tel-dan-inscription-the-first-historical-evidence-of-the-king-david-bible-story/, 11/08/2016.

Biltz, Mark. When Will the Tribulation Begin? Where are we on the Biblical Calendar?

Clear Signs of Trouble and Great Joy

(What Time is It?). Video, 1:05:11. 2021. El Shaddai Ministries.

Black, Jeremy and Green, Anthony. Gods, Demons and Symbols of Ancient Mesopotamia: An Illustrated Dictionary. 2003. University of Texas Press. Austin, Texas. 1992. Fifth printing,

Bostrom, Philippe. How Mice May Have Saved Jerusalem 2,700 Years Ago from the Terrifying Assyrians. April 18, 2018. https://www.haartz.com/archeology/.premium.MAGAZINE-how-mice-may-have-saved-Jerusalem-2700-years-ago-from-the-Assyrians-1.6011735

Brown, Simon. "Founding Fibs: The Religious Right Is Still Trying To Hijack George Washington's Legacy". May 9, 2014. https://www.au.org/blogs/wall-of-separation/founding-fibs-the-religious-right-is-still-trying-to-hijack-George-Washington's-Legacy

Bulst, Werner S.J. The Shroud of Turin. 1957. The Bruce Publishing Company. Milwaukee, Wisconsin.

Bullinger, E.W. Number in Scripture: Its Supernatural Design and Spiritual Significance. Alacrity Press, 2014. (Originally published by Eyre & Spottiswoode Ltd., 1921.)

Cahn, Jonathan. The Harbinger: The Ancient Mystery that Holds the Secret of America's Future. 2011. Frontline, Charisma Media/Charisma House Book Group, 600 Rinehart Rd, Lake Mary, Florida 32746.

Choi, Charles Q. 7 Theories on the Origin of Life. March 24, 2016. http://www.livescience.com/13363-7-theories-origin-life.html

Collins, Andrew. Gobekli Tepe: Genesis of the Gods: The Temple of the Watchers and the Discovery of Eden. 2014. Bear & Company, One Park Street, Rochester, Vermont 05767.

Crispe, Sara Esther. "The Secret of Elul". Chabad.org Magazine, a division of Chabad-Lubavitch Media Center. 1993-2017. http://www.chabad.org/theJewishWoman/article_cdo/aid/424441/jewish/The-Jewish-Heart.htm

Dever, William G. "Hershel's No. 2 Crusade For King and Country: Chronology and Minimalism". Biblical Archeology Review. March/April/May/June 2018, Vol. 44 Nos. 2&3. Biblical Archeological Society. 2018.

DeWitt, Richard. Worldviews: An Introduction to the History and Philosophy of Science. Second Edition. 2010. Wiley-Blackwell, John Wiley & Sons, Ltd, The Atrium, Southern

Bibliography

Gate, Chichester, West Sussex, PO19 8SQ, United Kingdom.

Dr. Phil. The Strange Beasts on Bishop Bell's Tomb. July 8, 2007. Stories from the Diogenes Club. http://storiesfromthediogenesclub.blogspot.com/2007/07/strange-beasts-on-bishop-bells-tomb.html

Encyclopedia Britannica: A New Survey of Universal Knowledge. 1956. Encyclopedia Britannica, Inc.

Eusebius Pamphilus. The Ecclesiastical History of Eusebius Pamphilus. Translated by Christian Frederick Cruse and Isaac Boyle. 1973. Baker Book House, Grand Rapids, Michigan. Sixth printing.

Fabricius, Karl. What the Lascaux Cave Paintings Tell Us About How Our Ancestors Understood the Stars. October 10, 2009. http://scribol.com/anthropology-and-history/archeology

Faulkner, Dr. Danny R. A Further Examination of the Gospel in the Stars. https://answersingenesis.org/astronomy/stars/a-further/examination-of-the-gospel-in-the-stars/ February 6, 2013.

Federer, William J. Samuel Adams knew what would overthrow America's liberties. 2016. American-minute.com.

Ford, Charles V., M.D. Lies! Lies! Lies! The Psychology of Deceit. 1996. American Psychiatric Press, Inc., 1400 K Street, N.W., Washington, DC. 2005.

Friedlander, Blaine P. Jr. "A Bright New Star Will Burst into the Sky in Five Years, Astronomers Predict". The Washington Post. https://washingtonpost.com/news/speaking-of-science/wp/2017/01/06/a-bright-new-star-will-burst-into-the-sky-in-five-years-astronomers-predict/

Gillispie, Charles Coulton. Genesis *and* Geology: A Study in the Relations of Scientific Thought, Natural Theology, and Social Opinion in Great Britain 1790-1850. 1969. Harvard University Press. Cam-bridge.

Holmes, David L. The Faiths of the Founding Fathers. 2006. Oxford University Press.

Gitt, Dr. Werner. In the Beginning was Information, A Scientist Explains the Incredible Design in Nature. 2005. Master Books, Inc. PO Box 726, Green Forest, AR 72638.

Glyn-Jones, William. Yima and his Bull: Gemini and Taurus in the Lascaux Caves. January 28, 2008. https://grahamhancock.com/glynjonesw1/ 2003-2017.

Clear Signs of Trouble and Great Joy

Goldberg, Jonah. Liberal Fascism: The Secret History of the American Left from Mussolini to the Politics of Meaning. 2007. Doubleday. The Doubleday Broadway Publishing Group, a division of Random House, Inc., New York. www.doubleday.com.

Halpern, Ben. The Idea of the Jewish State. Second Edition. 1969. Harvard University Press, Cambridge M, Massachusetts.

Hanson, Victor Davis. Why America Was Indispensable to the Allies' Winning World War II. National Review. May 14, 2015. http://www.nationalreview.com/article/ 418329/why-america-was-indispensable-allies-winning-world-war-ii-victor-davis-hanson.

Harrison, Allen F. and Bramson, Robert M. Ph.D. Styles of Thinking: Strategies for Asking Questions, Making Decisions, and Solving Problems. 1982. Anchor Press/Doubleday, Garden City, New York.

Hodder, Ian. The Leopard's Tale: Revealing the Mysteries of Catalhoyuk. 2006. Thames & Hudson Ltd, London.

Hooker, Heather Lee & Tim. The Ancient Aratta Civilization of Ukraine, Older than Sumeria. 2009. T&H Hooker & Megalithomania. (Video).

http://blog.chrisify.com/2017/08/the-seven-salems-of-eclipse-coincidence.html

http://www.cnn.com/2014/06/05/opinion/opinion-d-day-myth-reality/index.html.

http://en.wiktionary.org/wiki/rapio#Latin

http://www.haaretz.com/israel-news/business/u-s-aid-to-israel-totals-233-7b-over-six-decades

http://www.history.com/topics/ancient-history/byzantine-empire. Copyright 2016. A+E networks, Corp.

http://www.pbs.org/wgbh/pages/frontline/shows/whales/man/myth.html. 1995-2014 by WGBH educational foundation.

http://realhistoryww.com/world_history/ancient/Misc./The_Bible2.htm

http://www.valuesvotersnews.com/2008/12/george-washingtons-view-of-gods-role-in.html

https://aratta.wordpress.com/the-history-of-the-swastika. Cradle of Civilisation: A Blog About the Birth of our Civilisation and development.

Bibliography

https://godsamazingstarsecret.blogspot.com/2019/featured-teachers-of-mazzaroth-part-.htm

https://www.archives.gov/files/education/lessons/us-israel/images/recognition-telegram-1.jpg

https://en.wikipedia.org. Wikimedia Foundation, Inc.

https://www.newsinfo.inquirer.net/721613/mecca-crane-collapse-act-of-god-engineer. Agence France Presse. 12:35PM September 13, 2015.

https://www.timetoast.com/timelines/the-fall-of-the-roman-empire-3. Daniel Todd, founder and CEO. Copyright 2007-2016.

Ibrahim, Raymond. Exposed: Obama Helped Decade-Old Plan to Create IS: The Resurrection of the Caliphate was Planned and Exposed to the World Nearly Ten Years Ago. November 6, 2014. http://www.frontpagemag.com/fpm/244702/exposed-obama-helped-decade-old-plan-create-raymond-ibrahim David Horowitz Freedom Center.

Johnson, Ken Th.D. Ancient Book of Jasher, Biblefacts Annotated Edition. 2008. Biblefacts Ministries, biblefacts.org.

Johnson, Ken Th.D. The Ancient Book of Enoch. 2012. Biblefacts Ministries, biblefacts.org.

Josephus, Flavius. (Complete Works.) Translated by William Whiston, AM. 1960. Kregel Publications, Grand Rapids, Michigan 49501.

Kaiser, David. What Hitler and the Grand Mufti Really Said. http://time.com/4084301/hitler-grand-mufi-1941/. October 22, 2015.

Kent, Simon. Ottoman Empire Redux: Turkey's Erdogan Tells the World 'Jerusalem is Outs'. October 4, 2020. https://brietbart.com/middle-east/2020/10/04/turkeys-erdogan-tells-the-world-jerusalem-is-our-city.

Kuhn, Thomas S. The Structure of Scientific Revolutions. 50th Anniversary Edition being the 4th edition. 2012. The University of Chicago Press, Chicago, IL 60637.

Larson, Frederick A. The Star of Bethlehem. 2007. Presented by MPower Pictures. (video)

Leeming, David. The Oxford Companion to World Mythology. 2005. Oxford University Press, Inc. 198 Madison Ave., New York, NY 10016.

Clear Signs of Trouble and Great Joy

Lima, Pedro. Lascaux Planetarium Prehistorique? The Incredible discovery of a paleo-astronomer. Translation displayed at http://cassiopaea.org/forum/index.php?topic=238 33.0

Llafayette, Maxilillien de. Encyclopedia of Gods and Goddesses of Mesopotamia, Phoenicia, Ugarit, Canaan, Carthage, and the Ancient Middle East. 2015. Times Square Press. New York. Berlin.

Lumpkin, Joseph B. Introduction to The First Book of Enoch. The Books of Enoch: The Angels, The Watchers and The Nephilim. 2011. Fifthe State Publishers, Post Office Box 116, Blountsville, AL 35031.

McCoy, Dan. http://egyptianmythology.org/gods-and-goddesses/apophis/ 2014-2016

McCoy, Dan. http://egyptianmythology.org/gods-and-goddesses/ra/ 2014-2016

Montgomery, David R. The Rocks Don't Lie: A Geologist Investigates Noah's Flood. 2012. W.W. Norton & Company, Inc. 500 Fifth Avenue, New York, NY 10110.

Naeye, Robert. "A Stellar Explosion You Could See on Earth!" https://www.nasa.gov/mission_pages/ swift/bursts/brightest_grb.html NASA's Goddard Space Flight Center. May 26, 2010.

Okasha, Samir. Philosophy of Science: A Very Short Introduction. Second edition, 2016. Oxford University Press, 198 Madison Ave, New York, NY 10016.

Parks, Jake. "Two Stars Won't Collide Into a Red Nova in 2022 After All". D-Brief; discovermagazine.com. 09-07-2018. http://blogs.discovermagazine.com/d-brief/2018/09/07/two-stars-won't-merge-explode-red-nova-2022/#.XJi7knbow.

Patton, Don PhD. Man and Dinosaur Co-Existed: Evidence from South America. (video)

Patton, Don PhD. The Record of the Rocks. (video)

Posner, Menachem. "How Does the Spring Equinox Relate to the Timing of Passover? About the Jewish Leap Year". http://www.chabad.org/holidays/passover/pesach_cdo/aid/495531/jewish/How-Does-the-Spriing-Equinox-Relate-to-the-Timing-of-Passover?-About-the-Jewish-Leap-Year.html

Rogers, John H. Origins of the ancient constellations: I. The Mesopotamian traditions. 1998. Journal of the British Astronomical Association. Volume 108, 1. (This article can be accessed free at http://adsabs.harvard.edu/abs/1998JBAA..108....9R)

Bibliography

Rohl, David. From Eden to Exile: The Five Thousand Year History of the People of the Bible. Greenleaf Press, Lebanon, Tennessee.

Rohl, David. Exodus: Myth or History? Thinking Man Media, 6900 West Lake Street, St. Luis Park, MN 55426. 2015.

Rosenberg, Alex. Philosophy of Science: A Contemporary Introduction. 3rd edition. 2012. Routledge, 711 Third Avenue, New York, NY 10017.

Savage, Michael. Liberalism Is a Mental Disorder: Savage Solutions. 2005. Nelson Current, a division of a wholly-owned subsidiary (Nelson Communications, Inc) of Thomas Nelson, Inc.

Schaff, Philip. History of the Christian Church. Vol. III Nicene and Post Nicene Christianity; From Constantine the Great to Gregory the Great A.D. 311-600. 1910. 5th edition revised. Wm. B. Eerdmans Publishing Company. Grand Rapids, Michigan.

Seiss, Joseph A. The Gospel in the Stars. (Originally published by Kregel Publishing in 1882.) 2005. Cosimo Classics. Cosimo, P.O. Box 416, Old Chelsea Station, New York, NY, 10113.

Sharp, Doug & Bergman, Dr. Jerry. Persuaded by the Evidence: True Stories of Faith, Science, & the Power of a Creator. 2008. Master Books, PO Box 726, Green Forest, AR 72638.

Stanglin, Doug. New test dates Shroud of Turin to era of Christ. USA Today. Published 12:53 PM ET March 30, 2013/Updated 4:23 PM ET March 30, 2013. www.usatoday.com/story/news/world/2013/03/ 30/shroud-turin-display/2038295/

Strong, James. Strong's Hebrew and Greek Dictionaries. Electronic Edition STEP Files copyright 2003, Quickverse, a division of Findex.com, Inc.

The Ancient Book of Jubilees. Translation by Charles, RH. 1917. 2013. Introduction and appendices by Dr. Ken Johnson, TH.D. Biblefacts Ministries. Biblefacts.org.

The Ica Stones of Peru. Tracking Ancient Man. 2016. http://www.ancient-hebrew.org/1001.html

"Timeline for Abraham (Abram) from the promise given at age 70 (Gen 12:1-4) until Jacob Arrives in Egypt - 220 Years". Harp's Crossing Baptist Church. https://www.harpscrossing.com/wp-content/ uploads/2014/04/Timeline-from-abraham-to-exodus.pdf

Ullman, B. L. and Henry, Norman E. Latin For Americans. Second Book. 1962. The

Clear Signs of Trouble and Great Joy

Macmillan Company.

White, Chris. Four Blood Moons Debunked. 00:6:54. January 1, 2014. YouTube video.

Wilson, Ian. Before the Flood: The Biblical Flood as a Real Event and How it Changed the Course of Civilization. First St. Martin's Griffin Editon: March 2004. St. Martin's Press, 175 Fifth Ave, New York, N.Y. 10010.

Wilson, Ian. The Shroud: The 2000-Year-Old Mystery Solved. 2010. Transworld Publishers, 61-63 Uxbridge Road, London W5 5SA. A Random House Group Company. www.rbooks.co.uk.

Woodmorappe, John, M.A. Geology, B.A. Biology. The Mythology of Modern Dating Methods: Why million/billion year results are not credible. 1999. Institute for Creation Research, P.O. Box 2667, El Cajon, California 92021.

Wright, Ann. www.constellationsofwords.com/Constellations/lepus.htm 2008.

Made in the USA
Columbia, SC
22 April 2022